The Standard Model
of Particle Physics

by
Trevor G. Underwood

2nd Edition (August 7, 2025)

By the same author:

Quantum Electrodynamics – annotated sources. Volumes I and II. (April 2023);

Special Relativity. (June 2023);

General Relativity. (November 2023);

Gravity. (March 2024);

Electricity & Magnetism. (May 2024);

Quantum Entanglement. (June 2024);

The Standard Model. (September 2024);

New Physics. (October 2024);

The Cosmological Redshift of Light. (November 2024);

Cosmic Microwave Background Radiation. (January 2025);

Fundamental Physics. (May 2025).

all distributed by Lulu.com.

Published by Trevor G. Underwood
18 SE 10th Ave.
Fort Lauderdale, FL 33301

ISBN: 979-8-218-50364-2 (hardcover)
Library of Congress Control Number: 2024918424

Printed and distributed by Lulu Press, Inc.

700 Park Offices Dr
Ste 250
Durham, NC, 27709
http://www.lulu.com/shop

CONTENTS.

Page no.

17   **PREFACE**

30   **PART I   The Standard Model of Particle Physics.**

50   **The four "fundamental interactions or forces".**

56   **Symmetry in the Standard Model.**

59   **Timeline of discovery of subatomic particles.**

89   **Paul Adrien Maurice Dirac (August 8, 1902–October 20, 1984).**

97   **Dirac, P. A. M. (February, 1928). The Quantum Theory of the Electron.** *Roy. Soc. Proc., A*, 117, 778, 610–24; https://doi.org/10.1098/rspa.1928.0023; this work led Dirac to predict the existence of the *positron*, the *electron's antiparticle*, which he interpreted in terms of what came to be called the *Dirac sea*.

102   **Paul Dirac – 1933 Nobel Lecture, December 12, 1933.** *Theory of electrons and positrons*; https://www.nobelprize.org/prizes/physics/1933/dirac/lecture/; in his Nobel Lecture, Dirac described the current state of his theory of electrons and positrons.

107   **Werner Karl Heisenberg (December 5, 1901–February 1, 1976).**

114   **Heisenberg, W. (January, 1932). Über den Bau der Atomkerne. I. (About the construction of atomic nuclei. I.); (March, 1932). Über den Bau der Atomkerne. II. (About the construction of atomic nuclei. II.); (September, 1933). Über den Bau der Atomkerne. III.** *Zeit. Phys.*, 77, 1–11; https://doi.org/10.1007/BF01342433; *Ibid.*, 78, 156–64; https://doi.org/10.1007/BF01337585; *Ibid.*, 80, 587–96; https://doi.org/10.1007/BF01335696; three-part paper by Heisenberg, which attempted to treat the protons and neutrons on an equal footing by *considering protons and neutrons as different charge states of the same particle*, which he referred to as the *isotopic spin parameter*.

115   **Dirac, P. A. M. (1934). Discussion of the infinite distribution of electrons in the theory of the positron.** *Proc. Camb. Phil. Soc.*, 30, 2, 150-163; doi:10.1017/S030500410001656X; attempt by Dirac to address problem with his *relativistic* 'hole' theory which implies an infinite number of negative-energy *electrons* (per unit volume) with energies extending continuously from $- mc^2$ to $- \infty$. When an electromagnetic field is present positive- and negative-energy states

cannot be distinguished in a *relativistically* invariant way. Need to set up assumptions for production of *electromagnetic field* by the electron distribution such that any finite change in distribution produces a change in the field in agreement with Maxwell's equations and such that the infinite field which would be required by Maxwell's equations from an infinite density of electrons is in some way cut out. *Assumes each electron has its own individual wave function in space-time* and each electron moves in an *electromagnetic field* which is the same for all electrons part coming from external causes and part from the electron distribution itself. He introduces a *relativistic density matrix* referring to two points in space and two times, and separates the density distribution into two parts where one, R_a contains the *singularities*, and the other, R_b describes the *electric* and *current densities* physically present, so that any alteration one may make in the distribution of *electrons* and *positrons* will correspond to an alteration in R_b, which is *relativistically invariant and gauge invariant*, and that the *electric density* and *current density* corresponding to it satisfy the conservation law. *It therefore appears reasonable to make the assumption that the electric and current densities corresponding to R_b are those which are physically present, arising from the distribution of electrons and positrons. In this way we can remove the infinities.*

124   **Heisenberg, W. (September, 1934). Bemerkungen zur Diracschen Theorie des Positrons. (Remarks on the Dirac theory of positron.)** *Zeit. Phys.*, 90, 209–31; https://doi.org/10.1007/BF01333516; translation by D. H. Delphenich at https://s3.cern.ch/inspire-prod-files-5/5e5a144a00179c2f8e799c87f4c44be2; Heisenberg responded by presenting his thinking on Dirac's theory and further development of the theory in two papers.

126   **Heisenberg, W., Euler, H. (November, 1936). Folgerungen aus der Diracschen Theorie des Positrons. (Consequences of Dirac's theory of the positron.)** *Zeit. Phys.*, 98, 11-12, 714-732; https://doi.org/10.1007/BF01343663; translation by W. Korolevski and H. Kleinert, Institut fur Theoretische Physik, Freie Universitat Berlin, Arnimallee 14, D-14195 Berlin, Germany; https://arxiv.org/pdf/physics/0605038. According to Dirac's *theory of the positron*, an *electromagnetic field* tends to create pairs of particles which leads to a change of Maxwell's equations in the vacuum. This paper examines the consequence of the possibility of transforming *electromagnetic radiation* into *matter* in *quantum electrodynamics* on Maxwell equations. It is no longer possible to separate processes in the vacuum from those involving *matter* since *electromagnetic fields* can create *matter* if they are strong enough. Even if they are not strong enough to create *matter* they will, due to the virtual possibility of creating *matter*, *polarize the vacuum*.

129   **Enrico Fermi (September 29, 1901–November 28, 1954).**

143   **Fermi, E. (March, 1934) Versuch einer Theorie der β-Strahlen. I. (Attempt at a theory of β rays. I.)** *Zeit. Phys.,* 88, 161–77 (1934). https://doi.org/10.1007/ BF01351864; also at https://www.nssp.uni-saarland.de/lehre/Vorlesung/ Kernphysik_SS19/History/ Papers/Fermi_1.pdf.; (German); translation by T. G. Underwood; Fermi proposed the first theory of the *weak interaction,* known as *Fermi's interaction.* A quantitative theory of *β decay* is proposed, in which the existence of the *neutrino* is assumed, and the emission of *electrons* and *neutrinos* from a nucleus in the β case is treated with a method similar to that of the emission of a quantum of light from an excited atom in radiation theory. Formulas for the lifetime and for the shape of the emitted continuous β radiation spectrum are derived and compared with experience.

151   **Enrico Fermi – 1938 Nobel Prize in Physics.**

152   **Hideki Yukawa (January 23, 1907–September 8, 1981).**

154   **Yukawa, H. (1935). On the Interaction of Elementary Particles.** *Proceedings of the Physico-Mathematical Society of Japan,* 17, 48; also in (January, 1955). *Progr. Theoret. Phys. Suppl.* 1, 1–10, https://doi.org/10.1143/PTPS.1.1; the existence of *mesons* was predicted by Hideki Yukawa's 1935 *theory of mesons* that postulated the particle as mediating the *nuclear force.* The interaction between *elementary* particles can be described by means of a *field of force,* just as the interaction between the *charged* particles is described by the *electromagnetic field.* In the quantum theory this field should be accompanied by a new sort of quantum, just as the electromagnetic field is accompanied by the *photon.* In this paper the possible natures of this field and the quantum accompanying it will be discussed briefly and also their bearing on the nuclear structure will be considered.

165   **Hideki Yukawa - 1949 Nobel Lecture, December 12, 1949. *Meson theory in its developments.*** https://www.nobelprize.org/uploads/2018/06/yukawa-lecture.pdf. In his Nobel Prize lecture Yukawa provided a brief history of the development of the *theory of the meson.*

173   **Eugene Paul Wigner (November 17, 1902–January 1, 1995).**

177   **Wigner, E. (January, 1937). On the Consequences of the Symmetry of the Nuclear Hamiltonian on the Spectroscopy of Nuclei.** *Phys. Rev.* 51, 2, 106; https://journals.aps.org/pr/abstract/10.1103/PhysRev.51.106; also at https:// harvest.aps.org/v2/journals/articles/10.1103/PhysRev.51.106/fulltext; the structure of the *multiplets* of nuclear terms is investigated, using as *first approximation* a Hamiltonian which does not involve the ordinary *spin* and corresponds to equal forces between all nuclear constituents, *protons* and *neutrons.* The *multiplets* turn

out to have a rather complicated structure, instead of the S of atomic spectroscopy, one has three quantum numbers S, T, Y. The *second approximation* can either introduce *spin* forces (method 2), or else can discriminate between *protons* and *neutrons* (method 3). The *last approximation* discriminates between *protons* and *neutrons* as in method 2 and takes the spin forces into account as in method 3. The method 2 is worked out schematically and is shown to explain qualitatively the table of *stable nuclei* to about Mo.

184   **Eugene Wigner – the 1963 Nobel Prize in Physics.**

185   **Shoichi Sakata (January 18, 1911–October 16, 1970).**

188   **Sakata, S. & Inoue, T. (1942). On the Correlations between Mesons and Yukawa Particles.** Sakata, S. & Inoue, T. (1942). Chukanshi to Yukawa ryushi no Kankei ni tuite. (in Japanese). *Nippon Suugaku-Butsuri Gakkaishi (Journal of the Mathematical and Physical Society of Japan)*, 16, 4; https://doi.org/10.11429/ subutsukaishi1927.16.232; translation at (November, 1946). *Prog. Theor. Phys.*, 1, 4, 143–50; https://doi.org/ 10.1143/PTP.1.143; also at https://academic.oup. com/ptp/article/1/4/143/1846220; Sakata and Inoue proposed their *two-meson theory* in 1942. [Sakata, S. & Inoue, T. (1942). Chukanshi to Yukawa ryushi no Kankei ni tuite. (in Japanese). *Nippon Suugaku-Butsuri Gakkaishi*, 16; https://doi.org/10.11429/subutsukaishi1927.16.232.]; Sakata-Inoue's *two-meson theory*. At the time, a charged particle discovered in the hard component cosmic rays was misidentified as the Yukawa's *meson* ($\pi^{\pm}$, nuclear force carrier particle). The misinterpretation led to puzzles in the discovered cosmic ray particle. Sakata and Inoue solved these puzzles by identifying the cosmic ray particle as a daughter charged *fermion* produced in the $\pi^{\pm}$ decay. A new neutral *fermion* was also introduced to allow $\pi^{\pm}$ decay into *fermions*. We now know that these charged and neutral *fermions* correspond to the second-generation *leptons* μ and ν_{μ} in the modern language. They then discussed the decay of the *Yukawa particle*, $\pi^{+} \rightarrow \mu^{+} + \nu^{\mu}$. Sakata and Inoue predicted correct *spin* assignment for the *muon*, and they also introduced the second *neutrino*. They treated it as a distinct particle from the *beta decay neutrino*, and anticipated correctly the three body decay of the *muon*. As a result of World War II, the English printing of Sakata-Inoue's *two-meson theory* paper was delayed until 1946, one year before the experimental discovery of $\pi \rightarrow \mu\nu$ decay.

197   **Chen-Ning Yang (born 1 October 1922).**

200   **Yang, C. N. & Mills, R. (October, 1954). Conservation of Isotopic Spin and Isotopic Gauge Invariance.** *Phys. Rev.*, 96, 1, 191–5; https://journals.aps.org/pr/ pdf/10.1103/PhysRev.96.191; the *Yang–Mills theory*. Chen-Ning Yang and Robert

Mills extended the concept of *gauge theory* for *abelian* groups, e.g. *quantum electrodynamics*, to *nonabelian* groups to provide an explanation for *strong interactions*. The *Yang–Mills theory* is a *quantum field theory* for *nuclear binding*. It is a *gauge theory* based on a *special unitary group* SU(n), or more generally any compact Lie group. It seeks to describe the behavior of *elementary particles* using these non-abelian Lie groups and *is at the core of the unification of the electromagnetic force and weak forces* (i.e. U(1) × SU(2)) as well as *quantum chromodynamics*, the theory of the *strong force* (based on SU(3)). Thus, it forms the basis of the understanding of the *Standard Model* of particle physics.

215 **Sakata, S. (September, 1956). On a Composite Model for the New Particles.** *Progr. Theor. Phys.*, 16, 6, 686-8; https://doi.org/10.1143/PTP.16.686; Sakata proposed his *Sakata Model* which explains the physics behind the *Nakano-Nishijima-Gell-Mann (NNG) rule* by postulating that the fundamental building blocks of all strongly interacting particles are the *proton*, the *neutron* and the *lambda baryon*. The positively charged *pion* is made out of a *proton* and an *anti-neutron*, in a manner similar to the *Fermi-Yang composite Yukawa meson model*, while the positively charged *kaon* is composed of a *proton* and an *anti-lambda*. Aside from the integer *charges*, the *proton*, *neutron*, and *lambda* have similar properties as the *up quark*, *down quark*, and *strange quark* respectively.

218 **Tsung-Dao Lee (born November 24, 1926).**

220 **Lee, T. D. & Yang, C. N. (October, 1956). Question of Parity Conservation in Weak Interactions.** *Phys. Rev.*, 104, 254; https://journals.aps.org/pr/abstract/ 10.1103/PhysRev.104.254; in 1956 Chen Ning Yang and Tsung Dao Lee formulated a theory that the *left-right symmetry law* is violated by the *weak interaction*. Recent experimental data indicate closely identical *masses* and *lifetimes* of the θ^+ and the τ^+ *mesons*. On the other hand, analyses of the decay products of τ^+ strongly suggest on the grounds of *angular momentum* and *parity conservation* that the τ^+ and θ^+ are not the same particle. This poses a rather puzzling situation that has been extensively discussed. One way out of the difficulty is to assume that *parity is not strictly conserved*, so that θ^+ and τ^+ are two different decay modes of the same particle, which necessarily has a single *mass* value and a single *lifetime*. We wish to analyze this possibility in the present paper against the background of the existing experimental evidence of *parity conservation*. It will become clear that existing experiments do indicate *parity conservation* in *strong* and *electromagnetic interactions* to a high degree of accuracy, but that for the *weak interaction* actions (i.e., decay *interactions* for the *mesons* and *hyperons*, and various Fermi *interactions*) *parity conservation* is so far only an extrapolated hypothesis unsupported by experimental evidence. The question of *parity*

conservation in *β decays* and in *hyperon* and *meson decays* is examined. Possible experiments are suggested which might test *parity conservation* in these interactions.

234 **Chen Ning Yang – 1957 Nobel Lecture, December 11, 1957.** *The law of parity conservation and other symmetry laws of physics.* In his Nobel Prize lecture Yang described the developments that led to the disproof of *parity conservation.*

238 **Tsung-Dao Lee – 1957 Nobel Lecture, December 11, 1957.** *Weak Interactions and Non-conservation of Parity.* In his Nobel Prize lecture Lee reviewed the current knowledge about elementary particles and their *interactions.*

239 **Yoichiro Nambu (January 18, 1921–July 5, 2015).**

241 **Nambu, Y. (April, 1960). Axial Vector Current Conservation in Weak Interactions.** *Phys. Rev. Lett.*, 4, 7, 380-2; https://journals.aps.org/prl/abstract/ 10.1103/PhysRevLett. 4.380; in analogy to the conserved *vector current interaction* in the *beta decay* suggested by Feynman and Gell-Mann, some speculations have been made about a possibly conserved *axial vector current.* We would like to suggest that there may not be a strict *pseudovector current conservation*, but that we may have an approximate conservation which becomes rigorous in the limit q^2 » m_π^2, m_π being the *pion* mass and q^2 a *massless, pseudo scalar*, and *charged quantum* bridging the *nucleon* and *lepton currents.* We are tempted to extend this approximate conservation of the *axial vector current* (and naturally also the *vector current*) to the *strangeness-non conserving beta decays.*

249 **Sakurai, J. J. (September, 1960). Theory of strong interactions.** *Ann. Phys.*, 11, 1, 1-48; https://www.sciencedirect.com/science/article/abs/pii/ 0003491660901263; all the *symmetry* models of *strong interactions* which have been proposed up to the present are devoid of deep physical foundations. It is suggested that, instead of postulating artificial "higher" *symmetries* which must be broken anyway within the realm of *strong interactions*, we take the *existing exact* symmetries of *strong interactions* more seriously than before and exploit them to the utmost limit. A new theory of *strong interactions* is proposed on this basis.

253 **Jeffrey Goldstone (born September 3, 1933).**

254 **Goldstone, J. (January, 1961). Field theories with "Superconductor" solutions.** *Nuovo Cimento*, 19, 154–64; https://doi.org/10.1007/BF02812722; the conditions for the existence of non-perturbative type "superconductor" solutions of field theories are examined. A non-covariant canonical transformation method is used to find such solutions for a theory of a fermion interacting with a pseudoscalar

boson. A covariant renormalizable method using Feynman integrals is then given. A "superconductor" solution is found whenever in the normal perturbative-type solution the boson mass squared is negative and the coupling constants satisfy certain inequalities. The symmetry properties of such solutions are examined with the aid of a simple model of self-interacting boson fields. The solutions have lower symmetry than the Lagrangian, and contain mass zero bosons. *Goldstone's theorem* in *relativistic quantum field theory* states that if there is an exact *continuous symmetry* of the Hamiltonian or Lagrangian defining the system, and this is not a symmetry of the vacuum state (i.e. there is *broken symmetry*), then there must be at least one *spin-zero massless particle* called a *Goldstone boson*. In the *quantum theory of many-body systems Goldstone bosons* are collective excitations such as *spin waves*. An important exception to *Goldstone's theorem* is provided in *gauge theories* with the *Higgs mechanism*, whereby the *Goldstone bosons* gain *mass* and become *Higgs bosons*.

260 **Sheldon Lee Glashow (born December 5, 1932).**

262 **Glashow, S. L. (February, 1961). Partial-symmetries of weak interactions.** *Nuclear Physics*, 22, 4, 579–88; doi:10.1016/0029-5582(61)90469-2; the *W and Z bosons* were predicted in detail by Sheldon Glashow, Mohammad Abdus Salam, and Steven Weinberg. In 1961, Sheldon Glashow combined the *electromagnetic* and *weak interactions* and extended *electroweak unification models* due to Schwinger by including a short-range *neutral current*, the Z_0. The resulting symmetry structure that Glashow proposed, SU(2) × U(1), forms the basis of the accepted *theory of the electroweak interactions*. For this discovery, Glashow along with Steven Weinberg and Abdus Salam, was awarded the 1979 Nobel Prize in Physics.

265 **Murray Gell-Mann (September 15, 1929–May 24, 2019).**

268 **Gell-Mann, M. (March, 1961). The Eightfold Way: A Theory of Strong Interaction Symmetry (Report).** Pasadena, CA: California Inst. of Tech., Synchrotron Laboratory; https://www.osti.gov/biblio/4008239; Gell-Mann introduces his formulation of a particle classification system for *hadrons* known as *the Eightfold Way* – or, in more technical terms, SU(3) *flavor symmetry*, streamlining its structure. A new model of the higher symmetry of elementary particles is introduced in which the eight known *baryons* are treated as a *super-multiplet*, degenerate in the limit of *unitary symmetry* but split into isotopic *spin multiplets* by a *symmetry-breaking* term. The *symmetry violation* is ascribed phenomenologically to the *mass* differences. The *baryons* correspond to an eight-dimensional irreducible representation of the *unitary group*. The *pion* and *K meson*

fit into a similar set of eight particles along with a predicted *pseudoscalar meson* χ^0 having I = 0. A ninth *vector meson* coupled to the *baryon* current can be accommodated naturally in the scheme. It is predicted that the eight *baryons* should all have the same *spin* and *parity* and that *pseudoscalar* and *vector mesons* should form octets with possible additional singlets. The mathematics of the *unitary group* is described by considering three fictitious *leptons*, ν, e^-, and μ^-, which may throw light on the structure of *weak interactions*.

286 **Glashow, S. L. & Gell-Mann, M. (September, 1961). Gauge Theories of Vector Particles.** *Ann. Phys. (N.Y.)*, 15, 437-60; https://doi.org/10.1016/0003-4916(61) 90193-2; also at https://www.semanticscholar.org/paper/Gauge-Theories-of-Vector-Particles-Glashow-Gell-Mann/c0184d2962be36beb622736 b14e0484651ec0ba0; the possibility of generalizing the Yang-Mills trick is examined. Thus, *we seek theories of vector bosons invariant under continuous groups of coordinate-dependent linear transformations*. All such theories may be expressed as super-positions of certain "simple" theories; we show that each "simple" theory is associated with a simple *Lie algebra*. We may introduce *mass* terms for the *vector bosons* at the price of destroying the *gauge-invariance* for *coordinate-dependent gauge functions*.

291 **Schwinger, J. (January, 1962). Gauge Invariance and Mass.** *Phys. Rev.*, 125, 397; https://doi.org/10.1103/PhysRev.125.397; also at https://harvest. aps.org/v2/journals/articles/10.1103/PhysRev.125.397/fulltext; it is argued that the *gauge invariance* of a *vector field* does not necessarily imply zero *mass* for an associated particle if the *current vector coupling* is sufficiently strong. This situation may permit a deeper understanding of nucleonic *charge conservation* as a manifestation of a *gauge invariance*, without the obvious conflict with experience that a massless particle entails.

295 **Goldstone, J., Salam, A. & Weinberg, S. (August, 1962). Broken Symmetries.** *Phys. Rev.* 127, 965; https://doi.org/10.1103/PhysRev.127.965; some proofs are presented of Goldstone's Theorem, that "*in a manifestly Lorentz-invariant quantum field theory, if there is a continuous symmetry transformation under which the Lagrangian is invariant, then either the vacuum state is also invariant under the transformation, or there must exist spinless particles of zero mass*".

301 **Philip Warren Anderson (December 13, 1923–March 29, 2020).**

303 **Anderson, P. W. (April, 1963). Plasmons, Gauge Invariance, and Mass.** *Phys. Rev.*, 130, 1, 439; https://journals.aps.org/pr/abstract/10.1103/PhysRev.130.439; also at https://web.archive.org/web/20160307022433/https://www.physics.

rutgers.edu/grad/601/Anderson_Plasmons.pdf.; Schwinger had pointed out that the *Yang-Mills vector boson* implied by associating a generalized *gauge transformation* with a conservation law (of *baryonic charge*, for instance) does not necessarily have zero *mass*, if a certain criterion on the vacuum fluctuations of the generalized *current* is satisfied. We show that the theory of plasma oscillations is a simple *nonrelativistic* example exhibiting all of the features of Schwinger's idea. It is also shown that Schwinger's criterion that the *vector field* $m \neq 0$ implies that the *matter* spectrum before including the *Yang-Mills interaction* contains $m = 0$, but that the example of *superconductivity* illustrates that the physical spectrum need not. Some comments on the relationship between these ideas and the zero-*mass* difficulty in theories with *broken symmetries* are given.

312   **Gell-Mann. M. (February, 1964). A Schematic Model of Baryons and Mesons.** *Physics Letters*, 8, 3, 214–5; https://doi.org/10.1016/S0031-9163(64)92001-3; Gell-Mann, and separately George Zweig, proposed that *baryons*, which include *protons* and *neutrons*, and *mesons* were composed of elementary particles. [Zweig, G. (February 21, 1964). An SU(3) Model for Strong Interaction Symmetry and its Breaking.] Zweig called the elementary particles "aces" while Gell-Mann called them "*quarks*"; the theory came to be called the *quark model*. The bootstrap model for *strongly interacting particles* described in terms of the broken eightfold way is discussed to determine algebraic properties of the interactions with scattering amplitudes on the mass shell. A mathematical model based on *field theory* is described.

318   **Murray Gell-Mann – 1969 Nobel Prize in Physics. *Presentation Speech.***

322   **James Watson Cronin (September 29, 1931–August 25, 2016).**

324   **Val Logsdon Fitch (March 10, 1923–February 5, 2015).**

327   **James W. Cronin and Val L. Fitch - the 1980 Nobel Prize in Physics.** Demonstration in 1964 using *neutral K-mesons* of the violation of all three symmetry principles (1) that the laws of Nature are exactly alike for both *antimatter* and ordinary *matter*; (2) that the fundamental laws have exact *mirror symmetry*; and (3) that the fundamental laws have exact *time reflection symmetry* – symmetry under motion reversal.

331   **François Englert (born November 6, 1932).**

333   **Englert, F. & Brout, R. (August, 1964). Broken Symmetry and the Mass of Gauge Vector Mesons.** *Phys. Rev. Lett.*, 13, 9, 321–3; https://doi.org/10.1103/PhysRevLett.13.321; https://journals.aps.org/prl/abstract/10.1103/PhysRevLett.13.321; Brout and Englert showed that *gauge vector fields*, abelian and non-abelian, could acquire *mass* if empty space were endowed with a particular type of structure that one encounters in material systems. Other physicists, Peter Higgs and Gerald Guralnik, C. R. Hagen and Tom Kibble had reached similar conclusions at about the same time. The *Brout–Englert–Higgs* (BEH) *mechanism* is believed to give rise to the *masses* of all the elementary particles in the *Standard Model*. This includes the *masses* of the W and Z *bosons*, and the *masses* of the *fermions*, i.e. the *quarks* and *leptons*.

338   **Peter Ware Higgs (29 May 1929 – 8 April 2024).**

341   **Higgs, P. W. (October, 1964). Broken Symmetries and the Masses of Gauge Bosons.** *Phys. Rev. Let.*, 13, 16, 508–9; https://doi.org/10.1103/PhysRevLett.13.508; in a recent note it was shown that the *Goldstone theorem*, that Lorentz-covariant field theories in which spontaneous breakdown of *symmetry* under an internal Lie group occurs contain *zero-mass* particles, *fails if and only if the conserved currents associated with the internal group are coupled to gauge fields*. The purpose of the present note is to report that, *as a consequence of this coupling, the spin-one quanta of some of the gauge fields acquire mass*; the longitudinal degrees of freedom of these particles (which would be absent if their *mass* were zero) go over into the Goldstone bosons *when the coupling tends to zero. The model is discussed mainly in classical terms*; nothing is proved about the quantized theory. It should be understood, therefore, that *the conclusions which are presented concerning the masses of particles are conjectures based on the quantization of linearized classical field equations.*

345   **Gerald Stanford Guralnik (September 17, 1936–April 26, 2014).**

346   **Guralnik, G., Hagen, C. & Kibble, T. (November, 1964). Global Conservation Laws and Massless Particles.** *Phys. Rev. Lett.*, 13, 20, 585-7; https://doi.org/10.1103/PhysRevLett.13.585; https://journals.aps.org/prl/pdf/ 10.1103/PhysRevLett.13.585; a third paper on the subject was written later in the same year by Gerald Guralnik, C. R. Hagen, and Tom Kibble.

351   **François Englert – 2013 Nobel Lecture, December 8, 2013.** *The BEH Mechanism and its Scalar Boson.* In his Nobel Prize lecture Englert provided a brief history of the theory.

368   **Peter W. Higgs – 2013 Nobel Lecture, December 8, 2013.** *Evading the Goldstone Theorem.* In his Nobel Prize lecture Higgs provided an account of how he came to publish his paper.

373   **Abdus Salam (January 29, 1926–November 21, 1996).**

377   **Salam, A. & Ward, J. C. (November, 1964). Electromagnetic and weak interactions.** *Physics Letters,* 13, 2, 168–71; https://doi.org/10.1016/0031-9163(64)90711-5; in 1964, Salam and Ward worked on the synthesis of the *weak* and *electromagnetic interactio*n, obtaining a *gauge theory* based on the SU(2) × U(1) model. Salam was convinced that all the elementary particle interactions are actually the *gauge interactions*.

378   **Higgs, P. W. (May, 1966). Spontaneous Symmetry Breakdown without Massless Bosons.** *Phys. Rev.,* 145, 4, 1156; https://doi.org/10.1103/PhysRev. 145.1156; we examine a simple *relativistic* theory of two *scalar fields*, first discussed by Goldstone, in which as a result of *spontaneous breakdown* of U(1) *symmetry* one of the *scalar bosons* is *massless*, in conformity with the *Goldstone theorem.* When the *symmetry group* of the Lagrangian is extended from *global* to *local* U(1) *transformations* by the introduction of *coupling* with a *vector gauge field*, the *Goldstone boson* becomes the *longitudinal state* of a *massive vector boson* whose *transverse states* are the *quanta* of the *transverse gauge field.* A *perturbative* treatment of the model is developed in which the major features of these phenomena are present in zero order. *Transition amplitudes* for decay and scattering processes are evaluated in lowest order, and it is shown that they may be obtained more directly from an equivalent Lagrangian in which the original *symmetry* is no longer manifest. When the system is coupled to other systems in a *U(1) invariant* Lagrangian, the other systems display an induced *symmetry breakdown*, associated with a partially conserved *current* which interacts with itself via the *massive vector boson.*

379   **Steven Weinberg (May 3, 1933–July 23, 2021).**

382   **Weinberg, S. (November, 1967). Model of Leptons.** *Phys. Rev. Let.,* 19, 21, 1264–6;   https://doi.org/10.1103/PhysRevLett.19.1264;   Steven   Weinberg incorporated the *Brout–Englert–Higgs (BEH) mechanism* into Glashow's *electroweak interaction,* giving it its modern form. Weinberg proposed his *model of unification of electromagnetism and nuclear weak forces* with the *masses* of the force-carriers of the *weak* part of the *interaction* being explained by *spontaneous symmetry breaking,* in which the *symmetry* between the *electromagnetic* and *weak*

interactions is *spontaneously broken*, but in which the *Goldstone bosons* are avoided by introducing the *photon* and the *intermediate boson fields* as *gauge fields*.

389 **Gerardus (Gerard) 't Hooft (born July 5, 1946).**

392 **'t Hooft, G. (December, 1971). Renormalizable Lagrangians for Massive Yang-Mills Fields.** *Nucl. Phys.*, B35, 167-88; https://doi.org/10.1016/0550-3213(71)90139-8; *renormalizable* models are constructed in which *local gauge invariance* is *broken spontaneously*. *Feynman rules* and *Ward identities* can be found by means of a path integral method, and they can be checked by algebra. In one of these models, which is studied in more detail, *local* SU(2) is broken in such a way that *local* U(1) remains as a *symmetry*. A *renormalizable and unitary theory results*, with *photons*, *charged massive vector particles*, and additional *neutral scalar particles*. It has three independent parameters. Another model has local SU(2) x U(1) as a *symmetry* and may serve as a *renormalizable* theory for ρ-*mesons* and *photons*. In such models, *electromagnetic mass-differences* are finite and can be calculated in *perturbation theory*.

395 **Sheldon Lee Glashow, Abdus Salam and Steven Weinberg – the 1979 Nobel Prize in Physics.**

398 **Steven Weinberg – Nobel Lecture, December 8, 1979.** *Conceptual Foundations of the Unified Theory of Weak and Electromagnetic Interactions.* In his Nobel lecture, Weinberg attempted (but failed) to justify renormalization in the Standard Model.

409 **Abdus Salam – 1979 Nobel Lecture, December 8, 1979.** *Guage Unification of Fundamental Forces.* In his Nobel Prize lecture Salam provided a brief history of the development of *the gauge unification of the fundamental forces*.

423 **Sheldon Glashow – 1979 Nobel Lecture, December 8, 1979.** *Towards a Unified Theory – Threads in a Tapestry.* In his Nobel Prize lecture Glashow provided a brief history of his work.

436 **Makoto Kobayashi (born April 7, 1944, in Nagoya, Japan).**

437 **Toshihide Maskawa (February 7, 1940–July 23, 2021).**

438 **Kobayashi, M. & Maskawa, T. (February, 1973). CP-Violation in the Renormalizable Theory of Weak Interaction.** *Progress of Theoretical Physics*, 49, 2, 652–7; https://doi.org/10.1143/PTP.49.652; explained *broken symmetry* within the framework of the *Standard Model*, but *required that the Model be extended to three families of quarks to explain CP violation*, which ultimately led to the *six-quark model*.

445   **Yoichiro Nambu, Makoto Kobayashi and Toshihide Maskawa – the 2008 Nobel Prize in Physics.**

447   **Makoto Kobayashi – Nobel Lecture, December 8, 2008.** *CP Violation and Flavor Mixing.* In his Nobel Prize lecture, Kobayashi provided a brief history of the development of the *six-quark model*.

461   **PART II Unified Gravity** (Addition: June 12, 2025.)

468   **Partanen, M. & Tulkki, J. (March, 2024). QED based on an eight-dimensional spinorial wave equation of the electromagnetic field and the emergence of quantum gravity.** (Addition: June 12, 2025.) *Phys. Rev.*, 109, 032224; https://doi.org/10.1103/PhysRevA,109.03224; here we develop a coupling between the electromagnetic field, Dirac electron-positron field, and the gravitational field based on an eight-component spinorial representation of the electromagnetic field. Our spinorial representation is analogous to the well-known representation of particles in the Dirac theory but it is given in terms of 8×8 bosonic gamma matrices. In distinction from earlier works on the spinorial representations of the electromagnetic field, we reformulate QED using eight-component spinors. This enables us to introduce the generating Lagrangian density of gravity based on the special unitary symmetry of the eight-dimensional spinor space. The generating Lagrangian density of gravity plays, in the definition of the gauge theory of gravity and its symmetric stress energy-momentum tensor source term, a similar role as the conventional Lagrangian density of the free Dirac field plays in the definition of the gauge theory of QED and its electric four-current density source term. The fundamental consequence, the Yang-Mills gauge theory of *unified gravity*, is studied in a separate work [arXiv:2310.01460], where the theory is also extended to cover the other fundamental interactions of the Standard Model. We devote ample space for details of the eight-spinor QED to provide solid mathematical basis for the present work and the related work on the Yang-Mills gauge theory of *unified gravity*.

490   **Partanen, M. & Tulkki, J. (May, 2025). Gravity generated by four one-dimensional unitary gauge symmetries and the Standard Model.** (Addition: June 12, 2025.) *Rep. Prog. Phys.*, 88 057802; https://doi.org/10.1088/1361-6633/adc82e; this paper attempts to derive the gauge theory of gravity using compact, finite-dimensional symmetries in a way that resembles the formulation of the fundamental interactions of the Standard Model. For our eight-spinor representation of the Lagrangian, we define a quantity, called the space-time dimension field, which enables extracting four-dimensional space-time quantities from the eight-dimensional spinors. Four U(1) symmetries of the components of the space-time dimension field are used to derive a gauge theory, called *unified gravity*. The stress-energy-momentum tensor source term of gravity follows directly from these symmetries. The metric tensor enters in *unified gravity* through geometric conditions. Based on the Minkowski metric, *unified gravity* allows us to describe gravity within a single coherent mathematical framework together with the quantum fields of all fundamental interactions of the Standard Model. We present the Feynman rules for *unified gravity* and study the renormalizability and radiative corrections of the theory at one-loop order. The equivalence principle is formulated by requiring that the renormalized values of the inertial and gravitational masses are equal. In contrast to previous gauge theories of gravity, all infinities that are encountered in the calculations of loop diagrams can be absorbed by the redefinition of the small number of parameters of the theory in the same way as in the gauge theories of the Standard Model. This result and our observation that *unified gravity* fulfills the Becchi–Rouet–Stora–Tyutin (BRST) symmetry and its coupling constant is dimensionless suggest that *unified gravity* can provide the basis for a complete, renormalizable theory.

509   **PART III Supersymmetry.**

519   **PART IV String Theory.**

PREFACE

This is the fascinating story of the development of the *Standard Model* of particle physics between Dirac's prediction of the *positron* in 1928 and the introduction of the *six-quark model* in the mid 1970's, as described in the primary sources.

The *Standard Model of particle physics* is the theory describing three of the four known fundamental *forces* (*electromagnetic, weak and strong interactions* – excluding gravity) in the universe and *classifying all known elementary particles*.

It was developed in stages throughout the latter half of the 20th century, through the work of many scientists worldwide, with the current formulation being finalized in the mid-1970s upon experimental confirmation of the existence of *quarks*. Since then, proof of the *top quark* (1995), the *tau neutrino* (2000), and the *Higgs boson* (2012) added further credence to the *Standard Model*. In addition, the *Standard Model* has predicted various properties of *weak neutral currents* and the W and Z *bosons* with great accuracy.

The Standard Model is a paradigm of a *quantum field theory*, exhibiting a wide range of phenomena, including *spontaneous symmetry breaking, anomalies*, and *non-perturbative behavior*. It is used as a basis for building more exotic models that incorporate hypothetical particles, extra dimensions, and elaborate *symmetries* (such as *supersymmetry*) to explain experimental results at variance with the *Standard Model*, such as the existence of dark matter and neutrino oscillations.

Although the *Standard Model* is believed to be theoretically self-consistent there are mathematical issues regarding *quantum field theories* still under debate. It has demonstrated some success in providing experimental predictions, but *it leaves some physical phenomena unexplained and so falls short of being a complete theory of fundamental interactions*. Although the physics of *special relativity* is included, *general relativity is not*, and it will fail at energies or distances where the *graviton* is expected to emerge. It does not fully explain *baryon asymmetry*, or account for the *universe's accelerating expansion as possibly described by dark energy*. The model does *not contain any viable dark matter particle* that possesses all of the required properties deduced from observational cosmology. It also does not incorporate *neutrino oscillations* and their non-zero masses.

This analysis of the failures of the *Standard Model* suggest that they stem from the attempt to base it on *relativistic quantum field theory* and make it *Lorentz covariant*, and the reliance on *renormalization* to remove infinities and bring theoretical values in line with experimental ones. (See Underwood, T. G. (2023). *Quantum Electrodynamics – annotated sources*, Vol. II.)

[***Lorentz transformations*** are linear transformations from a coordinate frame in spacetime to another frame that moves at a constant velocity relative to the former in which observers moving at different velocities may measure different distances, elapsed times, and even different orderings of events, but always **such that the speed of light is the same in all inertial reference frames**. The invariance of light speed is one of the postulates of **special relativity**.

Frames of reference can be divided into two groups: **inertial** (relative motion with constant velocity) and **non-inertial** (accelerating, moving in curved paths, rotational motion with constant angular velocity, etc.). **"Lorentz transformations" only refer to transformations between *inertial* frames, in the context of special relativity.**

Lorentz transformations

A physical quantity is said to be ***Lorentz covariant*** if it transforms under a given representation of the Lorentz group. According to the representation theory of the Lorentz group, these quantities are built out of scalars, four-vectors, four-tensors, and spinors. Theories that are *consistent with the principle of Special Relativity* must have the same form in all Lorentz frames, that is, they must be *covariant*. The *covariant* formulation of *classical electromagnetism* refers to ways of writing the laws of *classical electromagnetism* in a form that is manifestly invariant under *Lorentz transformations*.]

Part I describes the development of the *Standard Model* from the Bohr model of the atom in 1913, based on what were believed to be 3 stable particles, *electrons* in orbit around a *nucleus* comprised of *protons* and *neutrons*, to its emergence in 1973 as the *six-quark model*, comprising 52 elementary particles and anti-particles, revealed largely by tracks in cloud chambers from high energy collisions between particles, or at high altitudes with cosmic rays, of which only the electron is stable.

It starts with an overview of the Standard Model. This is followed by introductions to the four "fundamental interactions or forces" and the role of symmetry in fundamental physics, which describes the existence of symmetries in classical dynamics and in quantum mechanics, on which the Standard Model is based. This is followed by a summary of the timeline of the discovery of subatomic particles; a valuable resource which describes the large number *elementary and composite particles* and their *classifications*.

The *classifications* include *fermions* (elementary and composite particles with ½ integer spin (including quarks and leptons); *leptons* (elementary particles with ½ integer spin), including *muons* (charged leptons) and *neutrinos* (neutral leptons); *bosons* (elementary and composite particles with integer spin), including *vector bosons* (spin 1), of which *gauge bosons* are elementary particles which act as force carriers) including *gluons* (which carry the strong interaction); *hadrons* (composite particles made of two or more quarks held together by the strong interaction), including *baryons* (composite particles that contains an odd number of valence quarks and antiquarks) and *mesons* (composite particles that contain an equal number of quarks and antiquarks, usually one of each, bound together by the strong interaction); *hyperons* are baryons containing one or more strange quarks, but no charm, bottom, or top quark.

Quarks are elementary particles and a fundamental constituent of matter, which combine to form *hadrons*, the most stable of which are *protons* and *neutrons*, the components of atomic nuclei; they are the only elementary particles in the *Standard Model* to experience all four fundamental interactions, also known as fundamental forces (*electromagnetism*, *gravitation*, *strong interaction*, and *weak interaction*), as well as the only known particles whose *electric charges* are not integer multiples of the elementary *charge*. There are six types (*flavors*) of *quarks*: *up*, *down*, *charm*, *strange*, *top*, and *bottom*.

The *Standard Model* includes 26 *elementary* particles: 12 *fermions* (6 *quarks* and 6 *leptons*) and 14 *bosons* (13 *gauge bosons* and 1 *scalar boson*, the *Higgs boson*). The 6 *quarks* comprise *up*, *down*, *charm*, *strange*, *top* and *bottom*; the 6 *leptons* comprise the *electron*, *electron neutrino*, *muon*, *muon neutrino*, *tau* and *tau neutrino*; and the 13 *gauge bosons* comprise the *photon*, 2 *W bosons*, the *Z boson*, 8 types of *gluon*, and the hypothetical *graviton*. Each has its own anti-particle, making a total of 52 *elementary particles and anti-particles*.

In his 1928 article [Dirac, P. A. M. (February, 1928). The Quantum Theory of the Electron] and his 1933 Nobel Lecture [Theory of electrons and positrons], Dirac describes how by subjecting quantum mechanics to *relativistic requirements* he was able to deduce the existence and properties of the *antiparticle* of the *electron*, the *positron*, a particle with the same *mass* but opposite *electric charge*. He noted that *there was a complete and perfect symmetry between positive and negative electric charge*.

In 1932 and 1933, Heisenberg published a three-part paper on atomic nuclei which concluded by treating *protons* and *neutrons* on an equal footing by *considering protons and neutrons as different charge states of the same particle, which he referred to as the isotopic spin parameter*. [Heisenberg, W. (January, 1932). Über den Bau der Atomkerne. I. (About the construction of atomic nuclei. I.); (March, 1932). Über den Bau der

Atomkerne. II. (About the construction of atomic nuclei. II.); (September, 1933). Über den Bau der Atomkerne. III.]

In 1932, from a cloud chamber photograph of cosmic rays, the American physicist Carl David Anderson identified a track as having been made by a *positron*.

In Dirac, P. A. M. (April, 1934) [Discussion of the infinite distribution of electrons in the theory of the positron], Dirac addressed the problem with his *relativistic* 'hole' theory which implied an infinite number of negative-energy *electrons* (per unit volume) with energies extending continuously from $- mc^2$ to $- \infty$, so that when an electromagnetic field is present positive- and negative-energy states cannot be distinguished in a *relativistically* invariant way. He saw the need to set up assumptions for production of *electromagnetic field* by the electron distribution such that any finite change in distribution produces a change in the field in agreement with Maxwell's equations and such that the infinite field which would be required by Maxwell's equations from an infinite density of electrons is in some way cut out. Dirac assumed that each *electron* has its own individual wave function in space-time and moved in an *electromagnetic field* which was the same for all *electrons* part coming from external causes and part from the *electron distribution* itself. He then introduced a *relativistic density matrix* referring to two points in space and two times, and separated the density distribution into two parts where one, R_a contained the *singularities*, and the other, R_b described the *electric* and *current densities* physically present, so that any alteration in the distribution of *electrons* and *positrons* would correspond to an alteration in R_b, which was *relativistically invariant and gauge invariant*, and the *electric density* and *current density* corresponding to it satisfied the conservation law.

> [A **gauge theory** is a type of field theory in which the **Lagrangian**, and hence the dynamics of the system itself, do not change under *local* transformations according to certain smooth families of operations (Lie groups). Formally, the **Lagrangian** is invariant under these transformations.
>
> **Lagrangian mechanics** is a formulation of classical mechanics founded on the d'Alembert principle of virtual work. It was introduced by the Italian-French mathematician and astronomer Joseph-Louis Lagrange in his presentation to the Turin Academy of Science in 1760 culminating in his 1788 grand opus, *Mécanique analytique*. Lagrangian mechanics describes a mechanical system as a pair (M, L) consisting of a configuration space M and a smooth function L within that space called a **Lagrangian**. For many systems, L = T − V, where T and V are the **kinetic** and **potential energy** of the system, respectively. The stationary **action principle** requires that the action functional of the system derived from L must remain at a stationary point (specifically, a maximum, minimum, or saddle point) throughout

the time evolution of the system. This constraint allows the calculation of the equations of motion of the system using Lagrange's equations.

> ***Newton's laws and the concept of forces are the usual starting point for teaching about mechanical systems***. This method works well for many problems, but for others the approach is nightmarishly complicated. Lagrangian mechanics adopts *energy* rather than *force* as its basic ingredient, leading to more abstract equations capable of tackling more complex problems. ***Lagrange's approach was to set up independent generalized coordinates for the position and speed of every object, which allows the writing down of a general form of Lagrangian (total kinetic energy minus potential energy of the system) and summing this over all possible paths of motion of the particles yielded a formula for the '<u>action</u>', which he minimized to give a generalized set of equations***. This summed quantity is minimized along the path that the particle actually takes. This choice eliminates the need for the constraint force to enter into the resultant generalized system of equations. There are fewer equations since one is not directly calculating the influence of the constraint on the particle at a given moment.]

In this way, he *removed the infinities* and assumed that the *electric and current densities corresponding to R_b were those which were physically present, arising from the distribution of electrons and positrons*.

Heisenberg responded by presenting his thinking on Dirac's theory and further development of the theory in two papers. [Heisenberg, W. (September, 1934). Bemerkungen zur Diracschen Theorie des Positrons. (Remarks on the Dirac theory of positron.); and Heisenberg, W., Euler, H. (1936). Folgerungen aus der Diracschen Theorie des Positrons. (Consequences of Dirac's theory of the positron.)].

In these papers Heisenberg was the first to reinterpret the Dirac equation as a "classical" field equation for any point particle of *spin* $\hbar/2$, itself subject to quantization conditions involving anti-commutators. Thus, reinterpreting it as a *quantum field equation* accurately describing *electrons*, Heisenberg put matter on the same footing as *electromagnetism*: as being described by *relativistic quantum field equations* which allowed the possibility of particle creation and destruction.

In their 1936 paper, [Heisenberg, W., Euler, H. (1936). Folgerungen aus der Diracschen Theorie des Positrons. (Consequences of Dirac's theory of the positron.)] Heisenberg and Euler noted that according to Dirac's *theory of the positron*, an *electromagnetic field* tends to create pairs of particles which leads to a change of Maxwell's equations in the vacuum. Their paper examined the consequence of the possibility of transforming *electromagnetic radiation* into *matter* in *quantum electrodynamics* on Maxwell equations. It is no longer possible to separate processes in the vacuum from those involving *matter* since *electromagnetic fields* can create *matter* if they are strong enough. Even if they are not strong enough to create *matter* they will, due to the virtual possibility of creating *matter*, *polarize the vacuum.*

In Fermi, E. (March, 1934) Versuch einer Theorie der β-Strahlen. I. (Attempt at a theory of β rays. I.), Enrico Fermi proposed the first theory of the *weak interaction*, known as *Fermi's interaction*. He proposed a quantitative theory of *β decay* in which the existence of the *neutrino* was assumed, and the emission of *electrons* and *neutrinos* from a nucleus in the β case was treated with a method similar to that of the emission of a quantum of light from an excited atom in radiation theory. Formulas for the lifetime and for the shape of the emitted continuous β radiation spectrum are derived and compared with experience.

In Yukawa, H. (1935). On the Interaction of Elementary Particles, Hideki Yukawa predicted the existence of *mesons* in his *theory of mesons* that postulated the particle as mediating the *nuclear force.* He noted that the interaction between *elementary* particles could be described by means of a *field of force*, just as the interaction between the *charged* particles is described by the *electromagnetic field.* He suggested that in quantum theory this field should be accompanied by a new sort of quantum, just as the electromagnetic field is accompanied by the *photon.* In this paper the possible natures of this field and the quantum accompanying it were discussed and their bearing on the nuclear structure considered.

In Wigner, E. (January, 1937). On the Consequences of the Symmetry of the Nuclear Hamiltonian on the Spectroscopy of Nuclei, Eugene Wigner investigated the structure of the *multiplets* of nuclear terms using as *first approximation* a Hamiltonian which does not involve the ordinary *spin* and corresponds to equal forces between all nuclear constituents, *protons* and *neutrons.*

> [A *multiplet* is the *state space* for 'internal' degrees of freedom of a particle, that is, degrees of freedom associated to a particle itself, as opposed to 'external' degrees of freedom such as the particle's position in space. Examples of such degrees of freedom are the *spin state* of a particle in *quantum mechanics*, or the *color, isospin* and *hypercharge state* of particles in the *Standard Model.* Formally, this state space is described by a *vector space* which carries the *action* of a group of *continuous symmetries.*]

The *multiplets* turn out to have a rather complicated structure, instead of the S of atomic spectroscopy, one has three quantum numbers S, T, Y. The *second approximation* can either introduce *spin* forces (method 2), or else can discriminate between *protons* and *neutrons* (method 3). The *last approximation* discriminates between *protons* and *neutrons* as in method 2 and takes the *spin* forces into account as in method 3. The method 2 is worked out schematically and is shown to explain qualitatively the table of *stable nuclei* to about Mo.

Sakata and Inoue proposed their *two-meson theory* in 1942. [Sakata, S. & Inoue, T. (1942). Chukanshi to Yukawa ryushi no Kankei ni tuite. (in Japanese). (On the Correlations between Mesons and Yukawa Particles.)] At the time, a charged particle discovered in the hard component cosmic rays was misidentified as the Yukawa's *meson* ($\pi^{\pm}$, nuclear force carrier particle). The misinterpretation led to puzzles in the discovered cosmic ray particle. Sakata and Inoue solved these puzzles by identifying the cosmic ray particle as a daughter charged *fermion* produced in the $\pi^{\pm}$ decay. A new neutral *fermion* was also introduced to allow $\pi^{\pm}$ decay into *fermions*. We now know that these charged and neutral *fermions* correspond to the second-generation *leptons* μ and ν_{μ} in the modern language. They then discussed the decay of the *Yukawa particle*, $\pi^{+} \rightarrow \mu^{+} + \nu^{\mu}$. Sakata and Inoue predicted correct *spin* assignment for the *muon*, and they also introduced the second *neutrino*. They treated it as a distinct particle from the *beta decay neutrino*, and anticipated correctly the three body decay of the *muon*. As a result of World War II, the English printing of Sakata-Inoue's *two-meson theory* paper was delayed until 1946, one year before the experimental discovery of $\pi \rightarrow \mu\nu$ decay.

In Yang, C. N. & Mills, R. (October, 1954). Conservation of Isotopic Spin and Isotopic Gauge Invariance, Chen-Ning Yang and Robert Mills extended the concept of *gauge theory* for *abelian* groups, e.g. *quantum electrodynamics*, to *nonabelian* groups to provide an explanation for *strong interactions*. The *Yang–Mills theory* is a *quantum field theory* for *nuclear binding*. It is a *gauge theory* based on a *special unitary group* SU(n), or more generally any compact Lie group.

> [The *special unitary group* of degree n, denoted SU(n), is the Lie group of n × n *unitary matrices* with *determinant* (real) 1.]

It seeks to describe the behavior of *elementary particles* using these non-abelian Lie groups and *is at the core of the unification of the electromagnetic force and weak forces* (i.e. U(1) × SU(2)) as well as *quantum chromodynamics*, the theory of the *strong force* (based on SU(3)). Thus, it forms the basis of the understanding of the *Standard Model* of particle physics.

In 1956, [Sakata, S. (September, 1956). On a Composite Model for the New Particles.] Sakata proposed his *Sakata Model* which explains the physics behind the *Nakano-Nishijima-Gell-Mann (NNG) rule* by postulating that the fundamental building blocks of all strongly interacting particles are the *proton*, the *neutron* and the *lambda baryon*. The positively charged *pion* is made out of a *proton* and an *anti-neutron*, in a manner similar to the *Fermi-Yang composite Yukawa meson model*, while the positively charged *kaon* is composed of a *proton* and an *anti-lambda*. Aside from the integer *charges*, the *proton*, *neutron*, and *lambda* have similar properties as the *up quark*, *down quark*, and *strange quark* respectively.

In 1956 [Lee, T. D. & Yang, C. N. (October, 1956). Question of Parity Conservation in Weak Interactions], Chen-Ning Yang and Tsung-Dao Lee formulated a theory that the *left-right symmetry law* is violated by the *weak interaction*. They noted that recent experimental data indicated closely identical *masses* and *lifetimes* of the θ^+ and the τ^+ *mesons*. On the other hand, analyses of the decay products of τ^+ strongly suggested on the grounds of *angular momentum* and *parity conservation* that the τ^+ and θ^+ were not the same particle. This posed a rather puzzling situation that has been extensively discussed. They suggested that one way out of the difficulty was to assume that *parity is not strictly conserved*, so that θ^+ and τ^+ are two different decay modes of the same particle, which necessarily has a single *mass* value and a single *lifetime*. They analyzed this possibility in this paper against the background of the existing experimental evidence of *parity conservation*, and concluded that existing experiments indicated *parity conservation* in *strong* and *electromagnetic interactions* to a high degree of accuracy, but that for the *weak interaction* actions (i.e., decay *interactions* for the *mesons* and *hyperons*, and various Fermi *interactions*) *parity conservation* was so far only an extrapolated hypothesis unsupported by experimental evidence. The question of *parity conservation* in *β decays* and in *hyperon* and *meson decays* was examined, and possible experiments were suggested which might test *parity conservation* in these *interactions*.

In Nambu, Y. (April, 1960). Axial Vector Current Conservation in Weak Interactions, Yoichiro Nambu noted that, in analogy to the conserved *vector current interaction* in the *beta decay* suggested by Feynman and Gell-Mann, some speculations have been made about a possibly conserved *axial vector current*. We would like to suggest that there may not be a strict *pseudovector current conservation*, but that we may have an approximate conservation which becomes rigorous in the limit $q^2 \gg m_\pi^2$, m_π being the *pion* mass and q^2 a *massless, pseudo scalar*, and *charged quantum* bridging the *nucleon* and *lepton currents*. We are tempted to extend this approximate conservation of the *axial vector current* (and naturally also the *vector current*) to the *strangeness-non conserving beta decays*.

In Sakurai, J. J. (September, 1960). Theory of strong interactions, Jun John Sakurai, a Japanese-American particle physicist, noted that all the *symmetry* models of *strong interactions* which had been proposed up to the present were devoid of deep physical foundations. He suggested that, instead of postulating artificial "higher" *symmetries* which must be broken anyway within the realm of *strong interactions*, we take the *existing exact* symmetries of *strong interactions* more seriously than before and exploit them to the utmost limit. A new theory of *strong interactions* was proposed on this basis.

In Goldstone, J. (January, 1961). Field theories with "Superconductor" solutions, Jeffrey Goldstone, a British theoretical physicist, examined the conditions for the existence of non-perturbative type "superconductor" solutions of field theories. A *non-covariant canonical transformation method* was used to find such solutions for a theory of a *fermion* interacting with a *pseudoscalar boson*. A *covariant renormalizable method* using Feynman integrals was then given. A *"superconductor" solution* was found whenever in the *normal perturbative-type solution* the *boson* mass squared was negative and the coupling constants satisfied certain inequalities. The *symmetry properties* of such solutions were examined with the aid of a simple model of self-interacting *boson* fields. The solutions had lower *symmetry* than the Lagrangian, and contain *mass zero bosons*.

Goldstone's theorem in *relativistic quantum field theory* states that if there is an exact *continuous symmetry* of the Hamiltonian or Lagrangian defining the system, and this is not a symmetry of the vacuum state (i.e. there is *broken symmetry*), then there must be at least one *spin-zero massless particle* called a *Goldstone boson*. In the *quantum theory of many-body systems Goldstone bosons* are collective excitations such as *spin waves*. An important exception to *Goldstone's theorem* is provided in *gauge theories* with the *Higgs mechanism*, whereby the *Goldstone bosons* gain *mass* and become *Higgs bosons*.

In Glashow, S. L. (February, 1961). Partial-symmetries of weak Interactions, Sheldon Glashow, Sheldon Glashow combined the *electromagnetic* and *weak interactions* and extended *electroweak unification models* due to Schwinger by including a short-range *neutral current*, the Z_0. The resulting *symmetry structure* that Glashow proposed, $SU(2) \times U(1)$, forms the basis of the accepted *theory of the electroweak interactions*. The *W and Z bosons* were predicted in detail by Sheldon Glashow, Mohammad Abdus Salam, and Steven Weinberg. For this discovery, Glashow along with Steven Weinberg and Abdus Salam, was awarded the 1979 Nobel Prize in Physics.

In Gell-Mann, M. (March, 1961). The Eightfold Way: A Theory of Strong Interaction Symmetry (Report). Murray Gell-Mann introduced his formulation of a particle classification system for *hadrons* known as *the Eightfold Way* – or, in more technical terms, $SU(3)$ *flavor symmetry*, streamlining its structure. A new model of the higher symmetry of elementary particles was introduced in which the eight known *baryons* were treated as a *super-multiplet*, degenerate in the limit of *unitary symmetry* but split into isotopic *spin multiplets* by a *symmetry-breaking* term.

The *symmetry violation* was ascribed phenomenologically to the *mass* differences. The *baryons* corresponded to an eight-dimensional irreducible *representation* of the *unitary group*. The *pion* and *K meson* fit into a similar set of eight particles along with a predicted *pseudoscalar meson* χ^0 having I = 0. A ninth *vector meson* coupled to the *baryon* current could be accommodated naturally in the scheme. Gell-Mann predicted that the eight *baryons* should all have the same *spin* and *parity* and that *pseudoscalar* and *vector mesons* should form *octets* with possible additional *singlets*. The mathematics of the *unitary group* was described by considering three fictitious *leptons*, v , e⁻, and μ⁻, which might throw light on the structure of *weak interactions*.

In Schwinger, J. (January, 1962). Gauge Invariance and Mass, Julian Schwinger argued that the *gauge invariance* of a *vector field* did not necessarily imply zero *mass* for an associated particle if the *current vector coupling* was sufficiently strong. He suggested that this situation might permit a deeper understanding of nucleonic *charge conservation* as a manifestation of a *gauge invariance*, without the obvious conflict with experience that a massless particle entailed.

In Goldstone, J., Salam, A. & Weinberg, S. (August, 1962). Broken Symmetries, some proofs were presented of Goldstone's Theorem, that *"in a manifestly Lorentz-invariant quantum field theory, if there is a continuous symmetry transformation under which the Lagrangian is invariant, then either the vacuum state is also invariant under the transformation, or there must exist spinless particles of zero mass"*.

In Anderson, P. W. (April, 1963). Plasmons, Gauge Invariance, and Mass, Philip Anderson noted that Schwinger had pointed out that the *Yang-Mills vector boson* implied by associating a generalized *gauge transformation* with a conservation law (of *baryonic charge*, for instance) did not necessarily have zero *mass*, if a certain criterion on the vacuum fluctuations of the generalized *current* was satisfied. He showed that the theory of plasma oscillations was a simple *nonrelativistic* example exhibiting all of the features of Schwinger's idea. He also showed that Schwinger's criterion that the *vector field* $m \neq 0$ implied that the *matter* spectrum before including the *Yang-Mills interaction* contained $m = 0$, but that the example of *superconductivity* illustrated that the physical spectrum need not. Some comments on the relationship between these ideas and the zero-*mass* difficulty in theories with *broken symmetries* were given.

In Gell-Mann. M. (February, 1964). A Schematic Model of Baryons and Mesons, Murray Gell-Mann proposed that *baryons*, which include *protons* and *neutrons*, and *mesons* were composed of elementary particles, which Gell-Mann called "*quarks*". The theory came to be called the *quark model*. The bootstrap model for *strongly interacting particles* described in terms of the *broken eightfold way* was discussed to determine algebraic properties of the interactions with scattering amplitudes on the mass shell. A mathematical model based on *field theory* was described.

In 1964, James Cronin and Val Fitch, American particle and nuclear physicists, demonstrated, using *neutral K-mesons*, the violation of all three symmetry principles (1) that the laws of Nature are exactly alike for both *antimatter* and ordinary *matter*; (2) that the fundamental laws have exact *mirror symmetry*; and (3) that the fundamental laws have exact *time reflection symmetry* – symmetry under motion reversal. For this discovery, they received the 1980 Nobel Prize in Physics.

In Englert, F. & Brout, R. (August, 1964). Broken Symmetry and the Mass of Gauge Vector Mesons, François Englert and Robert Brout showed that *gauge vector fields*, abelian and non-abelian, could acquire *mass* if empty space were endowed with a particular type of structure that one encounters in material systems. Other physicists, Peter Higgs and Gerald Guralnik, C. R. Hagen and Tom Kibble had reached similar conclusions at about the same time. The *Brout–Englert–Higgs* (BEH) *mechanism* is believed to give rise to the *masses* of all the elementary particles in the *Standard Model*. This includes the *masses* of the W and Z *bosons*, and the *masses* of the *fermions*, i.e. the *quarks* and *leptons*.

In Higgs, P. W. (October, 1964). Broken Symmetries and the Masses of Gauge Bosons. Peter Higgs observed that in a recent note he had shown that the *Goldstone theorem*, that Lorentz-covariant field theories in which spontaneous breakdown of *symmetry* under an internal Lie group occurs contain *zero-mass* particles, *fails if and only if the conserved currents associated with the internal group are coupled to gauge fields*. The purpose of the present note was to report that, *as a consequence of this coupling, the spin-one quanta of some of the gauge fields acquire mass*; the longitudinal degrees of freedom of these particles (which would be absent if their *mass* were zero) go over into the Goldstone bosons *when the coupling tends to zero. The model was discussed mainly in classical terms*; nothing was proved about the quantized theory. Higgs commented that it should be understood, therefore, that *the conclusions which were presented concerning the masses of particles were conjectures based on the quantization of linearized classical field equations.*

In Salam, A. & Ward, J. C. (November, 1964). Electromagnetic and Weak Interactions, Abdus Salam and John Ward worked on the synthesis of the *weak* and *electromagnetic interaction*, obtaining a *gauge theory* based on the SU(2) × U(1) model.

In Weinberg, S. (November, 1967). Model of Leptons, Steven Weinberg incorporated the *Brout–Englert–Higgs (BEH) mechanism* into Glashow's *electroweak interaction*, giving it its modern form. Weinberg proposed his *model of unification of electromagnetism and nuclear weak forces* with the *masses* of the force-carriers of the *weak* part of the *interaction* being explained by *spontaneous symmetry breaking*, in which the *symmetry* between the *electromagnetic* and *weak interactions* was *spontaneously broken*, but in which the *Goldstone bosons* were avoided by introducing the *photon* and the *intermediate boson fields* as *gauge fields*.

In 't Hooft, G. (December, 1971). Renormalizable Lagrangians for Massive Yang-Mills Fields, Gerard 't Hooft showed how *renormalizable* models are constructed in which *local gauge invariance* was *broken spontaneously*. He noted that *Feynman rules* and *Ward identities* could be found by means of a path integral method, and they could be checked by algebra. In one of these models, which was studied in more detail, *local* SU(2) was broken in such a way that *local* U(1) remained as a *symmetry*. This resulted in a *renormalizable and unitary theory*, with *photons*, *charged massive vector particles*, and additional *neutral scalar particles*. It had three independent parameters. Another model had local SU(2) x U(1) as a *symmetry* and might serve as a *renormalizable* theory for ρ-mesons and *photons*. In such models, *electromagnetic mass-differences* were finite and could be calculated in *perturbation theory*.

In Kobayashi, M. & Maskawa, T. (February, 1973). CP-Violation in the Renormalizable Theory of Weak Interaction, Makoto Kobayashi and Toshihde Maskawa explained *broken symmetry* within the framework of the *Standard Model*, but *required that the Model be extended to three families of quarks to explain CP violation,* which ultimately led to the *six-quark model*.

PART II is an addition on ***Unified Gravity*** in the second edition**,** describing two recent papers by Partanen, M. & Tulkki, J. The introduction focusses on the innovation of most interest, which is Partanen, M. & Tulkki, J. (March, 2024)'s introduction of an eight-component spinorial wave equation of the electromagnetic field and the application of this to produce an eight-component spinorial (non-relativistic) formulation of Maxwell's equations in which the four equations are represented by a single eight-component spinorial equation.

Partanen, M. & Tulkki, J. (March, 2024). QED based on an eight-dimensional spinorial wave equation of the electromagnetic field and the emergence of quantum gravity, the applied this to *relativistic* quantum electrodynamics (QED). A coupling between the electromagnetic field, Dirac electron-positron field, and the gravitational field was developed based in an eight-component spinorial representation of the electromagnetic field which is analogous to the well-known representation of particles in the Dirac theory but it is given in terms of 8×8 bosonic gamma matrices. This enabled the introduction of the generating Lagrangian density of gravity based on the special unitary symmetry of the eight-dimensional spinor space. The generating Lagrangian density of gravity played, in the definition of the gauge theory of gravity and its symmetric stress energy-momentum tensor source term, a similar role as the conventional Lagrangian density of the free Dirac field plays in the definition of the gauge theory of QED and its electric four-current density source term.

Partanen, M. & Tulkki, J. (May, 2025). Gravity generated by four one-dimensional unitary gauge symmetries and the Standard Model, attempted to derive the gauge theory of gravity using compact, finite-dimensional symmetries in a way that resembled the formulation of

28

the fundamental interactions of the Standard Model. For our eight-spinor representation of the Lagrangian, a quantity, called the space-time dimension field, was defined which enabled the extraction of four-dimensional space-time quantities from the eight-dimensional spinors. Four $U(1)$ symmetries of the components of the space-time dimension field were used to derive a gauge theory, called **unified gravity**. The stress-energy-momentum tensor source term of gravity follows directly from these symmetries. The metric tensor entered in unified gravity through geometric conditions. Based on the Minkowski metric, unified gravity allowed gravity to be described within a single coherent mathematical framework together with the quantum fields of all fundamental interactions of the Standard Model. The equivalence principle was formulated by requiring that the renormalized values of the inertial and gravitational masses were equal. In contrast to previous gauge theories of gravity, all infinities that were encountered in the calculations of loop diagrams could be absorbed by the redefinition of the small number of parameters of the theory in the same way as in the gauge theories of the Standard Model. This result and the observation that unified gravity fulfills the Becchi–Rouet–Stora–Tyutin (BRST) symmetry and its coupling constant is dimensionless suggest that unified gravity can provide the basis for a complete, renormalizable theory.

Part III, which was previously Part II, introduces an alternative to the *Standard Model, Supersymmetry*; and **Part IV**, which was previously Part II, introduces, *String Theory*, neither of which fare any better.

The second edition also includes a description of the Lagrangian on pages 20-1.

I would like to acknowledge Wikipedia, in particular, which provided much of this material, as well as other referenced sources.

Trevor G. Underwood
18 SE 10th Ave
Fort Lauderdale, FL33301.

August 7, 2025.

PART I   The Standard Model of Particle Physics.

The *Standard Model of particle physics* is the theory describing three of the four known fundamental *forces* (*electromagnetic, weak and strong interactions* – excluding gravity) in the universe and *classifying all known elementary particles*.

It was developed in stages throughout the latter half of the 20th century, through the work of many scientists worldwide, with the current formulation being finalized in the mid-1970s upon experimental confirmation of the existence of *quarks*. Since then, proof of the *top quark* (1995), the *tau neutrino* (2000), and the *Higgs boson* (2012) have added further credence to the *Standard Model*. In addition, the *Standard Model* has predicted various properties of *weak neutral currents* and the W and Z *bosons* with great accuracy.

Although the *Standard Model* is believed to be theoretically self-consistent*

> * There are mathematical issues regarding quantum field theories still under debate (see e.g. Landau pole), but the predictions extracted from the Standard Model by current methods applicable to current experiments are all self-consistent.

and has demonstrated some success in providing experimental predictions, *it leaves some physical phenomena unexplained and so falls short of being a complete theory of fundamental interactions*. Although the physics of *special relativity* is included, *general relativity is not*, and it will fail at energies or distances where the *graviton* is expected to emerge. It does not fully explain *baryon asymmetry*, or account for the *universe's accelerating expansion as possibly described by dark energy*. The model does *not contain any viable dark matter particle* that possesses all of the required properties deduced from observational cosmology. It also does not incorporate *neutrino oscillations* and their non-zero masses.

The development of the *Standard Model* was driven by theoretical and experimental particle physicists alike. *The Standard Model* is a paradigm of a *quantum field theory* for theorists, exhibiting a wide range of phenomena, including spontaneous symmetry breaking, anomalies, and non-perturbative behavior. It is used as a basis for building more exotic models that incorporate hypothetical particles, extra dimensions, and elaborate symmetries (such as *supersymmetry*) to explain experimental results at variance with the *Standard Model*, such as the existence of dark matter and neutrino oscillations.

The *Standard Model* is a *Yang–Mills theory*, a *quantum field theory for nuclear binding* devised by Chen Ning Yang and Robert Mills in 1953, who extended the concept of *gauge theory* for *abelian* groups, e.g. *quantum electrodynamics*, to *nonabelian* groups to provide an explanation for *strong interactions*. [Yang, C. N. & Mills, R. (1954). Conservation of Isotopic Spin and Isotopic Gauge Invariance. *Phys. Rev.*, 96, 1, 191–5. See below.]

> [A Yang–Mills theory is a *gauge theory* based on a special unitary group SU(n), or more generally any compact *Lie group*. A Yang–Mills theory seeks to describe the behavior of elementary particles using *non-abelian Lie groups* and is at the core of the *unification of the electromagnetic force and weak forces* (i.e. U(1) × SU(2)) as well as *quantum chromodynamics*, the theory of the *strong force* (based on SU(3)).]

In 1961, Sheldon Glashow combined the *electromagnetic* and *weak interactions*. [Glashow, S. L. (1961). Partial-symmetries of weak interactions. *Nuclear Physics*, 22, 4, 579–88. See below.] In 1964, a theory, subsequently known as the *Higgs mechanism*, which is believed to give rise to the *masses* of all the elementary particles in the *Standard Model*, includes the masses of the W and Z *bosons*, and the masses of the *fermions*, was published almost simultaneously by three independent groups; by Robert Brout and François Englert; by Peter Higgs; and by Gerald Guralnik, C. R. Hagen, and Tom Kibble. [Higgs, P.W. (1964). Broken Symmetries and the Masses of Gauge Bosons. See below.] In 1967 Steven Weinberg and Abdus Salam incorporated the *Higgs mechanism* [Weinberg, S. (1967). Model of Leptons. See below.] into Glashow's *electroweak interaction*, giving it its modern form.

After the *neutral weak currents* caused by *Z boson exchange* were discovered at CERN in 1973, [Hasert, F.J. et al. (1973). Observation of neutrino-like interactions without muon or electron in the Gargamelle neutrino experiment. *Phys. Rev. Let.* B, 46, 1, 138; https://doi.org/10.1016/0370-2693(73)90499-1] the *electroweak theory* became widely accepted and Glashow, Salam, and Weinberg shared the 1979 Nobel Prize in Physics for discovering it. The $W^{\pm}$ and Z^0 *bosons* were discovered experimentally in 1983; and the ratio of their masses was found to be as the *Standard Model* predicted.

The theory of the *strong interaction* (i.e. *quantum chromodynamics*), to which many contributed, acquired its modern form in 1973–74 when asymptotic freedom was proposed (a development which made *quantum chromodynamics* the main focus of theoretical research) and experiments confirmed that the *hadrons* were composed of fractionally charged *quarks*.

> [The term "*Standard Model*" was introduced by Abraham Pais and Sam Treiman in 1975, [Pais, A. & Treiman, S. B. (1975). How Many Charm Quantum Numbers are There? *Phys. Rev. Let.*, 35, 23, 1556–9; https://doi.org/10.1103/PhysRevLett.35.1556.] with reference to the electroweak theory with four quarks. Steven Weinberg, has since claimed priority, explaining that he chose the term

Standard Model out of a sense of modesty and used it in 1973 during a talk in Aix-en-Provence in France.]

Elementary particles (26)

In the *Standard Model*, there is a total of 26 *elementary particles*, including the hypothetical graviton. These can be summarized as follows.

Standard Model of Elementary Particles

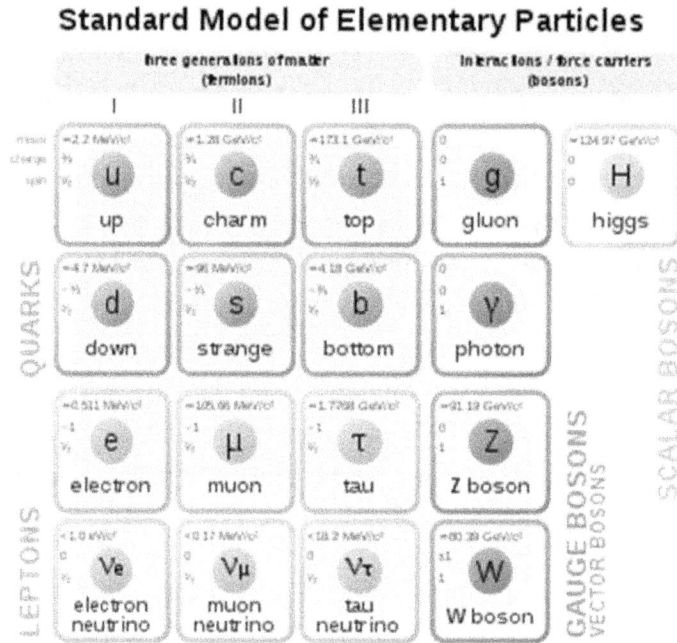

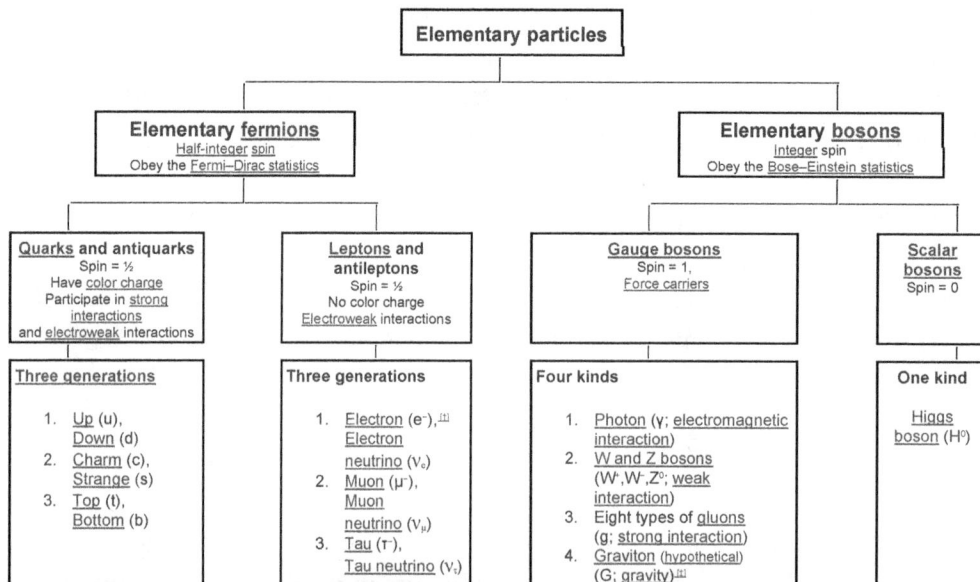

[†] An anti-electron (e⁺) is conventionally called a "positron".

[†] An anti-electron (e^+) is conventionally called a "positron".

Fermions – half-integer spin (12)

The *Standard Model* includes 12 elementary particles of spin 1/2, known as *fermions*. *Fermions* respect the *Pauli exclusion principle*, meaning that *two identical fermions cannot simultaneously occupy the same quantum state in the same atom*. Each *fermion* has a corresponding *antiparticle*, which are particles that have corresponding properties with the exception of *opposite charges*. *Fermions* are classified based on how they interact, which is determined by the charges they carry, into two groups: *quarks* and *leptons*.

Within each group, pairs of particles that exhibit similar physical behaviors are then grouped into *generations* (see the table). Each member of a *generation* has a greater *mass* than the corresponding particle of generations prior. Thus, *there are three generations of quarks and leptons. As first-generation particles do not decay, they comprise all of ordinary (baryonic) matter.*

Specifically, *all atoms consist of electrons orbiting around the atomic nucleus*, ultimately constituted of *up* and *down quarks*. On the other hand, *second- and third-generation charged particles decay with very short half-lives and can only be observed in high-energy environments*. *Neutrinos* of all generations also do not decay, and pervade the universe, but rarely interact with *baryonic* matter.

Quarks – spin ½; color charge (6)

There are six quarks: up, down, charm, strange, top, and bottom. Quarks carry *color charge*, and hence interact via the *strong interaction*. *Quarks* also carry *electric charge* and *weak isospin*, and *thus interact with other fermions through the electromagnetism and weak interaction*. The *color confinement* phenomenon results in *quarks* being strongly bound together such that they form color-neutral composite particles called *hadrons*; *quarks cannot individually exist* and must always bind with other quarks.

Hadrons - mesons and baryons

Hadrons can contain either a *quark-antiquark pair* (*mesons*) or *three quarks* (*baryons*). *The lightest baryons are the nucleons: the proton and neutron.*

Leptons – spin ½; no color charge (6)

The six leptons consist of the electron, electron neutrino, muon, muon neutrino, tau, and tau neutrino. The leptons do not carry *color charge*. They interact via the *electroweak interaction*, and do not respond to the *strong interaction*.

The main *leptons* carry an *electric charge* of –1 e. The three *neutrinos* carry a *neutral electric charge*. Thus, the *neutrinos'* motion is only influenced by *weak interaction* and *gravity*, making them difficult to observe.

Bosons – integer spin (14)

Gauge bosons – spin 1 (13)

The *Standard Model* includes 4 kinds of *gauge bosons* of *spin* 1, with *bosons* being *quantum particles* containing an *integer spin*. The *gauge bosons* are defined as *force carriers*, as they are responsible for *mediating* the *fundamental interactions. The Standard Model explains the four fundamental forces as arising from the interactions, with fermions exchanging virtual force carrier particles, thus mediating the forces.* At a macroscopic scale, this manifests as a *force*. As a result, *they do not follow the Pauli exclusion principle* that constrains fermions; *bosons* do not have a theoretical limit on their *spatial density*. The types of *gauge bosons* are described below.

Photons (1)

Photons mediate the *electromagnetic force*, responsible for interactions between *electrically charged particles*. The *photon* is *massless* and is *described by the theory of quantum electrodynamics* (QED).

Gluons (8)

Gluons mediate the *strong interactions*, which *binds quarks to each* other by influencing the *color charge*, with the interactions being *described in the theory of quantum chromodynamics* (QCD). They have *no mass*, and there are *eight distinct gluons*, with each being denoted through a *color-anticolor charge combination* (e.g. red–antigreen).

> [Although nine color–anticolor combinations mathematically exist, gluons form color octet particles. As one color-symmetric combination is linear and forms a color singlet particle, there are eight possible gluons.]

As *gluons* have an effective color charge, they can also interact amongst themselves.

W⁺, W⁻, and Z gauge bosons (3)

The W⁺, W⁻, and Z *gauge bosons* mediate the *weak interactions* between all *fermions*, being responsible for *radioactivity*. They contain *mass*, with the Z having more *mass* than the W$^{\pm}$. The *weak interactions* involving the W$^{\pm}$ act only on left-handed *particles* and right-handed *antiparticles*. The W$^{\pm}$ carries an *electric charge* of +1 and −1 and couples to

34

the *electromagnetic interaction*. The electrically *neutral Z boson* interacts with both left-handed *particles* and right-handed *antiparticles*. These three *gauge bosons* along with the *photons* are grouped together, as collectively mediating the *electroweak interaction*.

Graviton – hypothetical (1)

Gravity is currently unexplained in the *Standard Model*, as the hypothetical mediating particle the *graviton* has been proposed, but not observed. This is due to the *incompatibility of quantum mechanics and Einstein's theory of general relativity*, regarded as being the best explanation for *gravity*. In *general relativity*, gravity is explained as being the geometric curving of spacetime.

Scalar bosons – spin 0 (1)

Higgs boson (1)

The *Higgs particle* is a massive scalar elementary particle theorized by Peter Higgs (and others) in 1964, when he showed that Goldstone's 1962 theorem (generic continuous symmetry, which is spontaneously broken) provides a third polarization of a massive vector field. Hence, Goldstone's original scalar doublet, the massive spin-zero particle, was proposed as the *Higgs boson*, and is a key building block in the *Standard Model*. It has *no intrinsic spin*, and for that reason is classified as a boson with spin-0.

The *Higgs boson* plays a unique role in the *Standard Model*, by explaining why the other elementary particles, except the *photon* and *gluon*, have *mass*. In *particular, the Higgs boson* explains why the *photon* has *no mass*, while the W and Z *bosons* are very heavy. Elementary-particle *masses* and the differences between *electromagnetism* (mediated by the *photon*) and the *weak force* (mediated by the W and Z *bosons*) are critical to many aspects of the structure of microscopic (and hence macroscopic) *matter*. In *electroweak theory*, the *Higgs boson* generates the *masses* of the *leptons* (electron, muon, and tau) and *quarks*. As the *Higgs boson* has *mass*, it must interact with itself.

Because the *Higgs boson* is a *very massive* particle and also decays almost immediately when created, only a very high-energy particle accelerator can observe and record it. Experiments to confirm and determine the nature of the *Higgs boson* using the Large Hadron Collider (LHC) at CERN began in early 2010 and were performed at Fermilab's Tevatron until its closure in late 2011. Mathematical consistency of the *Standard Model* requires that any mechanism capable of generating the masses of elementary particles must become visible at energies above 1.4 TeV; therefore, the LHC (designed to collide two 7 TeV proton beams) was built to answer the question of whether the *Higgs boson* actually exists.

On July 4, 2012, two of the experiments at the LHC (ATLAS and CMS) both reported independently that they had found a new particle with a mass of about 125 GeV/c2 (about 133 proton masses, on the order of 10^{-25} kg), which is "consistent with the *Higgs boson*". On March 13, 2013, it was confirmed to be the searched-for *Higgs boson.*

Technically, *quantum field theory* provides the mathematical framework for the *Standard Model*, in which a *Lagrangian* controls the dynamics and kinematics of the theory. Each kind of particle is described in terms of a dynamical field that pervades space-time. [Jaeger, G. (2021). The Elementary Particles of Quantum Fields. *Entropy*, 23, 11, 1416; https://doi.org/10.3390/e23111416.] The construction of the *Standard Model* proceeds following the modern method of constructing most field theories: by first postulating a set of symmetries of the system, and then by writing down the most general renormalizable Lagrangian from its particle (field) content that observes these symmetries.

The local SU(3) × SU(2) × U(1) *gauge symmetry* is an *internal symmetry* that essentially defines the *Standard Model*.

> [The *symmetry group* of a geometric object is the group of all *transformations under which the object is invariant*, endowed with the group operation of composition.
>
> A *gauge theory* is a type of field theory in which the Lagrangian, and hence the dynamics of the system itself, do not change under local transformations according to certain smooth families of operations (*Lie groups*). Formally, the Lagrangian is invariant.
>
> The term *gauge* refers to any specific mathematical formalism to regulate redundant degrees of freedom in the Lagrangian of a physical system. The transformations between possible *gauges*, called *gauge transformations*, form a *Lie group*— referred to as the *symmetry group* or the *gauge group* of the theory. Associated with any *Lie group* is the *Lie algebra* of *group generators*. For each *group generator* there necessarily arises a corresponding *field* (usually a *vector field*) called the *gauge field*. *Gauge fields are included in the Lagrangian to ensure its invariance under the local group transformations (called gauge invariance)*. When such a theory is quantized, the *quanta* of the *gauge fields* are called *gauge bosons*. If the symmetry group is *non-commutative*, then the *gauge theory* is referred to as *non-abelian gauge theory*, the usual example being the *Yang–Mills theory.*
>
> > [An **Abelian gauge theory** is a **quantum field theory** that explains the dynamics of elementary particles. It is a **gauge theory** with the symmetry group U(1) and has one **gauge field**, the electromagnetic four-potential,

with the photon being the gauge boson. Gauge theories are quantum theories of vector field Aa μ(x) whose interactions with each other and with other fields follows from a local symmetry.]

Many powerful theories in physics are described by Lagrangians that are invariant under some symmetry transformation groups. When they are invariant under a transformation identically performed at every point in the spacetime in which the physical processes occur, they are said to have a *global symmetry*. *Local symmetry*, the cornerstone of *gauge theories*, is a stronger constraint. In fact, a global symmetry is just a local symmetry whose group's parameters are fixed in spacetime (the same way a constant value can be understood as a function of a certain parameter, the output of which is always the same).

Gauge theories are important as the successful *field theories* explaining the dynamics of elementary particles. *Quantum electrodynamics is an abelian gauge theory with the symmetry group* U(1) *and has one gauge field, the electromagnetic four-potential, with the photon being the gauge boson. The Standard Model is a non-abelian gauge theory with the symmetry group U(1) × SU(2) × SU(3) and has a total of twelve gauge bosons: the photon, three weak bosons and eight gluons.*

The *unitary group* of degree n, denoted U(n), is the group of n × n *unitary matrices*, with the *group operation* of *matrix multiplication*. In the simple case n = 1, the group U(1) corresponds to the *circle group*, consisting of all *complex numbers* with absolute value 1, under multiplication.

> The *special unitary group* of degree n, denoted SU(n), is the Lie group of n × n *unitary matrices* with *determinant* (real) 1. The matrices of the more general unitary group may have complex determinants with absolute value 1, rather *than real 1 in the special case*. The *group operation is matrix multiplication*. The special unitary group is a normal subgroup of the unitary group U(n), consisting of all n × n unitary matrices. The SU(n) groups find wide application in the *Standard Model* of particle physics, especially SU(2) in the *electroweak interaction* and SU(3) in the *strong interaction* (*quantum chromodynamics*). The group SU(2) is a simple Lie group consisting on all 2 x 2 matrices of *determinant* 1. The group SU(3) is a simple Lie group consisting on all 3 x 3 matrices of *determinant* 1.]

Roughly, *the three factors of the gauge symmetry give rise to the three fundamental interactions*. The *fields* fall into different representations of the various *symmetry groups* of the *Standard Model*. Upon writing the most general Lagrangian, one finds that the

37

dynamics [of the 26 types of *elementary particles*] depends on 19 parameters, whose numerical values are established by experiment (see table below).

Parameters of the Standard Model

#	Symbol	Description	Value
1	m_e	Electron mass	0.511 MeV
2	m_μ	Muon mass	105.7 MeV
3	m_τ	Tau mass	1.78 GeV
4	m_u	Up quark mass	1.9 MeV
5	m_d	Down quark mass	4.4 MeV
6	m_s	Strange quark mass	87 MeV
7	m_c	Charm quark mass	1.32 GeV
8	m_b	Bottom quark mass	4.24 GeV
9	m_t	Top quark mass	173.5 GeV
10	θ_{12}	CKM 12-mixing angle	13.1°
11	θ_{23}	CKM 23-mixing angle	2.4°
12	θ_{13}	CKM 13-mixing angle	0.2°
13	δ	CKM CP violation Phase	0.995
14	g_1 or g'	U(1) gauge coupling	0.357
15	g_2 or g	SU(2) gauge coupling	0.652
16	g_3 or g_s	SU(3) gauge coupling	1.221
17	θ_{QCD}	QCD vacuum angle	~0
18	v	Higgs vacuum expectation value	246 GeV
19	m_H	Higgs mass	125.09 ± 0.24 GeV.

Mathematical formulation of the Standard Model

The Standard Model is a *quantum field theory*, meaning its fundamental objects are *quantum fields* which are defined at all points in spacetime. *Quantum field theory* treats particles as *excited states* (also called *quanta*) of their underlying *quantum fields*, which are more fundamental than the particles. These fields are

- the *fermion fields*, ψ, which account for "*matter particles*";
- the *electroweak boson fields* W_1, W_2, W_3, and B;
- the *gluon field*, G_a; and
- the *Higgs field*, φ.

That these are *quantum* rather than classical fields has the mathematical consequence that they are *operator-valued*. In particular, *values of the fields generally do not commute*. As *operators*, they act upon a *quantum state* (*ket vector*).

Much of the qualitative descriptions of the *Standard Model* in terms of "particles" and "forces" comes from the perturbative *quantum field theory* view of the model. In this, the Lagrangian is decomposed as $L = L_0 + L_I$ into separate *free field* and *interaction* Lagrangians. The *free fields* care for particles in isolation, whereas processes involving several particles arise through *interactions*. The idea is that the *state vector* should only change when particles *interact*, meaning a free particle is one whose quantum state is constant. This corresponds to the *interaction picture* in quantum mechanics.

In the more common *Schrödinger picture*, even the *states* of free particles change over time: typically the phase changes at a rate which depends on their energy. In the alternative *Heisenberg picture*, *state vectors* are kept constant, at the price of having the *operators* (in particular the observables) be time-dependent. The *interaction picture* constitutes an intermediate between the two, where some time dependence is placed in the *operators* (the *quantum fields*) and some in the *state vector*. In *quantum field theory*, the former is called the *free field* part of the model, and the latter is called the *interaction* part. The *free field model* can be solved exactly, and then the solutions to the full model can be expressed as perturbations of the *free field* solutions, for example using the Dyson series.

The decomposition into *free fields* and *interactions* is in principle arbitrary. For example, *renormalization* in *quantum electrodynamics* modifies the *mass* of the *free field* electron to match that of a physical electron (with an *electromagnetic field*), and will in doing so add a term to the *free field* Lagrangian which must be cancelled by a counterterm in the *interaction* Lagrangian, that then shows up as a two-line vertex in the Feynman diagrams. *This is also how the Higgs field is thought to give particles mass*: the part of the interaction term which corresponds to the nonzero vacuum expectation value of the Higgs field is moved from the *interaction* to the *free field* Lagrangian, *where it looks just like a mass term having nothing to do with the Higgs field.*

Free fields

Under the usual free/interaction decomposition, which is suitable for low energies, the free fields obey the following equations:

- The *fermion field* ψ satisfies the Dirac equation; $(i\hbar\gamma^\mu\partial_\mu - m_f c)\psi_f = 0$ for each type f of fermion.
- The *photon field* A satisfies the wave equation $\partial_\mu\partial\mu Av=0$.
- The *Higgs field* φ satisfies the Klein–Gordon equation.
- The *weak interaction fields* Z, $W^\pm$ satisfy the Proca equation.

These equations can be solved exactly. One usually does so by considering first solutions that are periodic with some period L along each spatial axis; later taking the limit: $L \to \infty$ will lift this periodicity restriction.

In the *periodic case*, the solution for a field F (any of the above) can be expressed as a Fourier series of the form

$$F(x) = \ldots\ldots\ldots$$

...

In the limit $L \to \infty$, the sum would turn into an integral ...

... For these derivations, one starts out with expressions for the operators in terms of the quantum fields. That the operators with † are *creation operators* and the one without *annihilation operators* is a convention, imposed by the sign of the commutation relations postulated for them.

An important step in preparation for calculating in *perturbative quantum field theory* is to separate the "operator" factors ... from their corresponding vector or spinor factors ... .

...

Interaction terms and the path integral approach

The Lagrangian can also be derived without using *creation* and *annihilation operators* (the "canonical" formalism) by using a *path integral formulation*, pioneered by Feynman building on the earlier work of Dirac. Feynman diagrams are pictorial representations of *interaction terms*. ...

Lagrangian formalism

We can now give some more detail about the aforementioned *free* and *interaction* terms appearing in the *Standard Model* Lagrangian *density*. Any such term must be both *gauge* and *reference-frame* invariant, otherwise the laws of physics would depend on an arbitrary choice or the frame of an observer. Therefore, the *global* Poincaré symmetry, consisting of *translational symmetry*, *rotational symmetry* and the *inertial reference frame invariance* central to *the theory of special relativity* must apply. [???] The *local* $SU(3) \times SU(2) \times U(1)$ *gauge symmetry* is the *internal* symmetry. The three factors of the *gauge symmetry* together give rise to the three fundamental interactions, after some appropriate relations have been defined.

Kinetic terms

A *free particle* can be represented by a *mass* term, and a *kinetic* term which relates to the "motion" of the fields.

Fermion fields

The kinetic term for a *Dirac fermion* is $i\bar{\psi}\gamma^{\mu}\partial_{\mu}\psi$ where the notations are carried from above. ψ can represent any, or all, *Dirac fermions* in the *Standard Model*. Generally, as below, this term is included within the *couplings* (creating an overall "dynamical" term).

Gauge fields

For the spin-1 fields, first define the *field strength tensor*

$$F_{\mu\nu}{}^{a} = \partial_{\mu}A_{\nu}{}^{a} - \partial_{\nu}A_{\mu}{}^{a} + g f^{abc}A_{\mu}{}^{b}A_{\nu}{}^{c}$$

for a given gauge field (here we use A), with *gauge coupling constant* g. The quantity f^{abc} is the *structure constant* of the particular *gauge group*, defined by the *commutator*

$$[ta, tb] = i f^{abc}t_{c},$$

where t_i are the *generators* of the group. In an *Abelian (commutative) group* (such as the U(1) we use here) the *structure constants* vanish, since the generators t_a all commute with each other. Of course, this is not the case in general – the *Standard Model* includes the *non-Abelian* SU(2) and SU(3) groups (such groups lead to what is called a *Yang–Mills gauge theory*).

> [*Yang–Mills gauge theory* is a quantum field theory for nuclear binding devised by Chen Ning Yang and Robert Mills in 1953, as well as a generic term for the class of similar theories. The *Yang–Mills theory* is a *gauge theory* based on a *special unitary group* SU(n), or more generally any compact Lie group. A Yang–Mills theory seeks to describe the behavior of elementary particles using these *non-abelian* Lie groups and is at the core of the *unification of the electromagnetic force and weak forces* (i.e. U(1) × SU(2)) as well as *quantum chromodynamics*, the theory of the strong force (based on SU(3)). Thus, it forms the basis of the understanding of the *Standard Model* of particle physics.]

We need to introduce three *gauge fields* corresponding to each of the subgroups SU(3) × SU(2) × U(1).

- The *gluon field tensor* will be denoted by $G_{\mu\nu}{}^a$, where the index *a* labels elements of the 8 representation of *color* SU(3). The *strong coupling constant* is conventionally labelled g_s (or simply g where there is no ambiguity
- The notation $W_{\mu\nu}{}^a$ will be used for the *gauge field tensor* of SU(2) where *a* runs over the 3 *generators* of this group. The *coupling* can be denoted g_w or again simply g. The *gauge field* will be denoted by $W_\mu{}^a$.
- The *gauge field tensor* for the U(1) of *weak hypercharge* will be denoted by $B_{\mu\nu}$, the *coupling* by g', and the *gauge field* by B_μ.

The *kinetic term* can now be written as

$$L_{\text{kin}} = -\tfrac{1}{4}\, B_{\mu\nu}B_{\mu\nu} - \tfrac{1}{2}\, \text{tr}W_{\mu\nu}W^{\mu\nu} - \tfrac{1}{2}\, \text{tr}G_{\mu\nu}G^{\mu\nu}$$

where the traces are over the SU(2) and SU(3) indices hidden in W and G respectively. The two-index objects are the *field strengths* derived from W and G the *vector fields*. There are also two extra *hidden parameters*: the *theta angles* for SU(2) and SU(3).

Coupling terms

The next step is to "couple" the gauge fields to the fermions, allowing for interactions.

Electroweak sector

The *electroweak sector* is a *Yang-Mills gauge theory* with the *symmetry group* U(1) × SU(2)$_L$, where the subscript L indicates coupling only to *left-handed fermions*.

$$L_{\text{EW}} = \sum_\psi \bar{\psi}\, \gamma^\mu \left(i\partial_\mu - g'\, \tfrac{1}{2}\, Y_W B_\mu - g\, \tfrac{1}{2}\, \tau W_\mu\right)$$

where B_μ is the U(1) *gauge field*; Y_W is the *weak hypercharge* (the *generator* of the U(1) group); W_μ is the three-component SU(2) *gauge field*; and the *components* of τ are the *Pauli matrices* (infinitesimal *generators* of the SU(2) group) whose *eigenvalues* give the *weak isospin*.

> [*Note that we have to redefine a new U(1) symmetry of the weak hypercharge, different from quantum electrodynamics, in order to achieve the unification with the weak force.*]

The *electric charge* Q, third *component* of *weak isospin* T_3 (also called T_z, I_3 or I_z) and weak hypercharge Y_W are related by $Q = T_3 + \tfrac{1}{2}\, Y_W$, (or by the alternative convention $Q = T_3 + Y_W$). The first convention, used here, is equivalent to the earlier *Gell-Mann–Nishijima formula*. It makes the hypercharge be twice the average charge of a given *isomultiplet*.

One may then define the *conserved current* for *weak isospin* as

$$\mathbf{j}_\mu = \tfrac{1}{2}\,\overline{\psi}_L\gamma_\mu\boldsymbol{\tau}\psi_L$$

and for *weak hypercharge* as

$$j_\mu{}^Y = 2(j_\mu{}^{em} - j_\mu{}^3),$$

where $j_\mu{}^{em}$ is the *electric current* and $j_\mu{}^3$ the *third weak isospin current*. As explained above, these currents mix to create the physically observed *bosons*, which also leads to testable relations between the *coupling constants*.

Quantum chromodynamics sector

The *quantum chromodynamics sector* defines the *interactions between quarks and gluons, with SU(3) symmetry, generated* by T_a. Since *leptons* do not interact with *gluons*, they are not affected by this sector. The *Dirac Lagrangian* of the *quarks* coupled to the *gluon fields* is given by

$$L_{QCD} = i\overline{U}\,(\partial_\mu - ig_s G_\mu{}^a T^a)\gamma^\mu U + i\overline{D}\,(\partial_\mu - ig_s G_\mu{}^a T^a)\gamma^\mu D.$$

where U and D are the *Dirac spinors* associated with *up and down-type quarks*, and other notations are continued from the previous section.

Mass terms and the Higgs mechanism

Mass terms

The *mass term* arising from the *Dirac Lagrangian* (for any *fermion* ψ) is $-m\overline{\psi}\,\psi$ which is not *invariant* under the *electroweak symmetry*. This can be seen by writing ψ in terms of left and right-handed components (skipping the actual calculation):

$$-m\overline{\psi}\,\psi = -m(\overline{\psi}_L\psi_R + \overline{\psi}_R\psi_L)$$

i.e. contribution from $\overline{\psi}_L\psi_L$ and $\overline{\psi}_R\psi_R$ terms do not appear. *We see that the mass-generating interaction is achieved by constant flipping of particle chirality.* The *spin-half* particles have *no right/left chirality pair* with the same SU(2) representations and *equal and opposite weak hypercharges*, so assuming these *gauge charges* are conserved in the vacuum, *none of the spin-half particles could ever swap chirality, and must remain massless.*

[In mathematics, a figure is *chiral* (and said to have *chirality*) if it cannot be mapped to its mirror image by rotations and translations alone. In physics, *chirality* may be found in the spin of a particle, where the handedness of the object is determined by the direction in which the particle spins.]

Additionally, we know *experimentally* that the W and Z bosons are *massive*, but a *boson mass term* contains the combination e.g. $A^\mu A_\mu$, which clearly depends on the choice of gauge. Therefore, *none of the Standard Model fermions or bosons can "begin" with mass, but must acquire it by some other mechanism.*

The Higgs mechanism

The solution to both these problems comes from the *Higgs mechanism*, which involves *scalar fields* (the number of which depend on the exact form of *Higgs mechanism*) which (to give the briefest possible description) are "absorbed" by the *massive bosons* as degrees of freedom, and which couple to the *fermions* via *Yukawa coupling* to create what looks like *mass terms*.

[In particle physics, *Yukawa's interaction* or *Yukawa coupling*, named after Hideki Yukawa, is an interaction between particles according to the Yukawa potential. Specifically, it is a *scalar field* (or *pseudoscalar field*) ϕ and a *Dirac field* ψ of the type $V \approx g\bar{\psi}\phi\psi$ (scalar) or $g\bar{\psi}i\gamma 5\phi\psi$ (pseudoscalar).
The Yukawa interaction was developed to model the *strong force* between *hadrons*. A *Yukawa interaction* is thus used to describe the *nuclear force* between *nucleons* mediated by *pions* (which are *pseudoscalar mesons*).
A *Yukawa interaction* is also used in the *Standard Model* to describe the *coupling* between the *Higgs field* and *massless quark* and *lepton fields* (i.e., the fundamental *fermion* particles). Through spontaneous *symmetry breaking*, these *fermions* acquire a *mass* proportional to the vacuum expectation value of the *Higgs field. This Higgs-fermion coupling was first described by Steven Weinberg in 1967 to model lepton masses.*]

In the *Standard Model*, the *Higgs field* is a *complex scalar field* of the group SU(2)L:

$$\phi = 1/\sqrt{2}\ (\phi^+ \phi 0),$$

where the superscripts + and 0 indicate the *electric charge* (Q) of the *components*. The *weak hypercharge* (Y_W) of both components is 1.

The Higgs part of the Lagrangian is

$$L_H = [(\partial_\mu - igW_\mu^a t^a - ig'Y_\phi B_\mu)\phi]^2 + \mu^2\phi^\dagger\phi - \lambda(\phi^\dagger\phi)^2,$$

where $\lambda > 0$ and $\mu^2 > 0$, so that the mechanism of *spontaneous symmetry breaking* can be used. *There is a parameter here*, at first hidden within the shape of the potential, that is very important. In a *unitarity gauge* one can set $\phi^+ = 0$ and make ϕ^0 real. Then $\langle\phi^0\rangle = v$ is the non-vanishing vacuum expectation value of the *Higgs field*. v has units of *mass*, and it is the only parameter in the *Standard Model* which is not dimensionless. It is also much smaller than the Planck scale and about twice the *Higgs mass*, setting the scale for the *mass* of all other particles in the *Standard Model*. This is the only real fine-tuning to a small nonzero value in the *Standard Model*. Quadratic terms in W_μ and B_μ arise, which give masses to the W and Z *bosons*:

$$M_W = \tfrac{1}{2}\,vg$$
$$M_Z = \tfrac{1}{2}\,v\sqrt{(g2 + g'2)}.$$

The *mass* of the *Higgs boson* itself is given by $M_H = \sqrt{(2\mu^2)} \equiv \sqrt{(2\lambda v^2)}$.

Neutrino masses

As previously mentioned, evidence shows *neutrinos* must have *mass*. But within the Standard Model, the right-handed neutrino does not exist, so, even with a *Yukawa coupling*, neutrinos remain massless. An obvious solution is to simply *add a right-handed neutrino* v_R, which requires the addition of a new *Dirac mass* term in the Yukawa sector:

This field however must be a *sterile neutrino*, since being right-handed it experimentally belongs to an *isospin* singlet ($T_3 = 0$) and also has charge $Q = 0$, implying $Y_W = 0$ i.e. it does not even participate in the *weak interaction*. *The experimental evidence for sterile neutrinos is currently inconclusive.*

Since in any case new fields must be postulated to explain the experimental results, *neutrinos* are an obvious gateway to searching physics beyond the *Standard Model*.

Detailed information

This section provides more detail on some aspects, and some reference material. Explicit Lagrangian terms are also provided here.

Field content in detail

The *Standard Model* has the following *fields*. These describe one generation of leptons and quarks, and there are three generations, so there are three copies of each *fermionic field*. By CPT symmetry, there is a set of *fermions* and *antifermions* with opposite *parity* and *charges*. If a *left-handed fermion* spans some representation its antiparticle (*right-handed antifermion*) spans the dual representation (note that $\mathbf{2}^- = \mathbf{2}$ for SU(2), because it is *pseudo-real*). The column "**representation**" indicates under which *representations* of the *gauge groups* that each *field* transforms, in the order (SU(3), SU(2), U(1)) and for the U(1) group, the value of the *weak hypercharge* is listed. There are twice as many left-handed lepton field components as right-handed lepton field components in each generation, but an equal number of left-handed quark and right-handed quark field components.

[Table: *Field content of the standard model*. For tables see https://en.wikipedia.org/wiki/ Mathematical_formulation_of_the_Standard_Model]

Fermion content

[Table: *Left-handed fermions in the Standard Model*.]

Free parameters

Upon writing, *the most general Lagrangian with massless neutrinos, one finds that the dynamics depend on 19 parameters, whose numerical values are established by experiment*. Straightforward extensions of the *Standard Model* with massive *neutrinos* need *7 more parameters* (3 masses and 4 PMNS matrix parameters) for *a total of 26 parameters*. *The neutrino parameter values are still uncertain*. The 19 certain parameters are summarized here.

[Table: *Parameters of the Standard Model*. See above.]

The choice of *free parameters* is somewhat arbitrary. Instead of *fermion masses*, dimensionless *Yukawa couplings* can be chosen as *free parameters*. The value of the vacuum energy (or more precisely, the *renormalization scale* used to calculate this energy) may also be treated as an additional *free parameter*. The *renormalization scale* may be identified with the Planck scale or fine-tuned to match the observed *cosmological constant*. However, *both options are problematic*.

Additional symmetries of the Standard Model

From the theoretical point of view, the *Standard Model* exhibits *four additional global symmetries*, not postulated at the outset of its construction, collectively denoted *accidental symmetries*, which are continuous U(1) *global symmetries*.

In addition to the accidental (but exact) symmetries described above, the *Standard Model* exhibits several *approximate symmetries*. These are the "SU(2) *custodial symmetry*" and the "SU(2) or SU(3) *quark flavor symmetry*."

[Table: *Symmetries of the Standard Model and associated conservation laws*]

The U(1) symmetry

For the *leptons*, the *gauge group* can be written $SU(2)_l \times U(1)_L \times U(1)_R$. The two U(1) factors can be combined into $U(1)_Y \times U(1)_l$ where l is the *lepton number*. Gauging of the *lepton number* is ruled out by experiment, leaving only the possible gauge group $SU(2)_L \times U(1)_Y$. A similar argument in the *quark* sector also gives the same result for the *electroweak theory*.

Challenges

Self-consistency of the *Standard Model* (currently formulated as a *non-abelian gauge theory quantized through path-integrals*) has not been mathematically proved. While regularized versions useful for approximate computations (for example *lattice gauge theory*) exist, it is not known whether they converge (in the sense of S-matrix elements) in the limit that the regulator is removed. A key question related to the consistency is the *Yang–Mills existence and mass gap problem*.

Experiments indicate that *neutrinos* have *mass*, which the classic *Standard Model* did not allow. To accommodate this finding, the classic *Standard Model* can be modified to include neutrino mass, although it is not obvious exactly how this should be done.

If one insists on using only *Standard Model* particles, this can be achieved by adding a *non-renormalizable interaction of leptons with the Higgs boson*. On a fundamental level, such an interaction emerges in the seesaw mechanism where heavy right-handed neutrinos are added to the theory. This is natural in the left-right symmetric extension of the Standard Model and in certain grand unified theories. As long as new physics appears below or around 10^{14} GeV, the neutrino masses can be of the right order of magnitude.

Another problem is referred to as the *strong CP problem* which brings up the following quandary: why does quantum chromodynamics seem to preserve *CP-symmetry*? CP

47

stands for the combination of *charge conjugation symmetry* (C) and *parity symmetry* (P). According to the current mathematical formulation of *quantum chromodynamics*, there is no known reason for *CP-symmetry* to be conserved so a violation of *CP-symmetry* in *strong interactions* could occur. However, no violation of the *CP-symmetry* has ever been seen in any experiment involving only the *strong interaction*.

Other inadequacies of the *Standard Model* include:

- *The model does not explain gravitation*, although physical confirmation of a theoretical particle known as a *graviton* would account for it to a degree. Though it addresses strong and electroweak interactions, the *Standard Model* does not consistently explain the canonical theory of gravitation, general relativity, in terms of quantum field theory. The reason for this is, among other things, that quantum field theories of gravity generally break down before reaching the Planck scale. As a consequence, we have no reliable theory for the very early universe.

- Some physicists consider it to be *ad hoc and inelegant*, requiring 19 numerical constants *whose values are unrelated and arbitrary*. Although the *Standard Model*, as it now stands, can explain why neutrinos have masses, the specifics of neutrino mass are still unclear. It is believed that explaining neutrino mass will require an additional 7 or 8 constants, which are also arbitrary parameters.

- *The Higgs mechanism gives rise to the hierarchy problem* if some new physics (coupled to the Higgs) is present at high energy scales. In these cases, in order for the weak scale to be much smaller than the Planck scale, severe fine tuning of the parameters is required; there are, however, other scenarios that include quantum gravity in which such fine tuning can be avoided. There are also issues of quantum triviality, which suggests that *it may not be possible to create a consistent quantum field theory involving elementary scalar particles*.

- *The model is inconsistent with the emerging Lambda-CDM model of cosmology.* Contentions include the absence of an explanation in the *Standard Model* of particle physics for the observed amount of *cold dark matter* (CDM) and its contributions to *dark energy*, which are many orders of magnitude too large. It is also difficult to accommodate the *observed predominance of matter over antimatter (matter/antimatter asymmetry)*. The isotropy and homogeneity of the visible universe over large distances seems to require a mechanism like cosmic inflation, which would also constitute an extension of the *Standard Model*.

Physics beyond the Standard Model

Physics beyond the *Standard Model* refers to the theoretical developments needed to explain the deficiencies of the *Standard Model*, such as *the inability to explain the fundamental parameters of the Standard Model, the strong CP problem, neutrino oscillations, matter–antimatter asymmetry*, and the *nature of dark matter and dark energy*. Another problem lies within the mathematical framework of the *Standard Model* itself: *the Standard Model is inconsistent with that of general relativity*, and one or both theories break down under certain conditions, such as spacetime singularities like the Big Bang and black hole event horizons.

Theoretical and experimental research has attempted to extend the *Standard Model* into a *unified field theory* or a theory of everything, a complete theory explaining all physical phenomena including *constants*. Theories that lie beyond the *Standard Model* include various extensions of the Standard Model through *supersymmetry*, such as the Minimal Supersymmetric Standard Model (MSSM) and *Next-to-Minimal Supersymmetric Standard Model* (NMSSM), and entirely novel explanations, such as *string theory, M-theory*, and *extra dimensions*. As these theories tend to reproduce the entirety of current phenomena, the question of which theory is the right one, or at least the "best step" towards a Theory of Everything, can only be settled via experiments, and is one of the most active areas of research in both theoretical and experimental physics.

Many of the current problems with the *Standard Model* can probably be resolved by replacing the requirement of *inertial reference frame invariance* (in order to comply with the *theory of special relativity*) with the more realistic assumption that the speed of light is constant relative to the frame of the emitter (Ritz's theory). See Underwood, T. G. (2023). *Special Relativity*. Even Einstein recognized that *special relativity* was inconsistent with *general relativity*. [Einstein, A. (July, 1912). Relativität und Gravitation: Erwiderung auf eine Bemerkung von M. Abraham. (Relativity and Gravitation. Reply to a Comment by M. Abraham.) *Phys. Zeit.*, 38, 10, 1059-64; see translation in Underwood, T. G. (2023). *General Relativity*, pp. 176-82.]

The four "fundamental interactions or forces".

In the *Standard Model of particle physics* four *fundamental interactions* or *forces* are assumed: *gravity*, and the *electromagnetic, weak and strong interactions*, of which the latter three are incorporated in the model.

Gravity (the *gravitational interaction* or *gravitational force*).

Gravity (from Latin gravitas 'weight') is a *fundamental interaction* primarily observed as mutual attraction between all things that have *mass*. *Gravity* is, by far, the weakest of the four *fundamental interactions*, approximately 10^{38} times weaker than the *strong interaction*, 10^{36} times weaker than the *electromagnetic force* and 10^{29} times weaker than the *weak interaction*. As a result, it has no significant influence at the level of subatomic particles. However, *gravity* is the most significant interaction between objects at the macroscopic scale, and it determines the motion of planets, stars, galaxies, and even light. [See Underwood, T. G. (2024). *Gravity.*]

On Earth, *gravity* gives weight to physical objects, and the Moon's *gravity* is responsible for sublunar tides in the oceans. The corresponding antipodal tide is caused by the inertia of the Earth and Moon orbiting one another. The *gravitational attraction* between the original gaseous matter in the universe caused it to coalesce and form stars which eventually condensed into galaxies, so *gravity* is responsible for many of the large-scale structures in the universe. *Gravity* has an infinite range, although its effects become weaker as objects get farther away. *Gravity* conforms to *Newton's law of universal gravitation*, which describes *gravity* as a force causing any two bodies to be attracted toward each other, with magnitude proportional to the product of their *masses* and inversely proportional to the square of the distance between them.

Current models of particle physics imply that the earliest instance of *gravity* in the universe, possibly in the form of *quantum gravity*, *supergravity* or a *gravitational singularity*, along with ordinary space and time, developed during the Planck epoch (up to 10^{-43} seconds after the birth of the universe), possibly from a primeval state, such as a false vacuum, quantum vacuum or virtual particle, in a currently unknown manner. Scientists are currently working to develop a *theory of gravity* consistent with *quantum mechanics*, a *quantum gravity theory*, which would allow *gravity* to be united in a common mathematical framework (a theory of everything) with the other three fundamental *interactions* of physics.

The *electromagnetic interaction* or *electromagnetic force*.

The *electromagnetic interaction* or *electromagnetic force* occurs between particles with *electric charge* via *electromagnetic fields*. It is the dominant *force* in the *interactions of atoms and molecules. Electromagnetism* can be thought of as a combination of *electrostatics* and *magnetism*, which are distinct but closely intertwined phenomena. *Electromagnetic forces* occur between any two *charged particles. Electric forces* cause an attraction between particles with opposite *charges* and repulsion between particles with the same *charge*, while *magnetism* is an *interaction* that occurs between *charged particles* in *relative motion*. These two forces are described in terms of *electromagnetic fields*. Macroscopic charged objects are described in terms of *Coulomb's law for electricity* and *Ampère's force law for magnetism*; the *Lorentz force* describes microscopic *charged particles*.

The *electromagnetic force* is responsible for many of the *chemical* and *physical* phenomena observed in daily life. The *electrostatic* attraction between *atomic nuclei* and their *electrons* holds *atoms* together. *Electric forces* also allow different *atoms* to combine into *molecules*. *Magnetic interactions* between the *spin* and *angular momentum magnetic moments* of *electrons* also play a role in *chemical reactivity*; such relationships are studied in *spin chemistry*.

In the 18th and 19th centuries, prominent scientists and mathematicians such as Coulomb, Gauss and Faraday developed namesake laws which helped to explain the formation and *interaction* of *electromagnetic fields*. This process culminated in the 1860s with the discovery of *Maxwell's equations*, a set of four partial differential equations which provide a complete description of classical electromagnetic fields.

> [**Maxwell's equations** are a set of coupled partial differential equations that, together with the **Lorentz force law**, form the foundation of classical electromagnetism, classical optics, electric and magnetic circuits. **They describe how electric and magnetic fields are generated by charges, currents, and changes of the fields.** The equations are named after the physicist and mathematician James Clerk Maxwell, who, in 1861 and 1862, published an early form of the equations that included the Lorentz force law. Maxwell's equations may be combined to demonstrate how fluctuations in **electromagnetic fields** (waves) propagate at a constant speed in vacuum, c.]

Maxwell's equations provided a sound mathematical basis for the relationships between electricity and magnetism that scientists had been exploring for centuries, and predicted the existence of self-sustaining *electromagnetic waves*. Maxwell postulated that such waves make up visible light, which was later shown to be true. Gamma-rays, x-rays, ultraviolet, visible, infrared radiation, microwaves and radio waves were all determined to be *electromagnetic radiation* differing only in their range of frequencies. [See Underwood, T. G. (2024). *Electricity & Magnetism*.]

51

In the modern era, the field of *quantum electrodynamics* (QED) has modified *Maxwell's equations* to be consistent with the quantized nature of matter. In QED, the changes in the *electromagnetic field* are expressed in terms of discrete excitations, particles known as *photons*, the quanta of light. The *valence bond* (in chemistry) between two uncharged atoms, and *ferromagnetism* have been shown to be due to *quantum entanglement* of the *spin* of *electrons*. [See Underwood, T. G. (2024). *Quantum Entanglement*.]

The *weak interaction* or *weak force*.

The *weak interaction*, also called the *weak force*, is the mechanism of *interaction* between subatomic particles that is responsible for the *radioactive decay* of atoms: the *weak interaction* participates in nuclear fission and nuclear fusion. The effective range of the *weak force* is limited to subatomic distances and is less than the diameter of a *proton*.

In 1933, Enrico Fermi proposed the first theory of the *weak interaction*, known as *Fermi's interaction*. He suggested that *beta decay* could be explained by a four-*fermion* interaction, involving a contact force with no range. [Fermi, E. (March, 1934) Versuch einer Theorie der β-Strahlen. I. (Attempt at a theory of β rays. I.). See below.]

> [A *fermion* is a particle that follows *Fermi–Dirac statistics*. *Fermions* have a half-odd-integer spin (spin 1/2, spin 3/2, etc.) and obey the *Pauli exclusion principle*. These particles include all *quarks* and *leptons* and all composite particles made of an odd number of these, such as all *baryons* and many atoms and nuclei. Some *fermions* are elementary particles (such as *electrons*), and some are composite particles (such as *protons*). *Fermions* differ from *bosons*, which have integer *spin* and obey *Bose–Einstein statistics*.]

In the mid-1950s, Chen-Ning Yang and Tsung-Dao Lee first suggested that the handedness of the *spins* of particles in *weak interaction* might violate the *conservation law* or *symmetry*. [Lee, T. D. & Yang, C. N. (October, 1956). Question of Parity Conservation in Weak Interactions. See below.]

In the 1960s, Sheldon Glashow, Abdus Salam and Steven Weinberg unified the *electromagnetic force* and the *weak interaction* by showing them to be two aspects of a single force, now termed the *electroweak force*. [Glashow, S. L. (February, 1961). Partial-symmetries of weak interactions. See below.]

The *Standard Model* of particle physics provides a uniform framework for understanding *electromagnetic*, *weak*, and *strong interactions*. *A weak interaction occurs when two particles (typically, but not necessarily, half-integer spin fermions) exchange integer-spin, force-carrying bosons.*

[This could be seen to be another example of *quantum entanglement*.]

The *fermions* involved in such *exchanges* can be either *elementary* (e.g. *electrons* or *quarks*) or *composite* (e.g. *protons* or *neutrons*), although at the deepest levels, all *weak interactions* ultimately are between *elementary particles*.

In the *weak interaction*, *fermions* can *exchange* three types of force carriers, namely W+, W−, and Z *bosons*. The *masses* of these *bosons* are far greater than the *mass* of a *proton* or *neutron*, which is consistent with the short range of the *weak force*. In fact, the force is termed *weak* because its field strength over any set distance is typically several orders of magnitude less than that of the *electromagnetic force*, which itself is further orders of magnitude less than the *strong nuclear force*.

The existence of the W and Z *bosons* was not directly confirmed until 1983.

The *weak interaction* is the only fundamental *interaction* that breaks *parity symmetry*, and similarly, but far more rarely, the only *interaction* to break *charge–parity symmetry*.

The *weak interaction* is unique in that it allows *quarks* to *swap* their flavor for another. *Quarks*, which make up composite particles like *neutrons* and *protons*, come in six "*flavors*" – *up*, *down*, *charm*, *strange*, *top* and *bottom* – which give those composite particles their properties. The *swapping* of those properties is mediated by the force carrier *bosons*. For example, during *beta-minus decay*, a *down quark* within a *neutron* is changed into an *up quark*, thus converting the *neutron* to a *proton* and resulting in the emission of an *electron* and an *electron antineutrino*.

Most *fermions* decay by a *weak interaction* over time. Such decay makes radiocarbon dating possible, as carbon-14 decays through the *weak interaction* to nitrogen-14. It can also create *radioluminescence*, commonly used in tritium luminescence, and in the related field of betavoltaics (but not similar radium luminescence).

The *electroweak force* is believed to have separated into the *electromagnetic* and *weak forces* during the *quark* epoch of the early universe.

The *strong interaction* or *strong force*.

The *strong interaction*, also called the *strong force* or *strong nuclear force*, is a fundamental *interaction* that confines *quarks* into *protons*, neutrons, and other *hadron* particles. The *strong interaction* also binds *neutrons* and *protons* to create *atomic nuclei*, where it is called the *nuclear force*.

Most of the *mass* of a *proton* or *neutron* is the result of the *strong interaction* energy; the individual *quarks* provide only about 1% of the *mass* of a *proton*.

Before 1971, physicists were uncertain as to how the *atomic nucleus* was bound together. It was known that the *nucleus* was composed of *protons* and *neutrons* and that *protons* possessed positive *electric charge*, while *neutrons* were electrically neutral. By the understanding of physics at that time, positive charges would repel one another and the positively charged *protons* should cause the *nucleus* to fly apart. However, this was never observed. A stronger attractive force was postulated to explain how the *atomic nucleus* was bound despite the *protons'* mutual *electromagnetic* repulsion. This hypothesized force was called the *strong force*, which was believed to be a fundamental force that acted on the *protons* and *neutrons* that make up the *nucleus*.

In 1964, Murray Gell-Mann, and separately George Zweig, proposed that *baryons*, which include *protons* and *neutrons*, and *mesons* were composed of elementary particles. Zweig called the elementary particles "aces" while Gell-Mann called them "*quarks*"; the theory came to be called the *quark model*. [Gell-Mann. M. (February, 1964). A Schematic Model of Baryons and Mesons. See below.]

The *strong attraction* between *nucleons* was the side-effect of a more fundamental force that bound the *quarks* together into *protons* and *neutrons*. The theory of *quantum chromodynamics* explains that *quarks* carry what is called a *color charge*, although it has no relation to visible *color*. *Quarks* with unlike *color charge* attract one another as a result of the *strong interaction*, and the particle that mediates this was called the *gluon*.

The *strong interaction* is observable at two ranges, and mediated by different force carriers in each one. On a scale less than about 0.8 fm (roughly the radius of a *nucleon*), the force is carried by *gluons* and holds *quarks* together to form *protons*, *neutrons*, and other *hadrons*. On a larger scale, up to about 3 fm, the force is carried by *mesons* and binds *nucleons* (*protons* and *neutrons*) together to form the nucleus of an atom. In the former context, it is often known as the *color force*, and is so strong that if *hadrons* are struck by high-energy particles, they produce jets of massive particles instead of emitting their constituents (*quarks* and *gluons*) as freely moving particles. This property of the *strong force* is called *color confinement*.

The force carrier particle of the *strong interaction* is the *gluon*, a massless *gauge boson*. *Gluons* are thought to interact with *quarks* and other *gluons* by way of a type of *charge* called *color charge*. *Color charge* is analogous to *electromagnetic charge*, but it comes in three types (± red, ± green, and ± blue) rather than one, which results in different rules of behavior. These rules are described by *quantum chromodynamics* (QCD), the theory of *quark–gluon interactions*. Unlike the *photon* in *electromagnetism*, which is neutral, the *gluon* carries a *color charge*. *Quarks* and *gluons* are the only fundamental particles that carry non-vanishing *color charge*, and hence they participate in *strong interactions* only with each other. The *strong force* is the expression of the *gluon interaction* with other *quark* and *gluon* particles. The strength of *interaction* is parameterized by the *strong coupling constant*. This strength is modified by the *gauge color charge* of the particle, a group-theoretical property.

The *strong force* acts between *quarks*. Unlike all other forces (*electromagnetic*, *weak*, and *gravitational*), the *strong force* does not diminish in strength with increasing distance between pairs of *quarks*. After a limiting distance (about the size of a *hadron*) has been reached, it remains at a strength of about 10,000 N, no matter how much farther the distance between the *quarks*. As the separation between the *quarks* grows, the *energy* added to the pair creates new pairs of matching *quarks* between the original two; hence it is impossible to isolate *quarks*. The explanation is that the amount of work done against a force of 10,000 N is enough to create *particle–antiparticle* pairs within a very short distance. The *energy* added to the system by pulling two *quarks* apart would create a pair of new *quarks* that will pair up with the original ones. In QCD, this phenomenon is called *color confinement*; as a result, only *hadrons*, not individual free *quarks*, can be observed. The failure of all experiments that have searched for free *quarks* is considered to be evidence of this phenomenon.

Symmetry in the Standard Model.

The Standard Model of particle physics is a *gauge quantum field theory* containing the *internal (local) symmetries* of the *unitary product group* SU(3) × SU(2) × U(1). Roughly, *the three factors of the gauge symmetry give rise to the three fundamental interactions*, the *strong, weak* and *electromagnetic interactions*.]

[*Gauge theories* are a type of *quantum field theory*. In a *gauge theory*, there is a group of transformations of the *field variables* (*gauge transformations*) that leaves the basic physics of the *quantum field* unchanged. This condition, called *gauge invariance*, gives the theory a certain *symmetry*, which governs its equations. *Gauge* theories constrain the laws of physics, because all the changes induced by a *gauge transformation* have to cancel each other out when written in terms of observable quantities.

Quantum field theory is *relativistic quantum electrodynamics*; a theoretical framework that combines *classical field theory*, *special relativity*, and *quantum mechanics*. [See Underwood, T. G. (2023). *Quantum Electrodynamics – annotated sources*, Volume II.]

The *unitary group* of degree n, denoted by U(n), is the group of n × n *unitary matrices*, with the *group operation of matrix multiplication*. In the simple case n = 1, the *group U(1)* corresponds to the one dimensional *circle group*, consisting of all complex numbers with absolute value 1, under multiplication. All the unitary groups contain copies of this group.

The *special unitary group* of degree n, denoted SU(n), is the Lie group of n × n *unitary matrices* with *determinant* 1. The SU(n) groups find wide application in the *Standard Model* of particle physics, especially SU(2) in the *electroweak interaction*, and SU(3) in the *strong interaction*, in *quantum chromodynamics*. The group SU(2) is a simple Lie group consisting on all 2 x 2 matrices of *determinant* 1. The group SU(3) is a simple Lie group consisting on all 3 x 3 matrices of *determinant* 1.

The simplest case, SU(1), is the trivial group, with the 1 x 1 matrix having only a single element. Representations of SU(2) describe *non-relativistic spin*, due to being a double covering of the rotation group of Euclidean 3-space. SU(2) symmetry also supports concepts of *isobaric spin* and *weak isospin*, collectively known as *isospin*. When an element of SU(2) is written as a complex 2 × 2 matrix, it is simply a multiplication of column 2-vectors. It is known in physics as the spin-1/2.

The group SU(3) is an 8-dimensional simple Lie group consisting of all 3 × 3 unitary matrices with determinant 1. The Gell-Mann matrices, developed by Murray Gell-Mann, are a set of eight linearly independent 3×3 traceless Hermitian matrices used in the study of the *strong interaction* in particle physics. These matrices are traceless, Hermitian, and obey the extra trace orthonormality relation, so they can generate unitary matrix group elements of SU(3) through exponentiation. These properties were chosen by Gell-Mann because they then naturally generalize the Pauli matrices for SU(2) to SU(3), which formed the basis for Gell-Mann's *quark model*.

Symmetries may be broadly classified as *global* or *local*. A *global symmetry* is one that keeps a property invariant for a transformation that is applied simultaneously at all points of spacetime, whereas a *local symmetry* is one that keeps a property invariant when a possibly different symmetry transformation is applied at each point of spacetime; specifically, a *local symmetry transformation* is parameterized by the spacetime coordinates, whereas a *global symmetry* is not. This implies that a *global symmetry* is also a *local symmetry*. *Local symmetries* play an important role in physics as they form the basis for *gauge theories*.

According to the *Standard Model*, there are three *local symmetries* in the universe in which we live, which should be indistinguishable where a certain type of change is introduced.

The three *local symmetries* addressed by the *Standard Model* are:

> *C-symmetry* (*charge symmetry*), a universe where every particle is replaced with its *antiparticle*.

> *P-symmetry* (*parity symmetry*), a universe where everything is mirrored along the three physical axes. This excludes *weak interactions*.

> *T-symmetry* (*time reversal symmetry*), a universe where the direction of time is reversed. *T-symmetry* is counterintuitive (the future and the past are not symmetrical) but explained by the fact that the *Standard Model* describes *local* properties, not *global* ones like entropy. To properly reverse the direction of time, one would have to put the Big Bang and the resulting low-entropy state in the "future". Since we perceive the "past" ("future") as having lower (higher) entropy than the present, the inhabitants of this hypothetical time-reversed universe would perceive the future in the same way as we perceive the past, and vice versa.

These *symmetries* are near-symmetries because each is broken in the present-day universe. However, the *Standard Model* predicts that the combination of the three (that is, the simultaneous application of all three transformations) must be a symmetry, called *CPT symmetry*. *CP violation*, the violation of the combination of *C-* and *P-symmetry*, is necessary for the presence of significant amounts of *baryonic matter* in the universe.

Symmetry breaking is a phenomenon where a disordered but symmetric state collapses into an ordered, but less symmetric state. This collapse is often one of many possible bifurcations that a particle can take as it approaches a lower energy state. Due to the many possibilities, an observer may assume the result of the collapse to be arbitrary. This phenomenon is fundamental to *quantum field theory*. Specifically, it plays a central role in the *Glashow–Weinberg–Salam model* which forms part of the *Standard Model* modelling the *electroweak* sector.

See Chen Ning Yang – 1957 Nobel Lecture, December 11, 1957 [*The law of parity conservation and other symmetry laws of physics*] below.

Timeline of discovery of subatomic particles.

The *Standard Model* consists of 26 elementary particles (indicated below in bold): 12 *fermions* (6 *quarks* and 6 *leptons*) and 14 *bosons* (13 *gauge bosons* and 1 *scalar boson*, the *Higgs boson*). The 6 *quarks* comprise *up, down, charm, strange, top* and *bottom*; the 6 *leptons* comprise the *electron, electron neutrino, muon, muon neutrino, tau* and *tau neutrino*; and the 13 *gauge bosons* comprise the *photon*, 2 *W bosons*, the *Z boson*, 8 types of *gluons*, and the hypothetical *graviton*. Each has its own anti-particle, making a total of 52 elementary particles and anti-particles.

Electron [elementary particle]

In 1897, Joseph John Thomson (December 18, 1856–August 30, 1940), a British physicist, was credited with the discovery of the *electron*, the first subatomic particle to be found. Thomson showed that cathode rays were composed of previously unknown negatively charged particles (now called *electrons*), which he calculated must have bodies much smaller than atoms and a very large charge-to-mass ratio. [Thomson, J. J. (1897). Cathode Rays. *Phil. Mag.*, 44, 269, 293–316; https://doi.org/10.1080/14786449708621070.] Thomson was awarded the 1906 Nobel Prize in Physics "in recognition of the great merits of his theoretical and experimental investigations on the conduction of electricity by gases". The *electron* is a *lepton* and a *fermion*.

Lepton [classification]

A *lepton* is an elementary particle of half-integer *spin* (*spin* ½) that does not undergo *strong interactions*. Two main classes of *leptons* exist: *charged leptons* (also known as the *electron-like leptons* or *muons*), including the *electron, muon,* and *tauon,* and *neutral leptons*, better known as *neutrinos*. *Charged leptons* can combine with other particles to form various composite particles such as atoms and *positronium*, while *neutrinos* rarely interact with anything, and are consequently rarely observed. The best known of all *leptons* is the *electron*.

There are six types of *leptons*, known as flavors, grouped in three generations. The *first-generation leptons*, also called electronic leptons, comprise the electron (e^-) and the electron neutrino (ν_e); the second are the *muonic leptons*, comprising the muon (μ^-) and the *muon neutrino* (ν_μ); and the third are the *tauonic leptons*, comprising the *tau* (τ^-) and the *tau neutrino* (ν_τ). *Electrons* have the least mass of all the *charged leptons*. The heavier *muons* and *taus* will rapidly change into *electrons* and *neutrinos* through a process of particle decay: the transformation from a higher mass state to a lower mass state. Thus, *electrons* are stable and the most common charged *lepton* in the universe, whereas *muons* and *taus* can only be produced in high-energy collisions (such as those involving cosmic rays and those carried out in particle accelerators).

Leptons have various intrinsic properties, including *electric charge*, *spin*, and *mass*. Unlike *quarks*, however, *leptons* are not subject to the *strong interaction*, but they are subject to the other three fundamental interactions: *gravitation*, the *weak interaction*, and to *electromagnetism*, of which the latter is proportional to charge, and is thus zero for the electrically neutral *neutrinos*.

For every *lepton flavor*, there is a corresponding type of *antiparticle*, known as an *antilepton*, that differs from the *lepton* only in that some of its properties have equal magnitude but opposite sign. According to certain theories, *neutrinos* may be their own antiparticle. It is not currently known whether this is the case.

The first *charged lepton*, the *electron*, was theorized in the mid-19th century by several scientists and was discovered in 1897 by J. J. Thomson. The next *lepton* to be observed was the *muon*, discovered by Carl D. Anderson in 1936, which was classified as a meson at the time. After investigation, it was realized that the *muon* did not have the expected properties of a meson, but rather behaved like an *electron*, only with higher mass. It took until 1947 for the concept of "*leptons*" as a family of particles to be proposed. The term *lepton* was first used by physicist Léon Rosenfeld in 1948.

The first *neutrino*, the *electron neutrino*, was proposed by Wolfgang Pauli in 1930 to explain certain characteristics of *beta decay*. It was first observed in the Cowan–Reines *neutrino* experiment conducted by Clyde Cowan and Frederick Reines in 1956. The *muon neutrino* was discovered in 1962 by Leon M. Lederman, Melvin Schwartz, and Jack Steinberger, and the *tau* discovered between 1974 and 1977 by Martin Lewis Perl and his colleagues from the Stanford Linear Accelerator Center and Lawrence Berkeley National Laboratory. The *tau neutrino* remained elusive until July 2000, when the DONUT collaboration from Fermilab announced its discovery.

Leptons are an important part of the *Standard Model*. Electrons are one of the components of atoms, alongside protons and neutrons. Exotic atoms with muons and *taus* instead of *electrons* can also be synthesized, as well as *lepton–antilepton* particles such as *positronium*.

Fermion [*classification*]

A *fermion* is a particle that follows *Fermi–Dirac statistics*. Fermions have a half-odd-integer spin (spin 1/2, spin 3/2, etc.) and obey the *Pauli exclusion principle*. These particles include all *quarks* and *leptons* and all composite particles made of an odd number of these, such as all *baryons* and many atoms and nuclei. Some *fermions* are elementary particles (such as *electrons*), and some are composite particles (such as *protons*). *Fermions* differ from *bosons*, which have integer *spin* and obey *Bose–Einstein statistics*.

In addition to the *spin* characteristic, *fermions* have another specific property: they possess conserved *baryon* or *lepton* quantum numbers. Therefore, what is usually referred to as the *spin-statistics relation* is, in fact, a *spin statistics-quantum number relation*.

As a consequence of the *Pauli exclusion principle*, only one *fermion* can occupy a particular *quantum state* at a given time. Suppose multiple *fermions* have the same spatial probability distribution. *Then, at least one property of each fermion, such as its spin, must be different. Fermions* are usually associated with *matter*, whereas *bosons* are generally *force carrier* particles. However, in the current state of particle physics, the distinction between the two concepts is unclear. Weakly interacting fermions can also display bosonic behavior under extreme conditions. For example, at low temperatures, *fermions* show superfluidity for uncharged particles and superconductivity for charged particles. Composite *fermions*, such as *protons* and *neutrons*, are the key building blocks of everyday matter.

Paul Dirac coined the name *fermion* from the surname of Italian physicist Enrico Fermi.

Photon [elementary particle]

The *photon* (from Ancient Greek φῶς, φωτός 'light') is an elementary particle that is a quantum of the *electromagnetic field*, including electromagnetic radiation such as light and radio waves, and the force carrier for the *electromagnetic force*. *Photons* are massless particles that always move at the speed of light measured in vacuum. The photon is a *gauge boson*, and also a *vector boson*, with *spin* = 1. The *photon* has no *electric charge*, is generally considered to have zero *rest mass* and is a stable particle. The experimental upper limit on the *photon mass* is very small, on the order of 10^{-50} kg; its lifetime would be more than 10^{18} years. For comparison the age of the universe is about 1.38×10^{10} years.

In a vacuum, a photon has two possible *polarization states*. The photon is the *gauge boson* for *electromagnetism,*and therefore all other quantum numbers of the *photon* (such as *lepton number, baryon number,* and *flavor quantum numbers*) are zero. Also, the *photon* obeys *Bose–Einstein statistics*, and not Fermi–Dirac statistics. That is, they do not obey the Pauli exclusion principle,and more than one can occupy the same bound *quantum state*.

As with other elementary particles, *photons* are best explained by quantum mechanics and exhibit wave–particle duality, their behavior featuring properties of both waves and particles. The modern photon concept originated during the first two decades of the 20th century with the work of Albert Einstein, who built upon the research of Max Planck. While Planck was trying to explain how matter and electromagnetic radiation could be in thermal equilibrium with one another, he proposed that the energy stored within a material object should be regarded as composed of an integer number of discrete, equal-sized parts. To explain the photoelectric effect, Einstein introduced the idea that light itself is made of discrete units of energy. In 1926, Gilbert N. Lewis popularized the term *photon* for these energy units.

In the *Standard Model* of particle physics, *photons* and other elementary particles are described as a necessary consequence of physical laws having a certain symmetry at every point in spacetime. The intrinsic properties of particles, such as *charge, mass,* and *spin*, are determined by *gauge symmetry*.

Boson [classification]

A *boson* is a subatomic particle whose *spin* quantum number has an integer value (0, 1, 2, ...). *Bosons* form one of the two fundamental classes of subatomic particle, the other being *fermions*, which have odd half-integer spin (1/2, 3/2, 5/2, ...). Every observed subatomic particle is either a *boson* or a *fermion*.

Some *bosons* are elementary particles occupying a special role in particle physics, distinct from the role of *fermions* (which are sometimes described as the constituents of "ordinary matter"). Certain elementary *bosons* (*gauge bosons*) act as force carriers, which give rise to forces between other particles, while one (the *Higgs boson*) contributes to the phenomenon of *mass*. Other *bosons*, such as *mesons*, are composite particles made up of smaller constituents.

Vector boson [classification]

A *vector boson* is a *boson* whose *spin* equals one. *Vector bosons* that are also elementary particles are *gauge bosons*, the force carriers of fundamental interactions. Some composite particles are *vector bosons*, for instance any *vector meson* (*quark* and *antiquark*). During the 1970s and 1980s, intermediate *vector bosons* (the *W* and *Z bosons*, which mediate the *weak interaction*) drew much attention in particle physics.

Gauge boson [*classification*]

A *gauge boson* is a bosonic elementary particle that acts as the force carrier for elementary *fermions*. Elementary particles whose interactions are described by a *gauge theory* interact with each other by the exchange of *gauge bosons*, usually as virtual particles. *Photons, W and Z bosons*, and *gluons* are *gauge bosons*. All known *gauge bosons* have a *spin* of 1; for comparison, the *Higgs boson* has *spin* zero and the hypothetical *graviton* has a *spin* of 2. Therefore, all known *gauge bosons* are *vector bosons*.

Gauge bosons are different from the other kinds of bosons: first, fundamental *scalar bosons* (the Higgs boson); second, *mesons*, which are *composite bosons*, made of *quarks*; third, larger composite, non-force-carrying bosons, such as certain atoms.

The *Standard Model* of particle physics recognizes four kinds of *gauge bosons*: *photons*, which carry the *electromagnetic interaction*; *W and Z bosons*, which carry the *weak interaction*; and *gluons*, which carry the *strong interaction*.

Paul Dirac coined the name *boson* to commemorate the contribution of Satyendra Nath Bose, an Indian physicist.

Proton [*composite particle*]

in 1919, Ernest Rutherford, (August 30, 1871–October 19, 1937), a New Zealand physicist and pioneering researcher in both atomic and nuclear physics, was credited with the discovery of the *proton*. He was awarded the 1908 Nobel Prize in Chemistry "for his investigations into the disintegration of the elements, and the chemistry of radioactive substances". In 1917, he performed the first artificially-induced nuclear reaction by conducting experiments where nitrogen nuclei were bombarded with alpha particles. As a result, he discovered the emission of a subatomic particle which he initially called the "hydrogen atom", but later (more accurately) named the *proton*. [Rutherford, E. (1899). Uranium Radiation and the Electrical Conduction Produced by it. *Phil. Mag.*, 47, 284, 109–63; https://doi.org/10.1080/14786449908621245.]

Neutron [*composite particle*]

In 1932, following a previously reported experiment in Paris, James Chadwick, (October 20, 1891–July 24, 1974) an English physicist, discovered the neutron. He aimed alpha radiation at paraffin wax, a hydrocarbon high in hydrogen content, offering a target dense with protons, and measured the range of these protons, and how the new radiation impacted the atoms of various gases. Measurements of the recoil energy showed that the mass of the radiation particles must be uncharged particles with about the same mass as the proton, which matched the properties Rutherford described in 1920 and which had later been called *neutrons*. [Chadwick, J. (1932). Possible Existence of a Neutron. *Nature*, 129, 3252, 312. https://doi.org/10.1038/129312a0.] Chadwick was awarded the 1935 Nobel Prize in Physics for his discovery of the neutron.

In 1920 Rutherford gave a Bakerian lecture at the Royal Society entitled the "*Nuclear Constitution of Atoms*", a summary of recent experiments on atomic nuclei and conclusions as to the structure of atomic nuclei. At that time, the existence of electrons within the atomic nucleus was widely assumed. It was assumed the nucleus consisted of hydrogen nuclei in number equal to the atomic mass. But since each hydrogen nucleus had charge +1, the nucleus required a smaller number of "internal electrons" each of charge −1 to give the nucleus its correct total charge. Such a model was consistent with the scattering of alpha particles from heavy nuclei, as well as the charge and mass of the many isotopes that had been identified. There were other motivations for the *proton–electron* model. As noted by Rutherford at the time, "We have strong reason for believing that the nuclei of atoms contain electrons as well as positively charged bodies...", namely, it was known that beta radiation was electrons emitted from the nucleus. [Rutherford, E. (1920). Nuclear Constitution of Atoms. *Proc. Roy. Soc.* A., 97, 686, 374–400; https://doi.org/10.1098/rspa.1920.0040.]

In that lecture, Rutherford conjectured the existence of new particles, including two new particles: one of two *protons* with a closely bound electron, and another of one *proton* and a closely bound *electron*. The former was the nucleus of *deuterium* [a stable isotope of hydrogen with a nucleus consisting of one *proton* and one *neutron*, which is double the mass of the nucleus of ordinary hydrogen] discovered in 1931 by Harold Urey. The mass of the hypothetical neutral particle would be little different from that of the *proton*. At about the time of Rutherford's lecture, other publications appeared with similar suggestions of a *proton–electron* composite in the nucleus, and in 1921 William Harkins, an American chemist, named the uncharged particle the *neutron*. In 1930, Walther Bothe and his collaborator Herbert Becker in Giessen, Germany found that if energetic alpha particles emitted from *polonium* fell on certain light elements, specifically beryllium ($^9_4$Be), boron ($^{11}_5$B), or lithium ($^7_3$Li), an unusually penetrating radiation was produced which was not influenced by an electric field.

Baryon [classification]

Protons and neutrons are *baryons*, a type of composite subatomic particle *that contains an odd number of valence quarks and antiquarks*, conventionally three. *Baryons* belong to the hadron family of particles; hadrons are composed of quarks. Because *quarks* have a *spin ½* , the difference in *quark* number results in being *fermions* because they have half-integer *spin*.

Each *baryon* has a corresponding antiparticle (*antibaryon*) where their corresponding *antiquarks* replace *quarks*. For example, a *proton* is made of two *up quarks* and one *down quark*; and its corresponding antiparticle, the *antiproton*, is made of two *up antiquarks* and one *down antiquark*.

Baryons participate in the residual *strong force*, which is mediated by particles known as *mesons*. The most familiar *baryons* are *protons* and *neutrons*, both of which contain three *quarks*, and for this reason they are sometimes called *triquarks*. These particles make up most of the mass of the visible matter in the universe and compose the nucleus of every atom (*electrons*, the other major component of the atom, are members of a different family of particles called *leptons*; leptons do not interact via the *strong force*). Exotic *baryons* containing five quarks, called *pentaquarks*, have also been discovered and studied.

Baryons are strongly interacting *fermions*; that is, they are acted on by the *strong nuclear force* and are described by *Fermi–Dirac statistics*, which apply to all particles obeying the Pauli exclusion principle. This is in contrast to the bosons, which do not obey the exclusion principle.

The name "*baryon*", introduced by Abraham Pais, comes from the Greek word for "heavy" (βαρύς, barýs), because, at the time of their naming, most known elementary particles had lower masses than the baryons.

Hadron [classification]

A *hadron* (from Ancient Greek ἁδρός (hadrós) 'stout, thick') is a *composite subatomic particle made of two or more quarks held together by the strong interaction*. They are analogous to molecules, which are held together by the electric force. Most of the *mass* of ordinary matter comes from two *hadrons*: the *proton* and the *neutron*, while most of the *mass* of the *protons* and *neutrons* is in turn due to the *binding energy* of their constituent *quarks*, due to the *strong force*.

Hadrons are categorized into two broad families: *baryons*, made of an odd number of *quarks* (usually three *quarks*) and *mesons*, made of an even number of *quarks* (usually two *quarks*: one *quark* and one *anti-quark*). *Protons* and *neutrons* (which make the majority of the *mass* of an atom) are examples of *baryons*; *pions* are an example of a *meson*.

"Exotic" *hadrons*, containing more than three *valence quarks*, have been discovered in recent years. A *tetraquark state* (an exotic *meson*), named the Z(4430)−, was discovered in 2007 by the Belle Collaboration and confirmed as a resonance in 2014 by the LHCb collaboration. Two *pentaquark states* (exotic *baryons*), named P+c(4380) and P+c(4450), were discovered in 2015 by the LHCb collaboration. There are several more exotic *hadron* candidates and other color-singlet *quark* combinations that may also exist.

Almost all "free" *hadrons* and *anti-hadrons* (meaning, in isolation and not bound within an atomic nucleus) are believed to be unstable and eventually decay into other particles. The only known possible exception is free *protons*, which appear to be stable, or at least, take immense amounts of time to decay (order of 10^{34}+ years). By way of comparison, free *neutrons* are the longest-lived unstable particle, and decay with a half-life of about 611 seconds, and have a mean lifetime of 879 seconds.

Hadron physics is studied by colliding *hadrons*, e.g. *protons*, with each other or the nuclei of dense, heavy elements, such as lead (Pb) or gold (Au), and detecting the debris in the produced particle showers. A similar process occurs in the natural environment, in the extreme upper-atmosphere, where *muons* and *mesons* such as *pions* are produced by the collisions of *cosmic rays* with rarefied gas particles in the outer atmosphere.

The term "*hadron*" is a new Greek word introduced by L.B. Okun in a plenary talk at the 1962 International Conference on High Energy Physics at CERN. He opened his talk with the definition of a new category term:

"Notwithstanding the fact that this report deals with *weak interactions*, we shall frequently have to speak of *strongly interacting* particles. These particles pose not only numerous scientific problems, but also a terminological problem. The point is that "*strongly interacting particles*" is a very clumsy term which does not yield itself to the formation of an adjective. For this reason, to take but one instance, decays into *strongly interacting particles* are called "*non-leptonic*". This definition is not exact because "*non-leptonic*" may also signify *photonic*. In this report I shall call strongly interacting particles "*hadrons*", and the corresponding decays "*hadronic*" (the Greek ἁδρός signifies "large", "massive", in contrast to λεπτός which means "small", "light"). I hope that this terminology will prove to be convenient." — L.B. Okun (1962).]

Positron (anti-electron) [elementary particle]

In 1932, Carl David Anderson (September 3, 1905 – January 11, 1991), an American physicist, discovered the *positron*. Under the supervision of Robert A. Millikan, he began investigations into cosmic rays during the course of which he encountered unexpected particle tracks in his cloud chamber photographs that he correctly interpreted as having been created by a particle with the same mass as the *electron*, but with *opposite electrical charge*.

Anderson first detected the particles in cosmic rays. He then produced more conclusive proof by shooting gamma rays produced by the natural radioactive nuclide ThC" (^{208}Tl) [Thallium] into other materials, resulting in the creation of *positron-electron* pairs. Anderson shared the 1936 Nobel Prize in Physics "for his discovery of the positron", with Victor Franz Hess, "for his discovery of cosmic radiation".

The *positron* (or *anti-electron*) is a particle with an *electric charge* of +1 e, a *spin* of 1/2 (the same as the *electron*), and the same mass as an *electron*. It is the antiparticle (antimatter counterpart) of the electron. When a *positron* collides with an *electron*, annihilation occurs. If this collision occurs at low energies, it results in the production of two or more *photons*.

Positrons can be created by *positron emission radioactive decay* (through *weak interactions*), or by *pair production* from a sufficiently energetic *photon* which is interacting with an atom in a material.

This discovery validated Paul Dirac's theoretical prediction in 1928 of the existence of the *positron*. [Dirac, P. A. M. (February, 1928). The Quantum Theory of the Electron. See below.]

Muon (mu lepton) [elementary particle]

In 1937, Seth Henry Neddermeyer (September 16, 1907 – January 29, 1988), an American physicist, together with Carl Anderson, co-discovered the *muon* (*mu lepton*), using cloud chamber measurements of cosmic rays. [Neddermeyer, S. H. & Anderson, C. D. (1937). Note on the nature of Cosmic-Ray Particles. *Phys. Rev.*, 51, 10, 884–6; https://doi.org/10.1103/PhysRev.51.884].

A *muon* is an elementary particle similar to the *electron*, with an *electric charge* of −1 e and *spin* ½, but with a *much greater mass*. It is classified as a *lepton*. As with other *leptons*, the *muon* is not thought to be composed of any simpler particles. The *muon* is an *unstable subatomic particle* with a *mean lifetime* of 2.2 μs, *much longer than many other subatomic particles*. As with the decay of the free *neutron* (with a lifetime around 15 minutes), *muon* decay is slow (by subatomic standards) because the decay is mediated only by the *weak interaction* (rather than the more powerful *strong interaction* or electromagnetic interaction), and because the mass difference between the muon and the set of its decay products is small, providing few kinetic degrees of freedom for decay. *Muon decay* almost always produces at least three particles, which must include an *electron* of the same *charge* as the muon and *two types of neutrinos*.

Like all elementary particles, the *muon* has a corresponding antiparticle of opposite *charge* (+1 e) but equal *mass* and *spin*: the *anti-muon* (also called a *positive muon*). Muons have a *mass* of 105.66 MeV/c2, which is approximately 206.8 times that of the electron. There is also a third lepton, the tau, approximately 17 times heavier than the *muon*.

Due to their greater mass, *muons accelerate slower than electrons in electromagnetic fields*, and emit less bremsstrahlung (deceleration radiation). This allows *muons* of a given energy to penetrate far deeper into matter because the deceleration of electrons and muons is primarily due to energy loss by the bremsstrahlung mechanism. For example, so-called secondary muons, created by cosmic rays hitting the atmosphere, can penetrate the atmosphere and reach Earth's land surface and even into deep mines.

Because *muons* have a greater mass and energy than the decay energy of radioactivity, *they are not produced by radioactive decay*. Nonetheless, they are produced in great amounts in high-energy interactions in normal matter, in certain particle accelerator experiments with hadrons, and in cosmic ray interactions with matter. These interactions usually produce *pi mesons* initially, which almost always decay to *muons*. [Formerly, *muons* were called *mu mesons*, but are not classified as *mesons* by modern particle physicists.]

As with the other *charged leptons*, the *muon* has an associated *muon neutrino*, which differs from the *electron neutrino* and participates in different nuclear reactions.

Pion (or pi meson) [composite particle]

A *pion* or *pi meson*, is any of three subatomic particles: π^0, π^+, and π^-. Each pion consists of a *quark* and an *antiquark* and is therefore a *meson*. *Pions* are the lightest *mesons* and, more generally, the lightest *hadrons*. They are unstable, with the *charged pions*, π^+ and π^-, decaying after a mean lifetime of 26.033 nanoseconds (2.6033×10^{-8} seconds), and the *neutral pion* π^0 decaying after a much shorter lifetime of 85 attoseconds (8.5×10^{-17} seconds). *Charged pions* most often decay into *muons* and *muon neutrinos*, while *neutral pions* generally decay into *gamma rays*.

The exchange of virtual *pions*, along with *vector, rho* and *omega mesons*, provides an explanation for the residual *strong force* between *nucleons*. *Pions* are not produced in radioactive decay, but commonly are in high-energy collisions between *hadrons*. *Pions* also result from some *matter–antimatter annihilation* events. All types of *pions* are also produced in natural processes when high-energy cosmic-ray *protons* and other hadronic cosmic-ray components interact with matter in Earth's atmosphere. In 2013, the detection of characteristic *gamma rays* originating from the decay of *neutral pions* in two supernova remnants has shown that *pions* are produced copiously after supernovas, most probably in conjunction with production of high-energy protons that are detected on Earth as cosmic rays

In 1947, the *pion* (or *pi meson*) was discovered by members of C. F. Powell's group at the University of Bristol, in England, including César Lattes, Giuseppe Occhialini and Hugh Muirhead. In the same year, Lattes, together with Powell and Occhialini, determined the new particle's mass. [Lattes, C. M. G., Muirhead, H., Occhialini, G. P. S. & Powell, C. F. (1947). Processes involving charged mesons. *Nature*. 159, 4047, 694–7; https://doi.org/ 10.1038/159694a0.]

Cecil Frank Powell, (December 5,1903 –August 9, 1969) was a British physicist, who headed the team that developed the photographic method of studying nuclear processes. The Nobel Prize in Physics 1950 was awarded to Powell "for his development of the photographic method of studying nuclear processes and his discoveries regarding mesons made with this method".

Since the advent of particle accelerators had not yet come, high-energy subatomic particles were only obtainable from atmospheric cosmic rays. Photographic emulsions based on the gelatin-silver process were placed for long periods of time in sites located at high-altitude mountains, first at Pic du Midi de Bigorre in the Pyrenees, and later at Chacaltaya in the Andes Mountains, where the plates were struck by cosmic rays. After development, the photographic plates were inspected under a microscope by a team of about a dozen women.

In 1948, Lattes, Eugene Gardner, and their team first artificially produced *pions* at the University of California's cyclotron in Berkeley, California, by bombarding carbon atoms with high-speed alpha particles.

Meson [*classification*]

A *meson* is a type of hadronic subatomic particle composed of *an equal number of quarks and antiquarks, usually one of each, bound together by the strong interaction*. Because *quarks* have a *spin ½*, the difference in quark number between *mesons* and *baryons* results in *mesons* being *bosons*, whereas *baryons,* the other members of the *hadron* family, composed of *odd numbers of valence quarks* (at least three), are *fermions*.

Some experiments show evidence of exotic *mesons*, which do not have the conventional *valence quark* content of two *quark*s (one *quark* and one *antiquark*), but four or more.

Because *mesons* are composed of *quark* subparticles, they have a meaningful physical size, a diameter of roughly one femtometre (10^{-15} m), which is about 0.6 times the size of a *proton* or *neutron*. *All mesons are unstable*, with the longest-lived lasting for only a few tenths of a nanosecond. Heavier *mesons* decay to lighter *mesons* and ultimately to stable *electrons*, *neutrinos* and *photons*.

Each type of *meson* has a corresponding antiparticle (*antimeson*) in which *quarks* are replaced by their corresponding *antiquarks* and vice versa.

Because *mesons* are composed of *quarks*, they participate in both the *weak interaction* and *strong interaction*. *Mesons* with net *electric charge* also participate in the *electromagnetic interaction*. *Mesons* are classified according to their *quark content*, *total angular momentum*, *parity* and various other properties, such as *C-parity* and *G-parity*. Although no *meson* is stable, those of lower mass are nonetheless more stable than the more massive, and hence are easier to observe and study in particle accelerators or in cosmic ray experiments. The lightest group of *mesons* is less massive than the lightest group of *baryons*, meaning that they are more easily produced in experiments, and thus exhibit certain higher-energy phenomena more readily than do *baryons*. But *mesons* can be quite massive: for example, the J/Psi meson (J/ψ) containing the charm quark, first seen 1974, is about three times as massive as a proton, and the upsilon meson (Υ) containing the bottom quark, first seen in 1977, is about ten times as massive as a proton.

The existence of *mesons* was predicted by Hideki Yukawa's 1935 *theory of mesons* that postulated the particle as mediating the nuclear force. [Yukawa, H. (1935). On the Interaction of Elementary Particles. See below.]

Kaon (or *K meson*) [*composite particle*]

In the same year, 1947, the *kaon* (or *K meson*), the first *strange* particle, was co-discovered by George Dixon Rochester, and Clifford Charles Butler, two British physicists at the University of Manchester. They published two cloud chamber photographs of cosmic ray-induced events, one showing what appeared to be a neutral particle decaying into two charged *pions*, and one which appeared to be a charged particle decaying into a charged *pion* and something neutral.

A *kaon*, also called a *K meson,* is any of a group of four *mesons* distinguished by a *quantum number* called *strangeness*. In the *quark model* they are understood to be bound states of a *strange quark* (or *antiquark*) and an *up or down antiquark* (or *quark*).

The four kaons are:

K⁻, *negatively charged* (containing a *strange quark* and an *up antiquark*) has *mass* 493.677 ± 0.013 MeV and *mean lifetime* $(1.2380 \pm 0.0020) \times 10^{-8}$ s;

K⁺ (*antiparticle* of above) *positively charged* (containing an *up quark* and a *strange antiquark*) must (by CPT invariance) have *mass* and *lifetime* equal to that of K⁻. Experimentally, the *mass* difference is 0.032 ± 0.090 MeV, consistent with zero; the difference in lifetimes is $(0.11 \pm 0.09) \times 10^{-8}$ s, also consistent with zero;

K⁰, *neutrally charged* (containing a *down quark* and a *strange antiquark*) has *mass* 497.648 ± 0.022 MeV. It has mean squared *charge radius* of -0.076 ± 0.01 fm²;

K⁰, *neutrally charged* (*antiparticle* of above) (containing a *strange quark* and a *down antiquark*) has the same *mass*.

As the *quark model* shows, assignments that the *kaons* form two doublets of *isospin*; that is, they belong to the fundamental representation of SU(2) called the 2. One doublet of *strangeness* +1 contains the K⁺ and the K⁰. The *antiparticles* form the other doublet (of *strangeness* −1).

Kaons have proved to be a copious source of information on the nature of fundamental interactions since their discovery in cosmic rays in 1947. They were essential in establishing the foundations of the *Standard Model* of particle physics, such as the quark model of *hadrons* and the theory of *quark* mixing.

Kaons have played a distinguished role in our understanding of fundamental conservation laws: *CP violation*, a phenomenon generating the observed *matter–antimatter asymmetry* of the universe, was discovered in the *kaon* system in 1964. [Christenson, J. H., Cronin, J. W., Fitch, V. L. & Turlay, R. (July 1964). Evidence for the 2π Decay of the K20 Meson. *Phys. Rev. Let.,* 13, 4, 138–40; https://doi.org/10.1103/PhysRevLett.13.13.] This was acknowledged by the award of the 1980 Nobel Prize in Physics jointly to James Watson Cronin and Val Logsdon Fitch "for the discovery of violations of fundamental symmetry principles in the decay of neutral K-mesons".

[*CP violation* is a violation of *CP-symmetry* (or *charge conjugation parity symmetry*): the combination of *C-symmetry* (charge conjugation symmetry) and *P-*

symmetry (*parity symmetry*). *CP-symmetry* states that the laws of physics should be the same if a particle is interchanged with its *antiparticle* (*C-symmetry*) while its *spatial coordinates* are inverted ("mirror" or *P-symmetry*).

A parity transformation (also called *parity inversion)* is the flip in the sign of one spatial coordinate.

Charge conjugation is a transformation that switches all particles with their corresponding *antiparticles*, thus changing the sign of all *charges*: not only *electric charge* but also the *charges* relevant to other forces.]

Direct *CP violation* was discovered in the *kaon* decays in the early 2000s by the NA48 experiment at CERN and the KTeV experiment at Fermilab. As the *quark model* shows, assignments that *the kaons form two doublets of isospin*; that is, they belong to the fundamental representation of SU(2) called the **2**. One doublet of strangeness +1 contains the K^+ and the K^0. The antiparticles form the other doublet (of strangeness −1).

$Λ^0$ (or lambda baryon) [composite particle]

In 1950, the $Λ^0$ (or *lambda baryon*) discovered during a study of cosmic-ray interactions in October 1950, by V. D. Hopper and S. Biswas of the University of Melbourne, as a neutral *V particle* with a *proton* as a decay product, thus correctly distinguishing it as a *baryon*, rather than a *meson*, i.e. different in kind from the K *meson* discovered in 1947 by Rochester and Butler.

> [In particle physics, V was a generic name for heavy, unstable subatomic particles that decay into a pair of particles, thereby producing a characteristic letter V in a bubble chamber or other particle detector.]

They were produced by cosmic rays and detected in photographic emulsions flown in a balloon at 70,000 feet (21,000 m). Though the particle was expected to live for ~10^{-23} s, it actually survived for ~10^{-10} s. The property that caused it to live so long was dubbed *strangeness* and led to the discovery of the *strange quark*. Furthermore, these discoveries led to a principle known as the *conservation of strangeness*, wherein lightweight particles do not decay as quickly if they exhibit *strangeness* (because *non-weak* methods of particle decay must preserve the *strangeness* of the decaying *baryon*).

The *lambda baryons* (Λ) are a family of subatomic *hadron* particles containing one *up quark*, one *down quark*, and a *third quark* from a *higher flavor generation*, in a combination where the *quantum wave function* changes *sign upon the flavor of any two quarks being swapped* (thus slightly different from a *neutral sigma baryon*, $Σ^0$). They are thus *baryons*, with *total isospin* of 0, and have either *neutral electric charge* or the *elementary charge* +1.

There are four types of *lambda baryons* represented by the symbols Λ^0, Λ^+_c, Λ^0_b, and Λ^+_t. In this notation, the superscript character indicates whether the particle is electrically *neutral* (^{0}) or carries a *positive charge* ($^+$). The subscript character, or its absence, indicates whether the third quark is a *strange quark* (Λ^0) (no subscript), a *charm quark* (Λ^+_c), a *bottom quark* (Λ^0_b), or a *top quark* (Λ^+_t). Physicists expect to not observe a *lambda baryon* with a *top quark*, because the *Standard Model* of particle physics predicts that the *mean lifetime* of *top quarks* is roughly 5×10^{-25} seconds; that is about 1/20 of the mean timescale for *strong interactions*, which indicates that the *top quark* would decay before a *lambda baryon* could form a *hadron*.

Ξ (*Xi*) *baryon* [*composite particle*]

Xi baryons or *cascade particles* are a family of subatomic *hadron* particles which may have an *electric charge* of +2 e, +1 e, 0, or −1 e, where e is the *elementary charge*. They are historically called the *cascade particles* because of their *unstable state*; they are typically observed to decay rapidly into lighter particles, through a chain of decays (cascading decays).

Like all conventional *baryons*, *Xi baryons* contain three *quarks*. *Xi baryons*, in particular, contain either one *up* or one *down quark* and two other, more massive *quarks*. The two more massive *quarks* are any two of *strange*, *charm*, or *bottom* (doubles allowed). For notation, the assumption is that the two heavy quarks in the *Xi baryon* are both *strange*; subscripts "c" and "b" are added for each even heavier *charm* or *bottom quark* that replaces one of the two presumed *strange quarks*.

The first discovery of a *charged Xi baryon* was in cosmic ray experiments by the Manchester group in 1952. [Armenteros, R. et al. (Manchester group) (1952). The properties of charged V-particles. *Phil. Mag.*, 43, 341, 597; https://doi.org/10.1080/14786440608520216. The first discovery of the *neutral Xi baryon* was at the Lawrence Berkeley Laboratory in 1959. [Alvarez, L.W. et al. (1959). Neutral Cascade Hyperon Event. *Phys. Rev. Let.*, 2, 5, 215; https://doi.org/10.1103/PhysRevLett.2.215.] It was later observed as a daughter product from the decay of the *omega baryon* (Ω^-) observed at Brookhaven National Laboratory in 1964.

The *Xi*^-_b particle, also known as the *cascade B* particle, was discovered by DØ and CDF experiments at Fermilab in 2007. It was the first known particle made of *quarks* from all three *quark generations* – namely, a *down quark*, a *strange quark*, and a *bottom quark*. It

The DØ and CDF collaborations reported the consistent masses of the new state. The Particle Data Group world average *mass* is 5.7924 ± 0.0030 GeV/c².

In 2012, the CMS experiment at the Large Hadron Collider detected a *Xi*$^{*0}_b$ *baryon* (reported mass 5945 ± 2.8 MeV/c2). (Here,"*" indicates a *baryon decuplet*.) The LHCb experiment at CERN discovered two new Xi *baryons* in 2014: *Xi*$^{'-}_b$ and *Xi*$^{*-}_b$. And, in 2017, the LHCb researchers reported yet another Xi *baryon*: the double charmed *Xi*$^{++}_{cc}$ *baryon*, consisting of two heavy *charm quarks* and one *up quark*. The *mass* of *Xi*$^{++}_{cc}$ is about 3.8 times that of a *proton*.

Antiproton (antiparticle) [*composite antiparticle*]

The *antiproton* was first experimentally confirmed in 1955 at the Bevatron particle accelerator by University of California, Berkeley, physicists Emilio Segrè and Owen Chamberlain, for which they were awarded the 1959 Nobel Prize in Physics "for their discovery of the antiproton".

The *antiproton*, is the *antiparticle* of the *proton*. *Antiprotons are stable*, but they are typically *short-lived*, since any collision with a *proton* will cause both particles to be annihilated in a burst of energy.

The existence of the *antiproton* with *electric charge* of −1 e, opposite to the *electric charge* of +1 e of the *proton*, was predicted by Paul Dirac in his 1933 Nobel Prize lecture. Dirac received the Nobel Prize for his 1928 publication of his Dirac equation that predicted the existence of positive and negative solutions to Einstein's energy equation ($E = mc^2$) and the existence of the *positron*, the *antimatter* analog of the *electron*, with *opposite charge and spin*.

In terms of *valence quarks*, an *antiproton* consists of two *up antiquarks* and one *down antiquark* (uud). The properties of the *antiproton* that have been measured all match the corresponding properties of the *proton, with the exception that the antiproton has electric charge and magnetic moment that are the opposites of those in the proton*, which is to be expected from the *antimatter* equivalent of a *proton*.

The questions of *how matter is different from antimatter*, and the relevance of *antimatter* in explaining how our universe survived the Big Bang, remain open problems—open, in part, due to the relative scarcity of *antimatter* in today's universe.

Electron neutrino [elementary particle]

In 1956, the *electron neutrino* was detected by Frederick Reines and Clyde Cowan. [F. Reines, F. & Cowan, C.L. (1956). The Neutrino. *Nature*, 178, 4531, 446–9; https://doi.org/10.1038/178446a0.] At the time it was simply referred to as the *neutrino* since there was only one known *neutrino*. Frederick Reines (March 16, 1918–August 26, 1998) and Clyde Cowan (December 6, 1919–May 24, 1974) were both American physicists. The Nobel Prize in Physics 1995 was awarded "for pioneering experimental contributions to *lepton* physics" jointly with one half to Martin L. Perl "for the discovery of the *tau lepton* [in 1975]" and with one half to Frederick Reines "for the detection of the *neutrino* [in 1956]".

In the early 1950s, working in Hanford and Savannah River Sites, Reines and Cowan developed the equipment and procedures with which they first detected the supposedly undetectable *neutrinos* in June 1956. Reines dedicated the major part of his career to the study of the *neutrino*'s properties and interactions, which work would influence study of the *neutrino* for many researchers to come. This included the detection of *neutrinos* created in the atmosphere by cosmic rays, and the 1987 detection of *neutrinos* emitted from Supernova SN1987A, which inaugurated the field of *neutrino* astronomy.

The existence of the *neutrino* had been proposed by Wolfgang Pauli in 1930 to explain the apparent violation of conservation of energy in beta decay. In a letter of 4 December to Lise Meitner et al., beginning, "Dear radioactive ladies and gentlemen", he proposed the existence of a hitherto unobserved neutral particle with a small mass, no greater than 1% the mass of a proton, to explain the continuous spectrum of beta decay. In 1934, Enrico Fermi incorporated the particle, which he called a *neutrino*, "little neutral one" in Fermi's native Italian, into his theory of beta decay. [Fermi, E. (1934). Radioattività indotta da bombardamento di neutroni. Ricerca Scientifica, 5, 1, 283; reprinted in Collected Papers (*Note e Memorie*), vol. 1 Italy 1921–1938 (Chicago: University of Chicago Press; Rome: Accademia Nazionale dei Lincei, 1962), 645–46; translated as "Radioactivity Induced by Neutron Bombardment. – I," in Collected Papers, 674–75. Also in Wilson, F. L. (1968). Fermi's Theory of Beta Decay. *Am. J. of Physics*, 36, 12, 1150–1160; https://doi.org/10.1119/1.1974382: (includes complete English translation of Fermi's 1934 paper published in *Zeitschrift für Physik* in 1934; https://pubs.aip.org/aapt/ajp/ article-abstract/36/12/1150/1047952/Fermi-s-Theory-of-Beta-Decay?redirected From=fulltext.]

Muon neutrino (or mu neutrino) [elementary particle]

In 1962, in a gargantuan "*neutrino experiment*" at the Brookhaven National Laboratory, New York, a group headed by Leon Lederman detected the *muon neutrino* (or *mu neutrino*) shown to be distinct from the *electron neutrino*. [Danby, G. et al. (1962). Observation of High-Energy Neutrino Reactions and the Existence of Two Kinds of Neutrinos. *Phys. Rev. Let.*, 9 ,1, 36–44; https://www.physik.uni-bielefeld.de/~yorks/pro13/PhysRevLett.9.36.pdf.]

Leon Max Lederman (July 15, 1922–October 3, 2018), an American experimental physicist, received the 1988 Nobel Prize in Physics, along with Melvin Schwartz, an American physicist, and Jack Steinberger, a German-born American physicist, "for the *neutrino* beam method and the demonstration of the doublet structure of the *leptons* through the discovery of the *muon neutrino*".

The *muon neutrino* is an elementary particle which has zero *electric charge* and no *color charge*. The *mass* of the *muon neutrino* is very small but non-zero. It has *spin* = ½, and *weak isospin* = ½. Together with the *muon* it forms the *second generation* of *leptons*.

The *muon neutrino*, or "*neutretto*", was hypothesized to exist by a number of physicists in the 1940s. The first paper on it may be Shoichi Sakata and Takesi Inoue's two-meson theory of 1942, which also involved two *neutrinos*. [Shoichi Sakata, S. & Inoue, T. (1942). Chukanshi to Yukawa ryushi no Kankei ni tuite. *Nippon Suugaku-Butsuri Gakkaishi*, 16; https://doi.org/10.11429/subutsukaishi1927.16.232].

Omega (Ω) baryon [composite particle]

In 1964, the *omega baryon* was discovered at Brookhaven National Laboratory. The first *omega baryon* discovered was the Ω^-, made of three *strange quarks*. [V. E. Barnes; et al. (1964). Observation of a Hyperon with Strangeness Minus Three. *Phys. Rev. Lett.*, 12, 8, 204; https://doi.org/10.1103/PhysRevLett.12.204.]

The *omega baryons* are a family of subatomic *hadron* (a *baryon*) particles that are either neutral or have a +2, +1 or −1 elementary *charge*. They are *baryons containing no up or down quarks*. Since *omega baryons* do not have any *up or down quarks*, they all have *isospin* 0. *Omega baryons* containing *top quarks* are not expected to be observed. This is because the *Standard Model* predicts the mean lifetime of *top quarks* to be roughly 5×10^{-25} s, which is about a twentieth of the timescale for strong interactions, and therefore that they do not form *hadrons*.

Besides the Ω^-, a charmed omega particle (Ω^0_c) was discovered in 1985, in which a *strange quark* is replaced by a *charm quark*. The $\Omega-$ decays only via the *weak interaction* and has therefore a relatively long lifetime. *Spin* and *parity* values for unobserved baryons are predicted by the *quark model*.

The existence of the *omega baryon*, and its mass, and decay products had been predicted in 1961 by the American physicist Murray Gell-Mann and, independently, by the Israeli physicist Yuval Ne'eman. In 1961, Gell-Mann introduced his formulation of a particle classification system for *hadrons* known as *the Eightfold Way* – or, in more technical terms, SU(3) *flavor symmetry*, streamlining its structure. [Gell-Mann, M. (March, 1961). *The Eightfold Way: A Theory of Strong Interaction Symmetry (Report)*. See below.] A similar scheme had been independently proposed by Yuval Ne'eman in the same year. [Ne'eman, Y. (August, 1961). Derivation of Strong Interactions from a Gauge Invariance. *Nuclear Physics*, 26, 2; https://doi.org/10.1016/0029-5582(61)90134-1.]

Up quark, down quark, and strange quark [elementary particles]

In 1969, what were referred to at that time as *partons* (internal constituents of *hadrons*) were observed in deep inelastic scattering experiments between *protons* and *electrons* at the Stanford Linear Accelerator Center (SLAC). [Bloom, E. D. et al. (1969). High-Energy Inelastic e–p Scattering at 6° and 10°". *Phys. Rev. Lett.*, 23, 16, 930–4; https://doi.org/10.1103/PhysRevLett.23.930; Breidenbach, M et al. (1969). Observed Behavior of Highly Inelastic Electron-Proton Scattering. *Phys. Rev. Lett.*, 23, 16, 935–9; https://doi.org/10.1103/PhysRevLett.23.935.] The objects that were observed at SLAC would later be identified as *up and down quarks* as the other flavors were discovered. Nevertheless, "parton" remains in use as a collective term for the constituents of *hadrons* (quarks, antiquarks, and gluons). Richard Taylor, Henry Kendall and Jerome Friedman received the 1990 Nobel Prize in physics "for their pioneering investigations concerning deep inelastic scattering of electrons on protons and bound neutrons, which have been of essential importance for the development of the *quark model* in particle physics".

The *strange quark*'s existence was indirectly validated by SLAC's scattering experiments: not only was it a necessary component of Gell-Mann and Zweig's three-quark model, but it provided an explanation for the *kaon* (K) and *pion* (π) hadrons discovered in cosmic rays in 1947. [See above.]

Quarks [*classification*]

A *quark* is a type of elementary particle and a fundamental constituent of matter. *Quarks* combine to form composite particles called *hadrons*, the most stable of which are *protons* and *neutrons*, the components of atomic nuclei. All commonly observable matter is composed of *up quarks*, *down quarks* and *electrons*. Owing to a phenomenon known as *color confinement*, *quarks* are never found in isolation; they can be found only within *hadrons*, which include *baryons* (such as protons and neutrons) and *mesons*, or in *quark–gluon* plasmas. For this reason, much of what is known about *quarks* has been drawn from observations of *hadrons*.

Quarks have various intrinsic properties, including *electric charge*, *mass*, *color charge*, and *spin*. They are the only elementary particles in the *Standard Model* of particle physics to experience all four fundamental interactions, also known as fundamental forces (*electromagnetism*, *gravitation*, *strong interaction*, and *weak interaction*), as well as the only known particles whose *electric charges* are not integer multiples of the elementary *charge*.

There are six types, known as *flavors*, of *quarks*: *up*, *down*, *charm*, *strange*, *top*, and *bottom*. *Up and down quarks* have the lowest *masses* of all *quarks*. The heavier *quarks* rapidly change into *up and down quarks* through a process of *particle decay*: the transformation from a higher *mass state* to a lower *mass state*. Because of this, *up and down quarks* are generally stable and the most common in the universe, whereas *strange*, *charm*, *bottom*, and *top quarks* can only be produced in high energy collisions (such as those involving cosmic rays and in particle accelerators). For every *quark flavor* there is a corresponding type of *antiparticle*, known as an *antiquark*, that differs from the *quark* only in that some of its properties (such as the *electric charge*) have equal magnitude but opposite sign.

The *quark model* was independently proposed by physicists Murray Gell-Mann and George Zweig in 1964. [Gell-Mann. M. (February, 1964). A Schematic Model of Baryons and Mesons. *Physics Letters*, 8, 3, 214–5] and George Zweig [Zweig, G. (February 21, 1964). An SU(3) Model for Strong Interaction Symmetry and its Breaking: II. *CERN Document Server*. CERN-TH-412]. *Quarks* were introduced as parts of an ordering scheme for *hadrons*, and there was little evidence for their physical existence until deep inelastic scattering experiments at the Stanford Linear Accelerator Center in 1968. Accelerator program experiments have provided evidence for all six *flavors*. The *top quark*, first observed at Fermilab in 1995, was the last to be discovered.

The name "*quark*" was coined by Gell-Mann, and is a reference to the novel Finnegans Wake, by James Joyce ("*Three quarks for Muster Mark!*" book 2, episode 4). Zweig had referred to the particles as "aces", but Gell-Mann's name caught on. *Quarks*, *antiquarks*, and *gluons* were soon established as the underlying elementary objects in the study of the structure of *hadrons*. The 1969 Nobel Prize in Physics was awarded to Gell-Mann "for his contributions and discoveries concerning the classification of elementary particles and their interactions".]

In a 1970 paper, Glashow, John Iliopoulos and Luciano Maiani presented the GIM mechanism (named from their initials) to explain the experimental non-observation of flavor-changing *neutral currents*. This theoretical model required the existence of the as-yet undiscovered *charm quark*. [See below.]

In the following years a number of suggestions appeared for extending the quark model to six quarks. Of these, the 1975 paper by Haim Harari was the first to coin the terms *top* and *bottom* for the additional *quarks*. The number of supposed *quark flavors* grew to the current six in 1973, when Makoto Kobayashi and Toshihide Maskawa noted that the experimental observation of CP violation could be explained if there were another pair of quarks. *Charm quarks* were produced almost simultaneously by two teams in November 1974 – one at SLAC under Burton Richter, and one at Brookhaven National Laboratory under Samuel Ting. The *charm quarks* were observed bound with *charm antiquarks* in *mesons*. The two parties had assigned the discovered *meson* two different symbols, J and ψ; thus, it became formally known as the J/ψ meson. [See below.] The discovery finally convinced the physics community of the validity of the *quark model*.

Hyperon [classification]

A *hyperon* is any *baryon* containing one or more *strange quarks*, but no *charm*, *bottom*, or *top quark*. Being *baryons*, all *hyperons* are *fermions*. That is, they have half-integer *spin* and obey Fermi–Dirac statistics. *Hyperons* all interact via the *strong nuclear force*, making them types of *hadron*. They are composed of three light *quarks*, at least one of which is a *strange quark*, which makes them *strange baryons*.

This form of matter may exist in a stable form within the core of some neutron stars. The first research into *hyperons* happened in the 1950s and spurred physicists on to the creation of an organized classification of particles. The term was coined by French physicist Louis Leprince-Ringuet in 1953.

Today, research in this area is carried out on data taken at many facilities around the world, including CERN, Fermilab, SLAC, JLAB, Brookhaven National Laboratory, KEK, GSI and others. Physics topics include searches for CP violation, measurements of spin, studies of excited states (commonly referred to as spectroscopy), and hunts for exotic forms such as pentaquarks and dibaryons.]

J/ψ (J/psi) meson [composite particle]; charm quark and charm anti-quark [elementary particles]

In 1974, the _J/ψ (J/psi) meson_ (a composite particle which is a _vector meson (quark_ and _antiquark_) was discovered by groups headed by Burton Richter and Samuel Ting, both American physicists, demonstrating the existence of the _charm quark and charm anti-quark_, for which they shared the 1976 Nobel Prize for Physics. They discovered that they had found the same particle, and both announced their discoveries on November 11, 1974. [Aubert, J. J. et al. (1974). Experimental Observation of a Heavy Particle J. _Phys. Rev. Let._, 33, 23, 1404–6; https://doi.org/10.1103/PhysRevLett.33.1404; Augustin, J. -E. et al. (1974). Discovery of a Narrow Resonance in e+e− Annihilation. _Phys. Rev. Let._, 33, 23, 1406–8; https://doi.org/10.1103/PhysRevLett.33.1406.]

Burton Richter (March 22, 1931–July 18, 2018) led the Stanford Linear Accelerator Center (SLAC) team, and Samuel Ting (born January 27, 1936) led the Brookhaven National Laboratory (BNL) team. The importance of this discovery is highlighted by the fact that the subsequent, rapid changes in high-energy physics at the time have become collectively known as the "_November Revolution_".

The _J/ψ meson_ is a subatomic particle, a _flavor-neutral meson_ consisting of a _charm quark_ and a _charm antiquark_. _Mesons_ formed by a bound state of a _charm quark_ and a _charm anti-quark_ are generally known as "_charmonium_" or _psions_. The J/ψ is the most common form of _charmonium_, due to its _spin_ of 1 and its low rest mass. The J/ψ has a rest mass of 3.0969 GeV/c^2, just above that of the η_c (2.9836 GeV/c^2), and a mean lifetime of 7.2×10^{-21} s. _This lifetime was about a thousand times longer than expected._

The _J/ψ meson_ had been proposed by James Bjorken and Sheldon Glashow in 1964) [Bjørken, B. J. & Glashow, S. L. (1964). Elementary Particles and SU(4). _Phys. Let._, 11, 3, 255–7; https://www.sciencedirect.com/science/article/abs/pii/0031916364904330.]

Vector meson [classification]

A _vector meson_ is a _meson_ with _total spin_ 1 and odd parity. _Vector mesons_ have been seen in experiments since the 1960s, and are well known for their spectroscopic pattern of masses.

The _vector mesons_ contrast with the _pseudovector mesons_, which also have a _total spin_ 1 but instead have _even parity_. The _vector_ and _pseudovector mesons_ are also dissimilar in that the spectroscopy of _vector mesons_ tends to show nearly pure states of constituent _quark flavors_, whereas _pseudovector mesons_ and _scalar mesons_ tend to be expressed as composites of mixed states.

Tau [elementary particle]

In 1975, the *tau* discovered by a group headed by Martin Perl. [Perl, M. L. (1975). Evidence for Anomalous Lepton Production in e+–e– Annihilation. *Phys. Rev. Let.*, 35, 22, 1489–92; doi:10.1103/PhysRevLett.35.1489.] Their equipment consisted of SLAC's then-new electron–positron colliding ring, called SPEAR, and the LBL magnetic detector. They could detect and distinguish between *leptons*, *hadrons*, and *photons*. They did not detect the *tau* directly, but rather discovered anomalous events.

The search for *tau* started in 1960 at CERN by the Bologna-CERN-Frascati (BCF) group led by Antonino Zichichi. Zichichi came up with the idea of a new sequential *heavy lepton*, now called *tau*, and invented a method of search. He performed the experiment at the ADONE facility in 1969 once its accelerator became operational; however, the accelerator he used did not have enough energy to search for the *tau* particle.

The *tau* was independently anticipated in a 1971 article by Yung-su Tsai. [Tsai, Y-S. (November, 1971). Decay correlations of heavy leptons in e+ + e− → ℓ+ + ℓ−. *Phys. Rev. D.*, 4, 9, 2821; https://doi.org/10.1103/PhysRevD.4.2821.] Providing the theory for this discovery, the tau was detected in a series of experiments between 1974 and 1977 by Perl with his and Tsai's colleagues at the Stanford Linear Accelerator Center (SLAC) and Lawrence Berkeley National Laboratory (LBL) group.

The *tau*, also called the *tau lepton*, *tau particle*, *tauon* or *tau electron*, is an elementary particle similar to the electron, with negative electric charge and a spin of ½. Like the *electron*, the *muon*, and the three *neutrinos*, the *tau* is a *lepton*, and like all elementary particles with *half-integer spin*, the *tau* has a corresponding *antiparticle* of *opposite charge but equal mass and spin*. In the *tau*'s case, this is the "*antitau*" (also called the *positive tau*). *Tau leptons* have a lifetime of 2.9×10^{-13} s and a *mass* of 1776.9 MeV/c^2 (compared to 105.66 MeV/c^2 for *muons* and 0.511 MeV/c^2 for electrons). Since their interactions are very similar to those of the *electron, a tau can be thought of as a much heavier version of the electron. Because of their greater mass, tau particles do not emit as much bremsstrahlung (braking radiation) as electrons; consequently, they are potentially much more highly penetrating than electrons.*

Because of its *short lifetime*, the *range* of the *tau* is mainly set by its decay length, which is too small for bremsstrahlung to be noticeable. *Its penetrating power appears only at ultra-high velocity and energy (above petaelectronvolt energies)*, when time dilation extends its otherwise very short path-length. As with the case of the other *charged leptons*, the *tau* has an associated *tau neutrino*.

Upsilon meson [composite particle]; **bottom quark and bottom antiquark [elementary particle and elementary antiparticle]**

In 1977, the *upsilon meson* was discovered at Fermilab, demonstrating the existence of the *bottom quark*. [Herb, S. W. et al. (1977). Observation of a Dimuon Resonance at 9.5 GeV in 400-GeV Proton-Nucleus Collisions. *Phys. Rev. Let.*, 39, 5, 252–5; https://doi.org/10.1103/PhysRevLett.39.252.] The evidence for the *bottom quark* was first obtained by the Fermilab E288 experiment team led by Leon M. Lederman, when *proton-nucleon* collisions produced bottomonium decaying to pairs of *muons*. The discovery was confirmed about a year later by the PLUTO and DASP2 Collaborations at the electron-positron collider DORIS at DESY.

The *upsilon meson* is a quarkonium state (i.e. flavorless *meson*) formed from a *bottom quark* and its *antiparticle*. It was the first particle containing a *bottom quark* to be discovered because it is the lightest that can be produced without additional massive particles. It has a *lifetime* of 1.21×10^{-20} s and a *mass* about 9.46 GeV/c^2 in the ground state.

The *bottom quark* is an elementary particle of the *third generation*. It is a *heavy quark* with a *charge* of $-1/3$ e. All *quarks* are described in a similar way by *electroweak interaction* and *quantum chromodynamics*, but the *bottom quark* has exceptionally low rates of transition to lower-mass quarks. The *bottom quark* is also notable because *it is a product in almost all top quark decays*, and is a frequent decay product of the *Higgs boson*.

The *bottom quark* was proposed by Kobayashi and Maskawa in 1973, for which they shared half of the 2008 Nobel Prize in Physics "for [their 1973] discovery of *the origin of the broken symmetry which predicts the existence of at least three families of quarks in nature*", with the other half going to Yoichiro Nambu "for the discovery of *the mechanism of spontaneous broken symmetry* in subatomic physics". [Kobayashi, M. & Maskawa, T. (February, 1973). CP-Violation in the Renormalizable Theory of Weak Interaction. *Progress of Theoretical Physics*, 49, 2, 652–7; https://doi.org/10.1143/PTP.49.652. See below.]

Gluons [elementary particles]

In 1979, the PLUTO detector at the DORIS electron-positron collider at the PETRA (Positron–Electron Tandem Ring Accelerator) storage ring (with a circumference of 2,304 m.) at DESY, produced the first evidence of hadronic decays of the very narrow resonance which could be interpreted as three-jet event topologies produced by three *gluons*. [Barber, D. P. et al. (1979). Discovery of Three-Jet Events and a Test of Quantum Chromodynamics at PETRA. *Phys. Rev. Let.*, 43, 12, 830–3; https://doi.org/10.1103/PhysRevLett.43.830.]

[DESY is a national research center for fundamental science located in Hamburg and Zeuthen near Berlin in Germany. It operates particle accelerators used to investigate the structure, dynamics and function of matter, and conducts a broad spectrum of interdisciplinary scientific research in four main areas: particle and high energy physics; photon science; astroparticle physics; and the development, construction and operation of particle accelerators. Its name refers to its first project, an electron synchrotron, the Deutsches Elektronen-Synchrotron.]

Quarks and *gluons* (*colored*) manifest themselves by fragmenting into more *quarks* and *gluons*, which in turn hadronize into normal (colorless) particles, correlated in *jets*. Later, published analyses by the same experiment confirmed this interpretation and also the *spin* = 1 nature of the *gluon*.

Gluon are *gauge bosons*, a type of massless elementary particle that *mediates the strong interaction* between *quarks, acting as the exchange particle for the interaction. Gluons are massless vector bosons*, thereby having a *spin* of 1. Through the *strong interaction, gluons* bind *quarks* into groups according to *quantum chromodynamics*, forming *hadrons* such as *protons* and *neutrons*. [Formally, *quantum chromodynamics* is a *gauge theory* with SU(3) gauge symmetry.]

Gluons carry the *color charge* of the *strong interaction, thereby participating in the strong interaction as well as mediating it*. Because *gluons* carry the *color charge, quantum chromodynamics* is more difficult to analyze compared to *quantum electrodynamics*, where the *photon* carries no *electric charge*.

The term was coined by Murray Gell-Mann in 1962 for being similar to an adhesive or glue that keeps the nucleus together. Together with the *quarks*, these particles were referred to as *partons* by Richard Feynman.

W and Z bosons [elementary particles]

In 1976, Carlo Rubbia (born March 31, 1934), an Italian particle physicist and inventor, suggested adapting CERN's Super Proton Synchrotron (SPS) to collide *protons* and *antiprotons* in the same ring – the Proton-Antiproton Collider. The collider started running in 1981 and, in early 1983, an international team of more than 100 physicists headed by Rubbia and Simon van der Meer, known as the UA1 Collaboration, detected the intermediate *vector bosons*, the *W and Z bosons*, which had become a cornerstone of modern theories of elementary particle physics long before this direct observation. [Aubert J. J. et al. (European Muon Collaboration) (1983). The ratio of the nucleon structure functions F2N for iron and deuterium. *Phys. Let. B.*, 123, 3–4, 275–8; https://doi.org/ 10.1016/0370-2693(83)90437-9; Arnison, G. et al. (UA1 collaboration) (1983).

Experimental observation of lepton pairs of invariant mass around 95 GeV/c2 at the CERN SPS collider. *Phys. Let. B.*, 126, 5, 398–410; https://doi.org/10.1016/0370-2693(83)90188-0.]

Carlo Rubbia and Simon van der Meer (November 24, 1925–March 4, 2011), a Dutch particle accelerator physicist, were awarded the 1984 Nobel Prize in Physics "for their decisive contributions to the large project, which led to the discovery of the field particles W and Z, communicators of weak interaction".

The *W and Z bosons* are *gauge bosons* and *vector bosons*, with of a spin = 1, the two fundamental communicators of the *weak interaction*. They carry *the weak force that causes radioactive decay in the atomic nucleus* and controls the combustion of the Sun, just as *photons*, massless particles of light, carry *the electromagnetic force which causes most physical and biochemical reactions*. The *weak force* also plays a fundamental role in the nucleosynthesis of the elements, as studied in theories of stars evolution. *The W and Z bosons have a mass almost 100 times greater than the proton.*

The *W and Z bosons* were predicted in detail by Sheldon Glashow, Abdus Salam, and Steven Weinberg. In 1961, Glashow extended *electroweak unification models* due to Schwinger by including a short-range *neutral current*, the Z_0. The resulting symmetry structure that Glashow proposed, SU(2) × U(1), forms the basis of the accepted *theory of the electroweak interactions*. For this discovery, Glashow along with Steven Weinberg and Abdus Salam, was awarded the 1979 Nobel Prize in Physics "for their contributions to the *theory of the unified weak and electromagnetic interaction* between elementary particles, including, inter alia, the prediction of the *weak neutral current*". [Glashow, S. L. (February, 1961). Partial-symmetries of weak interactions. See below.]

Top quark [elementary particle]

In 1977, the *bottom quark* was observed by a team at Fermilab led by Leon Lederman. This was a strong indicator of the *top quark*'s existence: without the *top quark*, the *bottom quark* would have been without a partner. It was not until 1995 that the *top quark* was finally observed at Fermilab [Abe, F. et al. (CDF collaboration) (1995). Observation of Top quark production in p–p Collisions with the Collider Detector at Fermilab. *Phys. Rev. Lett.*, 74, 14, 2626–31; arXiv:hep-ex/9503002; https://doi.org/10.1103/PhysRevLett.74.2626; S. Arabuchi et al. (D0 collaboration) (1995). Observation of the Top Quark. *Phys. Rev. Lett.*, 74, 14, 2632–7; arXiv:hep-ex/9503003; https://doi.org/10.1103/PhysRevLett.74.2632.]

[The DØ experiment (sometimes written D0 experiment) was a worldwide collaboration of scientists conducting research on the fundamental nature of matter. DØ was one of two major experiments (the other was the CDF experiment) both located at the Tevatron Collider at Fermilab in Batavia, Illinois. The Tevatron was the world's highest-energy accelerator from 1983 until 2009, when its energy was surpassed by the Large Hadron Collider. The DØ experiment stopped taking data in 2011, when the Tevatron shut down, but data analysis is still ongoing.]

It had a mass much larger than expected, almost as large as that of a gold atom.

The *top quark* is *the most massive of all observed elementary particles*. It derives its mass from its *coupling to the Higgs boson*. This coupling is very close to unity; in the *Standard Model* of particle physics, *it is the largest (strongest) coupling at the scale of the weak interactions and above*. Like all other quarks, the *top quark* is a fermion with *spin ½* and *participates in all four fundamental interactions: gravitation, electromagnetism, weak interactions, and strong interactions*. It has an *electric charge* of + 2/3 e. It has a mass of 172.76 ± 0.3 GeV/c2, which is close to the rhenium atom mass. The *antiparticle* of the *top quark* is the *top antiquark* (sometimes called *antitop quark* or simply *antitop*), which differs from it only in that some of its properties have equal magnitude but opposite sign.

The *top quark* interacts with *gluons* of the *strong interaction* and is typically produced in hadron colliders via this interaction. However, once produced, the *top* (or antitop) can decay only through the *weak force*. It decays to a *W boson* and either a *bottom quark* (most frequently), a *strange quark*, or, on the rarest of occasions, a *down quark*.

The *Standard Model* determines the *top quark*'s mean *lifetime* to be roughly 5×10^{-25} s. This is about a twentieth of the timescale for *strong interactions*, and therefore *it does not form hadrons*, giving physicists a unique opportunity to study a "bare" *quark* (all other *quarks* hadronize, meaning that they combine with other *quarks* to form *hadrons* and can only be observed as such).

Because the *top quark* is so massive, *its properties allowed indirect determination of the mass of the Higgs boson*. As such, the *top quark*'s properties are extensively studied as a means to discriminate between competing theories of new physics beyond the *Standard Model*. *The top quark is the only quark that has been directly observed* due to its decay time being shorter than the hadronization time.

Tau neutrino [elementary particle]

In 2000, the *tau neutrino* first observed directly at Fermilab ["*Physicists Find First Direct Evidence for Tau Neutrino at Fermilab*" (Press release). Fermilab. July 20, 2000.]

The *tau neutrino*, or *tauon neutrino*, is an elementary particle which has zero *electric charge*. Together with the *tau*, it forms the *third generation* of *leptons*, hence the name *tau neutrino*. Its existence was immediately implied after the *tau* particle was detected in a series of experiments between 1974 and 1977 by Martin Lewis Perl with his colleagues at the SLAC–LBL group. The discovery of the *tau neutrino* was announced in July 2000 by the DONUT collaboration (Direct Observation of the Nu Tau). [Kodama, K. et al. (DONUT collaboration) (2001). Observation of tau neutrino interactions. *Physics Letters B*, 504, 3, 218–24; arXiv:hep-ex/0012035; https://doi.org/10.1016/S0370-2693(01)00307-0.] The DONUT experiment from Fermilab was built during the 1990s to specifically detect the tau neutrino.

Higgs boson [elementary particle]

In 2012, after a 40-year search, a particle exhibiting most of the predicted characteristics of the *Higgs boson* was discovered by researchers conducting the Compact Muon Solenoid (CMS) and ATLAS experiments at the Large Hadron Collider at CERN near Geneva, Switzerland.

The *Higgs boson* is an elementary particle in the *Standard Model* of particle physics produced by the quantum excitation of the *Higgs field*, one of the fields in particle physics theory. Both the *field* and the *boson* are named after Scottish physicist Peter Higgs, who in 1964, along with five other scientists in three teams, proposed the *Higgs mechanism*, a way for some particles to acquire *mass*. All fundamental particles known at the time should be massless at very high energies, but *fully explaining how some particles gain mass at lower energies had been extremely difficult*. If these ideas were correct, a particle known as a *scalar boson* should exist with certain properties. This particle was called the *Higgs boson* and could be used to test whether the *Higgs field* was the correct explanation.

A theory able to finally explain *mass generation* without "breaking" *gauge theory* was published almost simultaneously by three independent groups in 1964: by Robert Brout and François Englert [Englert, F. & Brout, R. (1964). Broken symmetry and the mass of gauge vector mesons. See below.] by Peter Higgs [Higgs, P. W. (1964). Broken symmetries and the masses of gauge bosons. See below.] and by Gerald Guralnik, C. R. Hagen and Tom Kibble. [Guralnik, G. S., Hagen, C. R. & Kibble, T. W. B. (1964). Global conservation laws and massless particles. See below.]

Although the *Higgs field* would exist everywhere, proving its existence was far from easy. In principle, it can be proved to exist by detecting its excitations, which manifest as *Higgs particles* (the *Higgs boson*), but *these are extremely difficult to produce and detect due to the energy required to produce them and their very rare production even if the energy is sufficient*. It was, therefore, several decades before the first evidence of the *Higgs boson* could be found. Particle colliders, detectors, and computers capable of looking for *Higgs bosons* took more than 30 years (c. 1980–2010) to develop. The importance of this fundamental question led to a 40-year search, and the construction of one of the world's most expensive and complex experimental facilities to date, CERN's Large Hadron Collider, in an attempt to create *Higgs bosons* and other particles for observation and study.

On 4 July 2012, the discovery of a new particle with a mass between 125 and 127 GeV/c^2 was announced; physicists suspected that it was the Higgs boson. Since then, the particle has been shown to behave, interact, and decay in many of the ways predicted for Higgs particles by the *Standard Model*, as well as having even *parity* and zero *spin*, two fundamental attributes of a Higgs boson. [CMS Collaboration (2017). Constraints on anomalous Higgs boson couplings using production and decay information in the four-lepton final state. *Physics Letters B*. 775, 1–24; arXiv:1707.00541; https://doi.org/10.1016/j.physletb.2017.10.021.]

By March 2013, the existence of the *Higgs boson* was confirmed, and therefore, the concept of some type of *Higgs field* throughout space is strongly supported. The presence of the field, now confirmed by experimental investigation, explains why some fundamental particles have (a rest) mass, despite the symmetries controlling their interactions, implying that they should be "massless". It also resolves several other long-standing puzzles, such as the reason for the extremely short distance travelled by the *weak force bosons*, and, therefore, the *weak force*'s extremely short range. As of 2018, in-depth research shows the particle continuing to behave in line with predictions for the *Standard Model Higgs boson*. More studies are needed to verify with higher precision that the discovered particle has all of the properties predicted or whether, as described by some theories, multiple *Higgs bosons* exist.

Physicists from two of the three teams, Peter Higgs and François Englert, were awarded jointly the 2013 Nobel Prize in Physics "for the theoretical discovery of a mechanism that contributes to our understanding of the origin of mass of subatomic particles, and which recently was confirmed through the discovery of the predicted fundamental particle, by the ATLAS and CMS experiments at CERN's Large Hadron Collider". Although Higgs's name has come to be associated with this theory, several researchers between about 1960 and 1972 independently developed different parts of it.

In the *Standard Model*, the *Higgs particle* is a massive *scalar boson* with zero *spin*, even (positive) *parity*, no *electric charge*, and no *color charge* that couples to (interacts with) *mass. It is also very unstable, decaying into other particles almost immediately upon generation.* The *Higgs field* is a *scalar field* with *two neutral and two electrically charged components* that form a complex doublet of the *weak isospin* SU(2) *symmetry*. Its "*Sombrero potential*" leads it to take a nonzero value everywhere (including otherwise empty space), which *breaks the weak isospin symmetry of the electroweak interaction* and, via the *Higgs mechanism, gives a rest mass to all massive elementary particles* of the *Standard Model*, including the *Higgs boson* itself. The existence of the *Higgs field* became the last unverified part of the *Standard Model* of particle physics, and for several decades was considered "the central problem in particle physics".

Paul Adrien Maurice Dirac (August 8, 1902–October 20, 1984).

Dirac was an English theoretical physicist who is regarded as one of the most significant physicists of the 20th century. Dirac shared the 1933 Nobel Prize in Physics with Erwin Schrödinger "for the discovery of new productive forms of atomic theory". "During the intense period of 1925-26 quantum theories were proposed that accurately described the energy levels of electrons in atoms. These equations needed to be adapted to Einstein's theory of *relativity*, however. In 1928 Paul Dirac formulated a fully *relativistic* quantum theory. The equation gave solutions that he interpreted as being caused by a particle equivalent to the electron, but with a positive charge. This particle, the positron, was later confirmed through experiments." [Paul A. M. Dirac – Facts. NobelPrize.org. https://www.nobelprize.org/prizes/physics/1933/dirac/facts/.]

Dirac made fundamental contributions to the early development of both *quantum mechanics* and *quantum electrodynamics*. Among other discoveries, he formulated the *Dirac equation* which describes the behavior of fermions and predicted the existence of *antimatter*. The notion of an *antiparticle* to each fermion particle – e.g. the *positron* as *antiparticle* to the electron – stems from his equation. He was the first to develop *quantum field theory*, which underlies all theoretical work on sub-atomic or "elementary" particles today. He also made significant contributions to the reconciliation of *general relativity* with quantum mechanics. He proposed and investigated the concept of a magnetic monopole, an object not yet known empirically, as a means of bringing even greater symmetry to *Maxwell's Equations* of electromagnetism.

Dirac was born at his parents' home in Bristol, England, on August 8, 1902, and grew up in the Bishopston area of the city. His father, Charles, a Swiss national from Saint-Maurice, Switzerland, immigrated to London in 1890, where he worked as a teacher of French. In 1896 he moved to Bristol, where he was appointed Head of Modern Languages at the Merchant Venturers' School, where he supplemented his income with private language classes. His mother, Florence, née Holten, was the daughter of a ship's captain. Charles met her shortly after his arrival, when she was working as a librarian at the Bristol Central Library. Paul had a younger sister, Béatrice, known as Betty, and an older brother, Reginald, known as Felix, who died by suicide in March 1925.

Charles and the children were officially Swiss nationals until they became naturalized in 1919. Dirac's father was strict and authoritarian, although he disapproved of corporal punishment. Dirac had a strained relationship with his father. Charles forced Dirac to speak to him only in French so that he might learn the language. When Dirac found that he could not express what he wanted to say in French, he chose to remain silent. He grew to dislike eating, largely on account of his parents' insistence that he eat every morsel of food on his plate.

Dirac was educated first at Bishop Road Primary School, which was just around the corner from his home. Although initially he only just made the top third of his class, he steadily improved so that by the age of 10 he was consistently near the top of his class. At home he pursued his extra-curricular hobby of astronomy. The school did not teach science but gave classes in technical drawing, that provided Dirac with his unique way of thinking about science. Like many parents, Charles entered all his children for scholarship exams. Although Felix and Betty each failed one, Dirac passed every one, so was educated at minimal expense to his parents.

Dirac started his secondary education at the all-boys Merchant Venturers' School (later Cotham School), where his father worked, shortly after the outbreak of the 1st World War on August 4, 1914. For Charles it was a fifteen-minute cycle to the school, but he made his sons to walk there and back twice a day, as they had lunch at home, rather than taking the tram. The school was an institution attached to the University of Bristol, which shared grounds and staff. It emphasized technical subjects like bricklaying, shoemaking and metal work, technical drawing, and modern languages. This was unusual at a time when secondary education in Britain was still dedicated largely to the classics. It took only weeks for Dirac to establish himself as a stellar pupil. Except for history and German, he shone at every academic subject, and was usually ranked as the top student of his class. He excelled at science, including chemistry, where he learned about atoms; and he began mulling over the nature of space and time. In particular, it further advanced Dirac's ability to visualize objects and their movements in three dimensions.

Dirac's teacher, Arthur Pickering, gave up on teaching Dirac with the other boys, instead sending him to the library with a reading list. He suggested that he look beyond simple geometry to the theories of the German mathematician Bernhard Riemann.

Dirac was a workaholic, very quiet, and had no interest in sports. As the gap between the abilities of Felix and Dirac widened, their relationship deteriorated until they were no longer on speaking terms. In 1918, shortly before the end of the 1st World War, although Dirac could have taken his pick from dozens of science courses, and considered taking a degree in mathematics, he decided to follow his brother by studying engineering on a City of Bristol University Scholarship at the University of Bristol's engineering faculty, which was co-located with the Merchant Venturers' School.

On November 7, 1919, the London *Times* published its famous article about the "Revolution in Science", reporting the verification of Einstein's *Theory of General Relativity*, by Arthur Eddington's claim that they had verified the predicted bending of light by the Sun during the recent eclipse. *Relativity became Dirac's new passion*, but it was not easy to find an accessible technical account of the theory. It would also several decades before Einstein's *Theory of Special Relativity*, which applied to observers moving relative to each other at uniform speed in a straight line, could be convincingly demonstrated. In the meantime, Einstein's reasoning made it possible to amend the description of everything given by Newton's theory and produce a "special relativistic" version. Dirac began transcribing every bit of physics expressed in non-relativistic form to make it fit with special relativity. *This appears to be the origin of Dirac's unquestioning obsession with introducing special relativity into quantum theory.*

Shortly before he completed his degree in 1921, he sat for the entrance examination for St. John's College, Cambridge. He passed and was awarded a £70 scholarship, but this fell short of the amount of money required to live and study at Cambridge. Despite his having graduated with a first-class honors Bachelor of Science degree in engineering, the economic climate of the post-war depression was such that he was unable to find work as an engineer. Instead, he took up an offer to study for a Bachelor of Arts degree in mathematics at the University of Bristol free of charge. He was permitted to skip the first year of the course owing to his engineering degree.

In 1923, Dirac graduated, once again with first-class honors, and received a £140 scholarship from the Department of Scientific and Industrial Research. Along with his £70 scholarship from St John's College, this was enough to live at Cambridge. There, *Dirac pursued his interests in the theory of general relativity*, an interest he had gained earlier as a student in Bristol, and in the nascent field of *quantum physics*, under the supervision of Ralph Fowler. From 1925 to 1928 he held an 1851 Research Fellowship from the Royal Commission for the Exhibition of 1851.

Dirac's first step into a new *quantum theory* was taken late in September 1925. Fowler had received a proof copy of Heisenberg's paper [Heisenberg, W. (July, 1925). Über quantentheoretische Umdeutung kinematischer und mechanischer Beziehungen. (On the quantum-theoretical re-interpretation of kinematic and mechanical relations.) *Zeit. Phys.*, 33, 879-93], which Fowler sent on to Dirac, who was on vacation in Bristol, asking him to look into this paper carefully.

Dirac's attention was drawn to a mysterious mathematical relationship, at first sight unintelligible, that Heisenberg had established, between *non-commuting variables*. Several weeks later, back in Cambridge, Dirac suddenly recognized that this mathematical form had the same structure as the Poisson brackets that occur in the classical dynamics of particle motion. From this thought he restated Heisenberg's quantum theory in terms of *non-commuting dynamical variables* represented by Poisson brackets, and demonstrated mathematically some of the assumptions that Heisenberg had made by appealing to the *Correspondence Principle*. This led him at the age of 25 to a formulation of quantum mechanics that allowed him to obtain the *quantization rules* in a novel and illuminating manner. Dirac described the quantization of the *electromagnetic field* as an ensemble of harmonic oscillators with the introduction of the concept of creation and annihilation operators of particles. For this work, [Dirac, P. A. M. (December, 1925). The Fundamental Equations of Quantum Mechanics. *Roy. Soc. Proc., A*, 109, 752, 642-53; received November 7, 1925.] published in June 1926, the first thesis on quantum mechanics to be submitted anywhere, Dirac received a PhD from Cambridge.

Dirac was regarded by his friends and colleagues as unusual in character. In a 1926 letter to Paul Ehrenfest, Albert Einstein wrote of Dirac, "I have trouble with Dirac. This balancing on the dizzying path between genius and madness is awful." In another letter concerning the Compton effect he wrote, "I don't understand Dirac at all."

He wrote a series of papers, published mainly in the Proceedings of the Royal Society, leading up to his *relativistic* theory of the electron (1928) and the theory of *holes* (1930). This latter theory required the existence of a positive particle having the same mass and charge as the known (negative) electron. This, the positron was discovered experimentally at a later date (1932) by C. D. Anderson, while its existence was likewise proved by Blackett and Occhialini (1933) in the phenomena of "pair production" and "annihilation".

In 1928, building on 2×2 spin matrices, which Dirac purported to have discovered independently of Wolfgang Pauli's work on *non-relativistic* spin systems, he proposed the *Dirac equation* as a *relativistic equation of motion* for the *wave function* of the electron. This work led Dirac to predict the existence of the positron, the electron's *antiparticle*, which he interpreted in terms of what came to be called the *Dirac sea*. The *Dirac equation* also contributed to explaining the origin of *quantum spin* as a relativistic phenomenon. *However, introduction of special relativity into the wave equation resulted in a second class of solutions of the wave equation in which the energy of a free electron was negative.* [Dirac, P. A. M. (February, 1928). The Quantum Theory of the Electron. *Roy. Soc. Proc., A*, 117, 778, 610–24]; introduces vectors with *4 components* resulting in a *relativistic equation of motion* for the wave function of the electron referred to as the *Dirac equation* that describes all spin-½ particles with mass.

In the spring of 1929, he was a visiting professor at the University of Wisconsin–Madison. An anecdote recounted in a review of the 2009 biography [Pais, A. (2009). *Paul Dirac: The Man and His Work*. Cambridge University Press.] tells of Heisenberg and Dirac sailing on an ocean liner to a conference in Japan in August 1929:

> "Both still in their twenties, and unmarried, they made an odd couple. Heisenberg was a ladies' man who constantly flirted and danced, while Dirac—'an Edwardian geek', as biographer Graham Farmelo puts it—suffered agonies if forced into any kind of socializing or small talk. 'Why do you dance?' Dirac asked his companion. 'When there are nice girls, it is a pleasure,' Heisenberg replied. Dirac pondered this

notion, then blurted out: 'But, Heisenberg, how do you know beforehand that the girls are nice?'"

Dirac's *The Principles of Quantum Mechanics*, published in 1930, is a landmark in the history of science. It quickly became one of the standard textbooks on the subject and is still used today. In that book, Dirac incorporated the previous work of Heisenberg on matrix mechanics and of Schrödinger on wave mechanics into a single mathematical formalism that associates measurable quantities to operators acting on the Hilbert space of vectors that describe the state of a physical system. The book also introduced the *Dirac delta function*. Following his 1939 article, he also included the *bra–ket notation* in the third edition of his book, thereby contributing to its universal use nowadays.

Whilst Dirac was relaxing on the Crimean coast, during one of his visits to the Soviet Union, in July 1932, Carl Anderson, working on the effects of cosmic rays in his cloud chamber at Caltech, was the first to detect the positive electron (positron) predicted by Dirac. By the autumn of 1932 this was confirmed by Patrick Blackett and an Italian visitor, Guiseppe Occhialini at the Cavendish, Cambridge University.

In the autumn of 1932, Dirac returned to considering how quantum mechanics can be developed by analogy with classical mechanics, finding another way of doing this other than by using Newton's laws, by generalizing the property of classical physics that enables the path of any object to be calculated, using the **Lagrangian**, where the **Lagrangian** is the difference between an object's kinetic and potential energy.

> [*Lagrangian* mechanics is a formulation of classical mechanics founded on the d'Alembert principle of virtual work. It was introduced by the Italian-French mathematician and astronomer Joseph-Louis Lagrange in his presentation to the Turin Academy of Science in 1760 culminating in his 1788 grand opus, *Mécanique analytique*. Lagrangian mechanics describes a mechanical system as a pair (M, L) consisting of a configuration space M and a smooth function L within that space called a **Lagrangian**. For many systems, $L = T - V$, where T and V are the **kinetic** and **potential energy** of the system, respectively. The stationary *action principle* requires that the action functional of the system derived from L must remain at a stationary point (specifically, a maximum, minimum, or saddle point) throughout the time evolution of the system. This constraint allows the calculation of the equations of motion of the system using Lagrange's equations.

> **Newton's laws and the concept of forces are the usual starting point for teaching about mechanical systems.** This method works well for many problems, but for others the approach is nightmarishly complicated. Lagrangian mechanics adopts **energy** rather than **force** as its basic ingredient, leading to more abstract

equations capable of tackling more complex problems. **Lagrange's approach was to set up independent generalized coordinates for the position and speed of every object, which allows the writing down of a general form of Lagrangian (total kinetic energy minus potential energy of the system) and summing this over all possible paths of motion of the particles yielded a formula for the 'action', which he minimized to give a generalized set of equations.** This summed quantity is minimized along the path that the particle actually takes. This choice eliminates the need for the constraint force to enter into the resultant generalized system of equations. There are fewer equations since one is not directly calculating the influence of the constraint on the particle at a given moment.]

The path taken between two points in any specified time interval is the *path of least action*, where the *action* associated with the object's path is obtained by adding the values of the Lagrangian along the path. When he generalized the idea to quantum mechanics, he found that a quantum particle has an infinite number of paths centered around the path predicted by classical mechanics. He found a way of taking into account all of the available paths by calculating their probability. [Dirac, P. A. M. (1933). The Lagrangian in Quantum Mechanics. *Phys. Zeit. Sowjet.*, 3, 1, 64-72]; alternative formulation of quantum mechanics in terms of Lagrangian in place of Hamiltonian, "*many-time*" theory.

Normally, he would submit a paper like this to a British journal but this time he chose to demonstrate his support for Soviet physics by sending the paper to a new Soviet journal that was about to publish his collaborative paper on field theory. [Dirac, P. A. M., Fock, V. A., Podolsky, B. (1932). On quantum electrodynamics. *Phys. Zeit. Sowjet.*, 2, 468]; *relativistic* model in which a fixed number of electrons interact through a second-quantized electromagnetic field, applies Dirac's *interaction representation* formulation of quantum field theory to full electrodynamics. Dirac was quietly pleased with his "little paper". It was not until almost a decade later that a few theoreticians in the next generation recognized the significance of the paper. [Farmelo, G. (2009). *The Strangest Man. The hidden life of Paul Dirac*. Basic Books, New York.]

In 1933 Dirac was awarded the Nobel Prize in Physics.

In 1934, he published a paper showing how expressions for the *electric* and *current densities* can be separated into two parts, where one contains the singularities that result in an infinite number of negative-energy electrons with infinite energies, and other describes the densities physically present. [Dirac, P. A. M. (March, 1934). Discussion of the infinite distribution of electrons in the theory of the positron. *Proc. Camb. Phil. Soc.*, 30, 2, 150-63]; attempts to addresses problem of electrons with negative energy with relativistic 'hole' theory.

However, further studies by Felix Bloch with Arnold Nordsieck, and Victor Weisskopf, in 1937 and 1939, revealed that such computations were reliable only at a first order of perturbation theory, a problem already pointed out by Robert Oppenheimer. At higher orders in the series infinities emerged, making such computations meaningless and casting serious doubts on the internal consistency of the theory itself. With no solution for this problem known at the time, *it appeared that a fundamental incompatibility existed between special relativity and quantum mechanics.*

Meanwhile, after the Gamows fled the Soviet Union following the Solvay conference in 1933 and arrived in Cambridge in 1934, Dirac had a dalliance with George Gamow's wife Rho, a strikingly attractive brunette, who taught him Russian, Dirac's fourth language. Then, after the Gamows left for Copenhagen, he had another with the wife of Fellow of St. Johns, a Russian émigré poet, Lydia Jackson, who continued with his Russian tuition.

On the day after Dirac arrived in Princeton at the end of September, as a visitor at the Institute for Advanced Studies, he ran into one of his new colleagues, Eugene Wigner, having lunch with his sister, Margit, known as Manci, who was visiting from their native Hungary, Dirac, the "lonely-looking man at the next table" was invited to join them. In 1937, Dirac married Margit and adopted Margit's two children, Judith and Gabriel. Paul and Margit Dirac had two children together, both daughters, Mary Elizabeth and Florence Monica.

Einstein was at Princeton at the time of Dirac's visit, having arrived with his wife in in October 1933, who was fifty-four but looked older. The two men respected each other but there was no special warmth between them. Einstein admired the success of quantum theory but mistrusted it. During 1935, Einstein completed his collaboration with his younger research associates, Boris Podolsky and Nathan Rosen, on a paper that cast serious doubts on the conventional interpretation of the theory. [Einstein, A., Podolsky, B. & Rosen, N. (May, 1935). Can Quantum-Mechanical Description of Physical Reality Be Considered Complete? *Phys. Rev.*, 47, 777-80.]

In 1942, Dirac gave his Bakerian Lecture, which was well received. [Dirac, P. A. M. (March, 1942.) Bakerian Lecture - The physical interpretation of quantum mechanics. *Roy. Soc. Proc., A*, 180, 980, 1-40.] And, in 1945 and 1948 made two more important contributions. [Dirac, P. A. M. (April, 1945). On the Analogy Between Classical and Quantum Mechanics. *Rev. Mod. Phys.*, 17, 195]; Dirac's proposal of the *path integral formulation* of quantum mechanics, extensively developed in Feynman, R. P. (1948). Space-Time Approach to Non-Relativistic Quantum Mechanics; and [Dirac, P. A. M. (May, 1948). Quantum Theory of Localizable Dynamical Systems. *Phys. Rev.*, 73, 9, 1092-103]; *relativistic* quantum theory in terms of variables on a space-like surface in space-time, referenced in Schwinger (1948). Quantum Electrodynamics. I. A Covariant Formulation.

A possible way out of the difficulties facing quantum theory, was given by Hans Bethe in 1947, who made the first *non-relativistic* computation of the shift of the lines of the hydrogen atom as measured by Lamb and Rutherford. Despite the limitations of the computation, agreement was excellent. The idea was simply to attach infinities to corrections of *mass* and *charge* that were actually fixed to a finite value by experiments. In this way, the infinities get absorbed in those constants and yield a finite result in good agreement with experiments. This procedure was named *renormalization*.

Even though *renormalization* works very well in practice, Dirac never accepted it, Dirac commented in 1975: "I must say that I am very dissatisfied with the situation because this so-called 'good theory' does involve neglecting infinities which appear in its equations, neglecting them in an arbitrary way. This is just not sensible mathematics. Sensible mathematics involves neglecting a quantity when it is small – not neglecting it just because it is infinitely great and you do not want it!" [Kragh, Helge (1990). Dirac: A Scientific Biography. Cambridge: Cambridge University Press.]. His final judgment on quantum field theory in his last paper was that "These rules of *renormalization* give surprisingly, excessively good agreement with experiments. Most physicists say that these working rules are, therefore, correct. I feel that is not an adequate reason. Just because the results happen to be in agreement with observation does not prove that one's theory is correct." [Dirac, P. A. M. (1987). The inadequacies of quantum field theory. In *Paul Adrien Maurice Dirac*, page 194. B. N. Kursunoglu and E. P. Wigner, eds., Cambridge University Press.] Nor was Feynman entirely comfortable with its mathematical validity, even referring to *renormalization* as a "shell game" and "hocus pocus" [Feynman, Richard (1985). *QED: The Strange Theory of Light and Matter*, page 128. Princeton University Press.]

Dirac was the Lucasian Professor of Mathematics at the University of Cambridge, was a member of the Center for Theoretical Studies, University of Miami, and spent the last decade of his life at Florida State University. Dirac was also awarded the Royal Medal in 1939 and both the Copley Medal and the Max Planck Medal in 1952. He was elected a Fellow of the Royal Society in 1930, an Honorary Fellow of the American Physical Society in 1948, and an Honorary Fellow of the Institute of Physics, London in 1971. He received the inaugural J. Robert Oppenheimer Memorial Prize in 1969. Dirac became a member of the Order of Merit in 1973, having previously turned down a knighthood as he did not want to be addressed by his first name. In 1984, Dirac died in Tallahassee, Florida, and was buried at Tallahassee's Roselawn Cemetery.

Dirac, P. A. M. (February, 1928). The Quantum Theory of the Electron.

Roy. Soc. Proc., A, 117, 778, 610–24; https://doi.org/10.1098/rspa.1928.0023; also in Underwood, T. G. (2023). *Quantum Electrodynamics – annotated sources*, pp. 126-33.

Communicated by R. H. Fowler, F.R.S.

Received January 2, 1928.

St. John's College, Cambridge.

This work led Dirac to predict the existence of the *positron*, the *electron*'s *antiparticle*, which he interpreted in terms of what came to be called the *Dirac sea*.

[In 1932, Carl David Anderson (September 3, 1905 – January 11, 1991), an American physicist, discovered the *positron*. Under the supervision of Robert A. Millikan, he began investigations into cosmic rays during the course of which he encountered unexpected particle tracks in his cloud chamber photographs that he correctly interpreted as having been created by a particle with the same mass as the *electron*, but with *opposite electrical charge*.

Anderson first detected the particles in cosmic rays. He then produced more conclusive proof by shooting gamma rays produced by the natural radioactive nuclide ThC" (^{208}Tl) [Thallium] into other materials, resulting in the creation of *positron-electron* pairs. Anderson shared the 1936 Nobel Prize in Physics "for his discovery of the positron", with Victor Franz Hess, "for his discovery of cosmic radiation".

The *positron* (or *anti-electron*) is a particle with an *electric charge* of +1 e, a *spin* of 1/2 (the same as the *electron*), and the same mass as an *electron*. It is the *antiparticle* (*antimatter* counterpart) of the electron. When a *positron* collides with an *electron*, annihilation occurs. If this collision occurs at low energies, it results in the production of two or more *photons*.

Positrons can be created by *positron emission radioactive decay* (through *weak interactions*), or by *pair production* from a sufficiently energetic *photon* which is interacting with an atom in a material.

This discovery validated Paul Dirac's theoretical prediction in 1928 of the existence of the *positron*. [Dirac, P. A. M. (February, 1928). The Quantum Theory of the Electron.]

———————————

The new quantum mechanics, when applied to the problem of the *structure of the atom with point-charge electrons*, does not give results in agreement with experiment. The discrepancies consist of "duplexity" phenomena, the observed number of stationary states for an electron in an atom being twice the number given by the theory. To meet the difficulty, Goudsmit and Uhlenbeck have *introduced the idea of an electron with a spin angular momentum of half a quantum* and a *magnetic moment* of one Bohr magneton[1].

[1] Uhlenbeck, G.E. & Goudsmit, S.A. (1925). Ersetzung der Hypothese vom unmechanischen Zwang durch eine Forderung bezüglich des inneren Verhaltens jedes einzelnen Elektrons. (Replacement of the hypothesis of unmechanical coercion by a requirement regarding the internal behavior of each individual electron.) *Naturwiss.*, 13, 953-54; http://dx.doi.org/10.1007/BF01558878.

This model for the electron has been fitted into the new mechanics by Pauli[*], and Darwin[#], working with an equivalent theory, has shown that it gives results in agreement with experiment for hydrogen-like spectra to the first order of accuracy.

[*] Pauli, W. (September, 1927). Zur Quantenmechanik des magnetischen Elektrons. (On the quantum mechanics of magnetic electrons.) *Zeit. Phys.*, 43, 601-23[; shows how the *non-relativistic* formulation by Dirac [Dirac (January, 1927). The Physical Interpretation of the Quantum Dynamics] and Jordan using the general canonical transformations of the Schrödinger functions enables a quantum-mechanical representation of electrons by the method of *eigenfunctions*, the differential equations for the *eigenfunctions* of the magnetic electron that are given in the present paper can be regarded as only provisional and approximate, like the Heisenberg-Jordan matrix formulation they *are not written down in a relativistically-invariant way*, for the hydrogen atom they are valid only in the approximation in which the dynamical behavior of the proper moment can be considered to be a secular perturbation].

[#] Darwin, C. G. (September, 1927). The Electron as a Vector Wave. *Roy. Soc. Proc.*, A, 116, 773, 227-53[; difficulties in interpretation of the spinning electron in terms of wave theory, *wave functions with 2 components*, should be interpreted in terms of a *vector*, but vector found to be in some degree arbitrary, when *relativity* transformation is applied to identify the "doublet effect" with the Zeeman effect gives value for the doublet separation twice as great as it should be, not at present possible to see what form the Thomas correction should take in the wave theory, the trouble is no doubt connected with the fact that the hydrogen spectrum has only been verified to a first approximation and goes wrong in the second—a difficulty at present shared by all theories].

The question remains as to why Nature should have chosen this particular model for the electron instead of being satisfied with the *point-charge*. One would like to find some incompleteness in the previous methods of applying quantum mechanics to the *point-charge electron* such that, when removed, the whole of the duplexity phenomena follow without arbitrary assumptions. In the present paper it is shown that this is the case, *the incompleteness of the previous theories lying in their disagreement with relativity, or, alternatively, with the general transformation theory of quantum mechanics.* It appears that *the simplest Hamiltonian for a point-charge electron satisfying the requirements of both relativity and the general transformation theory leads to an explanation of all duplexity phenomena without further assumption.*

All the same there is a great deal of truth in the spinning electron model, at least as a first approximation. *The most important failure of the model seems to be that the magnitude of the resultant orbital angular momentum of an electron moving in an orbit in a central field of force is not a constant*, as the model leads one to expect.

§ 1. *Previous Relativity Treatments.*

The *relativity Hamiltonian* according to the classical theory *for a point electron moving in an arbitrary electro-magnetic field with scalar potential A_0 and vector potential A is*

$$F = (W/c + e/c\ A_0)^2 + (\mathbf{p} + e/c\ \mathbf{A})^2 + m^2c^2,$$

where **p** is the *momentum* vector. It has been suggested by Gordon*

> * Gordon, W. (January, 1927). Der Comptoneffekt nach der Schrödingerschen Theorie. (The Compton effect according to Schrödinger's theory.) *Zeit. Phys.*, 40, 117-33[; Heisenberg and Schrödinger provided alternative methods for determination of quantum *frequencies* and *intensities*, Compton effect already calculated by Dirac (June, 1926) using Heisenberg method, here the same problem treated by Schrödinger method, starts with the same *classic relativistic equation for kinetic energy* in terms of *momentum* and *energy*, which is *Hamiltonian equation* for the system, introduces same imaginary variables for *time* and *energy* to create same space-time symmetric form, applies in same way to *electron in electromagnetic field* described in terms of *vector potential* and *scalar potential*, and introduces same imaginary variable for scalar potential, adds the same *field energy* to the *kinetic energy* resulting in the same *classical relativistic Hamiltonian equations for a point electron moving in an electromagnetic field*, in accordance with Schrödinger's rules Gordon then substitutes the classical *quantum differential operators* for the momentum vector in the amended *Hamiltonian equation* and applies resulting differential operator to the *wave function* ψ to obtain the *Klein-Gordon equation*, $1/c^2\ \partial^2/\partial t^2\ \psi - \nabla^2\ \psi + m^2c^2/h^2\ \psi = 0$, (Dirac [February, 1928). The Quantum Theory of the Electron.] objected to this substitution on grounds of the interpretation of the wave function, and solutions with negative probabilities, negative energy, and positive charge for the electron); calculates radiation from *current density* and *charge density*, applies to Compton effect.

that *the operator of the wave equation of the quantum theory should be obtained from this F by the same procedure as in non-relativity theory, namely, by putting in it.*

$$W = ih\ \partial/\partial t$$
$$p_r = -ih\ \partial/\partial x_r, \qquad r = 1, 2, 3,$$

This gives the *wave equation*

$$F\psi = \{(ih\, \partial/c\partial t + e/c\, A_0)^2 + \Sigma_r\, (-ih\, \partial/\partial x_r + e/c\, A_r)^2 + m^2c^2\}\, \psi = 0, \quad (1)$$

the *wave function* ψ being a function of x_1, x_2, x_3, t. *This gives rise to two difficulties.*

The first is in connection with the physical interpretation of ψ. Gordon, and also independently Klein*,

* Klein, O. (October, 1927). Elektrodynamik und Wellenmechanik vom Standpunkt des Korrespondenzprinzips. (Electrodynamics and wave mechanics from the standpoint of the correspondence principle.) *Zeit. Phys.*, 41, 10, 407-42[; alternative calculation of Compton effect restricted to the *one-electron problem*, starts from Maxwell-Lorentz field equations, describes motion of an electron in an electromagnetic field by *four-potential* and *scalar potential*, regards *Hamilton-Jacobi differential equation* for the action function (Klein–Gordon equation) as expression for motion of the electron, following de Broglie and Schrodinger replaces this first order equation with a second-order linear equation representing *relativistic* generalization of Schrödinger's wave equation for one-electron problem, evaluates equations determining the electromagnetic field with the help of wave mechanics using the correspondence principle to determine wave-mechanical expressions for *electric density* and *current vector*, after neglecting relativity results in the same expressions as those obtained by Schrodinger, applies to a "bound" electron moving in an axially symmetric electrostatic field over which a weak homogeneous magnetic field is superimposed to derive normal Zeeman effect, applies to scattered radiation from a light wave on a "force-free" electron to obtain the Compton effect, five-dimensional wave mechanics].

from considerations of the conservation theorems, make the assumption that if ψ_m, ψ_n are two solutions

$$\rho_{mn} = -\, e/2mc^2\, \{ih\, (\psi_m \partial\psi_n/\partial t - \psi_n{}^- \partial\psi_m/\partial t) + 2eA_0\, \psi_m\psi_n{}^-\},$$

and

$$I_{mn} = -\, e/2m\, \{-ih\, (\psi_m\, \text{grad}\, \psi_n{}^- - \psi_n{}^-\, \text{grad}\, \psi_m) + 2\, e/c\, A_m\, \psi_m\psi_n{}^-\}$$

are to be interpreted as the *charge* and *current* associated with the transition $m \to n$.

This appears to be satisfactory so far as *emission* and *absorption* of radiation are concerned, but is not so general as the interpretation of the *non-relativity quantum mechanics*, which has been developed[#] sufficiently to enable one to answer the question: *What is the probability of any dynamical variable at any specified time having a value lying between any specified limits,* when the system is represented by a given *wave function* ψ_n?

[#] Jordan, P. (November, 1927). Über eine neue Begründung der Quantenmechanik. (On a new justification of quantum mechanics.) *Zeit. Phys.*, 40, 809-38; https://doi.org/10.1007/ BF01390903; Dirac, P. A. M. (January, 1927). The Physical Interpretation of the Quantum Dynamics. *Roy. Soc. Proc., A*, 113, 765, 621-41.

The Gordon-Klein interpretation can answer such questions if they refer to the position of the electron (by the use of ρ_{nm}), but not if they refer to its momentum, or angular momentum or any other dynamical variable. We should expect the interpretation of the *relativity* theory to be just as general as that of the *non-relativity* theory.

The general interpretation of *non-relativity quantum mechanics* is based on the *transformation theory*, and is made possible by the *wave equation* being of the form

$$(H - W)\, \psi = 0, \tag{2}$$

i.e., being linear in W or $\partial/\partial t$, so that the *wave function* at any time determines the *wave function* at any later time. *The wave equation of the relativity theory must also be linear in W if the general interpretation is to be possible.*

The second difficulty in Gordon's interpretation arises from the fact that if one takes the *conjugate imaginary* of equation (1), one gets

$$\{(-W/c + e/c\, A_0)^2 + (-\mathbf{p} + e/c\, \mathbf{A})^2 + m^2c^2\}\, \psi = 0,$$

which is the same as one would get if one put $-e$ for e. *The wave equation (1) thus refers equally well to an electron with charge e as to one with charge $-e$.* If one considers for definiteness the limiting case of large *quantum* numbers one would find that some of the solutions of the *wave equation* are *wave packets* moving in the way a particle of charge $-e$ would move on the classical theory, while others are *wave packets* moving in the way a particle of charge e would move classically. *For this second class of solutions W has a negative value.*

Introduction of *special relativity* into the wave equation for the electron results in solutions with negative energy.

One gets over the difficulty on the classical theory by arbitrarily excluding those solutions that have a negative W. *One cannot do this on the quantum theory, since in general a perturbation will cause transitions from states with W positive to states with W negative.* Such a transition would appear experimentally as the electron suddenly changing its charge from $-e$ to e, a phenomenon which has not been observed. *The true relativity wave equation should thus be such that its solutions split up into two non-combining sets, referring respectively to the charge $-e$ and the charge e.*

In the present paper we shall be concerned only with the removal of the first of these two difficulties. The resulting theory is therefore still only an approximation, but it appears to be good enough to account for all the duplexity phenomena without arbitrary assumptions. …

Paul Dirac – 1933 Nobel Lecture, December 12, 1933. *Theory of electrons and positrons.*

[https://www.nobelprize.org/prizes/physics/1933/dirac/lecture/.]

Dirac shared the 1933 Nobel Prize in Physics with Erwin Schrödinger "for the discovery of new productive forms of atomic theory".

During the intense period of 1925-26 quantum theories were proposed that accurately described the energy levels of electrons in atoms. These equations needed to be adapted to Einstein's theory of relativity, however. In 1928 Paul Dirac formulated a fully *relativistic quantum theory*. The equation gave solutions that he interpreted as being caused by a particle equivalent to the *electron*, but with a positive *charge*. This particle, the *positron*, was later confirmed through experiments. [Paul A.M. Dirac – Facts. NobelPrize.org. https://www.nobelprize.org/prizes/physics/1933/dirac/facts/.]

In his Nobel Lecture, Dirac described the current state of his theory of *electrons* and *positrons*.

———————————————

Matter has been found by experimental physicists to be made up of small particles of various kinds, the particles of each kind being all exactly alike. Some of these kinds have definitely been shown to be composite, that is, to be composed of other particles of a simpler nature. But there are other kinds which have not been shown to be composite and which one expects will never be shown to be composite, so that one considers them as elementary and fundamental.

From general philosophical grounds one would at first sight like to have as few kinds of elementary particles as possible, say only one kind, or at most two, and to have all matter built up of these elementary kinds. It appears from the experimental results, though, that there must be more than this. In fact, the number of kinds of elementary particle has shown a rather alarming tendency to increase during recent years.

The situation is perhaps not so bad, though, because on closer investigation it appears that the distinction between elementary and composite particles cannot be made rigorous. To get an interpretation of some modern experimental results one must suppose that particles can be created and annihilated. Thus, if a particle is observed to come out from another particle, one can no longer be sure that the latter is composite. The former may have been created. The distinction between elementary particles and composite particles now becomes a matter of convenience. This reason alone is sufficient to compel one to give up the attractive philosophical idea that all matter is made up of one kind, or perhaps two kinds of bricks.

I should like here to discuss the simpler kinds of particles and to consider what can be inferred about them from purely theoretical arguments. The simpler kinds of particle are:

(i) the *photons* or light-quanta, of which light is composed;
(ii) the *electrons*, and the recently discovered *positrons* (which appear to be a sort of mirror image of the *electrons*, differing from them only in the sign of their *electric charge*);
(iii) the heavier particles - *protons* and *neutrons*.

Of these, I shall deal almost entirely with the *electrons* and the *positrons* - not because they are the most interesting ones, but because in their case the theory has been developed further. There is, in fact, hardly anything that can be inferred theoretically about the properties of the others. The *photons*, on the one hand, are so simple that they can easily be fitted into any theoretical scheme, and the theory therefore does not put any restrictions on their properties. The *protons* and *neutrons*, on the other hand, seem to be too complicated and no reliable basis for a theory of them has yet been discovered.

The question that we must first consider is how theory can give any information at all about the properties of elementary particles. There exists at the present time a general quantum mechanics which can be used to describe the motion of any kind of particle, no matter what its properties are. The general quantum mechanics, however, is valid only when the particles have small velocities and fails for velocities comparable with the velocity of light, when effects of *relativity* come in. There exists no *relativistic quantum mechanics* (that is, one valid for large velocities) which can be applied to particles with arbitrary properties. Thus, when one subjects quantum mechanics to *relativistic requirements*, one imposes restrictions on the properties of the particle. In this way one can deduce information about the particles from purely theoretical considerations, based on general physical principles.

This procedure is successful in the case of *electrons* and *positrons*. It is to be hoped that in the future some such procedure will be found for the case of the other particles. I should like here to outline the method for *electrons* and *positrons*, showing how one can deduce the *spin* properties of the *electron*, and then how one can infer the existence of *positrons* with similar *spin* properties and with the possibility of being annihilated in collisions with *electrons*.

We begin with the equation connecting the *kinetic energy* W and *momentum* p_r (r = 1, 2, 3), of a particle in *relativistic* classical mechanics

$$W^2/c^2 - p_r^2 - m^2c^2 = 0. \tag{1}$$

From this we can get a *wave equation* of quantum mechanics, by letting the left-hand side operate on the wave function ψ and understanding W and p_r to be the operators $ih\partial/\partial t$ and $-ih\partial/\partial x_r$. With this understanding, the *wave equation* reads

$$[W^2/c^2 - p_r^2 - m^2c^2]\,\psi = 0 \tag{2}$$

Now it is a general requirement of quantum mechanics that its *wave equations* shall be linear in the operator W or $\partial/\partial t$ so this equation will not do. We must replace it by some equation linear in W, and in order that this equation may have *relativistic invariance* it must also be linear in the p's.

We are thus led to consider an equation of the type

$$[W/c - \alpha_r p_r - \alpha_0 mc]\,\psi = 0 \tag{3}$$

This involves four new variables α_r and α_0 which are operators that can operate on ψ. We assume they satisfy the following conditions,

$$\alpha_\mu^2 = I \qquad \alpha_\mu\alpha_\nu + \alpha_\nu\alpha_\mu = 0$$

for

$$\mu \neq \nu \text{ and } \mu, \nu = 0, 1, 2, 3$$

and also, the α's commute with the p's and W. These special properties for the α's make Eq. (3) to a certain extent equivalent to Eq. (2), since if we then multiply (3) on the left-hand side by $W/c + \alpha_r p_r + \alpha_0 mc$ we get exactly (2).

The new variables α, which we have to introduce to get a *relativistic wave equation* linear in W, give rise to the *spin* of the *electron*. From the general principles of quantum mechanics one can easily deduce that these variables a give the *electron* a *spin angular momentum* of half a quantum and a *magnetic moment* of one Bohr magneton in the reverse direction to the *angular momentum*. These results are in agreement with experiment. They were, in fact, first obtained from the experimental evidence provided by spectroscopy and afterwards confirmed by the theory.

The variables α also give rise to some rather unexpected phenomena concerning the motion of the *electron*. These have been fully worked out by Schrödinger. *It is found that an electron which seems to us to be moving slowly, must actually have a very high frequency oscillatory motion of small amplitude superposed on the regular motion which appears to us. As a result of this oscillatory motion, the velocity of the electron at any time equals the velocity of light.* This is a prediction which cannot be directly verified by experiment, since the frequency of the oscillatory motion is so high and its amplitude is so small. But one must believe in this consequence of the theory, since other consequences of the theory which are inseparably bound up with this one, such as the law of scattering of light by an *electron*, are confirmed by experiment.

There is one other feature of these equations which I should now like to discuss, *a feature which led to the prediction of the positron*. If one looks at Eq. (1), one sees that it allows the *kinetic energy* W to be either a positive quantity greater than mc^2 or a negative quantity less than – mc^2. This result is preserved when one passes over to the quantum equation (2) or (3). These quantum equations are such that, when interpreted according to the general scheme of quantum dynamics, they allow as the possible results of a measurement of W either something greater than mc^2 or something less than – mc^2.

Now in practice the *kinetic energy* of a particle is always positive. We thus see that our equations allow of two kinds of motion for an *electron*, only one of which corresponds to what we are familiar with. *The other corresponds to electrons with a very peculiar motion such that the faster they move, the less energy they have, and one must put energy into them to bring them to rest.*

One would thus be inclined to introduce, as a new assumption of the theory, that only one of the two kinds of motion occurs in practice. But this gives rise to a difficulty, since we find from the theory that if we disturb the *electron*, we may cause a transition from a positive-energy *state* of motion to a negative-energy one, so that, even if we suppose all the electrons in the world to be started off in positive-energy *states*, after a time some of them would be in negative-energy *states*.

Thus, in allowing negative-energy states, the theory gives something which appears not to correspond to anything known experimentally, but which we cannot simply reject by a new assumption. We must find some meaning for these *states*.

An examination of the behavior of these *states* in an *electromagnetic field* shows that they *correspond to the motion of an electron with a positive charge instead of the usual negative one* - what the experimenters now call a *positron*. One might, therefore, be inclined to assume that *electrons* in negative-energy *states* are just *positrons*, but this will not do, because *the observed positrons certainly do not have negative energies*. We can, however, establish a connection between *electrons* in negative-energy *states* and *positrons*, in a rather more indirect way.

We make use of the exclusion principle of Pauli, according to which there can be only one *electron* in any state of motion. We now make the assumptions that in the world as we know it, nearly all the *states* of negative *energy* for the *electrons* are occupied, with just one *electron* in each *state*, and that a uniform filling of all the negative-energy *states* is completely unobservable to us. Further, any unoccupied negative-energy *state*, being a departure from uniformity, is observable and is just a *positron*.

An unoccupied negative-energy *state*, or *hole*, as we may call it for brevity, will have a positive *energy*, since it is a place where there is a shortage of negative *energy*. *A hole is, in fact, just like an ordinary particle, and its identification with the positron seems the most reasonable way of getting over the difficulty of the appearance of negative energies in our equations*. On this view *the positron is just a mirror-image of the electron*, having exactly the same *mass* and opposite *charge*. This has already been roughly confirmed by experiment. The *positron* should also have similar *spin* properties to the *electron*, but this has not yet been confirmed by experiment.

From our theoretical picture, we should expect an ordinary *electron*, with positive *energy*, to be able to drop into a hole and fill up this hole, the energy being liberated in the form of *electromagnetic radiation*. This would mean a process in which an *electron* and a *positron* annihilate one another. The converse process, namely the creation of an *electron* and a *positron* from *electromagnetic radiation*, should also be able to take place. Such processes appear to have been found experimentally, and are at present being more closely investigated by experimenters.

The theory of *electrons* and *positrons* which I have just outlined is a self-consistent theory which fits the experimental facts so far as is yet known. One would like to have an equally satisfactory theory for *protons*. One might perhaps think that the same theory could be applied to *protons*. This would require the possibility of existence of negatively charged *protons* forming a mirror-image of the usual positively charged ones. There is, however, some recent experimental evidence obtained by Stern about the spin magnetic moment of the *proton*, which conflicts with this theory for the *proton*. As the *proton* is so much heavier than the *electron*, it is quite likely that it requires some more complicated theory, though one cannot at the present time say what this theory is.

In any case I think it is probable that negative *protons* can exist, since as far as the theory is yet definite, *there is a complete and perfect symmetry between positive and negative electric charge*, and if this *symmetry* is really fundamental in nature, it must be possible to reverse the *charge* on any kind of particle. The negative *protons* would of course be much harder to produce experimentally, since a much larger *energy* would be required, corresponding to the larger *mass*.

If we accept the view of complete symmetry between positive and negative electric charge so far as concerns the fundamental laws of Nature, we must regard it rather as an accident that the Earth (and presumably the whole solar system), contains a preponderance of negative *electrons* and positive *protons*. It is quite possible that for some of the stars it is the other way about, these stars being built up mainly of *positrons* and negative *protons*. In fact, there may be half the stars of each kind. The two kinds of stars would both show exactly the same spectra, and there would be no way of distinguishing them by present astronomical methods.

Werner Karl Heisenberg (December 5, 1901–February 1, 1976).

Heisenberg was a German theoretical physicist and one of the key pioneers of quantum mechanics. He published his seminal work in 1925 in a breakthrough paper. In the subsequent series of papers with Max Born and Pascual Jordan, during the same year, this matrix formulation of quantum mechanics was substantially elaborated. He is known for the uncertainty principle, which he published in 1927. Heisenberg was awarded the 1932 Nobel Prize in Physics "for the creation of quantum mechanics".

Heisenberg also made important contributions to the theories of the hydrodynamics of turbulent flows, the atomic nucleus, ferromagnetism, cosmic rays, and subatomic particles. He was a principal scientist in the German nuclear weapons program during World War II. He was also instrumental in planning the first West German nuclear reactor at Karlsruhe, together with a research reactor in Munich, in 1957.

Heisenberg was born in Würzburg, Germany, to Kaspar Ernst August Heisenberg, and his wife, Annie Wecklein. His father was a secondary school teacher of classical languages who became Germany's only ordentlicher Professor (ordinarius professor) of medieval and modern Greek studies in the university system.

In his youth Heisenberg was a member and Scoutleader of the Neupfadfinder, a German Scout association and part of the German Youth Movement.

From 1920 to 1923, he studied physics and mathematics at the Ludwig Maximilian University of Munich under Arnold Sommerfeld and Wilhelm Wien; and at the Georg-August University of Göttingen with Max Born and James Franck and mathematics with David Hilbert. In June 1922, Sommerfeld took Heisenberg to Göttingen to attend the Bohr Festival, because Sommerfeld knew of Heisenberg's interest in Niels Bohr's theories on atomic physics. At the event, Bohr was a guest lecturer and gave a series of comprehensive lectures on quantum atomic physics and Heisenberg met Bohr for the first time.

Heisenberg's doctoral thesis, the topic of which was suggested by Sommerfeld, was on turbulence; the thesis discussed both the stability of laminar flow and the nature of turbulent flow. The problem of stability was investigated by the use of the Orr–Sommerfeld equation, a fourth order linear differential equation for small disturbances from laminar flow. He received his doctorate in 1923.

At Göttingen, under Born, he completed his habilitation in 1924 with a Habilitationsschrift (habilitation thesis) on the anomalous Zeeman effect.

From 1924 to 1927, Heisenberg was a Privatdozent at Göttingen, meaning he was qualified to teach and examine independently, without having a chair. From September 17, 1924 to May 1, 1925, under an International Education Board Rockefeller Foundation fellowship,

Heisenberg went to do research with Niels Bohr, director of the Institute of Theoretical Physics at the University of Copenhagen.

In Copenhagen, Heisenberg and Hans Kramers collaborated on a paper on dispersion, or the scattering from atoms of radiation whose wavelength is larger than the atoms. They showed that the successful formula Kramers had developed earlier could not be based on Bohr orbits, because the *transition frequencies* are based on level spacings which are not constant. The frequencies which occur in the Fourier transform of sharp classical orbits, by contrast, are equally spaced. But these results could be explained by a semi-classical virtual state model: the incoming radiation excites the valence, or outer, electron to a virtual state from which it decays. In a subsequent paper Heisenberg showed that this virtual oscillator model could also explain the polarization of fluorescent radiation.

These two successes, and the continuing failure of the Bohr–Sommerfeld model to explain the outstanding problem of the anomalous Zeeman effect, led Heisenberg to use the virtual oscillator model to try to calculate *spectral frequencies*. The method proved too difficult to immediately apply to realistic problems, so Heisenberg turned to a simpler example, the anharmonic oscillator.

The dipole oscillator consists of a simple harmonic oscillator, which is thought of as a charged particle on a spring, perturbed by an external force, like an external charge. The motion of the oscillating charge can be expressed as a Fourier series in the frequency of the oscillator. Heisenberg solved for the quantum behavior by two different methods. First, he treated the system with the virtual oscillator method, calculating the transitions between the levels that would be produced by the external source. He then solved the same problem by treating the anharmonic potential term as a perturbation to the harmonic oscillator and using the perturbation methods that he and Born had developed. Both methods led to the same results for the first and the very complicated second order correction terms. This suggested that behind the very complicated calculations lay a consistent scheme. Heisenberg returned to Göttingen and, with Max Born and Pascual Jordan over a period of about six months, developed the *matrix mechanics formulation of quantum mechanics*.

In his 1925 paper, which assumes that the reader is familiar with Kramers-Heisenberg transition probability calculations, Heisenberg set out to *try to construct a theory of quantum mechanics in which only relationships among observable quantities occur*. In place of assigning to the electron a point in space as a function of time he *assigned to the electron an emitted radiation;* where the observables are the *energies W(n) of the (Bohr) stationary states*, together with the associated (Einstein-Bohr) *frequencies v* and *amplitudes* which characterize the radiation emitted in the transition between the stationary states. Recognizing that *quantum theory* describes transitions between two stationary states he substituted two variables in place of one in the classical theory. He justified this replacement by an appeal to *Bohr's correspondence principle* and the *Pauli doctrine* that *quantum mechanics* must be limited to observables. In order to calculate the *energy* of a harmonic oscillator, in which the *amplitudes* do not combine in the same way as the classical harmonics, but rather in accordance with the *Ritz combination principle,* instead of reinterpreting x(t) as a *sum* over transition components, he represented the position by the *set* of *transition components*, thereby introducing non-commutative multiplication of matrices by physical reasoning, based on the *correspondence principle*, despite the fact that he was not then familiar with the mathematical theory of matrices.

After addressing what he referred to as the kinematic of the *quantum theory*, Heisenberg turned to mechanical problems aiming at the determination of the *amplitudes, frequencies* and *energies in order to construct the line spectrum of an atom from the given force on the electron*. He achieved this by translating the old quantum condition that fixes the properties of the states to a new condition that fixes the properties of the *transitions between states*, replacing the differential in the equation for the *phase* integral by a difference, resulting in an equation that has a simple quantum theoretical connection to the *Kramer's dispersion theory*.

On July 9, Heisenberg gave Born his paper to review and submit for publication. Heisenberg's seminal paper was published in September 1925. [Heisenberg, W. (July, 1925). Über quantentheoretische Umdeutung kinematischer und mechanischer Beziehungen. (On the Quantum-Theoretical Re-interpretation of Kinematic and Mechanical Relations.) *Zeit. Physik*, 33, 879-93.] This is the first paper in the famous trilogy which launched the *matrix mechanics formulation of quantum mechanics*.

When Born read the paper, he recognized the formulation as one which could be transcribed and extended to the systematic language of matrices, which he had learned from his study under Jakob Rosanes at Breslau University. Up until this time, matrices were seldom used by physicists; they were considered to belong to the realm of pure mathematics. Gustav Mie had used them in a paper on electrodynamics in 1912; and Born had used them in his work on the lattice theory of crystals in 1921. While matrices were used in these cases, the algebra of matrices with their multiplication did not enter the picture as they did in the matrix formulation of quantum mechanics.

Born, with the help of his assistant and former student Pascual Jordan, began immediately to make the transcription and extension, and they submitted their results for publication; the paper was received for publication just 60 days after Heisenberg's paper. [Born, M. & Jordan, P. (December, 1925). Zur Quantenmechanik. (On Quantum Mechanics.) *Zeit. Phys.*, 34, 858-88.] A follow-on paper was submitted for publication before the end of the year by all three authors. [Born, M., Heisenberg, W. & Jordan, P. (August, 1926). Zur Quantenmechanik II. (On Quantum Mechanics II.) *Zeit. Phys.*, 35, 557-615.]

On May 1, 1926, Heisenberg began his appointment as a university lecturer and assistant to Bohr in Copenhagen. It was in Copenhagen, in 1927, that Heisenberg developed his *uncertainty principle*, while working on the mathematical foundations of quantum mechanics. On February 23, Heisenberg wrote a letter to fellow physicist Wolfgang Pauli, in which he first described his new principle. In his paper on the principle, Heisenberg used the word "Ungenauigkeit" (imprecision), not "uncertainty", to describe it.

In 1928, Heisenberg was appointed ordentlicher Professor (professor ordinarius) of theoretical physics and head of the department of physics at the University of Leipzig; he gave his inaugural lecture there on 1 February 1928. In his first paper published from Leipzig, Heisenberg used the *Pauli exclusion principle* to solve the mystery of *ferromagnetism*. [Heisenberg, W. (September, 1928). Zur Theory of Ferromagnetismus. (On the theory of ferromagnetism.) *Zeit. Phys.*, 49, 619–36 (in German); https://doi.org/ 10.1007/BF01328601; translation in Underwood, T. G. (2024). *Quantum Entanglement*.]

In 1928, the British mathematical physicist Paul Dirac had derived his relativistic wave equation of quantum mechanics, which implied the existence of positive electrons, later to be named *positrons*. In early 1929, Heisenberg and Pauli submitted the first of two papers laying the foundation for *relativistic quantum field theory*. [Heisenberg, W. & Pauli, W. (July, 1929). Zur Quantendynamik der Wellenfelder. (On the quantum dynamics of wave fields.); (January, 1930). Zur Quantendynamik der Wellenfelder II. (On the quantum dynamics of wave fields II.) See below.]

Also in 1929, Heisenberg went on a lecture tour of China, Japan, India, and the United States. In the spring of 1929, he was a visiting lecturer at the University of Chicago, where he lectured on quantum mechanics.

In 1932 and 1933, Heisenberg published a three-part paper on atomic nuclei which concluded by treating *protons* and neutrons on an equal footing by *considering protons and neutrons as different states of the same particle*. [Heisenberg, W. (January, 1932). Über den Bau der Atomkerne. I. (About the construction of atomic nuclei. I.); (March, 1932). Über den Bau der Atomkerne. II. (About the construction of atomic nuclei. II.); (September, 1933). Über den Bau der Atomkerne. III. See below.]

In 1932, from a cloud chamber photograph of cosmic rays, the American physicist Carl David Anderson identified a track as having been made by a *positron*. In mid-1933, Heisenberg presented his theory of the positron. His thinking on Dirac's theory and further development of the theory were set forth in two papers. The first, [Heisenberg, W. (March, 1934). Bemerkungen zur Diracschen Theorie des Positrons. (Remarks on the Dirac theory of positron.) See below.] was published in 1934, and the second, Heisenberg, W., Euler, H. (1936). Folgerungen aus der Diracschen Theorie des Positrons. (Consequences of Dirac's theory of the positron.) See below.] was published in 1936.

In these papers Heisenberg was the first to reinterpret the Dirac equation as a "classical" field equation for any point particle of spin $\hbar/2$, itself subject to quantization conditions involving anti-commutators. Thus, reinterpreting it as a quantum field equation accurately describing electrons, Heisenberg put *matter* on the same footing as *electromagnetism*: as being described by *relativistic quantum field equations* which allowed the possibility of particle creation and destruction. (Hermann Weyl had already described this in a 1929 letter to Albert Einstein.)

In 1928, Albert Einstein nominated Heisenberg, Born, and Jordan for the Nobel Prize in Physics. The announcement of the Nobel Prize in Physics for 1932 was delayed until November 1933. It was at that time that it was announced Heisenberg had won the Prize for 1932 "for the creation of *quantum mechanics*, the application of which has, inter alia, led to the discovery of the allotropic forms of hydrogen".

Heisenberg enjoyed classical music and was an accomplished pianist. His interest in music led to meeting his future wife. In January 1937, Heisenberg met Elisabeth Schumacher (1914–1998) at a private music recital. Elisabeth was the daughter of a well-known Berlin economics professor, and her brother was the economist E. F. Schumacher, author of Small Is Beautiful. Heisenberg married her on April 29. Fraternal twins Maria and Wolfgang were born in January 1938, whereupon Wolfgang Pauli congratulated Heisenberg on his "pair creation"—a word play on a process from elementary particle physics, pair production. They had five more children over the next 12 years: Barbara, Christine, Jochen, Martin and Verena. In 1936 he bought a summer home for his family in Urfeld am Walchensee, in southern Germany.

Heisenberg was involved in the German nuclear weapons program, known as *Uranverein*, which was formed on September 1, 1939, the day World War II began, The Kaiser-Wilhelm Institut für Physik (KWIP, Kaiser Wilhelm Institute for Physics) in Berlin-Dahlem, was placed under the authority of the Heereswaffenamt (HWA, Army Ordnance Office), and the military control of the nuclear research commenced. In February 1942, at a scientific conference called by the Army Weapons Office, Heisenberg presented a lecture to Reichs officials on energy acquisition from nuclear fission entitled *Die theoretischen Grundlagen für die Energiegewinning aus der Uranspaltung* (The theoretical basis for energy generation from uranium fission). He lectured on the enormous energy potential of nuclear fission, stating that 250 million electron volts could be released through the fission of an atomic nucleus. Heisenberg stressed that pure U-235 had to be obtained to achieve a chain reaction; and explored various ways of obtaining isotope $^{235}_{92}U$ in its pure form, including uranium enrichment and an alternative layered method of normal uranium and a moderator in a machine.

In April 1942, Reichs Minister Rust decided to move the nuclear project from the Physics Institute to the Reichs Research Council; returning the Physics Institute to the Kaiser Wilhelm Society, and naming Heisenberg as Director at the Institute. With this appointment, Heisenberg obtained his first professorship. Heisenberg still also had his department of physics at the University of Leipzig.

In February 1943, Heisenberg was appointed to the Chair for Theoretical Physics at the Friedrich-Wilhelms-Universität (today, the Humboldt-Universität zu Berlin). In April, his election to the Preußische Akademie der Wissenschaften (Prussian Academy of Sciences) was approved. That same month, he moved his family to their retreat in Urfeld as Allied bombing increased in Berlin.

The Alsos Mission, an Allied effort to determine if the Germans had an atomic bomb program and to exploit German atomic related facilities, research, material resources, and scientific personnel for the benefit of the US, generally moved into areas which had just come under control of the Allied military forces, but sometimes they operated in areas still under control by German forces. The Kaiser-Wilhelm-Institut für Physik (KWIP, Kaiser Wilhelm Institute for Physics) had been bombed so it had mostly been moved in 1943 and 1944 to Hechingen and its neighboring town of Haigerloch, on the edge of the Black Forest, which eventually became included in the French occupation zone. This allowed the American task force of the Alsos Mission to take into custody a large number of German scientists associated with nuclear research. In January 1945, Heisenberg, with most of the rest of his staff, moved from the Kaiser-Wilhelm Institut für Physik to the facilities in the Black Forest.

On 30 March, 1945, the Alsos Mission reached Heidelberg, where important scientists were captured. Their interrogation revealed that Otto Hahn was at his laboratory in Tailfingen, while Heisenberg and Max von Laue were at Heisenberg's laboratory in Hechingen, and that the experimental natural uranium reactor that Heisenberg's team had built in Berlin had been moved to Haigerloch. Thereafter, the main focus of the Alsos Mission was on these nuclear facilities in the Württemberg area. Heisenberg was captured and arrested in Urfeld, on May 3, 1945, in an alpine operation in territory still under control by German forces.

Germany surrendered on May 7. Heisenberg would not see his family again for eight months, as he was moved across France and Belgium and flown to England on July 3, 1945. Nine prominent German scientists, including Heisenberg, who were members of the *Uranverein* were captured and incarcerated at Farm Hall in England. The facility had been a safe house of the British foreign intelligence MI6. During their detention, their conversations were recorded. Conversations thought to be of intelligence value were transcribed and translated into English. The Farm Hall transcripts reveal that Heisenberg, along with other physicists interned at Farm Hall including Otto Hahn and Carl Friedrich von Weizsäcker, were glad the Allies had won the war. Heisenberg told other scientists that he had never contemplated a bomb, only an atomic pile to produce energy.

On 3 January 1946, the Operation Epsilon detainees were transported to Alswede in Germany. Heisenberg settled in Göttingen, which was in the British zone of Allied-occupied Germany. Following the Kaiser Wilhelm Society's obliteration by the Allied Control Council, and the establishment of the Max Planck Society in the British zone, the Kaiser Wilhelm Institute for Physics was renamed, and Heisenberg became the director of the Max Planck Institute for Physics.

In 1951, Heisenberg agreed to become the scientific representative of the Federal Republic of Germany at the UNESCO conference, with the aim of establishing a European laboratory for nuclear physics. Heisenberg's aim was to build a large particle accelerator, drawing on the resources and technical skills of scientists across the Western Bloc. On 1 July 1953 Heisenberg signed the convention that established CERN on behalf of the Federal Republic of Germany. Although he was asked to become CERN's founding scientific director, he declined. Instead, he was appointed chair of CERN's science policy committee and went on to determine the scientific program at CERN.

In 1958, the Max Planck Institute for Physics was moved to Munich and renamed Max Planck Institute for Physics und Astrophysics, of which Heisenberg was a co-director, and then sole director until he resigned his directorship on December 31, 1970.

> [Heisenberg gave a joint lecture with Dirac at the old Cavendish Laboratory on May 22, 1963, which the author attended. [Underwood, T. G. (1962-3). *Cambridge University lecture notebook 6.*]

Heisenberg died age 74 of kidney cancer at his home, on February 1, 1976. Heisenberg is buried in Munich Waldfriedhof.

Heisenberg, W. (January, 1932). Über den Bau der Atomkerne. I. (About the construction of atomic nuclei. I.); (March, 1932). Über den Bau der Atomkerne. II. (About the construction of atomic nuclei. II.); (September, 1933). Über den Bau der Atomkerne. III.[†]

Zeit. Phys., 77, 1–11; https://doi.org/10.1007/BF01342433; *Ibid.*, 78, 156–64; https://doi.org/10.1007/BF01337585; *Ibid.*, 80, 587–96; https://doi.org/10.1007/ BF01335696.

[†] Each part is hidden behind pay walls but can be purchased from the publisher.

Three-part paper by Heisenberg, which attempted to treat the *protons* and *neutrons* on an equal footing by *considering protons and neutrons as different charge states of the same particle, which he referred to as the isotopic spin parameter.*

Abstract of Part I.

The consequences of the assumption that the *atomic nuclei* are made up of *protons* and *neutrons* without the participation of *electrons* are discussed. § 1. The Hamiltonian function of the *nucleus*. § 2. The relationship between *charge* and *mass* and the special stability of the *Helium* (He) *nucleus*. § 3 to 5; Stability of the *nuclei* and radioactive decay series. § 6. Discussion of the basic physical assumptions.

Abstract of Part II.

§ 1. Stability of even and odd *neutron* nuclei. § 2. Scattering of γ-rays at the atomic *nucleus*. § 3. The properties of the *neutron*.

Abstract of Part III.

The experiments of Curie, Joliot and Chadwick on the existence and stability of the *neutron* prompted the attempt made in Parts I and II of this thesis to define the role played by *neutrons* in the structure of *atomic nuclei* in very specific physical assumptions and to test the usefulness of these assumptions on the factual material of nuclear physics. The incompleteness of the empirical results available so far leads to a great uncertainty even of the foundations of any theory, and only in very few cases do the experiments force a certain interpretation. For this reason, *it seemed necessary to first put a certain hypothesis at the top and see how it is suitable for ordering experience.* In the following, however, it will also be discussed in detail which consequences are characteristic of the *chosen hypothesis* and at which points a different choice of the basic assumptions would lead to the same results. Before this discussion, the considerations of the first two parts will be supplemented and corrected in some places.

Dirac, P. A. M. (April, 1934). Discussion of the infinite distribution of electrons in the theory of the positron.

Proc. Camb. Phil. Soc., 30, 2, 150-63; https://doi.org/10.1017/S030500410001656X; also in Underwood, T. G. (2023). *Quantum Electrodynamics – annotated sources*, pp. 126-33.

Received February 2, 1934, read March 5, 1934.

St. John's College, Cambridge.

Attempt by Dirac to address problem with his *relativistic* 'hole' theory which implies an infinite number of negative-energy *electrons* (per unit volume) with energies extending continuously from $-mc^2$ to $-\infty$. When an *electromagnetic field* is present positive- and negative-energy states cannot be distinguished in a *relativistically* invariant way. Need to set up assumptions for production of *electromagnetic field* by the electron distribution such that any finite change in distribution produces a change in the field in agreement with Maxwell's equations and such that the infinite field which would be required by Maxwell's equations from an infinite density of electrons is in some way cut out. *Assumes each electron has its own individual wave function in space-time* and each electron moves in an *electromagnetic field* which is the same for all electrons part coming from external causes and part from the electron distribution itself. He introduces a *relativistic density matrix* referring to two points in space and two times, and separates the density distribution into two parts where one, R$_a$ contains the *singularities*, and the other, R$_b$ describes the *electric* and *current densities* physically present, so that any alteration one may make in the distribution of *electrons* and *positrons* will correspond to an alteration in R$_b$, which is *relativistically invariant and gauge invariant*, and that the *electric density* and *current density* corresponding to it satisfy the conservation law. *It therefore appears reasonable to make the assumption that the electric and current densities corresponding to R$_b$ are those which are physically present, arising from the distribution of electrons and positrons. In this way we can remove the infinities.*

———————————————

1. *Use of the density matrix.*

The quantum theory of the electron allows states of negative *kinetic energy* as well as the usual states of positive *kinetic energy* and also allows transitions from one kind of *state* to the other. Now particles in *states* of negative *kinetic energy* are never observed in practice. *We can get over this discrepancy between theory and observation by assuming that, in the world as we know it, nearly all the states of negative kinetic energy are occupied, with one electron in each state in accordance with Pauli's exclusion principle, and that the distribution of negative-energy electrons is unobservable to us on account of its uniformity.* Any unoccupied negative-energy states would be observable to us, as holes in the distribution of negative-energy electrons, but *these holes would appear as particles with positive kinetic energy* and thus not as things foreign to all our experience. It seems reasonable and in agreement with all the facts known at present to identify these holes with the recently discovered positrons and thus to obtain a theory of the positron*.

* As this theory was first put forward, Dirac (January, 1930). [A theory of electrons and protons. *Roy. Soc. Proc., A*, 126, 801, 360-65; https://doi.org/10.1098/rspa.1930.0013] and (October, 1930). [On the Annihilation of Electrons and Protons. *Proc. Camb. Phil. Soc.*, 26, 361-75; https://doi.org/10.1017/S0305004100016091], the holes were assumed to be protons, but this assumption was afterwards seen to be untenable, since it was found that the holes must correspond to particles with the same rest-mass as electrons. See Dirac (September, 1931). [Quantized singularities in the electromagnetic field. *Roy. Soc. Proc., A*, 133, 821, 60-72; https://doi.org/10.1098/rspa.1931.0130]; page 61.

We now have a picture of the world *in which there are an infinite number of negative-energy electrons (in fact an infinite number per unit volume) having energies extending continuously from* $-mc^2$ *to* $-\infty$. The problem we have to consider is the way this infinity can be handled mathematically and the physical effects it produces. In particular, we must set up some assumptions for the production of *electromagnetic field* by the electron distribution, which assumptions must be such that any finite change in the distribution produces a change in the field in agreement with Maxwell's equations, *but such that the infinite field which would be required by Maxwell's equations from an infinite density of electrons is in some way cut out.*

These problems are quite simple when we suppose each electron to be moving in a space free of electromagnetic field. *They are not so simple when there is a field present, since the positive- and negative-energy states then get mixed together so intimately that one cannot in general distinguish accurately between them in a relativistically invariant way.* A careful investigation is then necessary, even for such an elementary problem as seeing that a precise meaning can be given to a distribution such as occurs in practice, in which nearly all the negative-energy states are occupied and nearly all the positive-energy ones unoccupied.

To make an exact treatment of the matter would be very complicated and in the present paper only an approximate treatment will be given, on the lines of Hartree's method of the self-consistent field. *We shall suppose that each electron has its own individual wave function in space-time* (instead of there being one wave function in an enormous number of variables to describe the whole distribution), and also, *we shall suppose that each electron moves in a definite electromagnetic field*, which is the same for all the electrons. *This field will consist of a part coming from external causes and a part coming from the electron distribution itself*, the precise way in which the latter part depends on the electron distribution being one of the problems we have to consider.

Let the normalized functions for the electrons at any time be $\psi_a(x)$, where x stands for three positional coordinates x_1, x_2, x_3 of an electron and the suffix *a* takes on different values for the different electrons. *With electron spin taken into account, each $\psi_a(x)$ must have four components*, which may be specified by $\psi_{ak}(x)$ with k = 1, 2, 3, 4. The whole distribution of electrons may now be described by the *density matrix* ρ defined by

$$(x' \mid \rho \mid x'')_{k'k''} = \sum_a \psi_{ak'}(x') \, \psi^*_{ak''}(x''), \qquad (1)$$

in which the sum is taken over all the electrons. This is a matrix in the spin variables k as well as in the positional variables x. It is, of course, a Hermitian matrix. Its properties have been studied previously*,

* Dirac, P. A. M. (January, 1929). The basis of statistical quantum mechanics. *Proc. Camb. Phil. Soc.*, 25, 1, 62-6; https://doi.org/10.1017/S0305004100018570; Dirac, P. A. M. (July, 1930). Note on Exchange Phenomena in the Thomas Atom. *Proc. Camb. Phil. Soc.*, 26, 3, 376-85; https://doi.org/10.1017/s0305004100016108; Dirac, P. A. M. (April, 1931). Note on the Interpretation of the Density Matrix in the Many-Electron Problem. *Proc. Camb. Phil. Soc.*, 27, 2, 240-3; https://doi.org/10.1017/S0305004100010343.

the chief ones being the equation

$$\rho^2 = \rho, \qquad (2)$$

which expresses that the electron distribution satisfies the *exclusion principle*, and the *equation of motion*

$$ih \, d\rho/dt = H\rho - \rho H. \qquad (3)$$

Here H is the Hamiltonian for the motion of a single electron in the field, thus

$$H = \alpha_s(p_s + eA_s) - eA_0 + \alpha_4 m, \qquad (4)$$

the velocity of light being made equal to unity and a summation being implied over the values s = 1, 2, 3.

An alternative way of regarding the sum on the right-hand side of (1) is as a sum over all the occupied states*.

* The word 'all' used in this connection means each of a set of orthogonal states which is made as large as possible, and does not include states formed by superposition of these orthogonal states.

It may then conveniently be written $\sum_{oc} \psi_{k'}(x') \, \psi^*_{k''}(x'')$. There will be a corresponding sum over the unoccupied states, which may be written $\sum_{un} \psi_{k'}(x') \, \psi^*_{k''}(x'')$. If we add these two sums, we get the sum over all states and this must give us the unit matrix, from the transformation theory of quantum mechanics. Thus

117

$$\sum_{oc} \psi_{k'}(x') \, \psi^*_{k''}(x'') + \sum_{un} \psi_{k'}(x') \, \psi^*_{k''}(x'') = \delta(x' - x'') \, \delta_{k'k''}.$$

Put $\rho = \frac{1}{2}(1 + \rho_1),$ (5)

so that

$$(x' \mid \rho_1 \mid x'')_{k'k''} = \sum_{oc} \psi_{k'}(x') \, \psi^*_{k''}(x'') - \sum_{un} \psi_{k'}(x') \, \psi^*_{k''}(x'').$$

We may now consider the electron distribution as specified by the matrix ρ_1 instead of the matrix ρ. *This has the advantage that it makes a closer symmetry between the electrons and the positrons and leads to neater mathematical expressions.* The *equation of motion* (3) holds unchanged with ρ_1 instead of ρ and equation (2) gets modified to

$$\rho_1{}^2 = 1.$$ (6)

The *density matrices* that we have been discussing up to the present are *non-relativistic* things, since their elements each refer to two points in space x' and x" but to only one time. *To get a relativistic theory, we must introduce two times, t' and t", and use instead of ρ the relativistic density matrix R defined by*

$$(x'\, t' \mid R \mid x''\, t'')_{k'k''} = \sum_{a} \psi_{ak'}(x'\, t') \, \psi^*_{ak''}(x''\, t'')$$

$$= \sum_{oc} \psi_{k'}(x'\, t') \, \psi^*_{k''}(x''\, t'').$$ (7)

Instead of ρ_1 we shall now have R_1, defined by

$$(x'\, t' \mid R_1 \mid x''\, t'')_{k'k''} = \sum_{oc} \psi_{k'}(x'\, t') \, \psi^*_{k''}(x''\, t'') - \sum_{un} \psi_{k'}(x'\, t') \, \psi^*_{k''}(x''\, t''),$$

and instead of equation (5) we shall have

$$R = \frac{1}{2}(R_F + R_1),$$

where

$$(x'\, t' \mid R_F \mid x''\, t'')_{k'k''} = \sum_{oc} \psi_{k'}(x'\, t') \, \psi^*_{k''}(x''\, t'') + \sum_{un} \psi_{k'}(x'\, t') \, \psi^*_{k''}(x''\, t''),$$

R_F, representing *the full distribution with all possible states occupied*, is no longer simply the unit matrix, but all the same we should expect it to play some fundamental part in the theory.

The new matrices R, R_1, R_F are also *Hermitian*

> [A *Hermitian matrix* (or self-adjoint matrix) is a *complex square matrix* that is equal to its own *conjugate transpose*—that is, the element in the i-th row and j-th column is equal to the *complex conjugate* (two complex numbers having their real parts identical and their imaginary parts of equal magnitude but opposite sign) of the element in the j-th row and i-th column, for all indices i and j:
>
> *A* Hermitian $\Leftrightarrow$ $a_{ij} = a^*_{ji}$

118

or in matrix form:

$$A \text{ Hermitian} \Leftrightarrow A = A^{*\mathrm{T}}.$$

Hermitian matrices can be understood as the *complex extension of real symmetric matrices*. If the *conjugate transpose* of a matrix A is denoted by A^{H}, then the Hermitian property can be written concisely as

$$A \text{ Hermitian} \Leftrightarrow A = A^{\mathrm{H}}.$$

Hermitian matrices are named after Charles Hermite, who demonstrated in 1855 that matrices of this form share a property with real *symmetric matrices* of always having real *eigenvalues*.]

and their *equation of motion* is

$$\mathscr{H}\mathrm{R} = 0, \qquad \mathscr{H}\mathrm{R}_1 = 0, \qquad \mathscr{H}\mathrm{R}_\mathrm{F} = 0, \tag{8}$$

where $\mathscr{H}$ is the total operator that operates on the wave function in the *wave equation* for one *electron*, i.e.

$$\mathscr{H} = \mathrm{W} - \mathrm{H}$$

W being the operator ih times time-differentiation. Equations (2)

[$$\rho^2 = \rho, \tag{2}$$

which expresses that the electron distribution satisfies the *exclusion principle*, and the *equation of motion*

$$\mathrm{ih}\, d\rho/dt = \mathrm{H}\rho - \rho\mathrm{H}]$$

and (6),

[The *equation of motion* (3) holds unchanged with ρ_1 instead of ρ and equation (2) gets modified to

$$\rho_1^2 = 1. \tag{6}]$$

cannot be concisely expressed in terms of the R's.

To obtain the field produced by the distribution of electrons, we must first get the *electric density* and *current density*. *For this purpose, we must, according to the usual theory for finite distributions, take a diagonal element of ρ in the positional variables, or a diagonal element of R in the positional and time variables, and form its diagonal sum over the spin variables.* The resulting expression, namely $\sum_k (\mathrm{x} \mid \rho \mid \mathrm{x})_{kk}$ or $\sum_k (\mathrm{xt} \mid \mathrm{R} \mid \mathrm{xt})_{kk}$ would then be the *electric density* (apart from the factor $-\mathrm{e}$). The corresponding *current density* would have for its sth component $\sum_k (\mathrm{x} \mid \alpha_s\rho \mid \mathrm{x})_{kk}$ or $\sum_k (\mathrm{xt} \mid \alpha_s\mathrm{R} \mid \mathrm{xt})_{kk}$.

We can easily verify that this *electric density* and *current density* satisfy the *conservation law of electricity*. In equation (3)

[the *equation of motion*

$$\mathrm{ih}\, d\rho/dt = \mathrm{H}\rho - \rho\mathrm{H}. \tag{3}]$$

119

let us take diagonal elements in the *positional* variables, but keep to the symbolic matrix notation for the *spin* variables, so that a symbol like $(x \mid \rho \mid x)$ denotes a matrix with four rows and columns, of the same nature as an α. This gives

$$\text{ih } d/dt \, (x \mid \rho \mid x) = (x \mid H\rho - \rho H \mid x)$$
$$= \alpha_s \, (x \mid p_s\rho - \rho p_s \mid x) + (x \mid \{\alpha_s\rho - \rho\alpha_s\}\{p_s + eA_s\} \mid x)$$
$$+ (x \mid \{\alpha_4\rho - \rho\alpha_4 \mid x) \, m.$$

If we now take the *diagonal sum* with respect to the *spin variables*, the last two terms will contribute nothing, from the rule that the diagonal sum of the product of two matrices is independent of their order, and we shall be left with

$$\text{ih } d/dt \sum_k (x \mid \rho \mid x)_{kk} = \sum_{kk'} \alpha_{s\,kk'} \, (x \mid p_s\rho - \rho p_s \mid x)_{k'k}$$
$$= - \text{ih} \sum_{kk'} \alpha_{s\,kk'} \, \partial/\partial x_s \, (x \mid \rho \mid x)_{k'k},$$

i.e. $$d/dt \sum_k (x \mid \rho \mid x)_{kk} = - \partial/\partial x_s \sum_k (x \mid \alpha_s\rho \mid x)_{kk}, \qquad (9)$$

which is the required conservation law.

In our present theory the electric density and current density given by these formulae would be infinite and some alteration of the assumptions is therefore necessary. The problem now presents itself of finding some natural way of removing the infinities from $\sum_k (xt \mid R \mid xt)_{kk}$ and $\sum_k (xt \mid \alpha_s R \mid xt)_{kk}$ so as to leave finite remainders, which we could then assume to be the *electric* and *current* densities. This problem requires us to make a detailed investigation of the singularities in the matrix elements $(x't' \mid R \mid x''t'')_{k'k''}$ near the diagonal $x_s' = x_s''$, $t' = t''$.

2. *Case of no field.*

We shall begin our investigation by taking the case of *no electromagnetic field*, when the Hamiltonian (4)

$$[H = \alpha_s(p_s + eA_s) - eA_0 + \alpha_4 m, \qquad (4)]$$

reduces to

$$H = \alpha_s p_s + \alpha_4 m = (\boldsymbol{\alpha}, \mathbf{p}) + \alpha_4 m. \qquad (10)$$

In this case we can calculate accurately the matrix elements $(x't' \mid R \mid x''t'')_{k'k''}$ *for the distribution of electrons in which all the negative-energy states are occupied and all the-positive-energy ones unoccupied,* and see exactly how these matrix elements behave near the diagonal.

If we try to work directly from the definition (7) we meet with some awkward calculations in taking the *spin variables* into account and summing over the two possible spin orientations. We can avoid these calculations by using *symbolic methods* and first obtaining ρ. *The condition that a wave function ψ contains only Fourier components belonging to negative-energy states may be expressed symbolically by*

$$\{H + \surd(P^2 + m^2)\}\ \psi = 0,$$

where P denotes the length of the vector **p** and the *positive square root is understood.* Similarly, *the condition that the distribution ρ contains electrons only in negative-energy states may be expressed by*

$$\{H + \surd(P^2 + m^2)\}\ \rho = 0. \tag{11}$$

The condition that in the distribution ρ every negative-energy state is occupied, is just the condition that the distribution 1 − ρ contains electrons only in positive-energy states and may thus be expressed by

$$\{H - \surd(P^2 + m^2)\}\ (1 - \rho) = 0.$$

Adding this equation to (11) we get

$$\{H + \surd(P^2 + m^2)\}\ (2\rho - 1) = 0$$

or $\quad \rho = \tfrac{1}{2}\,[1 - H/\surd(P^2 + m^2)] = \tfrac{1}{2}\,[1 - \{(\boldsymbol{\alpha}, \mathbf{p}) + \alpha_4 m\}/\surd(P^2 + m^2)]. \tag{12}$

Hence, from the transformation theory,

$$(x' \mid \rho \mid x'') = 1/2h^3 \iint e^{i(x',p')/h}\, dp'\, [1 - \{(\boldsymbol{\alpha}, \mathbf{p}) + \alpha_4 m\}/\surd(P^2 + m^2)]$$
$$\times\ \delta(p' - p'')\, dp''\, e^{-i(x'',p')/h} \tag{13}$$

where dp denotes the product $dp_1 dp_2 dp_3$.

…

… Introducing the notation

$$x_s' - x_s'' = x_s, \qquad\qquad t' - t'' = t,$$

which we shall keep through the rest of the paper, …

…

It is clear from these equations that there will be singularities, not only at the point $x_s = 0$, $t = 0$, but also everywhere on the light-cone $t^2 - r^2 = 0$. In order to determine these singularities, we may expand the Bessel functions in power series of $\surd(t^2 - r^2)$ and retain only the first few terms. …

…

The main result of this investigation for the case of no field is that there are two quite distinct kinds of singularity occurring in the matrices R_F [representing the full distribution with all possible states occupied] and R_1 [representing the relativistic density matrix] respectively. The singularities occurring in R_F are all associated with the δ function and those in R_1 with the reciprocal function and logarithm. From the generality of this result, we may expect it to hold also when there is a field present.

3. *Case of an arbitrary field.*

Let us now examine the singularities in (x't' | R_F | x"t") and (x't' | R_1 | x"t") when there is a general field present. Our method will be to suppose that the singularities are of the same form as in the case of no field, but have unknown coefficients. These coefficients must be functions of x_s', t', $x_s"$, t" which are free from singularities and can be expanded as Taylor series for small values of x_s and t. We must try to choose them so that the *equations of motion* (8)

$$[\mathcal{H}R = 0, \qquad \mathcal{H}R_1 = 0, \qquad \mathcal{H}R_F = 0, \qquad\qquad (8)]$$

are satisfied.

The application of the method follows a parallel course for R_F and R_1, and we need therefore treat in detail only R_1 which is the *density matrix* we are mainly interested in. ...

... In this way all our equations are satisfied and all our unknowns are determined, with the exception of g, which is not itself determined although $\mathcal{H}$g is. The final result is

$$(x't' | R_1 | x"t") = ... \qquad\qquad (48)$$

...

To do the corresponding work for R_F we put, analogously to (29),

...

... Thus, R_F is completely fixed.

4. *Conclusion.*

From the foregoing work we see that the following results must hold, at least to the accuracy of the Hartree method of approximation:

(i)   One can give a precise meaning to the *distribution of electrons* in which every state is occupied. This distribution may be defined as that described by the *density matrix* R_F given by (49), this matrix being completely fixed for any given field.

(ii)   One can give a precise meaning to a *distribution of electrons* in which nearly all (i.e. all but a finite number, or all but a finite number per unit volume) of the negative-energy states are occupied and nearly all of the positive-energy ones unoccupied. Such a distribution may be defined as one described by a *density matrix* $R = \frac{1}{2} (R_F + R_1)$, where R_1 is of the form (48). This definition is permissible because *the only possible variations in R_1, namely those due to g not being completely defined, are free from singularity and thus correspond to finite changes, or finite changes per unit volume, in the electron distribution.* Our method does not give any precise meaning to which negative-energy states are unoccupied or which positive-energy ones are occupied. It is sufficiently definite, though, to take as the basis of the theory of the position the *assumption that only distributions described by R = ½ (R$_F$ + R$_1$) with R1 of the form (48) occur in nature.*

(iii)   A [density] distribution R such as occurs in nature according to the above assumption *can be divided naturally into two parts*

R = R$_a$ + R$_b$,

where R$_a$ contains all the singularities and is also completely fixed for any given field, so that any alteration one may make in the distribution of *electrons* and *positrons* will correspond to an alteration in R$_b$ but to none in R$_a$. We get this division into two parts by putting the term containing g into R$_b$ and all the other terms into R$_a$. Thus,

R$_b$ = g/4ih.

It is easily seen that *R$_b$ is relativistically invariant and gauge invariant*, and it may be verified after some calculation that R$_b$ is Hermitian and that the *electric density* and *current density* corresponding to it satisfy the conservation law (9)

[d/dt $\sum_k$ (x | ρ | x)$_{kk}$ = $- \partial/\partial x_s$ $\sum_k$ (x | $\alpha_s\rho$ | x)$_{kk}$.   (9)]

It therefore appears reasonable to make the assumption that the electric and current densities corresponding to R$_b$ are those which are physically present, arising from the distribution of electrons and positrons. In this way we can remove the infinities mentioned at the end of § 1.

The present paper is incomplete in that the effect of the *exclusion principle*, equation (2) or (6)

[$\rho^2 = \rho$,   (2)

          $\rho_1^2 = 1$,   (6)]

on R$_b$ has not been investigated. Further work that remains to be done is to examine the physical consequences of the foregoing assumption and to see whether it leads to any phenomena of the nature of a *polarization of a vacuum* by an *electromagnetic field.*]

Heisenberg, W. (September, 1934). Bemerkungen zur Diracschen Theorie des Positrons. (Remarks on the Dirac theory of positron.)

Zeit. Phys., 90, 209–31; https://doi.org/10.1007/BF01333516; translation by D. H. Delphenich at https://s3.cern.ch/inspire-prod-files-5/5e5a144a00179c2f8e799c87 f4c44be2.

Leipzig.

Received June 21, 1934.

Heisenberg responded by presenting his thinking on Dirac's theory and further development of the theory in two papers.

————————————

I. Intuitive theory of matter waves.
1. The inhomogeneous differential equation of the density matrix.
2. The conservation laws.
3. Applications (polarization of the vacuum).

II. Quantum theory of the wave field.
1. Presentation of the field equations.
2. Applications (the self-energy of light quanta).

The purpose of the present paper[1] is to construct the *Dirac theory of the positron*[2] in the formalism of *quantum electrodynamics*.

[1] The paper originated in some discussions that I had with Herren Pauli, Dirac, and Weisskopf, in part written and in part oral, and to them I am deeply grateful.
[2] E.g.: Dirac, P. A. M. (1930). The Principles of Quantum Mechanics. Oxford, pp. 255.

Thus, we shall demand that the *symmetry* in nature between positive and negative *charge* should be expressed in the basic equations from the outset, and that in addition to the well-known difficulties with the divergences that quantum electrodynamics leads to, no new infinities should appear in the formalism, moreover; i.e., that the theory should provide an approximation for the treatment of the circle of problems that have been treated by quantum electrodynamics up to now. By the latter postulate, one distinguishes the present effort from the investigations of Fock[3], Oppenheimer and Furry[4], and Peierls[5], the last of which is similar to it; he is closely linked with the paper of Dirac[6], moreover.

[3] Fock, V. & Leningrad, C. R. (1933). (*N. S.*) no. 6, 267-71.

[4] Furry, W. H. & Oppenheimer, J. R. (1943). *Phys. Rev.*, 45, 245.

[5] Peierls, R. to appear.

[6] Dirac, P. A. M. (April, 1934). Discussion of the infinite distribution of electrons in the theory of the positron. See above (in what follows, this is always referred to by *loc. cit.*).

In contrast to the Dirac treatment, one has the work on the meaning of the *conservation law for the total system of radiation-matter* and the necessity of formulating the basic equations of the theory in a way that grows out of the Hartree approximation.

…

Heisenberg, W., Euler, H. (November, 1936). Folgerungen aus der Diracschen Theorie des Positrons. (Consequences of Dirac's theory of the positron.)

Zeit. Phys., 98, 11-12, 714-732; https://doi.org/10.1007/BF01343663; translation by W. Korolevski and H. Kleinert, Institut fur Theoretische Physik, Freie Universitat Berlin, Arnimallee 14, D-14195 Berlin, Germany; https://arxiv.org/pdf/physics/0605038.

Leipzig.

Received December 22, 1935.

According to Dirac's *theory of the positron*, an *electromagnetic field* tends to create pairs of particles which leads to a change of Maxwell's equations in the vacuum. This paper examines the consequence of the possibility of transforming *electromagnetic radiation* into *matter* in *quantum electrodynamics* on Maxwell equations. It is no longer possible to separate processes in the vacuum from those involving *matter* since *electromagnetic fields* can create *matter* if they are strong enough. Even if they are not strong enough to create *matter* they will, due to the virtual possibility of creating *matter*, *polarize the vacuum*.

Abstract

According to Dirac's *theory of the positron*, an *electromagnetic field* tends to create pairs of particles which leads to a change of Maxwell's equations in the vacuum. These changes are calculated in the special case that no real *electrons* or *positrons* are present and the field varies little over a Compton wavelength. The resulting effective Lagrangian of the *field* reads:

$$\mathscr{L} = \tfrac{1}{2}\,(\mathscr{E}^2 - \mathscr{B}^2) + e^2/\hbar c \int_0^\infty e^{-\eta}\,d\eta/\eta^3\,[[i\eta^2(\mathscr{E}\mathscr{B})\cdot\cos\,[\eta/|\mathscr{E}_k|\sqrt{\{\mathscr{E}^2 - \mathscr{B}^2 + 2i(\mathscr{E}\mathscr{B})\}} +$$
$$\text{conj.}]/\cos\,[(\eta/|\mathscr{E}_k|\sqrt{\{\mathscr{E}^2 - \mathscr{B}^2 + 2i(\mathscr{E}\mathscr{B})\}} - \text{conj.}] + |\mathscr{E}_k|^2 + \eta^2/3\,(\mathscr{B}^2 - \mathscr{E}^2)]]$$

where $\mathscr{E}$, $\mathscr{B}$ are the field strengths, and $|\mathscr{E}_k| = m^2c^3/e\hbar = 1/137\ e/(e^2/mc^2)^2 = $ critical field strengths.

The expansion terms in *small fields* (compared to $\mathscr{E}$) describe *light-light scattering*. The simplest term is already known from perturbation theory. For *large fields*, the equations derived here differ strongly from *Maxwell's equations*. Our equations will be compared to those proposed by Born.

126

The fact that *electromagnetic radiation* can be transformed into *matter* and vice versa leads to fundamentally new features in *quantum electrodynamics*. One of the most important consequences is that, even in the vacuum, the Maxwell equations have to be exchanged by more complicated formulas. In general, it will be not possible to separate processes in the vacuum from those involving *matter* since *electromagnetic fields* can create *matter* if they are strong enough. Even if they are not strong enough to create *matter* they will, due to the virtual possibility of creating *matter*, *polarize the vacuum* and therefore change the Maxwell equations. *This polarization of the vacuum to be studied below* will give rise to a distinction between the vectors $\mathscr{B}$, $\mathscr{E}$ on the one hand and D, $\mathscr{H}$ on the other, where

$$D = \mathscr{E} + 4\pi\,\mathscr{B}, \qquad (1)$$
$$\mathscr{H} = \mathscr{B} - 4\pi\mathscr{M}$$

The *polarizations* β and $\mathscr{M}$ can be arbitrary functions of the field strengths at the same place, their derivatives, and the field strengths in the surroundings of the observed position. If the field strengths are small (which means, as we shall see small compared to $e^2/\hbar c$ - times of the field strength at the boundary of the *electron*), then β and $\mathscr{M}$ can be approximately considered as linear functions of $\mathscr{E}$ and $\mathscr{B}$. In this approximation, Uehling[1] and Serber[2] have calculated the modifications of Maxwell's theory.

[1] Uehling, E. A. (1935). *Phys. Rev.*, 48, 55.
[2] Serber, R. (1935). *Phys. Rev.*, 48, 49.

Another interesting case is obtained by not assuming small field strengths but instead slowly varying fields (i.e., the fields are nearly constant over the length $\hbar/mc$). Then one obtains β and $\mathscr{M}$ as functions of $\mathscr{E}$ and $\mathscr{B}$ at the same position. The derivatives of $\mathscr{E}$ and $\mathscr{B}$ do not appear in the approximation. The expansion of β and $\mathscr{M}$ in powers of $\mathscr{E}$ and $\mathscr{B}$ will contain only odd powers, as will be seen in the calculation. The expansion terms of third order are phenomenologically related to light-light scattering and are already known[3]. The goal of this work is to find the functions $\beta(\mathscr{E},\mathscr{B})$ and $\mathscr{M}(\mathscr{E},\mathscr{B})$ for slowly varying field strength. It is sufficient to calculate the *energy density* $U(\mathscr{E},\mathscr{B})$. From the *energy density* one can derive the fields using the Hamiltonian method; one introduces the Lagrangian $\mathscr{L}(\mathscr{E},\mathscr{B})$ and obtains

$$D_i = \partial\mathscr{L}/\partial\mathscr{E}_i, \quad \mathscr{H}_i = -\partial\mathscr{L}/\partial\mathscr{E}_i, \qquad (2)$$
$$U(\mathscr{E},\mathscr{B}) = 1/4\pi\,[\Sigma i\,D_i\mathscr{E}_i - \partial\mathscr{L}] = 1/4\pi\,[\Sigma_i\,\mathscr{E}_i\,\partial\mathscr{L}/\partial\mathscr{E}_i - \mathscr{L}] \qquad (3)$$

The Lagrangian is determined by (3) and D and $\mathscr{H}$ are determined by (2). Due to relativistic invariant the Lagrangian can only depend on the two invariants $\mathscr{E}^2 - \mathscr{B}^2$ and $(\mathscr{E}\mathscr{B})^2$

(compare 3). The calculation of $U(\mathscr{E},\mathscr{B})$ can be reduced to the question of how much *energy density* is associated with the *matter fields* in a background of constant fields $\mathscr{E}$ and $\mathscr{B}$. Before solving this problem, the *mathematical scheme of the positron theory*[4] will be presented.

[4] Heisenberg, W. (March, 1934). Bemerkungen zur Diracschen Theorie des Positrons. (Remarks on the Dirac theory of positron.). See above.

This will also correct some errors in earlier formulas.

...

Enrico Fermi (September 29, 1901–November 28, 1954).

Fermi was an Italian and naturalized American physicist, renowned for being the creator of the world's first nuclear reactor, the Chicago Pile-1, and a member of the Manhattan Project. He has been called the "architect of the nuclear age" and the "architect of the atomic bomb". He was one of very few physicists to excel in both theoretical physics and experimental physics. Fermi did important work in particle physics, especially related to *pions* and *muons*, and he speculated that cosmic rays arose when material was accelerated by magnetic fields in interstellar space.

Fermi was awarded the 1938 Nobel Prize in Physics for his work on induced radioactivity by neutron bombardment and for his discovery of trans-uranium elements. With his colleagues, Fermi filed several patents related to the use of nuclear power, all of which were taken over by the US government. He made significant contributions to the development of statistical mechanics, quantum theory, and nuclear and particle physics.

Fermi led the team at the University of Chicago that designed and built Chicago Pile-1, which went critical on December 2, 1942, demonstrating the first human-created, self-sustaining nuclear chain reaction. He was on hand when the X-10 Graphite Reactor at Oak Ridge, Tennessee went critical in 1943, and when the B Reactor at the Hanford Site did so the next year. During World War II, he worked on the Manhattan Project at Los Alamos, where he headed F Division, part of which worked on Edward Teller's thermonuclear "Super" bomb. He was present at the Trinity test on July 16, 1945, the first test of a full nuclear bomb explosion, where he used his Fermi method to estimate the bomb's yield.

After the war, he helped establish the Institute for Nuclear Studies at Chicago, and served on the General Advisory Committee, chaired by J. Robert Oppenheimer, which advised the Atomic Energy Commission on nuclear matters. After the detonation of the first Soviet fission bomb in August 1949, he strongly opposed the development of a hydrogen bomb on both moral and technical grounds. He was among the scientists who testified on Oppenheimer's behalf at the 1954 hearing that resulted in the denial of Oppenheimer's security clearance.

Enrico Fermi was born in Rome, Italy, on September 29, 1901. He was the third child of Alberto Fermi, a division head in the Ministry of Railways, and Ida de Gattis, an elementary school teacher. His sister, Maria, was two years older, his brother Giulio a year older.

At a local market in Campo de' Fiori, Fermi found a physics book, the 900-page *"Elementorum physicae mathematicae"*. Written in Latin by Jesuit Father Andrea Caraffa, a professor at the Collegio Romano, it presented mathematics, classical mechanics, astronomy, optics, and acoustics as they were understood at the time of its 1840 publication.

Fermi graduated from high school in July 1918, having skipped the third year entirely. Fermi learned German to be able to read the many scientific papers that were published in that language at the time, and he applied to the Scuola Normale Superiore in Pisa. Fermi took first place in the difficult entrance exam, which included an essay on the theme of "Specific characteristics of Sounds"; the 17-year-old Fermi chose to use Fourier analysis to derive and solve the partial differential equation for a vibrating rod, and after interviewing Fermi the examiner declared he would become an outstanding physicist.

At the Scuola Normale Superiore, Fermi was advised by Luigi Puccianti, director of the physics laboratory, who said there was little he could teach Fermi and often asked Fermi to teach him something instead. Fermi's knowledge of quantum physics was such that Puccianti asked him to organize seminars on the topic. During this time Fermi learned tensor calculus, a technique key to general relativity. Fermi initially chose mathematics as his major, but soon switched to physics. He remained largely self-taught, studying general relativity, quantum mechanics, and atomic physics.

In September 1920, Fermi was admitted to the physics department. Since there were only three students in the department Puccianti let them freely use the laboratory for whatever purposes they chose. Fermi decided that they should research X-ray crystallography, and the three worked to produce a Laue photograph—an X-ray photograph of a crystal. During 1921, his third year at the university, Fermi published his first scientific works in the Italian journal *Nuovo Cimento*. The first was entitled "*On the dynamics of a rigid system of electrical charges in translational motion*" (*Sulla dinamica di un sistema rigido di cariche elettriche in moto traslatorio*). A sign of things to come was that the mass was expressed as a tensor—a mathematical construct commonly used to describe something moving and changing in three-dimensional space. In classical mechanics, mass is a scalar quantity, but in relativity, it changes with velocity.

The second paper was "On the electrostatics of a uniform gravitational field of electromagnetic charges and on the weight of electromagnetic charges" (*Sull'elettrostatica di un campo gravitazionale uniforme e sul peso delle masse elettromagnetiche*). Using *general relativity*, Fermi showed that a charge has a weight equal to U/c^2, where U was the electrostatic energy of the system, and c is the speed of light.

The first paper seemed to point out a contradiction between the electrodynamic theory and the relativistic one concerning the calculation of the electromagnetic masses, as the former predicted a value of $4/3 \, U/c^2$. Fermi addressed this the next year in a paper "*Concerning a contradiction between electrodynamic and the relativistic theory of electromagnetic mass*" in which he showed that the apparent contradiction was a consequence of *relativity*. This paper was sufficiently well-regarded that it was translated into German and published in the German scientific journal *Physikalische Zeitschrift* in 1922.

That year, Fermi submitted his article "*On the phenomena occurring near a world line*" (*Sopra i fenomeni che avvengono in vicinanza di una linea oraria*) to the Italian journal *I Rendiconti dell'Accademia dei Lincei*. In this article he examined the *Principle of Equivalence*, and introduced the so-called "Fermi coordinates". He proved that on a world line close to the timeline, space behaves as if it were a Euclidean space.

Fermi submitted his thesis, "*A theorem on probability and some of its applications*" (*Un teorema di calcolo delle probabilità ed alcune sue applicazioni*), to the Scuola Normale Superiore in July 1922, and received his laurea at the unusually young age of 20. The thesis was on X-ray diffraction images. Theoretical physics was not yet considered a discipline in Italy, and the only thesis that would have been accepted was experimental physics. For this reason, Italian physicists were slow in embracing the new ideas like relativity coming from Germany. Since Fermi was quite at home in the lab doing experimental work, this did not pose insurmountable problems for him.

While writing the appendix for the Italian edition of the book *Fundamentals of Einstein Relativity* by August Kopff in 1923, Fermi was the first to point out that hidden inside the Einstein equation ($E = mc^2$) was an enormous amount of nuclear potential energy to be exploited. "It does not seem possible, at least in the near future", he wrote, "to find a way to release these dreadful amounts of energy—which is all to the good because the first effect of an explosion of such a dreadful amount of energy would be to smash into smithereens the physicist who had the misfortune to find a way to do it."

In 1923–1924, Fermi spent a semester studying under Max Born at the University of Göttingen, where he met Werner Heisenberg and Pascual Jordan. Fermi then studied in Leiden with Paul Ehrenfest from September to December 1924 on a fellowship from the Rockefeller Foundation obtained through the intercession of the mathematician Vito Volterra. Here Fermi met Hendrik Lorentz and Albert Einstein, and became friends with Samuel Goudsmit and Jan Tinbergen.

Fermi's first major contribution involved the field of statistical mechanics. After Wolfgang Pauli formulated his *exclusion principle* in 1925, Fermi followed with a paper in which he applied the principle to an ideal gas, employing a statistical formulation now known as *Fermi–Dirac statistics*. Today, particles that obey the *exclusion principle* are called "*fermions*". Pauli later postulated the existence of an uncharged invisible particle emitted along with an electron during beta decay, to satisfy the law of conservation of energy. Fermi took up this idea, developing a model that incorporated the postulated particle, which he named the "*neutrino*".

His theory, later referred to as *Fermi's interaction* and now called *weak interaction*, described one of the four fundamental interactions in nature.

[The *weak interaction*, also called the *weak force*, is the mechanism of *interaction* between subatomic particles that is responsible for the *radioactive decay* of atoms: the *weak interaction* participates in nuclear fission and nuclear fusion. The effective range of the *weak force* is limited to subatomic distances and is less than the diameter of a *proton*.

A weak interaction occurs when two particles (typically, but not necessarily, half-integer spin fermions) exchange integer-spin, force-carrying bosons.

In 1933, Enrico Fermi proposed the first theory of the *weak interaction*, known as *Fermi's interaction*. He suggested that *beta decay* could be explained by a four-*fermion* interaction, involving a contact force with no range. [Fermi, E. (March, 1934) Versuch einer Theorie der β-Strahlen. I. (Attempt at a theory of β rays. I.). See below.]

[A *fermion* is a particle that follows *Fermi–Dirac statistics*. *Fermions* have a half-odd-integer spin (spin 1/2, spin 3/2, etc.) and obey the *Pauli exclusion principle*. These particles include all *quarks* and *leptons* and all composite particles made of an odd number of these, such as all *baryons* and many atoms and nuclei. Some *fermions* are elementary particles (such as *electrons*), and some are composite particles (such as *protons*). *Fermions* differ from *bosons*, which have integer *spin* and obey *Bose–Einstein statistics*.]

In the mid-1950s, Chen-Ning Yang and Tsung-Dao Lee first suggested that the handedness of the *spins* of particles in *weak interaction* might violate the *conservation law* or *symmetry*. [Lee, T. D. & Yang, C. N. (October, 1956). Question of Parity Conservation in Weak Interactions. See below.]

In the 1960s, Sheldon Glashow, Abdus Salam and Steven Weinberg unified the *electromagnetic force* and the *weak interaction* by showing them to be two aspects of a single force, now termed the *electroweak force*. [Glashow, S. L. (February, 1961). Partial-symmetries of weak interactions. See below.]

132

The *Standard Model* of particle physics provides a uniform framework for understanding *electromagnetic*, *weak*, and *strong interactions*. A *weak interaction* occurs when two particles (typically, but not necessarily, *half-integer spin fermions*) exchange *integer-spin*, force-carrying *bosons*. The *fermions* involved in such exchanges can be either *elementary* (e.g. *electrons* or *quarks*) or *composite* (e.g. *protons* or *neutrons*), although at the deepest levels, all *weak interactions* ultimately are between *elementary particles*.

In the *weak interaction*, *fermions* can exchange three types of force carriers, namely W^+, W^-, and Z *bosons*. The *masses* of these *bosons* are far greater than the *mass* of a *proton* or *neutron*, which is consistent with the short range of the *weak force*. In fact, the force is termed *weak* because its field strength over any set distance is typically several orders of magnitude less than that of the *electromagnetic force*, which itself is further orders of magnitude less than the *strong nuclear force*.

The existence of the W and Z *bosons* was not directly confirmed until 1983.

The *weak interaction* is the only fundamental *interaction* that breaks *parity symmetry*, and similarly, but far more rarely, the only *interaction* to break *charge–parity symmetry*.

The *weak interaction* is unique in that it allows *quarks* to swap their flavor for another. *Quarks*, which make up composite particles like *neutrons* and *protons*, come in six "*flavors*" – *up*, *down*, *charm*, *strange*, *top* and *bottom* – which give those composite particles their properties. The swapping of those properties is mediated by the force carrier *bosons*. For example, during *beta-minus decay*, a *down quark* within a *neutron* is changed into an *up quark*, thus converting the *neutron* to a *proton* and resulting in the emission of an *electron* and an *electron antineutrino*.

Most *fermions* decay by a *weak interaction* over time. Such decay makes radiocarbon dating possible, as carbon-14 decays through the *weak interaction* to nitrogen-14. It can also create *radioluminescence*, commonly used in tritium luminescence, and in the related field of beta-voltaics (but not similar radium luminescence).

The *electroweak force* is believed to have separated into the *electromagnetic* and *weak forces* during the *quark* epoch of the early universe.]

Through experiments inducing radioactivity with the recently discovered *neutron*, Fermi discovered that slow *neutrons* were more easily captured by *atomic nuclei* than fast ones, and he developed the Fermi age equation to describe this. After bombarding thorium and uranium with slow *neutrons*, he concluded that he had created new elements. Although he was awarded the Nobel Prize for this discovery, the new elements were later revealed to be nuclear fission products.

From January 1925 to late 1926, Fermi taught mathematical physics and theoretical mechanics at the University of Florence. He also participated in seminars at the Sapienza University of Rome, giving lectures on quantum mechanics and solid-state physics.
After Wolfgang Pauli announced his exclusion principle in 1925, Fermi responded with a paper "*On the quantization of the perfect monoatomic gas*" (*Sulla quantizzazione del gas perfetto monoatomico*), in which he applied the exclusion principle to an ideal gas. The paper was especially notable for Fermi's statistical formulation, which describes the distribution of particles in systems of many identical particles that obey the exclusion *principle*. This was independently developed soon after by Paul Dirac, who also showed how it was related to the Bose–Einstein statistics. Accordingly, it is now known as *Fermi–Dirac statistics*. After Dirac, particles that obey the exclusion principle are today called "*fermions*", while those that do not are called "*bosons*".

Professorships in Italy were granted by competition (*concorso*) for a vacant chair, the applicants being rated on their publications by a committee of professors. Fermi applied for a chair of mathematical physics at the University of Cagliari on Sardinia, but was narrowly passed over in favor of Giovanni Giorgi. In 1926, at the age of 24, he applied for a professorship at the Sapienza University of Rome. This was a new chair, one of the first three in theoretical physics in Italy, that had been created by the Minister of Education at the urging of professor Orso Mario Corbino, who was the university's professor of experimental physics, the director of the Institute of Physics, and a member of Benito Mussolini's cabinet. Corbino, who also chaired the selection committee, hoped that the new chair would raise the standard and reputation of physics in Italy. The committee chose Fermi ahead of Enrico Persico and Aldo Pontremoli, and Corbino helped Fermi recruit his team, which was soon joined by notable students such as Edoardo Amaldi, Bruno Pontecorvo, Ettore Majorana and Emilio Segrè, and by Franco Rasetti, whom Fermi had appointed as his assistant. They were soon nicknamed the "Via Panisperna boys" after the street where the Institute of Physics was located.

Fermi married Laura Capon, a science student at the university, on July 19, 1928. They had two children: Nella, born in January 1931, and Giulio, born in February 1936. On March 18, 1929, Fermi was appointed a member of the Royal Academy of Italy by Mussolini, and on April 27 he joined the Fascist Party. He later opposed Fascism when the 1938 racial laws were promulgated by Mussolini in order to bring Italian Fascism ideologically closer to German Nazism. These laws threatened Laura, who was Jewish, and put many of Fermi's research assistants out of work.

During their time in Rome, Fermi and his group made important contributions to many practical and theoretical aspects of physics. In 1928, he published his *"Introduction to Atomic Physics"* (*Introduzione alla fisica atomica*), which provided Italian university students with an up-to-date and accessible text. Fermi also conducted public lectures and wrote popular articles for scientists and teachers in order to spread knowledge of the new physics as widely as possible. Part of his teaching method was to gather his colleagues and graduate students together at the end of the day and go over a problem, often from his own research. A sign of success was that foreign students now began to come to Italy. The most notable of these was the German physicist Hans Bethe, who came to Rome as a Rockefeller Foundation fellow, and collaborated with Fermi on a 1932 paper *"On the Interaction between Two Electrons"* (German: *Über die Wechselwirkung von Zwei Elektronen*).

At this time, physicists were puzzled by *beta decay*, in which an electron was emitted from the atomic nucleus. To satisfy the law of conservation of energy, Pauli postulated the existence of an invisible particle with no charge and little or no mass that was also emitted at the same time. Fermi took up this idea, which he developed in a tentative paper in 1933, and then a longer paper the next year that incorporated the postulated particle, which Fermi called a "*neutrino*". His theory, later referred to as Fermi's interaction, and still later as the theory of the *weak interaction*, described one of the four fundamental forces of nature. The *neutrino* was detected after his death, and his interaction theory showed why it was so difficult to detect. When he submitted his paper to the British journal *Nature*, that journal's editor turned it down because it contained speculations which were "too remote from physical reality to be of interest to readers".

According to Fermi's biographer David N. Schwartz, it is at least strange that Fermi seriously requested publication from the journal, since at that time *Nature* only published short notes on articles of this kind, and was not suitable for the publication of even a new physical theory. More suitable, if anything, would have been the *Proceedings of the Royal Society of London*. He agrees with some scholars' hypothesis, according to which the rejection of the British magazine convinced his young colleagues (some of them Jews and leftist) to give up the boycott of German scientific magazines, after Hitler came to power in January 1933. Thus, Fermi saw the theory published in Italian and German before it was published in English. [Fermi, E. (March, 1934) Versuch einer Theorie der β-Strahlen. I. (Attempt at a theory of β rays. I.) *Zeit. Phys.*, 88, 161–77; https://doi.org/10.1007/BF01351864; also at https://www.nssp. uni-saarland.de/lehre/Vorlesung/Kernphysik _SS19/History/Papers/Fermi_1.pdf. (German); translation by T. G. Underwood. See below.]

On the basis of his theory, the capture of an orbital *electron* by a *nucleus* was predicted and eventually observed. The consequences of the Fermi theory were vast. For example, β spectroscopy was established as a powerful tool for the study of nuclear structure. But perhaps the most influential aspect of this work was that his particular form of the β *interaction* established a pattern that has been appropriate for the study of other types of *interactions*. It was the first successful theory of the creation and annihilation of material particles. Previously, only *photons* had been known to be created and destroyed.

In 1934, Fermi decided to switch to experimental physics, using the *neutron*, which James Chadwick had discovered in 1932. In March 1934, Fermi wanted to see if he could induce radioactivity with Rasetti's polonium-beryllium neutron source. *Neutrons* had no *electric charge*, and so would not be deflected by the positively charged *nucleus*. This meant that they needed much less energy to penetrate the nucleus than charged particles, and so would not require a particle accelerator, which the Via Panisperna boys did not have.

Fermi then had the idea to resort to replacing the polonium-beryllium neutron source with a radon-beryllium one, which he created by filling a glass bulb with beryllium powder, evacuating the air, and then adding 50 mCi of radon gas. This created a much stronger *neutron* source, the effectiveness of which declined with the 3.8-day half-life of radon. He knew that this source would also emit gamma rays, but, on the basis of his theory, he believed that this would not affect the results of the experiment. He started by bombarding platinum, an element with a high atomic number that was readily available, without success. He turned to aluminum, which emitted an alpha particle and produced sodium, which then decayed into magnesium by *beta particle emission*. He tried lead, without success, and then fluorine in the form of calcium fluoride, which emitted an *alpha particle* and produced nitrogen, decaying into oxygen by *beta particle emission*. In all, he induced radioactivity in 22 different elements. Fermi rapidly reported the discovery of neutron-induced radioactivity in the Italian journal *La Ricerca Scientifica* on March 23, 1934.

The Via Panisperna boys also noticed some unexplained effects. The experiment seemed to work better on a wooden table than on a marble tabletop. Fermi remembered that Joliot-Curie and Chadwick had noted that paraffin wax was effective at slowing *neutrons*, so he decided to try that. When *neutrons* were passed through paraffin wax, they induced a hundred times as much radioactivity in silver compared with when it was bombarded without the paraffin. Fermi guessed that this was due to the hydrogen atoms in the paraffin. Those in wood similarly explained the difference between the wooden and the marble tabletops. This was confirmed by repeating the effect with water. He concluded that collisions with *hydrogen atoms* slowed the *neutrons*. The lower the atomic number of the nucleus it collides with, the more energy a neutron loses per collision, and therefore the fewer collisions that are required to slow a *neutron* down by a given amount. Fermi realized that this induced more radioactivity because slow *neutrons* were more easily captured than fast ones. He developed a diffusion equation to describe this, which became known as the Fermi age equation.

In 1938, Fermi received the Nobel Prize in Physics at the age of 37 for his "demonstrations of the existence of new radioactive elements produced by neutron irradiation, and for his related discovery of nuclear reactions brought about by slow neutrons".

After Fermi received the prize in Stockholm, he did not return home to Italy but rather continued to New York City with his family in December 1938, to escape new Italian racial laws that affected his Jewish wife, Laura Capon, where they applied for permanent residency. Fermi arrived in New York City on January 2, 1939. He was immediately offered positions at five universities, and accepted one at Columbia University, where he had already given summer lectures in 1936.

In December 1938, he received the news that the German chemists Otto Hahn and Fritz Strassmann had detected the element barium after bombarding uranium with neutrons, which Lise Meitner and her nephew Otto Frisch correctly interpreted as the result of nuclear fission. Frisch confirmed this experimentally on January 13, 1939. Fermi had dismissed the possibility of fission on the basis of his calculations, but he had not taken into account the binding energy that would appear when a nuclide with an odd number of neutrons absorbed an extra neutron. For Fermi, the news came as a profound embarrassment, as the transuranic elements that he had partly been awarded the Nobel Prize for discovering had not been transuranic elements at all, but fission products. He added a footnote to this effect to his Nobel Prize acceptance speech.

The scientists at Columbia decided that they should try to detect the energy released in the nuclear fission of uranium when bombarded by neutrons. On January 25, 1939, in the basement of Pupin Hall at Columbia, an experimental team including Fermi conducted the first nuclear fission experiment in the United States. French scientists Hans von Halban, Lew Kowarski, and Frédéric Joliot-Curie had demonstrated that uranium bombarded by neutrons emitted more neutrons than it absorbed, suggesting the possibility of a chain reaction. Fermi and Anderson did so too a few weeks later. Leó Szilárd obtained 200 kilograms (440 lb) of uranium oxide from Canadian radium producer Eldorado Gold Mines Limited, allowing Fermi and Anderson to conduct experiments with fission on a much larger scale. Fermi and Szilárd collaborated on the design of a device to achieve a self-sustaining nuclear reaction—a nuclear reactor.

Owing to the rate of absorption of neutrons by the hydrogen in water, it was unlikely that a self-sustaining reaction could be achieved with natural uranium and water as a neutron moderator. Fermi suggested, based on his work with neutrons, that the reaction could be achieved with uranium oxide blocks and graphite as a moderator instead of water. This would reduce the neutron capture rate, and in theory make a self-sustaining chain reaction possible. Szilárd came up with a workable design: a pile of uranium oxide blocks interspersed with graphite bricks. Szilárd, Anderson, and Fermi published a paper on *"Neutron Production in Uranium"*.

Fermi was among the first to warn military leaders about the potential impact of nuclear energy, giving a lecture on the subject at the Navy Department on March 18, 1939. The response fell short of what he had hoped for, although the Navy agreed to provide $1,500 towards further research at Columbia. Later that year, Szilárd, Eugene Wigner, and Edward Teller sent the letter signed by Einstein to US president Franklin D. Roosevelt, warning that Nazi Germany was likely to build an atomic bomb. In response, Roosevelt formed the Advisory Committee on Uranium to investigate the matter.

The Advisory Committee on Uranium provided money for Fermi to buy graphite, and he built a pile of graphite bricks on the seventh floor of the Pupin Hall laboratory. By August 1941, he had six tons of uranium oxide and thirty tons of graphite, which he used to build a still larger pile in Schermerhorn Hall at Columbia.

The S-1 Section of the Office of Scientific Research and Development, as the Advisory Committee on Uranium was now known, met on December 18, 1941, with the US now engaged in World War II, making its work urgent. Most of the effort sponsored by the committee had been directed at producing enriched uranium, but Committee member Arthur Compton determined that a feasible alternative was plutonium, which could be mass-produced in nuclear reactors by the end of 1944. He decided to concentrate the plutonium work at the University of Chicago. Fermi reluctantly moved, and his team became part of the new Metallurgical Laboratory there.

The possible results of a self-sustaining nuclear reaction were unknown, so it seemed inadvisable to build the first nuclear reactor on the University of Chicago campus in the middle of the city. Compton found a location in the Argonne Woods Forest Preserve, about 20 miles (32 km) from Chicago. Stone & Webster was contracted to develop the site, but the work was halted by an industrial dispute. Fermi then persuaded Compton that he could build the reactor in the squash court under the stands of the University of Chicago's Stagg Field. Construction of the pile began on 6 November 1942, and Chicago Pile-1 went critical on December 2. The shape of the pile was intended to be roughly spherical, but as work proceeded Fermi calculated that criticality could be achieved without finishing the entire pile as planned.

This experiment was a landmark in the quest for energy, and it was typical of Fermi's approach. Every step was carefully planned, and every calculation was meticulously done. When the first self-sustained nuclear chain reaction was achieved, Compton made a coded phone call to James B. Conant, the chairman of the National Defense Research Committee.

To continue the research where it would not pose a public health hazard, the reactor was disassembled and moved to the Argonne Woods site. There Fermi directed experiments on nuclear reactions, reveling in the opportunities provided by the reactor's abundant production of free neutrons. The laboratory soon branched out from physics and engineering into using the reactor for biological and medical research. Initially, Argonne was run by Fermi as part of the University of Chicago, but it became a separate entity with Fermi as its director in May 1944.

When the air-cooled X-10 Graphite Reactor at Oak Ridge went critical on November 4, 1943, Fermi was on hand just in case something went wrong. The technicians woke him early so that he could see it happen. Getting X-10 operational was another milestone in the plutonium project. It provided data on reactor design, training for DuPont staff in reactor operation, and produced the first small quantities of reactor-bred plutonium. Fermi became an American citizen in July 1944, the earliest date the law allowed.

In September 1944, Fermi inserted the first uranium fuel slug into the B Reactor at the Hanford Site, the production reactor designed to breed plutonium in large quantities. Like X-10, it had been designed by Fermi's team at the Metallurgical Laboratory and built by DuPont, but it was much larger and was water-cooled. Over the next few days, 838 tubes were loaded, and the reactor went critical. Shortly after midnight on September 27, the operators began to withdraw the control rods to initiate production. At first, all appeared to be well, but around 03:00, the power level started to drop and by 06:30 the reactor had shut down completely. The Army and DuPont turned to Fermi's team for answers. The cooling water was investigated to see if there was a leak or contamination. The next day the reactor suddenly started up again, only to shut down once more a few hours later. The problem was traced to neutron poisoning from xenon-135 or Xe-135, a fission product with a half-life of 9.1 to 9.4 hours. Fermi and John Wheeler both deduced that Xe-135 was responsible for absorbing neutrons in the reactor, thereby sabotaging the fission process. DuPont had deviated from the Metallurgical Laboratory's original design in which the reactor had 1,500 tubes arranged in a circle, and had added 504 tubes to fill in the corners. The scientists had originally considered this over-engineering a waste of time and money, but Fermi realized that if all 2,004 tubes were loaded, the reactor could reach the required power level and efficiently produce plutonium.

In April 1943, Fermi raised with Robert Oppenheimer the possibility of using the radioactive byproducts from enrichment to contaminate the German food supply. The background was fear that the German atomic bomb project was already at an advanced stage, and Fermi was also skeptical at the time that an atomic bomb could be developed quickly enough. Oppenheimer discussed the "promising" proposal with Edward Teller, who suggested the use of strontium-90. James B. Conant and Leslie Groves were also briefed, but Oppenheimer wanted to proceed with the plan only if enough food could be contaminated with the weapon to kill half a million people.

In mid-1944, Oppenheimer persuaded Fermi to join his Project Y at Los Alamos, New Mexico. Arriving in September, Fermi was appointed an associate director of the laboratory, with broad responsibility for nuclear and theoretical physics, and was placed in charge of F Division, which was named after him. F Division had four branches: F-1 Super and General Theory under Teller, which investigated the "Super" (thermonuclear) bomb; F-2 Water Boiler under L. D. P. King, which looked after the "water boiler" aqueous homogeneous research reactor; F-3 Super Experimentation under Egon Bretscher; and F-4 Fission Studies under Anderson. Fermi observed the Trinity test on July 16, 1945 and conducted an experiment to estimate the bomb's yield by dropping strips of paper into the blast wave. He paced off the distance they were blown by the explosion, and calculated the yield as ten kilotons of TNT; the actual yield was about 18.6 kilotons.

Along with Oppenheimer, Compton, and Ernest Lawrence, Fermi was part of the scientific panel that advised the Interim Committee on target selection. The panel agreed with the committee that atomic bombs would be used without warning against an industrial target. Like others at the Los Alamos Laboratory, Fermi found out about the atomic bombings of Hiroshima and Nagasaki from the public address system in the technical area. Fermi did not believe that atomic bombs would deter nations from starting wars, nor did he think that the time was ripe for world government. He therefore did not join the Association of Los Alamos Scientists.

Fermi became the Charles H. Swift Distinguished Professor of Physics at the University of Chicago on July 1, 1945, although he did not depart the Los Alamos Laboratory with his family until December 31, 1945. The Metallurgical Laboratory became the Argonne National Laboratory on July 1, 1946, the first of the national laboratories established by the Manhattan Project. The short distance between Chicago and Argonne allowed Fermi to work at both places.

The Manhattan Project was replaced by the Atomic Energy Commission (AEC) on January 1, 1947. Fermi served on the AEC General Advisory Committee; an influential scientific committee chaired by Robert Oppenheimer. After the detonation of the first Soviet fission bomb in August 1949, Fermi, along with Isidor Rabi, wrote a strongly worded report for the committee, opposing the development of a hydrogen bomb on moral and technical grounds. Nonetheless, Fermi continued to participate in work on the hydrogen bomb at Los Alamos as a consultant. Fermi was among the scientists who testified on Oppenheimer's behalf at the Oppenheimer security hearing in 1954 that resulted in the denial of Oppenheimer's security clearance.

In his later years, Fermi continued teaching at the University of Chicago, where he was a founder of what later became the Enrico Fermi Institute. Fermi conducted important research in particle physics, especially related to *pions* and *muons*. He made the first predictions of *pion-nucleon resonance*, relying on statistical methods, since he reasoned that exact answers were not required when the theory was wrong anyway. In a paper coauthored with Chen Ning Yang, he speculated that *pions* might actually be composite particles. The idea was elaborated by Shoichi Sakata. It has since been supplanted by the *quark model*, in which the *pion* is made up of *quarks*, which completed Fermi's model, and vindicated his approach.

Toward the end of his life, Fermi questioned his faith in society at large to make wise choices about nuclear technology. He said:

> "Some of you may ask, what is the good of working so hard merely to collect a few facts which will bring no pleasure except to a few long-haired professors who love to collect such things and will be of no use to anybody because only few specialists at best will be able to understand them? In answer to such questions, I may venture a fairly safe prediction.

> The history of science and technology has consistently taught us that scientific advances in basic understanding have sooner or later led to technical and industrial applications that have revolutionized our way of life. It seems to me improbable that this effort to get at the structure of matter should be an exception to this rule. What is less certain, and what we all fervently hope, is that man will soon grow sufficiently adult to make good use of the powers that he acquires over nature."

Fermi underwent what was called an "exploratory" operation in Billings Memorial Hospital in October 1954, after which he returned home. Fifty days later he died of inoperable stomach cancer in his home in Chicago. He was 53. Fermi suspected working near the nuclear pile involved great risk but he pressed on because he felt the benefits outweighed the risks to his personal safety. Two of his graduate student assistants working near the pile also died of cancer.

A memorial service was held at the University of Chicago chapel, where colleagues spoke to mourn the loss of one of the world's "most brilliant and productive physicists". His body was interred at Oak Woods Cemetery.

Fermi, E. (March, 1934) Versuch einer Theorie der β-Strahlen. I. (Attempt at a theory of β rays. I.)[1]

Zeit. Phys., 88, 161–77; https://doi.org/10.1007/BF01351864; also at https://www.nssp.uni-saarland.de/lehre/Vorlesung/Kernphysik_SS19/History/Papers/ Fermi_1.pdf (German); translation by T. G. Underwood.

Rome.

[1] Cf. the preliminary communication: (1933). *La Ricerca Scientifica*, 2, 12.

With 3 illustrations.

Received January 16, 1934.

Abstract

A quantitative theory of *β decay* is proposed, in which the existence of the *neutrino* is assumed, and the emission of *electrons* and *neutrinos* from a *nucleus* in the β case is treated with a method similar to that of the emission of a quantum of light from an excited atom in radiation theory. Formulas for the lifetime and for the shape of the emitted continuous β radiation spectrum are derived and compared with experience.

[The existence of the *neutrino* had been proposed by Wolfgang Pauli in 1930 to explain the apparent violation of conservation of energy in *beta decay*. In a letter of 4 December to Lise Meitner et al., beginning, "Dear radioactive ladies and gentlemen", he proposed the existence of a hitherto unobserved neutral particle with a small mass, no greater than 1% the mass of a proton, to explain the continuous spectrum of beta decay. In 1934, Enrico Fermi incorporated the particle, which he called a *neutrino*, "little neutral one" in Fermi's native Italian, into his theory of beta decay. [Fermi, E. (1934). Radioattività indotta da bombardamento di neutroni. Ricerca Scientifica, 5, 1, 283; reprinted in Collected Papers (*Note e Memorie*), vol. 1 Italy 1921–1938 (Chicago: University of Chicago Press; Rome: Accademia Nazionale dei Lincei, 1962), 645–46; translated as "Radioactivity Induced by Neutron Bombardment. – I," in Collected Papers, 674–75. Also in Wilson, F. L. (1968). Fermi's Theory of Beta Decay. *Am. J. of Physics*, 36, 12, 1150–1160; https://doi.org/10.1119/1.1974382: (includes complete English translation of Fermi's 1934 paper published in *Zeitschrift für Physik* in 1934; https://pubs.aip.org/aapt/ajp/article-abstract/36/12/1150/1047952/Fermi-s-Theory-of-Beta-Decay?redirected From=fulltext.

In 1956, the *electron neutrino* was detected by Frederick Reines and Clyde Cowan. [F. Reines, F. & Cowan, C.L. (1956). The Neutrino. *Nature*, 178, 4531, 446–9; https://doi.org/10.1038/178446a0.] At the time it was simply referred to as the *neutrino* since there was only one known *neutrino*. Frederick Reines (March 16, 1918–August 26, 1998) and Clyde Cowan (December 6, 1919–May 24, 1974) were both American physicists. The Nobel Prize in Physics 1995 was awarded "for pioneering experimental contributions to *lepton* physics" jointly with one half to Martin L. Perl "for the discovery of the *tau lepton* [in 1975]" and with one half to Frederick Reines "for the detection of the *neutrino* [in 1956]".]

---

1. Basic assumptions of the theory.

In the attempt to build up a theory of nuclear *electrons* and *β emission*, one encounters two visions. The first is due to the continuous β radiation spectrum. If the law of conservation of energy is to remain valid, one must assume that a fraction of the energy released during the β decay escapes our previous observation possibilities. According to W. Pauli's proposal, one can assume, for example, that not only an electron but also a new particle, the so-called "neutrino" (mass of the order of magnitude or less than the electron mass; no electric charge) is emitted during β decay. In the present theory, we will use the hypothesis of the neutrino.

Another difficulty for the theory of nuclear electrons is that the current relativistic theories of light particles (electrons or neutrinos) are not able to explain in a conclusive way how such particles can be bound in orbits of nuclear dimensions.

It therefore seems more appropriate to assume with Heisenberg[2] that a nucleus consists only of heavy particles, protons and neutrons.

> [2] Heisenberg, W. (January, 1932). Über den Bau der Atomkerne. I. (About the construction of atomic nuclei. I.). *Zeit. Phys.*, 77, 1–11; https://doi.org/10.1007/BF01342433; see above.

Nevertheless, in order to understand the possibility of *β emission*, we will try to construct a theory of the emission of light particles from a nucleus by analogy to the theory of the emission of a light quantum from an excited atom in the ordinary radiation process. In radiation theory, the total number of light quanta is not a constant: light quanta are created when they are emitted by an atom and disappear when they are absorbed. By analogy to this, we want to base the *β radiation* theory on the following assumptions:

a) The total number of *electrons*, as well as *neutrinos*, is never necessarily constant. *Electrons* (or *neutrinos*) can form and disappear. This possibility has no analogy with the emergence or disappearance of a pair of an *electron* and a positron; if one interprets the positron as Dirac's "hole", this last process can indeed be understood as a quantum leap of an *electron* between a *state* with negative energy and a *state* with positive energy with conservation of the total (infinitely large) number of *electrons*.

b) The heavy particles, *neutrons* and *protons*, can be regarded as two *inner quantum states* of the heavy particle, as in Heisenberg. We formulate this by introducing an inner coordinate ρ of the visual particle, which can only take two values: $\rho = 1$ if the particle is a *neutron*; $\rho = -1$ if the particle is a *proton*.

c) The Hamilton function of the system consisting of heavy and light particles must be chosen in such a way that each transition from *neutron* to *proton* is associated with the formation of an *electron* and a *neutrino*. The reverse process, the transformation of a *proton* into a *neutron*, on the other hand, is said to be associated with the disappearance of an *electron* and a *neutrino*. It should be noted that this ensures the preservation of the *charge*.

2. The operators that occur in theory.

A mathematical formalism of the theory in accordance with these three requirements can be most easily constructed with the help of the Dirac-Jordan-Klein method[1] of the "*second quantization*".

[1] Cf. e.g. Jordan, B. P. & Klein, O. (November, 1927). Zum Mehrkörperproblem der Quantentheorie. (On the multibody problem of quantum theory.) *Zeit. Phys.*, 45, 751–65; https://doi.org/10.1007/BF01329553; Heisenberg, W. (1931). Zum Paulischen Ausschließungsprinzip. (On the Paulian exclusion principle.) *Ann. Phys.*, 10, 888; https://doi.org/10.1002/andp.19314020710.

We will therefore consider the *probability amplitudes* ψ and φ of *electrons* and *neutrinos* as well as the *complex conjugated quantities* ψ^* and φ^* as *operators*; for the description of the heavy particles, on the other hand, we will use the usual *representation* in the *configuration space*, whereby of course ρ must also be counted as a coordinate.

We first introduce two *operators* Q and Q*, which act on the functions of the two-valued variable ρ as the linear substitutions.

$$Q = \begin{vmatrix} 0 & 1 \\ 0 & 0 \end{vmatrix}; \qquad Q^* = \begin{vmatrix} 0 & 0 \\ 1 & 0 \end{vmatrix} \qquad\qquad (1)$$

It is easy to see that Q corresponds to a transition from *proton* to *neutron* and Q* to a transition from *neutron* to *proton*.

...

3. Setting up the Hamiltonian function.

The *energy* of the entire system, consisting of heavy and light particles, is the sum of the *energies* H_{heavy} of the heavy particles + H_{light} of the light particles + the *interaction energy* H between heavy and light particles.

We write the first member, considering for the time being only a single visible particle, in the form

$$H_{heavy} = (1 + \rho)/2 \, N + (1 - \rho)/2 \, P, \tag{6}$$

where N and P represent the energy operators of the *neutron* and *proton*, respectively. For $\rho = 1$ (*neutron*), (6) is indeed reduced to N; for $\rho = -1$ (*proton*) (6) reduces to P.

The *energy* H_{light} of the light particles assumes the simplest form if one takes as *quantum states* $\psi_1 \psi_2 .. \psi_s ...$ and $\varphi_1 \varphi_2 .. \varphi_u ...$ as the *stationary states* for the *electrons* or *neutrinos*. For the *electrons*, for example, the *stationary states* in the *Coulomb field* of the *nucleus* should be chosen, taking into account the *electron displacement*. For *neutrinos*, one can simply assume plane de Broglie waves, since the forces acting on the *neutrinos* do not play a significant role. Let $H_1 H_2 ... H_s...$ and $K_1 K_2 ... K_\sigma ...$ the *energies* of the *stationary states* of *electrons* and *neutrinos*; Then we have:

$$H_{light} = \Sigma s \, H_s N_s + \Sigma_\sigma K_\sigma M_\sigma. \tag{7}$$

All that remains to be written is the *energy of interaction*. First, this consists of the *Coulomb energy* between *proton* and *electrons*; in heavy *nuclei*, however, the attraction of a single *proton* plays only a subordinate role[1] and in no case contributes to the process of *β decay*.

> [1] The Coulombian effect of the numerous other *protons* must of course be taken into account as a *static field*.

For the sake of simplicity, therefore, we will not consider this link. On the other hand, we must add to the Hamiltonian function a term that satisfies the condition c) for digit 1.
...

4. The perturbation matrix.

The *theory of β decay* can be carried out with the help of the established Hamiltonian function in full analogy to radiation theory. In the latter, as is well known, the Hamiltonian function consists of the sum: *energy of the atom + energy of the pure radiation field + coupling energy*. This last link is understood as a disturbance of the other two.

146

By analogy to this, in our case the sum

$$H_{heavy} + H_{light} \tag{16}$$

as an unperturbed Hamiltonian function; in addition, there is the disturbance represented by the *coupling element* (13).

$$[H = g[Q\psi\tilde{}^*\delta\phi + Q^*\psi\tilde{}\delta\phi^*], \tag{13}$$

where ψ and ϕ are to be written as vertical matrix columns.]

The *quantum states* of the unperturbed system can be numbered as follows:

$$(\rho, n, N_1 N_2 \ldots N_S \ldots M_1 M_2 \ldots M_\sigma \ldots), \tag{17}$$

where the first number ρ takes one of the two values ± 1 and indicates whether the heavy particle is a *neutron* or a *proton*. The second number n numbers the *quantum state* of the *neutron* or *proton*. For $\rho = 1$ (*neutron*), let the corresponding *eigenfunction*

$$\upsilon_n(x), \tag{18}$$

where x represents the coordinates of the heavy particles, except for ρ. For $\rho = -1$ (*proton*) let the *eigenfunction* be

$$v_n(x). \tag{19}$$

The remaining numbers $N_1 N_2 \ldots N_S \ldots M_1 M_2 \ldots M_\sigma \ldots$ are only capable of the two values 0 and 1 and indicate whether the respective state of the *electron* or the *neutrino* is occupied.

If we now consider the general form (9)

$$[H = Q \Sigma_{s\sigma} c_{s\sigma} a_s b_\sigma + Q^* \Sigma_{s\sigma} c^*_{s\sigma} a^*_s b^*_\sigma \tag{9}$$

where $c_{s\sigma}$ and $c^*_{s\sigma}$ represent quantities that depend on the coordinates, momenta, etc. of the heavy particle.]

of the *perturbation energy*, we see that it has different elements from zero only for such *transitions* in which either the heavy particle changes from a *neutron* to a *proton state* and at the same time an *electron* and a *neutrino* are produced, or vice versa.

With the help of (1), (3), (5), (9), (18), (19) you can easily find the matrix element in question

$$H_{-1mN1N2\ldots1s\ldots M1M2\ldots1\sigma\ldots}^{1nN1N2\ldots0s\ldots M1M2\ldots0\sigma\ldots} = \pm \int v^*_m c^*_{s\sigma} \upsilon_n \, d\tau \tag{20}$$

147

where the integration must be extended over the *configuration space* of the heavy particle (except for the coordinate ρ). The ± sign means more precisely

$$(-1)^{N1 + N2 + \ldots + Ns-1 + M1 + M2 + \ldots + M\sigma-1}$$

and, by the way, will fall out of the following calculations. The opposite *transition* corresponds to a *complex conjugate matrix element*.

If you introduce the value (15) for $c^*_{s\sigma}$, you get

$$H_{-1mls\sigma}{}^{1n0s0\sigma} = \pm\, g \int v^*_m \, \upsilon_n \, \tilde{\psi}_s \, \delta \, \varphi^*_\sigma \, d\tau \qquad (21)$$

where, because of the brevity, all the constant indices have been omitted in the first term.

5. Theory of β decay.

A *β decay* consists of a process in which a *nuclear neutron* turns into a *proton* and at the same time an *electron*, which is observed as a *β ray*, and a *neutrino* are emitted with the mechanism described. In order to calculate the probability of this process, let us assume that at the time t = 0 a *neutron* exists in a *nuclear state* with *eigenfunction* $\upsilon_n(x)$ and $N_s = M_\sigma = 0$, i.e. the *electron state* s and the *neutrino state* σ are empty. Then for t = 0 is the *probability amplitude* of the *state* $(1, n, 0_s, 0_\sigma)$

$$a_1 n 0_s 0_\sigma = 1 \qquad (22)$$

and that of the *state* $(-1, m, 1_s, 1_\sigma)$ where the *neutron* has changed into a *proton* with the *eigenfunction* $v_m(x)$ under emission of an *electron* and a *neutrino*, equal to zero.

…

6. Determinants of the transition probability.

(31) indicates the probability that a *β decay* with the transition of the *electron* to the *state* s will take place during time t. As it is, this probability is proportional to the time t (t has been assumed to be small in terms of lifetime); the coefficient of t indicates the *transition probability* for the process described.

…

7. The mass of the neutrino.

The *transition probability* (32)

$$[Ps = 8\pi^3 g^2/h^4 \mid \int v^*_m \, u_n \, d\tau \mid^2 \, p_\sigma^2/\upsilon_\sigma \, (\tilde{\psi}_s \, \psi_s - \mu c^2/K_\sigma \, \tilde{\psi}_s \, \beta \, \psi_s). \qquad (32)]$$

148

determines the shape of the continuous β spectrum. We first want to discuss how this form depends on the *rest mass* μ the *neutrino* in order to determine this constant by comparing it with the empirical curves. The *mass* μ is contained in the factor p_σ^2/v_σ. The dependence of the shape of the *energy distribution curve* on μ is most pronounced near the end point of the *distribution curve*. If E_0 is the *limiting energy* of the *β rays*, it is not difficult to see that the *distribution curve* for *energies* E in the vicinity of E_0 behaves as follows, except for one factor independent of E.

$$p_\sigma^2/v_\sigma = 1/c^3 \ (\mu c^2 + E_0 - E) \ \sqrt{\{(E_0 - E)^2 + 2\mu c^2(E_0 - E)\}} \tag{36}$$

...

8. Lifetime and shape of the distribution curve for "permitted" transitions.

From (39) one can derive a formula which indicates how many *β transitions* take place in the unit of time for which the β particle receives *momentum* between mcη and mc (η + dη). To do this, one must derive a formula for the sum of $\psi^\sim{}_s\psi_s$ at the location of the *nucleus* over all *quantum states* of the respective interval in the continuous spectrum.

It should be noted that the *relativistic eigenfunctions* in the *Coulomb field* for the *states* with j = ½ ($^2s_{1/2}$ and $^2p_{1/2}$) for r = 0 become infinitely large. Now, however, the nuclear attraction for the *electrons* obeys *Coulomb's law* only up to r > ρ, where ρ here means the *nuclear radius*. A rough calculation shows that if one makes plausible assumptions about the course of the *electric field* within the *nucleus*, the value of $\psi^\sim{}_s\psi_s$ at the center has a value that is very close to the value that $\psi^\sim{}_s\psi_s$ would assume in the case of *Coulomb's law* at the distance ρ from the center.

...

The reciprocal *lifetime* is obtained from (44) by integrating η = 0 to η = η_0; one finds:

$$1/\tau = 1.75 \ . \ 10^{95} \ g^2 \ | \int v^*{}_m \ u_n \ d\tau \ |^2 \ F(\eta_0), \tag{45}$$

where $F(\eta_0) = \dots$. $\tag{46}$

9. The forbidden transitions.

Before we move on to the comparison with experience, let's discuss some characteristics of the forbidden *β transitions*. As already mentioned, a *transition* is forbidden if the corresponding matrix element (35) disappears. If the *representation* of the *nucleus* with individual *quantum states* of the *neutrons* and *protons* is a good approximation, $Q^*{}_{mn}$ always disappears for symmetry reasons, if not

$$i = i', \tag{47}$$

149

where i and i' represent the *momentum* moments (in units $h/2\pi$) of the *neutron state* υ_n and the *proton state* υ_m.

The *selection rule* (47), if the individual *states* are not a good approximation, corresponds to the more general

$$l = l', \tag{48}$$

where l and l' mean the *momentum* moments of the *nucleus* before and after the *β decay*.
...

10. Comparison with experience.

Formula (45)
$$[1/\tau = 1.75 \cdot 10^{95} \, g^2 \, | \int \upsilon^*_m \, u_n \, d\tau \, |^2 \, F(\eta_0), \tag{45}]$$
gives a relationship between the maximum *momentum* of the emitted *β-rays* and the *lifetime* of the β-radiating substance: In this relationship, an unknown element still occurs, namely the integral

$$\int v^*_m \, \upsilon_n \, d\tau, \tag{50}$$

for the evaluation of which a knowledge of the *eigenfunctions* of the *proton* and the *neutron* in the nucleus would be necessary. However, in the case of *permitted transitions*, (50) is of order 1. So, you can expect that the product

$$\tau F (\eta_0) \tag{51}$$

has the same order of magnitude for all *permitted transitions*. However, if the *transition* in question is *prohibited*, the *lifetime* is about 100 times longer than in the normal case and the product (51) is also correspondingly larger. Table 2 lists the products (51) for the radioactive elements for which sufficient data are available on the continuous β spectrum.
...

To sum up, it should be said that the theory as given here is in agreement with the experimental data, which are not always particularly accurate. If, moreover, a closer comparison of theory and experience should lead to contradictions, it would still be possible to modify the theory without touching its conceptual foundations. One could keep equation (9) and make a different choice of the $c_{s\sigma}$. In particular, this could lead to a modification of the selection rule (48) and result in a different form of the *energy distribution curve* and the dependence of the *lifetime* on the *maximum energy*. Whether such a change will be necessary, however, can only be shown by a further development of the theory and possibly also by a tightening of the experimental data.

Enrico Fermi – 1938 Nobel Prize in Physics.

The Nobel Prize in Physics 1938 was awarded to Enrico Fermi "for his demonstrations of the existence of new radioactive elements produced by *neutron* irradiation, and for his related discovery of nuclear reactions brought about by slow *neutrons*".

Discovered in 1932, the *neutron* proved to be a powerful new tool for studying atoms. When Enrico Fermi irradiated heavy atoms with *neutrons*, these were captured by the atomic nuclei, creating new and often radioactive isotopes. In 1934, Fermi and his colleagues discovered that when *neutrons* are slowed down, e.g. by paraffin shielding, the *interaction rate* with *nuclei* increases. This revelation led to the discovery of many hitherto-unknown radioactive isotopes. [Enrico Fermi – Facts. NobelPrize.org. https://www.nobelprize.org/prizes/physics/1938/fermi/facts/.]

Hideki Yukawa (January 23, 1907–September 8, 1981).

Yukawa was a Japanese theoretical physicist and the first Japanese Nobel laureate for his prediction of the *pi meson*, or *pion*.

He was born as Hideki Ogawa in Tokyo and grew up in Kyoto with two older brothers, two older sisters, and two younger brothers. His father, for a time, considered sending him to technical college rather than university since he was "not as outstanding a student as his older brothers". However, when his father broached the idea with his middle school principal, the principal praised his "high potential" in mathematics and offered to adopt Ogawa himself in order to keep him on a scholarly career. At that, his father relented.

Ogawa decided against becoming a mathematician when in high school; his teacher marked his exam answer as incorrect when Ogawa proved a theorem but in a different manner than the teacher expected. He decided against a career in experimental physics in college when he demonstrated clumsiness in glassblowing, a requirement for experiments in spectroscopy.

In 1929, after receiving his bachelor's degree at Kyoto Imperial University, he stayed on as a lecturer for four years. After graduation, he was interested in theoretical physics, particularly in the theory of elementary particles.

In 1932, he married Sumi Yukawa. In accordance with Japanese customs of the time, since he came from a family with many sons but his father-in-law Genyo had none, he was adopted by Genyo and changed his family name from Ogawa to Yukawa. The couple had two sons, Harumi and Takaaki. In 1933 he became a lecturer at Osaka Imperial University, at 26 years old.

In 1935 he published his *theory of mesons*, which explained the interaction between *protons* and *neutrons* at Osaka Imperial University, and was a major influence on research into elementary particles. [Yukawa, H. (1935). On the Interaction of Elementary Particles. *Proceedings of the Physico-Mathematical Society of Japan*, 17, 48. See below.]

In 1938, he received his Ph.D. degree at Osaka Imperial University for his predictions regarding the existence of *mesons* and his theoretical work on the nature of nuclear forces. In 1940 he became a professor in Kyoto Imperial University.

In 1949 he became a professor at Columbia University, the same year he received the 1949 Nobel Prize in Physics "for his prediction of the existence of *mesons* on the basis of theoretical work on nuclear forces", after the discovery by Cecil Frank Powell, Giuseppe Occhialini and César Lattes of Yukawa's predicted *pi meson* in 1947.

Yukawa was an editor of *Progress of Theoretical Physics*, and published the books *Introduction to Quantum Mechanics* (1946) and *Introduction to the Theory of Elementary Particles* (1948).

Yukawa retired from Kyoto University in 1970 as a Professor Emeritus. Owing to increasing infirmity, in his final years he appeared in public in a wheelchair. He died at his home in Sakyo-ku, Kyoto, on September 8, 1981 from pneumonia and heart failure, aged 74. His tomb is in Higashiyama-ku, Kyoto.

Yukawa, H. (1935). On the Interaction of Elementary Particles.

Proceedings of the Physico-Mathematical Society of Japan, 17, 48; translation in (January, 1955). *Progr. Theoret. Phys. Suppl.* 1, 1–10, https://doi.org/10.1143/PTPS.1.1.

Received November 30, 1934.

The existence of *mesons* was predicted by Hideki Yukawa's 1935 *theory of mesons* that postulated the particle [referred to as the *Yukawa particle*] as mediating the *nuclear force*.

[A *meson* is a type of hadronic subatomic particle composed of *an equal number of quarks and antiquarks, usually one of each, bound together by the strong interaction*. Because *mesons* are composed of *quark* sub-particles, they have a meaningful physical size, a diameter of roughly one femtometre (10^{-15} m), which is about 0.6 times the size of a *proton* or *neutron*. *All mesons are unstable*, with the longest-lived lasting for only a few tenths of a nanosecond. Heavier *mesons* decay to lighter *mesons* and ultimately to stable *electrons*, *neutrinos* and *photons*.

The *Yukawa particle* is now identified as the *pion* (*pi meson*).

Pion (or pi meson) [*composite particle*]

A *pion* or *pi meson*, is any of three subatomic particles: π^0, π^+, and π^-. Each *pion* consists of a *quark* and an *antiquark* and is therefore a *meson*. *Pions* are the lightest *mesons* and, more generally, the lightest *hadrons*. They are unstable, with the *charged pions*, π^+ and π^-, decaying after a mean lifetime of 26.033 nanoseconds (2.6033×10^{-8} seconds), and the *neutral pion* π^0 decaying after a much shorter lifetime of 85 attoseconds (8.5×10^{-17} seconds). *Charged pions* most often decay into *muons* and *muon neutrinos*, while *neutral pions* generally decay into *gamma rays*.

The exchange of virtual *pions*, along with *vector, rho* and *omega mesons*, provides an explanation for the residual *strong force* between *nucleons*. *Pions* are not produced in radioactive decay, but commonly are in high-energy collisions between *hadrons*. *Pions* also result from some *matter–antimatter annihilation* events. All types of *pions* are also produced in natural processes when high-energy cosmic-ray *protons* and other hadronic cosmic-ray components interact with matter in Earth's atmosphere. In 2013, the detection of characteristic *gamma rays* originating from the decay of *neutral pions* in two supernova remnants has shown that *pions* are produced copiously after supernovas, most probably in conjunction with production of high-energy protons that are detected on Earth as cosmic rays.]

The *interaction* between *elementary* particles can be described by means of a *field of force*, just as the *interaction* between the *charged* particles is described by the *electromagnetic field*. In the *quantum theory* this field should be accompanied by a new sort of *quantum*, just as the *electromagnetic field* is accompanied by the *photon*. In this paper the possible natures of this *field* and the *quantum* accompanying it will be discussed briefly and also their bearing on the nuclear structure will be considered.

---

§ 1. Introduction

At the present stage of the quantum theory little is known about the nature of *interaction* of elementary particles, Heisenberg considered the interaction of *"Platzwechsel"* between the *neutron* and the *proton* to be of importance to the nuclear structure[1].

[1] Heisenberg, W. (January, 1932). Über den Bau der Atomkerne. I. (About the construction of atomic nuclei. I.); (March, 1932). Über den Bau der Atomkerne. II. (About the construction of atomic nuclei. II.); (September, 1933). Über den Bau der Atomkerne. III. *Zeit. Phys.*, 77, 1–11; https://doi.org/10.1007/ BF01342433; *Ibid.*, 78, 156–64; https://doi.org/10.1007/ BF01337585; *Ibid.*, 80, 587–96; https://doi.org/10.1007/ BF01335696; see below. We shall denote the first item by I.

Recently Fermi treated the problem of *β-disintegration* on the hypothesis of *"neutrino"*[2].

[2] Fermi, E. (March, 1934). Versuch einer Theorie der β-Strahlen. I. (Attempt at a theory of β rays. I .) *Zeit. Phys.*, 88, 161–77 (1934). https://doi.org/10.1007/BF01351864; see above.

According to this theory, the *neutron* and the *proton* can interact by emitting and absorbing a pair of *neutrino* and *electron*. Unfortunately, the *interaction energy* calculated on such assumption is much too small to account for the binding energies of *neutrons* and *protons* in the *nucleus*[3].

[3] Tamm, I. (June, 1934). Exchange Forces between Neutrons and Protons, and Fermi's Theory. *Nature*, 133, 981; https://doi.org/10.1038/133981a0; Iwanenko, D. (June, 1934). Interaction of Neutrons and Protons. *Idem.*, 981–2; https://doi.org/10.1038/133981b0.

To remove this defect, it seems natural to modify the theory of Heisenberg and Fermi in the following way. The transition of a heavy particle from *neutron* state to *proton* state is not always accompanied by the emission of light particles, i.e., a *neutrino* and an *electron*, but the energy liberated by the transition is taken up sometimes by another heavy particle, which in turn will be transformed from *proton* state into *neutron* state. If the probability of occurrence of the latter process is much larger than that of the former, the *interaction* between the *neutron* and the *proton* will be much larger than in the case of Fermi, whereas the probability of emission of light particles is not affected essentially.

Now such *interaction* between the *elementary* particles can be described by means of a *field of force*, just as the *interaction* between the *charged* particles is described by the *electromagnetic field*. The above considerations show that the *interaction* of heavy particles with this *field* is much larger than that of light particles with it.

In the *quantum theory* this field should be accompanied by a new sort of *quantum*, just as the *electromagnetic field* is accompanied by the *photon*.

In this paper the possible natures of this *field* and the *quantum* accompanying it will be discussed briefly and also their bearing on the nuclear structure will be considered.

Besides such an *exchange force* and the ordinary *electric* and *magnetic forces* there may be other forces between the elementary particles, but we disregard the   latter   for   the moment.

Fuller account will be made in the next paper.

§ 2. Field Describing the Interaction

In analogy with the *scalar potential* of the *electromagnetic field*, a function U(x,y,z,t) is introduced to describe the *field* between the *neutron* and the *proton*. This function will satisfy an equation similar to the *wave equation* for the *electromagnetic potential*.

Now the equation

$$\{\Delta - 1/c^2 \, \partial^2/\partial t^2\} \, U = 0 \tag{1}$$
$$[\text{where } \Delta = \partial^2/\partial x^2 + \partial^2/\partial y^2 + \partial^2/\partial z^2],$$

has only a static solution with central symmetry 1/r, except the additive and the multiplicative constants. The *potential of force* between the *neutron* and *proton* should, however, not be of Coulomb type, but decrease more rapidly with distance. It can be expressed, for example by

$$+ \text{ or } - g^2 e^{-\lambda r}/r, \tag{2}$$

where g is a constant with the dimension of *electric charge*, i.e., cm.$^{3/2}$ sec.$^{-1}$ gr.$^{1/2}$ and λ with the dimension cm.$^{-1}$.

Since this function is a *static solution* with a *central symmetry* of the *wave equation*

156

$$\{\Delta - 1/c^2\, \partial^2/\partial t^2 - \lambda^2\}\, U = 0 \qquad\qquad (3)$$

let this equation be assumed to be the correct equation for U in vacuum. In the presence of the heavy particles, the U–field interacts with them and causes the transition from *neutron* state to *proton* state.

Now, if we introduce the matrices[4]

[4] Heisenberg, *loc. cit.* I.

$$\tau_1 = \begin{pmatrix} 0 & 1 \\ 1 & 0 \end{pmatrix}, \qquad \tau_2 = \begin{pmatrix} 0 & -i \\ i & 0 \end{pmatrix}, \qquad \tau_3 = \begin{pmatrix} 1 & 0 \\ 0 & -1 \end{pmatrix}$$

and denote the *neutron* state and the *proton* state by $\tau_3 = 1$ and $\tau_3 = -1$ respectively, the *wave equation* is given by

$$\{\Delta - 1/c^2\, \partial^2/\partial t^2 - \lambda^2\}\, U = -4\pi g \Psi\tilde{}(\tau_1 - i\tau_2)/2\ \Psi, \qquad\qquad (4)$$

where Ψ denoted the *wave function* of the heavy particles, being a function of time, position, *spin* as well as τ'_3, which takes the value either 1 or -1.

Next, the *conjugate complex function* $U\tilde{}(x, y, z, t)$, satisfying the equation

$$\{\Delta - 1/c^2\, \partial^2/\partial t^2 - \lambda^2\}\, U\tilde{} = -4\pi g \Psi\tilde{}(\tau_1 + i\tau_2)/2\ \Psi, \qquad\qquad (5)$$

is introduced, corresponding to the inverse transition from *proton* to *neutron* state. Similar equation will hold for the *vector function*, which is the analogue of the *vector potential* of the *electromagnetic field*. However, we disregard it for the moment, *as there is no correct relativistic theory for the heavy particles*. Hence the simple *non–relativistic wave equation* neglecting *spin* will be used for the heavy particle, it the following way

$$[h^2/4\ \{(1 + \tau_3)/M_N + (1 - \tau_3)/M_P\}\ \Delta + ih\ \partial/\partial t - (1 + \tau_3)/2\ M_N c^2 - (1 - \tau_3)/2\ M_P c^2$$
$$-g\ \{U\tilde{}\ (\tau_1 - i\tau_2)/2 + U\ (\tau_1 + i\tau_2)/2\}]\ \Psi = 0, \qquad\qquad (6)$$

where h is Planck's constant divided by 2π and M_N, M_P are the *masses* of the *neutron* and the *proton* respectively. The reason for taking the negative sign in front of g will be mentioned later.

The equation (6) corresponds to the Hamiltonian

$$H = \{(1 + \tau_3)/4M_N + (1 - \tau_3)/4M_P\}\ \vec{p}^{\,2} + (1 + \tau_3)/2\ M_N c^2 + (1 - \tau_3)/2\ M_P c^2$$
$$+ g\ \{U\tilde{}\ (\tau_1 - i\tau_2)/2 + U\ (\tau_1 + i\tau_2)/2\} \qquad\qquad (7)$$

157

where $\vec{p}$ is the *momentum* of the particle. If we put $M_N c^2 - M_P c^2 = D$ and $M_N + M_P = 2M$, the equation (7) becomes approximately

$$H = \vec{p}^2/2M + g/2 \{U^{\sim} (\tau_1 - i\tau_2) + U (\tau_1 + i\tau_2)\} + D/2 \; \tau_3, \qquad (8)$$

where the constant term Mc^2 omitted.

Now consider two heavy particles at the point (x_1, y_1, z_1) and (x_2, y_2, z_2) respectively and assume their relative velocity to be small. The field at (x_1, y_1, z_1) due to the particle at (x_2, y_2, z_2) are, from (4) and (5),

$$U(x_1, y_1, z_1) = g \; e^{-\lambda\tau 12}/\tau_{12} \; (\tau_1^{(1)} - i\tau_2^{(1)})/2 \qquad (9)$$

and

$$U^{\sim}(x_1, y_1, z_1) = g \; e^{-\lambda\tau 12}/\tau_{12} \; (\tau_1^{(2)} - i\tau_2^{(2)})/2,$$

where $(\tau_1^{(1)}, \tau_2^{(1)}, \tau_3^{(1)})$ and $(\tau_1^{(2)}, \tau_2^{(2)}, \tau_3^{(2)})$ are the matrices relating to the first and the second particles respectively, and τ_{12} is the distance between them.

Hence the Hamiltonian for the system is given, in the absence of the external fields by,

$$H = \ldots = \vec{p_1}^2/2M + \vec{p_2}^2/2M + g^2/2 \; \{\tau_1^{(1)} \tau_1^{(2)} + \tau_2^{(1)} \tau_2^{(2)}\} \; e^{-\lambda\tau 12}/\tau_{12}$$
$$+ \{\tau_3^{(1)} + \tau_3^{(2)}\} \; D, \qquad (10)$$

where $\vec{p_1}$, $\vec{p_2}$ are the *momenta* of the particles.

This Hamiltonian is equivalent to Heisenberg's Hamiltonian (1)[5], if we take for "*Platzwechselintegral*"

[5] Heisenberg, I.

$$J(\tau) = -g^2 \; e^{-\lambda r}/r, \qquad (11)$$

Except that the interaction between the *neutrons* and the *electrostatic* repulsion between the *protons* are not taken into account. Heisenberg took the positive sign for $J(r)$, so that the *spin* of the lowest *energy state* of H^2 was 0, whereas in our case, owing to the negative sign in front of g^2, the lowest *energy state* has the *spin* 1, which is required from the experiment.

Two constants g and λ appearing in the above equations should be determined by comparison with experiment. For example, using the Hamiltonian (10) for heavy particles, we can calculate the *mass* defect of H^2 and the *probability of scattering* of a *neutron* by a *proton* provided that the relative velocity is small compared with the light velocity[6].

[6] These calculations were made previously, according to the theory of Heisenberg, by Mr. Tomonaga, to whom the writer owes much. A little modification is needed in our case. Detailed accounts will be made in the next paper.

Rough estimation shows that the calculated values agree with the experimental results, if we take for λ the value between 10^{12} cm^{-1}. And 10^{13} cm^{-1}. and for g a few times of the elementary *charge* e, although no direct relation between g and e was suggested in the above considerations.

§ 3. Nature of the Quanta Accompanying the Field

The *U–field* considered above should be quantized according to the general method of the quantum theory. Since the *neutron* and the *proton* both obey fermi's statistics, the *quanta* accompanying the *U–field* should obey Bose's statistics and the quantization can be carried out the lines similar to that of the *electromagnetic field*.

The *law of conservation of the electric charge* demands that the *quantum* should have charge either $+$ e or $-$ e. The *field* quantity U corresponds to the *operator* which increases the number of *negatively charged quanta* and decreases the number of *positively charged quanta* by one respectively. U$^{\sim}$, which is the *complex conjugate* of U, corresponds to the *inverse operator*.

Next, denoting

$$p_x = -ih\,\partial/\partial x \text{ , etc., } W = ih\,\partial/\partial t \text{ ,}$$
$$m_U c = \lambda h,$$

the *wave equation* for U in free space can be written in the form

$$\{p_x{}^2 + p_y{}^2 + p_z{}^2 - W^2/c^2 + m_U c^2\}\, U = 0, \tag{12}$$

so that the *quantum* accompanying the *field* has the proper *mass* $m_U = \lambda h/c$.

Assuming $\lambda = 5 \times 10^{12}$ cm^{-1}., we obtain for m_U a value 2×10^2 times as large as the *electron mass*. As such a quantum with large *mass* and positive or negative *charge* has never been found by the experiment, the above theory seems to be on a wrong line. We can show, however, that, in the ordinary nuclear transformation, such a *quantum* cannot be emitted into outer space.

Let us consider, for example, the *transition* from a *neutron* state of *energy* W_N to a *proton* state of *energy* W_P, both of which include the proper energies. These states can be expressed by the *wave function*

$$\Psi_N(x,y,z,t, 1) = u(x,y,z)e^{-iW_N t/h}, \qquad \Psi_N(x,y,z,t, -1) = 0$$

and

$$\Psi_P(x,y,z,t, 1) = 0, \qquad \Psi_P(x,y,z,t, -1) = v(x,y,z)e^{-iW_P t/h},$$

so that, on the right-hand side of the equation (4), the term

$$-4\pi g \tilde{v} u e^{-it(W_N - W_P)/h}$$

appears.

Putting $U = U'(x,y,z)e^{i\omega t}$, we have from (4)

$$\{\Delta - (\lambda^2 - \omega^2/c^2)\} U' = -4\pi g \tilde{v} u, \tag{13}$$

where $\omega = (W_N - W_P)/h$. Integrating this, we obtain a solution

$$U'(\vec{r}) = g \iiint e^{-\mu|r-r'|}/|\vec{r} - \vec{r}'| \, \tilde{v}(\vec{r}')u(\vec{r}')dv', \tag{14}$$

where $\mu = \sqrt{(\lambda^2 - \omega^2/c^2)}$

If $\lambda > |\omega|/c$ or $m_U c^2 > |W_N - W_P|$, μ is real and the function $J(r)$ of Heisenberg has the form $-g^2 e^{-\mu r}/r$, in which μ, however, depends on $|W_N - W_P|$, becoming smaller and smaller as the latter approaches $m_U c^2$. This means that the *range of interaction* between a *neutron* and a *proton* increases as $|W_N - W_P|$ increases.

Now the *scattering* (elastic or inelastic) of a *neutron* by a *nucleus* can be considered as the result of the following double process: the *neutron* falls into a *proton* level in the *nucleus* and a *proton* in the latter jumps to a *neutron* state of positive *kinetic energy*, the *total energy* being conserved throughout the process. The above argument, then shows that the *probability of scattering* may in some cases increase with the velocity of the *neutron*.

According to the experiment of Bonner[7], the *collision cross section* of the *neutron* increases, in fact, with the velocity in the case of lead whereas it decreases in the case of carbon and hydrogen, the rate of decrease being slower in the former than the latter.

[7] Bonner, T. W. (1934). *Phys. Rev.*, 45, 606.

The origin of this effect is not clear, but the above considerations do not, at least, contradict it. For, if the *binding energy* of the *proton* in the *nucleus* becomes comparable with $m_U c^2$, the range of *interaction* of the *neutron* with the former will increase considerably with the velocity of the *neutron*, so that the *cross section* will decrease slower in such case than in the case of hydrogen, i.e., free *proton*. Now the *binding energy* of the *proton* in C^{12}, which is estimated from the difference of *masses* of C^{12} and B^{11}, is

$$12, 0036 - 11, 0110 = 0, 9926.$$

This corresponds to a *binding energy* 0.0152 in *mass* unit, being thirty times the *electron mass*. Thus, in the case of carbon we can expect the effect observed by Bonner. The arguments are only tentative, other explanations being, of course, not excluded.

Next if $\lambda < |\omega|/c$ or $m_U c^2 <| W_N - W_P |$, μ becomes pure imaginary and U expresses spherical undamped wave, implying that a *quantum* with *energy* greater than $m_U c^2$ can be emitted in outer space by the transition of the heavy particle from *neutron* state to *proton* state, provided that $| W_N - W_P |> m_U c^2$.

The velocity of *U–wave* is greater but the group velocity is smaller than the light velocity c, as in the case of the *electron wave*.

The reason why such massive *quanta*, if they ever exist, are not yet discovered may be ascribed to the fact that the *mass* m_U is so large that condition $| W_N - W_P |> m_U c^2$ is not fulfilled in ordinary nuclear transformations.

§ 4. Theory of β – Disintegration

Hitherto we have considered only the *interaction* of U–quanta with heavy particles. Now, according to our theory, the quantum emitted when a heavy particle jumps from a *neutron* state to a *proton* state can be absorbed by a light particle which will then in consequence of energy absorption rise from a *neutrino* state of negative *energy* to an *electron* state of positive *energy*. Thus, an *anti–neutrino* and an *electron* are emitted simultaneously from the *nucleus*. Such *intervention* of a massive *quantum* does not alter essentially the probability of *β-disintegration*, which has been calculated on the hypothesis of direct coupling of a heavy particle and a light particle, just as, in the theory of *internal conversion of γ-ray*, the *intervention* of the *proton* does not affect the final result[8].

[8] Taylor, H. A. & Mott, N. F. (1932). *Proc. Roy. Soc. A*, 138, 665.

Our theory, therefore, does not differ essentially from Fermi's theory.

Fermi considered that an *electron* and a *neutrino* are emitted simultaneously from the radioactive *nucleus*, but this is formally equivalent to the assumption that a light particle jumps from a *neutrino state* of negative *energy* to an *electron state* of positive *energy*.

For, if the *eigenfunctions* of the *electron* and the *neutrino* be Ψ_k, φ_k respectively, where k = 1, 2, 3, 4, a term of the form

$$-4\pi g' \sum_{k=1}^{4} \tilde{\psi}_k \varphi_k \qquad (15)$$

should be added to the right-hand side of the equation (5) for $\tilde{U}$, where g' is a new constant with the same dimension as g.

Now the *eigenfunctions* of the *neutrino state* with *energy* and *momentum* just opposite to those of the *state* φ_k is given by $\varphi'_k = -\delta_{kl}\tilde{\varphi}_l$ and conversely $\varphi_k = \delta_{kl}\tilde{\varphi}_{l'}$, where

$$\delta = \begin{pmatrix} 0 & -1 & 0 & 0 \\ 1 & 0 & 0 & 0 \\ 0 & 0 & 0 & 1 \\ 0 & 0 & -1 & 0 \end{pmatrix},$$

so that (15) becomes

$$-4\pi g' \sum_{k,l=1}^{4} \tilde{\psi}_k \delta_{kl} \tilde{\varphi}_l \qquad (16)$$

From equations (13) and (15), we obtain for the matrix element of the *interaction energy* of the heavy particle and the light particle an expression

$$gg' \int \ldots \int v(\vec{r}_1)u(\vec{r}_1) \sum_{k=1}^{4} \tilde{\psi}_k(\vec{r}_2)\varphi_k(\vec{r}_2) e^{-\lambda r_{12}}/r_{12}dv_1dv_2, \qquad (17)$$

corresponding to the following double process: a heavy particle falls from the *neutron state* with the *eigenfunction* $u(\vec{r})$ into the *proton state* with the *eigenfunction* $v(\vec{r})$ and simultaneously a light particle jumps from the *neutrino state* $\varphi_k(\vec{r})$ of negative *energy* to the *electron state* $\psi_k(\vec{r})$ of positive *energy*. In (17) λ is taken instead of μ, since the difference of *energies* of the *neutron state* and the *proton state*, which is equal to the sum of the upper limit of the *energy* spectrum of *β-rays* and the proper *energies* of the *electron* and the *neutrino*, is always small compared with m_Uc^2.

As λ is much larger than the *wave numbers* of the *electron state* and the *neutrino state*, the function $e^{-\lambda r_{12}}/r_{12}$ can be regards approximately as a δ-function multiplied by $4\pi/\lambda^2$ for the integrations with respect to x_2, y_2, z_2.

The factor $4\pi/\lambda^2$ comes from

$$\iiint e^{-\lambda r_{12}}/r_{12} \, dv_2 = 4\pi/\lambda^2.$$

Hence (17) becomes

$$4\pi gg'/\lambda^2 \iiint v(\vec{r})u(\vec{r}) \sum_k \tilde{\psi}_k(\vec{r})\varphi_k(\vec{r}) \, dv \qquad (18)$$

162

or by (16)

$$4\pi gg'/\lambda^2 \iiint \tilde{v}(\vec{r})u(\vec{r}) \, \Sigma_{k,l} \, \tilde{\psi}_k(\vec{r})\delta_{kl} \, \tilde{\varphi}_l(\vec{r}) \, dv, \qquad (19)$$

which is the same as the expression (21) of Fermi, corresponding to the emission of a *neutrino* and an *electron* of positive *energy states* $\varphi'_k(\vec{r})$ and $\psi_k(\vec{r})$, except that the factor $4\pi gg'/\lambda^2$ is substituted for Fermi's g.

> [[Fermi, E. (March, 1934). Versuch einer Theorie der β-Strahlen. I. (Attempt at a theory of β rays. I.). See above]:
> "$H_{-1m1s1\sigma}^{1n0s0\sigma} = \pm g \int v^*_m \upsilon_n \, \tilde{\psi}_s \delta \, \varphi^*_\sigma \, d\tau$ (21)
> where, because of the brevity, all the constant indices have been omitted in the first term."]]

Thus, the result is the same as that of Fermi's theory, in this approximation, if we take

$$4\pi gg'/\lambda^2 = 4 \times 10^{-50} \text{cm}^3.\text{erg},$$

from which the constant g' can be determined. Taking, for example, $\lambda = 5 \times 10^{12}$ and $g = 2 \times 10^{-9}$, we obtain $g' \simeq 4 \times 10^{-17}$, which is about 10^{-8} times as small as g.

This means that the *interaction* between the *neutrino* and the *electron* is much smaller than between the *neutron* and the *proton* so that the *neutrino* will be far more penetrating than the *neutron* and consequently more difficult to observe. The difference of g and g' may be due to the difference of *masses* of heavy and light particles.

§ 4. Summary

The *interactions* of *elementary particles* are described by considering a hypothetical *quantum* which has the elementary *charge* and the proper *mass* and which obeys Bose's statistics. The *interaction* of such a *quantum* with the heavy particle should be far greater than that with the light particle in order to account for the large *interaction* of the *neutron* and the *proton* as well as the small probability of β-*disintegration*.

Such quanta, if they ever exist and approach the matter close enough to be absorbed, will deliver their *charge* and *energy* to the latter. If, then, the *quanta* with negative charge come out in excess, the matter will be charged to a negative *potential*[9].

[9] Huxley, G. H. (September, 1934). Origin of the Cosmic Corpuscles. *Nature*, 134, 418–9; 571; https://doi.org/10.1038/134418b0; Johnson, T. H. (May, 1934). Coincidence Counter Studies of the Corpuscular Component of the Cosmic Radiation. *Phys. Rev.*, 45, 569.

These arguments, of course, of merely speculative character, agree with the view that the high-speed positive particles in the *cosmic rays* are generated by the *electrostatic field* of the earth, which is charged to a negative *potential*.

The massive *quanta* may also have some bearing on the shower produced by *cosmic rays*.

Hideki Yukawa - 1949 Nobel Lecture, December 12, 1949. *Meson theory in its developments.*

https://www.nobelprize.org/uploads/2018/06/yukawa-lecture.pdf.

The Nobel Prize in Physics 1949 was awarded to Hideki Yukawa "for his prediction of the existence of *mesons* on the basis of theoretical work on nuclear forces".

Atomic nuclei consist of *protons* and *neutrons* held together by a *strong force*. Hideki Yukawa assumed that this force is borne by particles and that there is a relationship between the range of the force and the mass of the force-bearing particle. In 1934, Yukawa predicted that this particle should have a mass about 200 times that of an electron. He called this particle a *"meson"*. Mesons' existence was verified in later experiments. [Hideki Yukawa – Facts. NobelPrize.org. https://www.nobelprize.org/prizes/ physics/1949/yukawa/facts/.]

In his Nobel Prize lecture Yukawa provided a brief history of the development of the *theory of the meson.*

The *meson* theory started from the extension of the concept of the *field of force* so as to include the *nuclear forces* in addition to the *gravitational* and *electromagnetic forces*.

[A *meson* is a type of hadronic subatomic particle composed of *an equal number quarks and antiquarks, usually one of each, bound together by the strong interaction*. Because *mesons* are composed of *quark* subparticles, they have a meaningful physical size, a diameter of roughly one femtometre (10^{-15} m), which is about 0.6 times the size of a *proton* or *neutron*. *All mesons are unstable*, with the longest-lived lasting for only a few tenths of a nanosecond. Heavier *mesons* decay to lighter *mesons* and ultimately to stable *electrons, neutrinos* and *photons*.

The *strong interaction*, also called the *strong force* or *strong nuclear force*, is a fundamental *interaction* that confines *quarks* into *protons*, neutrons, and other *hadron* particles. The *strong interaction* also binds *neutrons* and *protons* to create *atomic nuclei*, where it is called the *nuclear force*.

Most of the *mass* of a *proton* or *neutron* is the result of the *strong interaction* energy; the individual *quarks* provide only about 1% of the *mass* of a *proton*.

Before 1971, physicists were uncertain as to how the *atomic nucleus* was bound together. It was known that the *nucleus* was composed of *protons* and *neutrons* and that *protons* possessed positive *electric charge*, while *neutrons* were electrically neutral. By the understanding of physics at that time, positive charges would repel one another and the positively charged *protons* should cause the *nucleus* to fly apart. However, this was never observed. A stronger attractive force was postulated to explain how the *atomic nucleus* was bound despite the *protons'* mutual *electromagnetic* repulsion. This hypothesized force was called the *strong force*, which was believed to be a fundamental force that acted on the *protons* and *neutrons* that make up the *nucleus*.

In 1964, Murray Gell-Mann, and separately George Zweig, proposed that *baryons*, which include *protons* and *neutrons*, and *mesons* were composed of elementary particles. Zweig called the elementary particles "aces" while Gell-Mann called them "*quarks*"; the theory came to be called the *quark model*. [Gell-Mann. M. (February, 1964). A Schematic Model of Baryons and Mesons. See below.]

The *strong attraction* between *nucleons* was the side-effect of a more fundamental force that bound the *quarks* together into *protons* and *neutrons*. The theory of *quantum chromodynamics* explains that *quarks* carry what is called a *color charge*, although it has no relation to visible *color*. *Quarks* with unlike *color charge* attract one another as a result of the *strong interaction*, and the particle that mediates this was called the *gluon*.]

The necessity of introduction of specific *nuclear forces*, which could not be reduced to *electromagnetic interactions* between *charged* particles, was realized soon after the discovery of the *neutron*, which was to be bound strongly to the *protons* and other *neutrons* in the *atomic nucleus*. As pointed out by Wigner[1],

[1] Wigner, E. (February, 1932). On the Mass Defect of Helium. *Phys. Rev.*, 43, 252; https://doi.org/10.1103/PhysRev.43.252.

specific *nuclear forces* between two *nucleons*, each of which can be either in the *neutron state* or the *proton state*, must have a very short range of the order of 10^{-13} cm, in order to account for the rapid increase of the *binding energy* from the deuteron to the alpha particle. The *binding energies* of *nuclei* heavier than the alpha-particle do not increase as rapidly as if they were proportional to the square of the *mass number* A, i.e. the number of *nucleons* in each *nucleus*, but they are in fact approximately proportional to A. This indicates that *nuclear forces* are saturated for some reason. Heisenberg[2]

[2] Heisenberg, W. (January, 1932). Über den Bau der Atomkerne. I. (About the construction of atomic nuclei. I.); (March, 1932). Über den Bau der Atomkerne. II. (About the construction of atomic nuclei. II.); (September, 1933). Über den Bau der Atomkerne. III.

Zeit. Phys., 77, 1–11; https://doi.org/10.1007/BF01342433; *Ibid.*, 78, 156–64; https://doi.org/10.1007/BF01337585; *Ibid.*, 80, 587–96; https://doi.org/10.1007/BF01335696; see above.

suggested that this could be accounted for, *if we assumed a force between a neutron and a proton*, for instance, due to the *exchange of the electron* or, more generally, due to the *exchange of the electric charge*, as in the case of the *chemical bond* between a *hydrogen* atom and a *proton*. Soon afterwards, Fermi[3]

[3] Fermi, E. (March, 1934) Versuch einer Theorie der β-Strahlen. I. (Attempt at a theory of β rays. I.) *Zeit. Phys.*, 88, 161–77 (1934). https://doi.org/10.1007/BF01351864. See above.

developed a theory of *beta-decay* based on the hypothesis by Pauli, according to which a *neutron*, for instance, could decay into a *proton*, an *electron*, and a *neutrino*, which was supposed to be a very penetrating neutral particle with a very small *mass*.

This gave rise, in turn, to the expectation that *nuclear forces* could be reduced to the *exchange of a pair of an electron and a neutrino* between two *nucleons*, just as *electromagnetic forces* were regarded as due to the *exchange of photons* between *charged* particles. It turned out, however, that the *nuclear force* thus obtained was much too small[4], because the *beta-decay* was a very slow process compared with the supposed rapid exchange of the *electric charge* responsible for the actual *nuclear forces*.

[4] Tamm, I. (June, 1934). Exchange Forces between Neutrons and Protons, and Fermi's Theory. *Nature*, 133, 981; https://doi.org/10.1038/133981a0; Iwanenko, D. (June, 1934). Interaction of Neutrons and Protons. *Idem.*, 981–2; https://doi.org/10.1038/133981b0.

The idea of the *meson field* was introduced in 1935 in order to make up this difference. The original assumptions of the *meson theory* were as follows:

I. The *nuclear forces* are described by a *scalar field* U, which satisfies the *wave equation*

$$\{\partial^2/\partial x^2 + \partial^2/\partial y^2 + \partial^2/\partial z^2 - 1/c^2\ \partial^2/\partial t^2 - \lambda^2\}\ U = 0 \qquad (1)$$

in vacuum, where λ is a constant with the dimension of reciprocal length[5].

[5] Yukawa, H. (1935). On the Interaction of Elementary Particles. *Proceedings of the Physico-Mathematical Society of Japan*, 17, 48; see above; Yukawa H. & Sakata, S. (1937). *Ibid.*, 19, 1084.

Thus, the *static potential* between two *nucleons* at a distance r is proportional to exp $(-\lambda r)/r$, the *range of forces* being given by t/λ.

II. According to the general principle of quantum theory, the *field* U is inevitably accompanied by new particles or quanta, which have the *mass*

$$\mu = \lambda\hbar/c \qquad\qquad\qquad (2)$$

and the *spin* 0, obeying Bose-Einstein statistics. The *mass* of these particles can be inferred from the range of *nuclear forces*. If we assume, for instance, $\lambda = 5 \times 10^{12}$ cm^{-1}, we obtain $\mu \cong 200\ m_e$, where m_e is the *mass* of the *electron*.

III. In order to obtain *exchange forces*, we must assume that these *mesons* have the *electric charge* + e or - e, and that a positive (negative) *meson* is emitted (absorbed) when the *nucleon* jumps from the *proton state* to the *neutron state*, whereas a negative (positive) *meson* is emitted (absorbed) when the *nucleon* jumps from the *neutron* to the *proton*. Thus, a *neutron* and a *proton* can interact with each other *by exchanging mesons* just as two *charged* particles interact by *exchanging photons*. In fact, we obtain an *exchange force* of Heisenberg type between the *neutron* and the *proton* of the correct magnitude, if we assume that the *coupling constant* g between the *nucleon* and the *meson field*, which has the same dimension as the *elementary charge* e, is a few times larger than e.

However, the above simple theory was incomplete in various respects. For one thing, the *exchange force* thus obtained was repulsive for the triplet S-state of the deuteron in contradiction to the experiment, and moreover we could not deduce the *exchange force* of Majorana type, which was necessary in order to account for the saturation of *nuclear forces* just at the alpha-particle. In order to remove these defects, more general types of *meson fields* including vector, pseudoscalar and pseudovector fields in addition to the scalar fields, were considered by various authors[6].

> [6] Kemmer, N. (1938). *Proc. Roy. Soc. London, A*, 166, 127; H. Fröhlich, H., Heitler, W. & Kemmer, N. (1938). *Ibid.*, 166, 154; Bhabha, H. J. (1938). *Ibid.*, 166, 501; Stueckelberg, E. C. G. (1938). *Helv. Phys. Acta*, II, 299; Yukawa, H., Sakata, S. & Taketani, M. (1938). *Proc. Phys.-Math. Soc. Japan*, 20, 319; Yukawa, H., Sakata, S., Kobayasi, M. & Taketani, M. (1938). *Ibid.*, 20, 720.

In particular, the *vector field* was investigated in detail, because it could give a combination of *exchange forces* of Heisenberg and Majorana types with correct signs and could further account for the *anomalous magnetic moments* of the *neutron* and the *proton* qualitatively. Furthermore, the *vector theory* predicted the existence of non-central forces between a *neutron* and a *proton*, so that the deuteron might have the *electric quadripole moment*. However, the actual *electric quadripole moment* turned out to be positive in sign, whereas the *vector theory* anticipated the sign to be negative. The only *meson field*, which gives the correct signs both for *nuclear forces* and for the *electric quadripole moment* of the *deuteron*, was the *pseudoscalar field*[7].

[7] Rarita, W. & Schwinger, J. (March, 1941). On the Neutron-Proton Interaction. *Phys. Rev.*, 59, 436, 556; https://doi.org/10.1103/PhysRev.59.436.

There was, however, another feature of *nuclear forces*, which was to be accounted for as a consequence of the *meson theory*. Namely, the results of experiments on the *scattering* of *protons* by *protons* indicated that the type and magnitude of *interaction* between two *protons* were, at least approximately, the same as those between a *neutron* and a *proton*, apart from the Coulomb force. Now the *interaction* between two *protons* or two *neutrons* was obtained only if we took into account the terms proportional to g^4, whereas that between a *neutron* and a *proton* was proportional to g^2, as long as we were considering *charged mesons* alone. Thus, it seemed necessary to assume further:

IV. In addition to *charged mesons*, there are *neutral mesons* with the *mass* either exactly or approximately equal to that of *charged mesons*. They must also have the integer *spin*, obey Bose-Einstein statistics and interact with *nucleons* as strongly as *charged mesons*.

This assumption obviously increased the number of arbitrary constants in *meson theory*, which could be so adjusted as to agree with a variety of experimental facts. These experimental facts could-not be restricted to those of nuclear physics in the narrow sense, but was to include those related to *cosmic rays*, because we expected that *mesons* could be created and annihilated due to the interaction of *cosmic ray* particles with *energies* much larger than μc^2 with *matter*. In fact, the discovery of particles of intermediate *mass* in *cosmic rays* in 1937[8] was a great encouragement to further developments of *meson theory*.

[8] Anderson, C. D. & Neddermeyer, S. H. (1937). *Phys. Rev.*, 51, 884; Street, J. C. & Stevenson, E. C. (1937). *Ibid.*, 51, 1005; Nishina, Y., Takeuchi, M. & Ichimiya, T. (1937). *Ibid.*, 52, 1193.

At that time, we came naturally to the conclusion that the *mesons* which constituted the main part of the hard component of *cosmic rays* at sea level were to be identified with the *mesons* which were responsible for the *nuclear force*[9].

[9] Yukawa, H. (1937). *Proc. Phys. Math. Soc. Japan*, 19, 712; Oppenheimer, J. R. & Serber, R. (1937). *Phys. Rev.*, 51, 1113; Stueckelberg, E. C. G. (1937). *Ibid.*, 53, 41.

Indeed, *cosmic ray mesons* had the *mass* around 200 m_e as predicted and moreover, there was the definite evidence for the spontaneous decay, which was the consequence of the following assumption of the original *meson theory*:

V. *Mesons* interact also with light particles, i.e. *electrons* and *neutrinos*, just as they interact with *nucleons*, the only difference being the smallness of the *coupling constant* g' in this case compared with g. Thus, a positive (negative) *meson* can change spontaneously into a positive (negative) *electron* and a *neutrino*, as pointed out first by Bhabha[10].

[10] Bhabha, H. J. (1938). Nuclear Forces, Heavy Electrons and the β-Decay. *Nature*, 141, 117–8; https://doi.org/10.1038/141117a0Nature, 141, 117.

The *proper lifetime*, i.e. the mean lifetime at rest, of the *charged scalar meson*, for example, is given by

$$\tau_0 = 2 \ \{\hbar c/(g')^2\} \ (\hbar/\mu c^2) \qquad\qquad (3)$$

For the *meson* moving with velocity v, the lifetime increases by a factor $1/\sqrt{\{1 - (v/c)^2\}}$ due to the well-known *relativistic* delay of the moving clock. [???] Although the spontaneous decay and the velocity dependence of the *lifetime* of *cosmic ray mesons* were remarkably confirmed by various experiments[11],

[11] Euler, H. & Heisenberg, W. (1938). *Ergeb. Exakt. Naturw.*, I, 1; Blackett, P. M. S. (1938). *Nature*, 142, 992; Rossi, B. (1938). Nature, 142, 993; P. Ehrenfest, P. Jr. & A. Freon, A. (1938). *Coopt. Rend.*, 207, 853; Williams, E. J. & Roberts, G. E. (1940). *Nature*, 145, 102.

there was an undeniable discrepancy between theoretical and experimental values for the *lifetime*. The original intention of *meson theory* was to account for the *beta-decay* by combining the assumptions III and V together. However, the *coupling constant* g', which was so adjusted as to give the correct result for the *beta-decay*, turned out to be too large in that it gave the *lifetime* τ_0 of *mesons* of the order of 10^{-8} sec, which was much smaller than the observed *lifetime* 2×10^{-6} sec. Moreover, there were indications, which were by no means in favor of the expectation that *cosmic-ray mesons* interacted strongly with *nucleons*. For example, the observed *cross section of scattering* of *cosmic-ray mesons* by *nuclei* was much smaller than that obtained theoretically. Thus, already in 1941, the identification of the *cosmic-ray meson* with the *meson*, which was supposed to be responsible for *nuclear forces*, became doubtful. In fact, Tanikawa and Sakata[12]

[12] Tanikawa, Y. (1947). *Progr. Theoret. Phys. Kyoto*, 2, 220; S. Sakata, S. & Inouye, K. (1946). *Ibid.*, 1, 143.

proposed in 1942 a new hypothesis as follows: The *mesons* which constitute the hard component of *cosmic rays* at sea level are not directly connected with *nuclear forces*, but are produced by the decay of heavier *mesons* which interacted strongly with *nucleons*.

However, we had to wait for a few years before this *two-meson hypothesis* was confirmed, until 1947, when two very important facts were discovered. First, it was discovered by Italian physicists that the negative *mesons* in *cosmic rays*, which were captured by lighter atoms, did not disappear instantly, but very often decayed into *electrons* in a mean time interval of the order of 10^{-6} sec.[13]

[13] Conversi, M., Pancini, E. & Piccioni, O. (1947). *Phys. Rev.*, 71, 209.

This could be understood only if we supposed that ordinary *mesons* in *cosmic rays* interacted very weakly with *nucleons*. Soon afterwards, Powell and others[14]

[14] Lattes, C. M. G., Muirhead, H., Occhialini, G. P. S. & Powell, C. F. (1947). *Nature*, 159, 694; Lattes, C. M. G., Occhialini, G. P. S. & Powell, C. F. (1947). *Nature*, 160, 453; 486.

discovered two types of *mesons* in *cosmic rays*, the heavier *mesons* decaying in a very short time into lighter *mesons*. Just before the latter discovery, the *two-meson hypothesis* was proposed by Marshak and Bethe[15] independent of the Japanese physicists above mentioned.

[15] R. E. Marshak, R. E. & Bethe, H. A. (1947). *Phys. Rev.*, 72, 506.

In 1948, *mesons* were created artificially in Berkeley[16] and subsequent experiments confirmed the general picture of *two-meson theory*.

[16] Gardner, E. & Lattes, C. M. G. (1948). *Science*, 107, 270; Barkas, W. H., Gardner, E. & Lattes, C. M. G. (1948). *Phys. Rev.*, 74, 1558.

The fundamental assumptions are now[17]

[17] As for further details, see Yukawa, H. (1949). *Rev. Mod. Phys.*, 21, 474.

(i) The heavier *mesons*, i.e. *n-mesons* with the *mass* about 280 m_e *interact strongly* with *nucleons* and can decay into lighter *mesons*, i.e. π-*mesons* and *neutrinos* with a lifetime of the order of 10^{-8} sec; π-*mesons* have integer *spin* (very probably spin 0) and obey Bose-Einstein statistics. *They are responsible for, at least, a part of nuclear forces.* In fact, the shape of *nuclear potential* at a distance of the order of or larger could be accounted for as due to *the exchange of π-mesons between nucleons.*

(ii) The lighter *mesons*, i.e. μ-*mesons* with the *mass* about 210 m_e are the main constituent of the hard component of *cosmic rays* at sea level and can decay into *electrons* and *neutrinos* with the lifetime 2 x 10^{-6} sec. They have very probably *spin* ½ and obey Fermi-Dirac statistics. As they interact only weakly with *nucleons*, *they have nothing to do with nuclear forces.* Now, if we accept the view that π-*mesons* are the *mesons* that have been anticipated from the beginning, then we may expect the existence of *neutral π-mesons* in addition to *charged π-mesons*. Such neutral *mesons*, which have integer *spin* and interact as strongly as *charged mesons* with *nucleons*, must be very unstable, because each of them can decay into two or three *photons*[18].

[18] Sakata, S. & Tanikawa, Y. (1940). *Phys. Rev.*, 57, 548; R. J. Finkelstein, R. J. (1947). *Ibid.*, 72, 415.

In particular, a *neutral meson* with *spin* 0 can decay into two *photons* and the lifetime is of the order of 10^{-14} sec or even less than that. Very recently, it became clear that some of the experimental results obtained in Berkeley could be accounted for consistently by considering that, in addition to *charged n-mesons*, *neutral n-mesons* with the *mass* approximately equal to that of *charged π-mesons* were created by collisions of high-energy *protons* with *atomic nuclei* and that each of these *neutral mesons* decayed into two *mesons* with the lifetime of the order of 10^{-13} sec or less[19].

[19] York, H. F., Moyer, B. J. & Bjorklund, R. (1949). *Phys. Rev.*, 76, 187.

Thus, the *neutral mesons* must have *spin* 0. In this way, *meson theory* has changed a great deal during these fifteen years. Nevertheless, there remain still many questions unanswered. Among other things, we know very little about *mesons* heavier than π-*mesons*. We do not know yet whether some of the heavier *mesons* are responsible for *nuclear forces* at very short distances. *The present form of meson theory is not free from the divergence difficulties*, although recent development of *relativistic field theory* has succeeded in removing some of them. We do not yet know whether the remaining divergence difficulties are due to our ignorance of the structure of elementary particles themselves[20].

[20] Yukawa, H. (1950). *Phys. Rev.*, 77, 219.

We shall probably have to go through another change of the theory, before we shall be able to arrive at the complete understanding of the nuclear structure and of various phenomena, which will occur in high energy regions.

Eugene Paul Wigner (November 17, 1902–January 1, 1995).

Wigner was a Hungarian-American theoretical physicist who also contributed to mathematical physics. He received a half share of the 1963 Nobel Prize in Physics "for his contributions to the theory of the atomic nucleus and the elementary particles, particularly through the *discovery and application of fundamental symmetry principles*".

Jenő Pál Wigner was born in Budapest, Austria-Hungary on November 17, 1902, to middle class Jewish parents, Elisabeth Elsa Einhorn and Antal Anton Wigner, a leather tanner. He had an older sister, Berta, known as Biri, and a younger sister Margit, known as Manci, who later married British theoretical physicist Paul Dirac. He was home schooled by a professional teacher until the age of 9, when he started school at the third grade. During this period, Wigner developed an interest in mathematical problems. At the age of 11, Wigner contracted what his doctors believed to be tuberculosis. His parents sent him to live for six weeks in a sanatorium in the Austrian mountains, before the doctors concluded that the diagnosis was mistaken.

Wigner's family was Jewish, but not religiously observant, and his Bar Mitzvah was a secular one. From 1915 through 1919, he studied at the secondary grammar school called Fasori Evangélikus Gimnázium, the school his father had attended. Religious education was compulsory, and he attended classes in Judaism taught by a rabbi. A fellow student was János von Neumann, who was a year behind Wigner. They both benefited from the instruction of the noted mathematics teacher László Rátz. In 1919, to escape the Béla Kun communist regime, the Wigner family briefly fled to Austria, returning to Hungary after Kun's downfall. Partly as a reaction to the prominence of Jews in the Kun regime, the family converted to Lutheranism. Wigner explained later in his life that his family decision to convert to Lutheranism "was not at heart a religious decision but an anti-communist one".

After graduating from the secondary school in 1920, Wigner enrolled at the Budapest University of Technical Sciences, known as the Műegyetem. He was not happy with the courses on offer, and in 1921 enrolled at the Technische Hochschule Berlin (now Technical University of Berlin), where he studied chemical engineering. He also attended the Wednesday afternoon colloquia of the German Physical Society. These colloquia featured leading researchers including Max Planck, Max von Laue, Rudolf Ladenburg, Werner Heisenberg, Walther Nernst, Wolfgang Pauli, and Albert Einstein. Wigner also met the physicist Leó Szilárd, who at once became Wigner's closest friend. A third experience in Berlin was formative. Wigner worked at the Kaiser Wilhelm Institute for Physical Chemistry and Electrochemistry (now the Fritz Haber Institute), and there he met Michael Polanyi, who became, after László Rátz, Wigner's greatest teacher. Polanyi supervised

Wigner's DSc thesis, *Bildung und Zerfall von Molekülen* ("Formation and Decay of Molecules").

After graduating at the Technical University of Berlin, Wigner returned to Budapest, where he went to work at his father's tannery, but in 1926, he accepted an offer from Karl Weissenberg at the Kaiser Wilhelm Institute in Berlin. Wigner worked as an assistant to Karl Weissenberg and Richard Becker at the Kaiser Wilhelm Institute in Berlin. Weissenberg wanted someone to assist him with his work on X-ray crystallography, and Polanyi had recommended Wigner. After six months as Weissenberg's assistant, Wigner went to work for Richard Becker for two semesters. Wigner explored quantum mechanics, studying the work of Erwin Schrödinger. He also delved into the group theory of Ferdinand Frobenius and Eduard Ritter von Weber.

Wigner received a request from Arnold Sommerfeld to work at the University of Göttingen as an assistant to the great mathematician David Hilbert. This proved a disappointment, as the aged Hilbert's abilities were failing, and his interests had shifted to logic. Wigner nonetheless studied independently. He laid the foundation for *the theory of symmetries in quantum mechanics* and in 1927 introduced what is now known as the *Wigner D-matrix*. *Wigner and Hermann Weyl were responsible for introducing group theory into quantum mechanics*. The latter had written a standard text, *Group Theory and Quantum Mechanics* (1928), but it was not easy to understand, especially for younger physicists. Wigner's *Group Theory and Its Application to the Quantum Mechanics of Atomic Spectra* (1931) made group theory accessible to a wider audience.

Along the way he performed ground-breaking work in pure mathematics, in which he authored a number of mathematical theorems. He is also known for his research into the *structure of the atomic nucleus*. In particular, in *Wigner's theorem*, proven by him in 1931, he laid the foundation for the *theory of symmetries in quantum mechanics*.

Wigner's theorem specifies how physical symmetries such as rotations, translations, and CPT symmetry are represented on the Hilbert space of states.

> [*CPT symmetry* is a fundamental symmetry of physical laws under the simultaneous transformations of *charge conjugation* (C), *parity transformation* (P), and *time reversal* (T). It is the only combination of C, P, and T that is observed to be an exact symmetry of nature at the fundamental level. CPT symmetry means that a mirror-image of our universe where we reverse all momenta (corresponding to the reversal of time) and with all matter replaced by anti-matter would evolve under exactly the same physical laws.

174

Charge conjugation is a transformation that switches all particles with their corresponding antiparticles, thus changing the sign of all charges: not only electric charge but also the charges relevant to other forces.

A *parity transformation* (also called parity inversion) is the flip in the sign of one spatial coordinate. In three dimensions, it can also refer to the simultaneous flip in the sign of all three spatial coordinates (a point reflection). It can also be thought of as a test for chirality of a physical phenomenon, in that *a parity inversion transforms a phenomenon into its mirror image.*

A *physical state* is represented mathematically by a vector in a *Hilbert space* (that is, vector spaces on which a positive-definite scalar product is defined); this is called the *space of states.* Physical properties like momentum, position, energy, and so on will be represented by *operators* acting in the *space of states.*]

According to the theorem, any symmetry transformation is represented by a linear and unitary or antilinear and antiunitary transformation of Hilbert space. The representation of a symmetry group on a Hilbert space is either an ordinary representation or a projective representation.

In the late 1930s, Wigner extended his research into *atomic nuclei.* By 1929, his papers were drawing notice in the world of physics. In 1930, Princeton University recruited Wigner for a one-year lectureship, at 7 times the salary that he had been drawing in Europe. Princeton recruited von Neumann at the same time. Jenő Pál Wigner and János von Neumann had collaborated on three papers together in 1928 and two in 1929. They anglicized their first names to "Eugene" and "John", respectively. When their year was up, Princeton offered a five-year contract as visiting professors for half the year. The Technische Hochschule responded with a teaching assignment for the other half of the year. This was very timely, since the Nazis soon rose to power in Germany.

At Princeton in 1934, Wigner introduced his sister Margit "Manci" Wigner to the physicist Paul Dirac, whom she married in 1937, and had two children. [See Underwood, T. G. (2023). *Quantum Electrodynamics – annotated sources.* Volume I, p. 292.]

In September 1936, Wigner delivered a paper at the Tercentenary Conference of Arts and Sciences at Harvard University, in which he was the first to introduce *total isotopic spin* [Wigner, E. (January, 1937). On the Consequences of the Symmetry of the Nuclear Hamiltonian on the Spectroscopy of Nuclei. See below.]

Princeton did not rehire Wigner when his contract ran out in 1936. Through Gregory Breit, Wigner found new employment at the University of Wisconsin. There, he met his first wife,

Amelia Frank, who was a physics student there. However, she died unexpectedly in 1937, leaving Wigner distraught. He accepted an offer in 1938 from Princeton to return there. Wigner became a naturalized citizen of the United States on January 8, 1937, and he brought his parents to the United States.

Although he was a professed political amateur, on August 2, 1939, Wigner participated in a meeting with Leo Szilard and Albert Einstein that resulted in the Einstein–Szilard letter, which prompted President Franklin D. Roosevelt to authorize the creation of the Advisory Committee on Uranium with the purpose of investigating the feasibility of nuclear weapons. Wigner was afraid that the German nuclear weapon project would develop an atomic bomb first.

On June 4, 1941, Wigner married his second wife, Mary Annette Wheeler, a professor of physics at Vassar College, who had completed her Ph.D. at Yale University in 1932. After the war she taught physics on the faculty of Rutgers University's Douglass College in New Jersey until her retirement in 1964. They remained married until her death in November 1977. They had two children, David Wigner and Martha Wigner Upton.

During the Manhattan Project, he led a team whose task was to design nuclear reactors to convert uranium into weapons grade plutonium. At the time, reactors existed only on paper, and no reactor had yet gone critical. Wigner was disappointed that DuPont was given responsibility for the detailed design of the reactors, not just their construction. He became director of research and development at the Clinton Laboratory (now the Oak Ridge National Laboratory) in early 1946, but became frustrated with bureaucratic interference by the Atomic Energy Commission, and returned to Princeton.

In the postwar period, he served on a number of government bodies, including the National Bureau of Standards from 1947 to 1951, the mathematics panel of the National Research Council from 1951 to 1954, the physics panel of the National Science Foundation, and the influential General Advisory Committee of the Atomic Energy Commission from 1952 to 1957 and again from 1959 to 1964. In later life, he became more philosophical, and published *The Unreasonable Effectiveness of Mathematics in the Natural Sciences*, his best-known work outside technical mathematics and physics.

His second wife, Mary, died in November 1977. In 1979, Wigner married his third wife, Eileen Clare-Patton (Pat) Hamilton, the widow of physicist Donald Ross Hamilton, the dean of the graduate school at Princeton University, who had died in 1972. In 1992, at the age of 90, he published his memoirs, *The Recollections of Eugene P. Wigner*.

Wigner died of pneumonia at the University Medical Center in Princeton, New Jersey on January 1, 1995.

Wigner, E. (January, 1937). On the Consequences of the Symmetry of the Nuclear Hamiltonian on the Spectroscopy of Nuclei.

Phys. Rev. 51, 2, 106 (1937); https://journals.aps.org/pr/abstract/10.1103/PhysRev.51.106; also at https://harvest.aps.org/v2/journals/articles/10.1103/PhysRev.51.106/fulltext.

* A paper delivered at the Tercentenary Conference of Arts and Sciences at Harvard University, September, 1936.

Princeton University, Princeton, New Jersey.

Received October 23, 1936.

Abstract.

The structure of the *multiplets* of nuclear terms is investigated, using as *first approximation* a Hamiltonian which does not involve the ordinary *spin* and corresponds to equal forces between all nuclear constituents, *protons* and *neutrons*.

[A *multiplet* is the *state space* for 'internal' degrees of freedom of a particle, that is, degrees of freedom associated to a particle itself, as opposed to 'external' degrees of freedom such as the particle's position in space. Examples of such degrees of freedom are the *spin state* of a particle in *quantum mechanics*, or the *color, isospin* and *hypercharge state* of particles in the *Standard Model* of particle physics. Formally, this state space is described by a *vector space* which carries the *action* of a group of *continuous symmetries*.]

The *multiplets* turn out to have a rather complicated structure, instead of the S of atomic spectroscopy, one has three quantum numbers S, T, Y. The *second approximation* can either introduce *spin forces* (method 2), or else can discriminate between *protons* and *neutrons* (method 3). The *last approximation* discriminates between *protons* and *neutrons* as in method 2 and takes the *spin forces* into account as in method 3. The method 2 is worked out schematically and is shown to explain qualitatively the table of stable *nuclei* to about Mo.

---

1.   Recent investigations[1] appear to show that the *forces* between all pairs of constituents of the *nucleus* are approximately equal.

[1] Tuve, M. A., Heydenburg, N. P. & Hafstad, L. R. (1936). *Phys. Rev.*, 50, 806; Breit, G., Condon, E. U., & Present, R. D. (November, 1936). Theory of Scattering of Protons by Protons. *Phys. Rev.*, 50, 825; https://journals.aps.org/pr/abstract/10.1103/ PhysRev.50.825.

This makes it desirable to treat the *protons* and *neutrons* on an equal footing. A scheme for this was devised in his original paper by W. Heisenberg[2] *who considered protons and neutrons as different states of the same particle.*

[2] Heisenberg, W. (January, 1932). Über den Bau der Atomkerne. I. (About the construction of atomic nuclei. I.) *Zeit. Phys.*, 77, 1–11; see above.

Heisenberg introduced a variable τ which we shall call the *isotopic spin*, the value -1 of this variable can be assigned to the *proton state* of the particle, the value $+1$ to the *neutron state*. The assumption that the forces between all pairs of particles are equal is equivalent, then, to the assumption that they do not depend on τ or that the Hamiltonian does not involve the *isotopic spin*.

In addition to this *isotopic spin* τ, we must keep, of course, the *ordinary spin* variable s also; s also can assume the two values $+1$ and -1. It has been pointed out lately[3] that the Pauli principle requires that the *wave function*

$$\Psi \left(r_1 s_1 \tau_1, \ r_2 s_2 \tau_2, \ldots \ r_n s_n \tau_n \right) \tag{1}$$

be *antisymmetric* with respect to the simultaneous interchange of Cartesian, *spin* and *isotopic spin* variables of any pair of *heavy particles*.

[3] Bartlett, J. H. (1936). *Phys. Rev.*, 49, 102; Elsasser, W. (1936). *J. de Phys. et Rad.* I, 312; and especially Cassen, B. & Condon, E. U. (1936). *Phys. Rev.*, 50, 846.

This fact is quite analogous to the similar statement for *ordinary spin*.

Of course, if Eq. (1) is to represent the state of a given *nucleus*, say with n_P *protons* and n_N *neutrons*, it must vanish at every place where the sum of the τ's

$$\tau_1 + \tau_2 + \ldots + \tau_n \neq r_N - n_P \tag{2}$$

is not equal to the "*isotopic number*" of this element. All *wave functions* which are finite for several sums of the τ's, refer to *states* which can be different elements with finite probabilities. No such *states* are known to be of any importance and the mathematical apparatus of the *isotopic spin* is, hence, somewhat redundant. It will turn out that it is very useful in spite of this.

178

In addition to the assumption of the approximate equality of *forces* between all pairs of particles, *it appears to be a useful approximation to neglect the forces involving the ordinary spin*. The Hamiltonian depends then *on the space coordinates alone*. By keeping both, one or none of these assumptions, one comes to four possible schemes;

(1) Take into account *forces* depending on *space coordinates alone*.

(2) Take into account *forces* depending on *space and ordinary spin coordinates*, assuming, however, *interactions* between all kinds of pairs to be equal.

(3) *Neglect ordinary spin forces*, take into account *forces* depending on space *coordinates and isotopic spin*, i.e., discriminate between proton-proton, proton-neutron and neutron-neutron interactions.

(4) Take *all kinds of interaction* into account.

The first is the roughest method, the last the most exact and it is probable that (2) is more accurate for light elements, (3) for heavy elements. On the other hand, of course, one can obtain most results from symmetry considerations for 1, fewest for 4. Approximation (1) is identical with the "all orbital forces equal" model[4].

[4] Feenberg, E. & Wigner, E. *Phys. Rev.* This issue.

The statement that an *operator* involves only one or another set of variables needs further amplification. As used in the ordinary theory of spectra, this expression means that the *operator* can be written in terms of these variables alone. It did not mean that it cannot be written in some other way as well. Thus, e.g., the *interchange* P of the space coordinates acts only on space coordinates, although it can be written by *Dirac's identity*,

$$P = -\tfrac{1}{2} - \tfrac{1}{2}\,(s_1 \cdot s_2)$$

entirely in terms of *spin operators* for *antisymmetric* functions. We shall keep this definition for the forces depending on Cartesian and ordinary *spin* coordinates for nuclei also.

The *operators* which involve τ are, however, somewhat specialized to begin with. Using Heisenberg's notation for *isotopic spin operators*

$$\tau = \tau_\zeta = \left\| \begin{matrix} -1 & 0 \\ 0 & -1 \end{matrix} \right\|, \qquad \tau_\xi = \left\| \begin{matrix} 0 & i \\ -i & 0 \end{matrix} \right\|, \qquad \tau_\eta = \left\| \begin{matrix} 0 & 1 \\ 1 & 0 \end{matrix} \right\|, \qquad (4)$$

the *conservation law for electric charge* requires that all *operators* commute with

$$\tau_{\zeta 1} + \tau_{\zeta 2} + \ldots + \tau_{\zeta n} = n_N - n_P = 2T_\zeta. \qquad (3)$$

179

In addition to this, one hardly would say that

$$\tau_{\xi 1}\tau_{\xi 2} + \tau_{\eta 1}\tau_{\eta 2} + \tau_{\zeta 1}\,\tau_{\zeta 2} = -1 - 2PQ. \qquad (5)$$

(P *interchange of space*, Q *interchange of spin coordinates*) does not involve the Cartesian or *spin* coordinates, since Eq. (5) is a rather artificial expression, τ_ξ and τ_η having no immediate physical significance. We shall assume hence for approximation (3) *only such operators which are equivalent to operators acting on the Cartesian coordinates alone, but in a different way for protons and neutrons.* This is equivalent to using only *operators* involving the space coordinates and the τ_ζ's. If we do this, the results of method (3) must become equivalent to the usual theory (without τ's) which neglects the *spin*. As a matter of fact, for approximation (3), the introduction of τ is entirely useless and it is taken up here only in order to establish the transition from approximation (1) to (3).

2.   The *interaction* in the *electronic shells* of atoms is a sum of terms containing two particles only and the *momenta* is no higher than the second power. The reason for the first is, that the *interaction* occurs through a *field* and this gives in first approximation only *interaction between two particles*. The reason that one can stop with the second power of the *momenta* is that these always enter in the combination p/mc which is a small quantity.

An advantage of introducing the variable τ is[2,3] *that one can take over these assumptions to nuclei.* If one does not use the variable τ the *interchange* of two particles if expressed as a power series of the *momenta* is an infinite series[5]

[5] Wheeler, J. A. (1936). The Dependence of Nuclear Forces on Velocity. *Phys. Rev.*, 50, 643.

$$\Sigma\, n_1 n_2 n_3 \{ \dots \} \times \{ \dots \}$$

However, it can be expressed by means of *Dirac's identity* also entirely without the *momenta* by means of Eq. (5). It must be admitted, however, that the *spin* cannot be considered to be small as in the atomic theory. We shall determine here all *interaction* forms between two particles which do not contain higher than first power terms of *momenta*[6] as far as the dependence on s and τ goes.

[6] Some of these were given previously by Cassen and Condon, reference 3. The expressions given here are invariant only under Galilei transformations. G. Breit has shown that, in order to ensure *relativistic invariance*, correction terms must be added to the expressions derived here.

Nothing can be said, of course, on the dependence on the distance, and this factor will be omitted hence. It seems to be of lesser importance for the present.

The *interaction* must have *spherical symmetry*, depending on the differences of *coordinates* and *momenta* only, be *invariant under inversion*, substitution of $-t$ for t and also be *symmetric in the particles*. The first requirements determine the dependence on s, x and p. From the two triples of *spin operators*, one can form two invariants

(i)   1; and (i') $\frac{1}{2} + \frac{1}{2} (s_{x1}s_{x2} + s_{y1}s_{y2} + s_{z1}s_{z2}) = Q_{12}$,

three axial vectors with Z components

(v) $s_{z1} + s_{z2}$; $s_{z1} - s_{z2}$; $s_{x1}s_{y2} - s_{y1}s_{x2}$

respectively, and one axial tensor, with components

$$s_{x1}s_{y2} + s_{y1}s_{x2}; \qquad s_{y1}s_{z2} + s_{z1}s_{y2}; \qquad s_{z1}s_{x2} + s_{x1}s_{z2};$$
$$s_{x1}s_{x2} - s_{y1}s_{y2}; \qquad s_{x1}s_{x2} + s_{y1}s_{y2} - 2s_{z1}s_{z2}$$

The **first two of these**, (i) and (i'), can be used as they stand, cannot be combined with first power expressions of p, however, since these change sign under the $t' = -t$ substitution. The last one (t) gives the familiar expression

(i") $(s_1 . r_{12})(s_2 . r_{12}) - 3(s_1 . s_2)r_{12}^2$

if combined with the similar tensor of the coordinates[7].

[7] (ii) has the property that it is identical with Q_{12}(ii). It is an interaction which shows saturation.

It cannot be combined with the p either. The middle one must be combined with the vector $p_1 - p_2$ which gives a useless axial *invariant* and tensor and an ordinary vector. This combined with the distance vector gives the familiar

$$(ia)\ (ib)\ (ic) \quad \begin{vmatrix} s_x & s_y & s_z \\ x_1 - x_2 & y_1 - y_2 & z_1 - z_2 \\ p_{x1} - p_{x2} & p_{y1} - p_{y2} & p_{z1} - p_{z2} \end{vmatrix}$$

Here s_x, s_y, s_z, can be the components of one of the three vectors (v). On the whole, we have 6 invariants. These *invariants* can be multiplied with one of the six expressions in τ which commute with $\tau_{\xi1} + \tau_{\xi2}$. These are, first of all

$(\tau_0)\ 1$ and $(\tau_0') - \frac{1}{2} - \frac{1}{2} (\tau_1 . \tau_2) = P_{12}Q_{12}$,

which give the same *interaction* between all pairs of particles. In addition to these, we have

181

(τ_1) $\frac{1}{2} + \frac{1}{2}\,\tau_{\xi 1}\tau_{\xi 2}$ and (τ_2') $\frac{1}{2}\,(\tau_{\xi 1} = \tau_{\xi 2})$.

The first of these gives ordinary *interaction* but only between like particles; the second gives a negative *interaction* for *proton* pairs, a positive for *neutron* pairs, none for unlike particles. These *interactions* are *symmetric* in the particles and can be combined with (i), (i'), (i") and (ia), giving in the whole 16 different forms.

Finally, we have

$(\tau_{\xi 2})$ $\tau_{\xi 1} - \tau_{\xi 2}$ and $(\tau_{\xi 2}')$ $\frac{1}{2}\,(\tau_{\xi 1}\tau_{\eta 2} - \tau_{\eta 1}\tau_{\xi 2})$,

which can be combined with (ib) and (ic) giving 4 more types of interaction.

In *approximation* (1) we can have only (i)(τ_0) and (i')(τ_0'), i.e., *ordinary* and *Majorana exchange forces*.

> [There are three types of exchange forces; the *Majorana exchange force*, which arises from the *spatial exchange*; the *Bartlett force*, arising from the *spin exchange*; and the *Heisenberg force*, arising from the *space-spin exchange*. Ettore Majorana (born 5 August 5, 1906 – likely dying in or after 1959) was an Italian theoretical physicist who worked on neutrino masses. On 25 March 1938, he disappeared under mysterious circumstances after purchasing a ticket to travel by ship from Naples to Palermo.]

In *approximation* (2), all 8 forms derived from (τ_0), (τ_0') and (i) (i') (i") and (ia). These are, in addition to the previous ones, *spin-spin* (i")(τ_0), *spin-orbit* (242) (τ_0) *ordinary forces*, *Heisenberg forces* (i)(τ_0'). Furthermore *spin-spin exchange forces* (i")(τ_0') and *spin-orbit exchange forces* (ia)(τ_0') of the *Heisenberg type*. The *Majorana exchange forces* of these types are identical with the *ordinary forces*. Finally, we have the *spin-exchange forces* (2') (τ_0) of *Bartlett*[8].

> [8] The content of this section is based on the fundamental mathematical works of Cartan, E. (1913). *Bull. Soc. Math. de France*, 41, 43; (1914). *J. de Math.*, 10, 149; Schur, I. (1924). *Berl. Ber.*, pp. 189, 297, 346; and particularly, Weyl, H. (1925). *Math. Zs.*, 23, 271. I attempted to compile in this section—often without giving rigorous proofs— those results which suffice for the discussion of the physical problems in question.

In *approximation* (3) we must permit according to the preceding section, in addition to those of 1, only (i)(τ_1) and (i)(τ_1'), allowing for different interactions between different kinds of pairs. The coefficient of (i)(τ_1') is certainly very small, the *proton-proton interaction* being very nearly equal to the *neutron-neutron interaction*.

In *approximation* (4), all 20 types become possible.

3.   We next go over to *approximation* (1), and try to define the analog of the *multiplet* system.

> [A *multiplet* is the *state space* for 'internal' degrees of freedom of a particle, that is, *degrees of freedom associated to a particle itself*, as opposed to 'external' degrees of freedom such as the particle's position in space. Examples of such degrees of freedom are the *spin state* of a particle in quantum mechanics, or the *color, isospin and hypercharge state* of particles in the *Standard Model* of particle physics. Formally, we describe this *state space* by a *vector space* which carries the action of a group of continuous symmetries.]

This can be defined in two ways: either by considering the functional dependence of the *wave functions* on the *spins* or else by considering their dependence on the space coordinates. We shall first consider the *spin function*.

The great difference between the ordinary *spin* and the spin considered here is that we have, for every particle, two *spin* coordinates s and τ giving in the whole four different sets of values $-1, -1; -1, 1; 1, -1; 1, 1$. Instead of two two-valued *spins*, one can introduce one four-valued *spin* η, which has the values 1, 2, 3, 4 for the four different doublets of values of s and τ, respectively. This η plays the same role which the two-valued *spin* plays in the ordinary *spin* theory. However, because of the four-valuedness of η, instead of the *representations* of the two-dimensional *unitary group* (or the equivalent three-dimensional *rotation group*), the *representations* of the four-dimensional *unitary group* will characterize the *multiplet* systems.

Since the Hamiltonian does not contain the *spin* coordinates, any transformation which affects only these, will bring an *eigenfunction* into an *eigenfunction*. We can consider first, the *permutations* of the η_i and second, simultaneous *unitary transformations* of all the η …

…

Eugene Wigner – 1963 Nobel Prize in Physics.

The Nobel Prize in Physics 1963 was awarded to Eugene Wigner "for his contributions to the theory of the atomic nucleus and the elementary particles, particularly through the discovery and application of fundamental *symmetry principles*".

After discovery of the *neutron*, it became evident that the atomic nucleus is made up of nucleons—*protons* and *neutrons*—that are affected by a cohesive *force*. In 1933 Eugene Wigner discovered that the *force* binding the *nucleons* together is very weak when the distance between them is great, but very strong when the *nucleons* are close to one another as in the *atomic nucleus*. Wigner also described several characteristics of the *nucleons* and the *nuclear force*, including the fact that the *force* between two *nucleons* is the same, regardless of whether they are *protons* or *neutrons*. [Eugene Wigner – Facts. NobelPrize.org. https://www. nobelprize.org/prizes/physics/1963/wigner/facts/]

Shoichi Sakata (January 18, 1911–October 16, 1970).

Shoichi Sakata was a Japanese physicist who was internationally known for theoretical work on the subatomic particles. He proposed the *two-meson theory* and the *Sakata model* (an early precursor to the *quark model*), and the Pontecorvo–Maki–Nakagawa–Sakata neutrino mixing matrix.

> [The *Sakata model* of *hadrons* was a precursor to the *quark model*. It proposed that the *proton*, *neutron*, and *Lambda baryon* were elementary particles, and that all other known *hadrons* were made of them. The model was proposed by Shoichi Sakata in 1956. The model was successful in explaining many features of *hadrons*, but was supplanted by the *quark model* as the understanding of *hadrons* progressed.]

Sakata was born in Tokyo, Japan on January 18, 1911, to a family that held a tradition of public service. He was the eldest of six children of Tatsue Sakata and Mikita Sakata. At the time of Sakata's birth, Mikita was secretary to Prime Minister Katsura Tarō, who became Sakata's godfather. While attending Kōnan Middle School in Hyōgo Prefecture in 1924, Sakata was taught by the physicist Bunsaku Arakatsu. During his time as a student at Kōnan High School from 1926 to 1929, Sakata attended a lecture given by the influential physicist Jun Ishiwara. Sakata also became closely acquainted with Katō Tadashi, who would later co-translate Friedrich Engels's 1883 unfinished work Dialectics of Nature into Japanese. According to Sakata, Dialectics of Nature and Vladimir Lenin's 1909 work Materialism and Empirio-criticism became formative works for his thinking.

Sakata got in to the Kyoto Imperial University in 1930. When he was a second-year student, Yoshio Nishina, a granduncle-in-law of Sakata, gave a lecture on quantum mechanics at the Kyoto Imperial University. Sakata became acquainted with Hideki Yukawa and Shin'ichirō Tomonaga, the first and the second Japanese Nobel laureates, through the lecture. After graduating from the university, Sakata worked with Tomonaga and Nishina at Rikagaku Kenkyusho (RIKEN) in 1933 and moved to Osaka Imperial University in 1934 to work with Yukawa. Yukawa published his first paper on the *meson theory* in 1935 and Sakata closely collaborated with him for the developments of the *meson theory*. Possible existence of the neutral *nuclear force carrier* particle π^0 was postulated by them. Accompanied by Yukawa, Sakata moved to Kyoto Imperial University as a lecturer in 1939.

Sakata and Inoue proposed their *two-meson theory* in 1942. [Sakata, S. & Inoue, T. (1942). Chukanshi to Yukawa ryushi no Kankei ni tuite. (in Japanese). *Nippon Suugaku-Butsuri Gakkaishi*, 16; https://doi.org/10.11429/subutsukaishi1927.16.232.] At the time, a charged particle discovered in the hard component cosmic rays was misidentified as the Yukawa's *meson* ($\pi^{\pm}$, nuclear force carrier particle). The misinterpretation led to puzzles in the discovered cosmic ray particle. Sakata and Inoue solved these puzzles by identifying the cosmic ray particle as a daughter charged *fermion* produced in the $\pi^{\pm}$ decay. A new neutral *fermion* was also introduced to allow $\pi^{\pm}$ decay into *fermions*.

We now know that these charged and neutral *fermions* correspond to the second-generation *leptons* μ and ν_{μ} in the modern language. They then discussed the decay of the *Yukawa particle*, $\pi^{+} \rightarrow \mu^{+} + \nu^{\mu}$.

Sakata and Inoue predicted correct *spin* assignment for the *muon*, and they also introduced the second *neutrino*. They treated it as a distinct particle from the *beta decay neutrino*, and anticipated correctly the three body decay of the *muon*. The English printing of Sakata-Inoue's *two-meson theory* paper was delayed until 1946, [Sakata, S. & Inoue, T. (1946). On the Correlations between Mesons and Yukawa Particles. *Prog. Theor. Phys.*, 1, 4, 143–50; https://doi.org/10.1143/PTP.1.143. See below.] one year before the experimental discovery of $\pi \rightarrow \mu\nu$ decay.

Sakata moved to Nagoya Imperial University as a professor in October 1942 and remained there until his death. The name of the university was changed to Nagoya University in October 1947 after the end of the Pacific War (1945). Sakata reorganized his research group in Nagoya to be administrated under the democracy principle after the War.

Sakata stayed at the Niels Bohr Institute from May to October 1954 at the invitation of N. Bohr and C. Møller. During his stay, Sakata gave a talk introducing works of young Japanese particle physics researchers, especially emphasizing an empirical relation among the strongly interacting particles (*hadrons*) found by Nakano and Nishijima, which is now known as the *Nakano-Nishijima-Gell-Mann (NNG) rule*.

After Sakata returned to Nagoya, Sakata and his Nagoya group started researches trying to uncover the physics behind the NNG rule. In 1956, he proposed his *Sakata Model* [Sakata, S. (September, 1956). On a Composite Model for the New Particles. *Progr. Theor. Phys.*, 16, 686-8. See below.] which explains the NNG rule by postulating the fundamental building blocks of all strongly interacting particles are the *proton*, the *neutron* and the *lambda baryon*. The positively charged *pion* is made out of a *proton* and an *anti-neutron*, in a manner similar to the *Fermi-Yang composite Yukawa meson model*, [E. Fermi, W. & Yang, C. N. (1949). Are Mesons Elementary Particles? *Phys. Rev.*, 76, 12, 1739–43; https://doi.org/10.1103/PhysRev.76.1739.] while the positively charged *kaon* is composed of a *proton* and an *anti-lambda*, succeeding to explain the NNG rule in the *Sakata model*.

Aside from the integer charges, the *proton*, *neutron*, and *lambda* have similar properties as the *up quark*, *down quark*, and *strange quark* respectively.

In 1959, Ikeda, Ogawa and Ohnuki and, independently, Yamaguchi found out the *U(3) symmetry* in the *Sakata model*. The *U(3) symmetry* provides a mathematical descriptions of *hadrons* in the Eightfold Way idea of Murray Gell-Mann [Gell-Mann, M. (March, 1961). *The Eightfold Way: A Theory of Strong Interaction Symmetry (Report)*. Pasadena, CA: California Inst. of Tech., Synchrotron Laboratory. See below.]. *Sakata's model* was superseded by the *quark model*, proposed by Gell-Mann and George Zweig in 1964, which keeps the U(3) symmetry, but made the constituents *fractionally charged* and rejected the idea that they could be identified with observed particles. Still, within Japan, *integer charged quark models* parallel to Sakata's were used until the 1970s, and are still used as effective descriptions in certain domains.

The 2008 physics Nobel laureates Yoichiro Nambu, Toshihide Maskawa and Makoto Kobayashi, who received their awards for work on *symmetry breaking*, all came under his tutelage and influence. The *neutrino oscillation* phenomena, as predicted by Maki, Nakagawa and Sakata, has been experimentally confirmed (2015 Nobel prize in physics).

Shoichi Sakata's "Sakata model" inspired Murray Gell-Mann and George Zweig's *quark* model, but the 1969 prize was only awarded to Murray Gell-Mann.

Sakata, S. & Inoue, T. (1942). On the Correlations between Mesons and Yukawa Particles*.

Sakata, S. & Inoue, T. (1942). Chukanshi to Yukawa ryushi no Kankei ni tuite. (in Japanese). *Nippon Suugaku-Butsuri Gakkaishi* (*Journal of the Mathematical and Physical Society of Japan*), 16, 4; https://doi.org/10.11429/subutsukaishi1927.16.232; translation at (November, 1946). *Prog. Theor. Phys.*, 1, 4, 143–50; https://doi.org/10.1143/PTP.1.143; also at https://academic.oup.com/ptp/article/1/4/143/1846220.

Institute of Theoretical Physics, Nagoya University.

* The content of this paper was read before the symposium of the meson theory held on September, 1943. The printing was, however, delayed owing to the war circumstances.

Sakata and Inoue proposed their *two-meson theory* in 1942. At the time, a charged particle discovered in the hard component cosmic rays was misidentified as the Yukawa's *meson* ($\pi^{\pm}$, nuclear force carrier particle). The misinterpretation led to puzzles in the discovered cosmic ray particle. Sakata and Inoue solved these puzzles by identifying the cosmic ray particle as a daughter charged *fermion* produced in the $\pi^{\pm}$ decay. A new neutral *fermion* was also introduced to allow $\pi^{\pm}$ decay into *fermions*. We now know that these charged and neutral *fermions* correspond to the second-generation *leptons* μ and ν_{μ} in the modern language. They then discussed the decay of the *Yukawa particle*, $\pi^{+} \rightarrow \mu^{+} + \nu^{\mu}$. Sakata and Inoue predicted correct *spin* assignment for the *muon*, and they also introduced the second *neutrino*. They treated it as a distinct particle from the *beta decay neutrino*, and anticipated correctly the three body decay of the *muon*. As a result of World War II, the English printing of Sakata-Inoue's *two-meson theory* paper was delayed until 1946, one year before the experimental discovery of $\pi \rightarrow \mu\nu$ decay.

Introduction

1.   At the present stage of its development, *meson theory*, which generated from the original idea of Yukawa, is confronted with several grave difficulties. Perhaps some of these difficulties, as has on many occasions been pointed out by Heisenberg[1], have close connections with the existence of a "*universal length*" that limits the validity of the present *relativistic quantum mechanics*.

[1] Heisenberg, W. (1938). Uber die in der Theorie der Elementarteilohen aujtretende universelle Lange. (About the universal length occurring in the theory of elementary particles.) *Ann. Phys.*, 5, 32, 20-33; https://link.springer.com/chapter/10.1007/978-3-642-70078-1_19 (1938). *Zeit. Phys.*, 110, 251; (1939). *ibid.*, 118, 61.

But on the other hand, it seems very likely too, that some of them are not of such essential nature, but of complicated character. In fact, it was recently shown that certain problems are of purely mathematical nature which results from an inappropriate application of perturbation theory as had been insisted on by Bhabha[2] from earlier times, and if the reaction of field is properly considered, can be removed[3,4].

[2] Bhabha, H. J. (1939). General classical theory of spinning particles in a meson field. *Proc. Roy. Soc.* A, 172, 384-409; https://royalsocietypublishing.org/doi/pdf/10.1098/rspa.1941.0057.
[3] Heitler. (1941). *Proc. Camb. Phil. Soc.*, 37, 291; Wilson. (1941). *Proc. Camb. Phil. Soc.*, 37, 301; Sokolow. (1941). *Jour. of Phys. U.S.S.R.*, 5, 231.
[4] Tomonaga. (1941). *Sci. Pap. I.P.C.R.*, 37, 247.

The object of the present paper is to show that a sequence of difficulties can be removed by the modification of the fundamental nature of elementary particles from a new stand-point.

Initially, the *Yukawa theory* was proposed in order to explain the problems of *nuclear forces* and *beta decay* phenomena in unification, and the identification of the particle introduced in this theory (*Yukawa particle*) with a new particle discovered in the hard component of cosmic rays (*meson*), at once led to theoretical stand-point from which very clear and reasonable accounts could be given for many problems on cosmic ray phenomena.

[The *Yukawa particle* is now identified as the *pion* (*pi meson*).

Pion (or pi meson) [*composite particle*]

A *pion* or *pi meson*, is any of three subatomic particles: π^0, π^+, and π^-. Each pion consists of a *quark* and an *antiquark* and is therefore a *meson*. *Pions* are the lightest *mesons* and, more generally, the lightest *hadrons*. They are unstable, with the *charged pions*, π^+ and π^-, decaying after a mean lifetime of 26.033 nanoseconds (2.6033×10^{-8} seconds), and the *neutral pion* π^0 decaying after a much shorter lifetime of 85 attoseconds (8.5×10^{-17} seconds). *Charged pions* most often decay into *muons* and *muon neutrinos*, while *neutral pions* generally decay into *gamma rays*.

The exchange of virtual *pions*, along with *vector, rho* and *omega mesons*, provides an explanation for the residual *strong force* between *nucleons*. *Pions* are not produced in radioactive decay, but commonly are in high-energy collisions between *hadrons*. *Pions* also result from some *matter–antimatter annihilation* events. All types of *pions* are also produced in natural processes when high-energy cosmic-ray *protons* and other hadronic cosmic-ray components interact with matter in Earth's atmosphere. In 2013, the detection of characteristic *gamma rays* originating from the decay of *neutral pions* in two supernova remnants has shown that *pions* are produced copiously after supernovas, most probably in conjunction with production of high-energy protons that are detected on Earth as cosmic rays.

In 1947, the *pion* (or *pi meson*) was discovered by members of C. F. Powell's group at the University of Bristol, in England, including César Lattes, Giuseppe Occhialini and Hugh Muirhead. In the same year, Lattes, together with Powell and Occhialini, determined the new particle's mass. [Lattes, C. M. G., Muirhead, H., Occhialini, G. P. S. & Powell, C. F. (1947). Processes involving charged mesons. *Nature*. 159, 4047, 694–7; https://doi.org/10.1038/ 159694a0.]

Cecil Frank Powell, (December 5,1903 –August 9, 1969) was a British physicist, who headed the team that developed the photographic method of studying nuclear processes. The Nobel Prize in Physics 1950 was awarded to Powell "for his development of the photographic method of studying nuclear processes and his discoveries regarding mesons made with this method".

Since the advent of particle accelerators had not yet come, high-energy subatomic particles were only obtainable from atmospheric cosmic rays. Photographic emulsions based on the gelatin-silver process were placed for long periods of time in sites located at high-altitude mountains, first at Pic du Midi de Bigorre in the Pyrenees, and later at Chacaltaya in the Andes Mountains, where the plates were struck by cosmic rays. After development, the photographic plates were inspected under a microscope by a team of about a dozen women.

In 1948, Lattes, Eugene Gardner, and their team first artificially produced *pions* at the University of California's cyclotron in Berkeley, California, by bombarding carbon atoms with high-speed alpha particles.]

For example, the theoretically anticipated instability of the *Yukawa particle* gave reasonable accounts for various effects on the variation of intensity of the hard component in cosmic rays (temperature effect, density effect etc.). But the more precise and quantitative the comparison between theory and experiment became, discrepancies manifested themselves so much the more markedly. For the *lifetime* of *mesons*, experimental results exceed the theoretical by 10^2 in magnitude[5].

[5] Sakata, S. (1941). On Yukawa's Theory of the Beta-Disintegration and the Lifetime of the Meson. *Proc. Phys. Math. Soc. Japan*, 23, 291; https://www.jstage.jst.go.jp/

article/ppmsj1919/23/0/23_0_291/_pdf; also in (January, 1955). *Progress of Theoretical Physics Supplement*, 1, 118–36, https://doi.org/10.1143/PTPS.1.118.

On the other hand, for the *cross section* of slow *mesons* with *energy* 2-20 x 10^8 eV, maximum estimation of experimental results[6] gave a *cross section* smaller than the calculated one by 10^{-2} in magnitude.

[6] Heitler, W. (1938). Showers Produced by the Penetrating Cosmic Radiation. *Proc. Roy. Soc. A*, 166, 927, 529-43; https://doi.org/10.1098/rspa.1938.0108.

In order to remove these difficulties, many trials[7] have been made by several authors, but they still seem unsatisfactory.

[7] Heitler & Ma. (194). *Proc. Roy. Soc.*, 176, 368; Bhabha. (1941). *Phys. Rev.*, 59, 100; Kobayasi. (1941). *Proc. Phys. Math. Soc. Japan*, 23, 891; Sakata. (1941). *Proc. Phys. Math. Soc. Japan*, 23, 283, 291.

In this paper, with the main aim to solve the above-mentioned two difficulties simultaneously, we propose a new *theory of mesons* developed from the stand-point based on the following assumption: *the meson is an elementary particle which has close correlations to the Yukawa particle, but it should be considered as an elementary particle of a different sort.*

For a *meson theory* from such a stand-point, two alternatives are possible: *Bose meson* and *Fermi meson theory*. In the following, we adopt the latter alternative[8].

[8] Another alternative which involves Bose-meson distinguished from Yukawa particle are adopted by Y. Tanikawa.

Interaction between Meson and Yukawa Particle.

2. According to the *Yukawa theory*, heavy particles (*nucleon*) and light particles *interact* with *Yukawa particles* by the following process (schematically represented) respectively:

$$P \leftrightarrow N + Y^+, \qquad N \leftrightarrow P + Y^-, \qquad \text{(I)}$$
$$e^- \leftrightarrow \nu + Y^-, \qquad \nu \leftrightarrow e^- + Y^+, \qquad \text{(II)}$$

(P: *proton*, N: *neutron*, e^-: *electron*, ν: *neutrino*, $Y^\pm$: positively and negatively charged *Yukawa particles*).

In our theory, it is assumed that the *meson* is a *Fermi particle* with spin $\hbar/2$, and furthermore corresponding to (I) and (II) the following interactions are introduced:

$$m^{\pm} \leftrightarrow n + Y^{\pm}, \qquad n \leftrightarrow m^{\pm} + Y^{\mp}, \qquad\qquad\qquad \text{(III)}$$

($m^{\pm}$ positively and negatively charged *meson*, n: neutral *meson* which is assumed in the following discussions to have a negligible *mass*, and consequently may be regarded as equivalent with the *neutrino*).

Furthermore. if we introduce a neutral *Yukawa particle* (Y^0) in order to explain the *proton-proton* and *neutron-neutron forces* and set up the following *interactions*:

$$P \leftrightarrow P + Y^0, \qquad N \leftrightarrow N + Y^0, \qquad\qquad\qquad \text{(I')}$$

it would be natural in our theory to introduce the following *interactions*

$$m^{\pm} \leftrightarrow m^{\pm} + Y^0, \qquad n \leftrightarrow n + Y^0, \qquad\qquad\qquad \text{(III')}$$

In the following, results obtained by the introduction of new *interactions* (III) and (III') will be discussed. Consequently, it is concluded that if g, g' and γ are the natural constants which represent the strength of *interactions* (I), (II) and (III) or (III') respectively and we adopt as their value $g^2/\hbar c \sim 10^{-1}$, $g'^2/\hbar c \sim 10^{-15}$, $\gamma^2/\hbar c \sim 10^{-2}$, it is possible to account for the phenomena in *atomic nuclei* and *cosmic rays* consistently, *without aiming to touch the inherent difficulties of field theory.*

Mass and Lifetime of Yukawa Particle.

3. For phenomena in *atomic nuclei* (nuclear forces, beta-decay etc.), *Yukawa theory* is conserved in its original form. But it is to be noted that the particle with *mass* determined from the range of *nuclear forces*, is the *Yukawa particle* and not the *meson* found in *cosmic rays*. This point is advantageous to explain the experimental results[9] about the *nuclear force range* ($\hbar/m_u c$) which gives half the value ($\hbar/\mu c$) obtained from the *meson mass* data (m_u: *mass* of *Yukawa particle*, μ: *mass* of *meson*).

[9] Aoki. (1939). *Phys. Rev.*, 65, 794; (1939). *Proc. Phys. Math. Soc. Japan*, 21, 282. Hoisington, Share & Breit. (1939). *Phys. Rev.*, 56, 884.

In order to account for these results, $m_u = 2\mu$ is a reasonable assumption. More generally speaking, it is allowable to assume $m_u > \mu$.

4. As the consequence of the above assumption $m_u > \mu$, it becomes possible that a *Yukawa particle* transforms into a *meson* in vacuum by the following process (*spontaneous disintegration of the Yukawa particle*):

$$Y^{\pm} \rightarrow m^{\pm} + n \qquad (m_u > \mu). \qquad\qquad\qquad \text{(IV)}$$

The reciprocal of the *proper lifetime* of *Yukawa particles* (t_0) calculated in vector theory is given by the following expression

$$1/t_0 = \ldots \tag{1}$$

where γ_1 and γ_2 are the constants of vector and tensor interactions respectively. If we assume … (as later shown, these figures are compatible with the considerations of *meson* decay), we obtain $t_0 \sim 10^{-21}$ sec. for $m_u/\mu \sim 2$. (*Mean free path* of the *Yukawa particle* with the *energy* 10^{10} eV. is. about 10^{-8} cm). Naturally these values depend on the ratio m_u/μ. As m_u approaches μ, the *lifetime* of *Yukawa particles* is prolonged by the last factor in (1); and the limit $m_u = \mu$, it becomes infinite, But, in this case, the *Yukawa particle* decays by the original process $Y \rightarrow e + v$. Consequently, the *lifetime* of the *Yukawa particle* does not become greater than 10^{-8} sec.

5. Process (IV) represents the creation of the hard component in *cosmic rays*. For the analysis of *cosmic ray* phenomena, we shall take up the proton primary hypothesis which has been proposed by Schein et al[10].

[10] Schein, Wollan & Jesse. (1941). *Phys. Rev.*, 69, 615.

Then primary incidental *protons* at first create *Yukawa particles* by collisions with *nuclei* of N or 0 atoms existing in the atmosphere. These *Yukawa particles* transform into *mesons* instantaneously by the above process. The *interaction* of the latter with *matter* is smaller than that of the former ($\gamma < g$). Thus, we observe these *mesons* as hard components of *cosmic-rays*. Previously, Nordheim[11] in his analysis on *cosmic-rays* suggested that the absorption process of hard component must have *cross sections* smaller by a factor of order 10 compared to the creation processes.

[11] Nordheim. (1939). *Phys. Rev.*, 66, 502.

This difficulty, which indicates an *asymmetry* inconsistent with the original Yukawa theory, is overwhelmed by the insertion of process (IV).

Decay and Scattering of Meson.

6. According to our theory, the decay of mesons occurs by the following process:

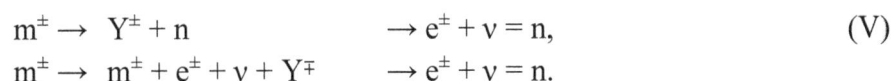

$$\begin{array}{lll}
m^{\pm} \rightarrow Y^{\pm} + n & \rightarrow e^{\pm} + v = n, & \text{(V)} \\
m^{\pm} \rightarrow m^{\pm} + e^{\pm} + v + Y^{\mp} & \rightarrow e^{\pm} + v = n.
\end{array}$$

The reciprocal of the proper lifetime of mesons (r_0) in vector and pseudoscalar theory are given as follows, …

$$1/t_0 = \ldots, \qquad \text{(vector theory)} \qquad\qquad (2)$$
$$1/t_0 = \ldots, \qquad \text{(pseudoscalar theory)} \qquad (2')$$
$$\ldots, \qquad\qquad\qquad\qquad\qquad\qquad\qquad\qquad (3)$$
$$\ldots,$$
$$\ldots,$$
$$\ldots.$$

$\ldots.$

7. Introducing the neutral *Yukawa particle* and taking *interaction* (III'), the *scattering* of *mesons* occurs by the following process:

$$m^+ + N \to m^+ + Y^0 + N \to m^{+\prime} + N'. \qquad\qquad \text{(VI)}$$

The *cross section* of these processes is determined by the *interaction constant* of (III'). Taking the same constant γ of (III) for this (analogous to Kemmer's *symmetrical theory of nuclear forces*), we have to examine whether the above-determined γ leads to a *cross section* consistent with scattering experiments or not. Furthermore, we must take into account the following process which has a probability of the same order of magnitude as that of the *scattering process*.

$$m^+ + N \to m^+ + Y^- + P \to n + P. \qquad\qquad \text{(VI')}$$

If the *cross section* of the latter process becomes appreciably large, it is provable that an attempt to account for the creation of *neutral mesons* by this process may lead to a contradiction with experimental results. From these considerations, it is desirable to take as small a value for γ as possible[*,**]. ...

> [*] Naturally, we cannot take too small a value for γ, in order that it be consistent with the decay probability of *Yukawa particles*. Eventually, it seems appropriate to take $\gamma^2 \sim 10^{-2}\,\hbar c$ (or a slightly smaller value).
>
> [**] If we take the formalism which excludes the neutral *Yukawa particle*, the scattering of *mesons* takes place only by a process of the fourth order, which results in a much smaller *cross section* than that of (VI). But, in this case, process (VI') is not excluded.

...

In the above consideration we took up the symmetrical formalism. But even if we assume that for the *interaction* with heavy particles the *interaction* constants of Y^0 are greater than that of $Y^\pm$ as in Heitler's theory or Bethe's neutral theory, it is still possible to make γ_1, γ_2 sufficiently small.

Some Remark on the Results.

8. As shown above, there is certainly a possibility that the contradiction between *Yukawa theory* and experimental results be removed. In the following, some related problems which are characteristic to our theory will be discussed.

(i) One of characteristics of our theory is that the *meson* has spin ħ/2. This point is interesting in relation with the results of analysis on radiative effects of the *meson* which are sensitively affected by the *spin* value. According to the calculation by Christy and Kusaka[12] of the sizes of bursts produced by mesons, our choice of spin ħ/2 for the *meson* is supported by experiment, in contrast to the usual theory with spin ħ.

[12] Christy & Kusaka. (1941). *Phys. Rev.*, 59, 414.

(ii) The role of *neutral* particles in the original *Yukawa theory* has been frequently discussed. If we take the *symmetrical theory* on *nuclear force* and furthermore assume $m_u > 2\mu$, the *neutral Yukawa particle* decays by the following process: (*spontaneous disintegration of the neutral Yukawa particle*)

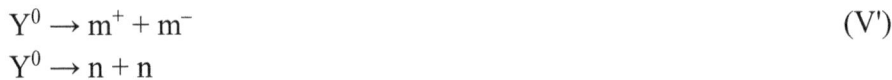

$$Y^0 \rightarrow m^+ + m^- \qquad\qquad (V')$$
$$Y^0 \rightarrow n + n$$

and its *proper lifetime* is of the same order as that of charged *Yukawa particles* calculated above ($\sim 10^{-21}$ sec). This value is smaller than that obtained from the previously discussed process[13], in which the neutral *Yukawa particle* (*neutretto*) transforms into *photons*.

[13] Sakata & Tanikawa. (1940). *Phys. Rev.*, 57, 548.

As a result, the *Yukawa particle* transforms into *mesons* (penetrating component) with a greater probability than into *photons* (shower producing component). This is very favorable in interpreting the results of experiment by Schein *et al*[10] which denied the existence of shower producing components in upper atmosphere and suggested, as the possible alternatives which was at first suggested by Taketani[14], the introduction of some process in which the *neutral Yukawa particle* transforms into the penetrating component with a greater probability than into shower-producing components, or otherwise the elimination of *neutral Yukawa particle* from the formalism.

[14] Taketani. (1941). *Kagaku*, 11, 523.

(iii) Also, it is to be noted that as the decay product of one *meson*, three particles are generated and two of them are *neutral particles* which have smaller *interaction* with matter. This consequence seems to be in agreement with the fact[15] that when *mesons* are stopped in matter, decayed *electrons* have rarely been observed, or that when *mesons* decay in a Wilson chamber, only slow *electrons* have been observed.

[15] Maier-Leibnitz. (1939). *Zeit. Phys.*, 112, 569; Neddermever & Anderson. (1938). *Phys. Rev.*, 54, 88.

Furthermore, the fraction of the total mesonic *energy* given to the *electron* by decay in this theory is smaller than that in the usual theory. This point is, too, supported by Nordheim's analysis[16] on the *intensity* of cascade showers.

[16] Nordheim. (1941). *Phys. Rev.*, 59, 554.

But eventually, the validity of introduction of such processes should be fully discussed in future, based on further experimental and theoretical investigations on *cosmic rays*.

Finally, the authors wish to express their appreciation to Prof. H. Yukawa for his continuous interest in this work, and to Y. Tanikawa and S. Nakamura for constant collaboration. They are also much indebted to M. Taketani for his valuable discussion.

Chen-Ning Yang (born 1 October 1922).

Yang is a Chinese theoretical physicist who made significant contributions to statistical mechanics, integrable systems, gauge theory, and both particle physics and condensed matter physics. He and Tsung-Dao Lee received the 1957 Nobel Prize in Physics for their work on *parity non-conservation of the weak interaction*. The two proposed that the *conservation of parity*, a physical law observed to hold in all other physical processes, *is violated in the so-called weak nuclear reactions*, those nuclear processes that result in the emission of beta or alpha particles. Yang is also well known for his collaboration with Robert Mills in developing *non-abelian gauge theory*, widely known as the *Yang–Mills theory*.

Yang was born in Hefei, Anhui, China. His father, Ko-Chuen Yang (1896–1973), was a mathematician, and his mother, Meng Hwa Loh Yang, was a housewife. Yang attended elementary school and high school in Beijing, and in the autumn of 1937 his family moved to Hefei after the Japanese invaded China. In 1938 they moved to Kunming, Yunnan, where National Southwestern Associated University was located. In the same year, as a second-year student, Yang passed the entrance examination and studied at National Southwestern Associated University. He received a Bachelor of Science in 1942, with his thesis on the application of group theory to molecular spectra.

Yang continued to study graduate courses there for two years, working on statistical mechanics. In 1944, he received a Master of Science from Tsinghua University, which had moved to Kunming during the Sino-Japanese War (1937–1945). Yang was then awarded a scholarship from the Boxer Indemnity Scholarship Program, set up by the United States government using part of the money China had been forced to pay following the Boxer Rebellion. His departure for the United States was delayed for one year, during which time he taught in a middle school as a teacher and studied field theory.

Yang entered the University of Chicago in January 1946 and studied with Edward Teller. He received a Doctor of Philosophy in 1948. Yang remained at the University of Chicago for a year as an assistant to Enrico Fermi.

At the University of Chicago, Yang first spent twenty months working in an accelerator lab, but he later found he was not as good as an experimentalist and switched back to theory. His doctoral thesis was about angular distribution in nuclear reactions.

In 1949 he was invited to do his research at the Institute for Advanced Study in Princeton, New Jersey, where he began a period of fruitful collaboration with Tsung-Dao Lee. He was made a permanent member of the Institute in 1952, and full professor in 1955. In 1963, Princeton University Press published his textbook, *Elementary Particles*.

Yang married Chih-li Tu, a teacher, in 1950 and has two sons and a daughter with her: Franklin Jr., Gilbert and Eulee. His father-in-law was the Kuomintang general Du Yuming. Some scholars suspect that Du was promoted to a high-ranking position in the Chinese People's Political Consultative Conference in order to convince Yang to return to China after seeking refuge in the US.

Yang is well known for his 1953 collaboration with Robert Mills in extending the concept of *gauge theory* for *abelian* groups, e.g. *quantum electrodynamics*, to *nonabelian* groups to provide an explanation for *strong interactions*. The idea was generally conceived by Yang, and the novice scientist Mills assisted him. [Yang, C. N. & Mills, R. (October, 1954). Conservation of Isotopic Spin and Isotopic Gauge Invariance. See below.]

In 1956, he and Tsung Dao Lee proposed that *in the weak interaction the parity symmetry was not conserved*. Chien-shiung Wu's team at the National Bureau of Standards in Washington experimentally verified the theory. Yang and Lee received the 1957 Nobel Prize in Physics "for their penetrating investigation of the so-called parity laws which has led to important discoveries regarding the elementary particles".

In 1965 he moved to Stony Brook University, where he was named the Albert Einstein Professor of Physics and the first director of the newly founded Institute for Theoretical Physics. Today this institute is known as the C. N. Yang Institute for Theoretical Physics. Yang retired from Stony Brook University in 1999, assuming the title Emeritus Professor.

In the 1970s Yang worked on the topological properties of gauge theory, collaborating with Wu Tai-Tsun to elucidate the Wu–Yang monopole. Unlike the Dirac monopole, it has no singular Dirac string. Also devised by the Wu–Yang dictionary, the *Yang-Mills theory* set the template for the *Standard Model* and modern physics in general, as well as the work towards a Grand Unified Theory.

Yang visited the Chinese mainland in 1971 for the first time after the thaw in China–US relations, and has subsequently worked to help the Chinese physics community rebuild the research atmosphere which was destroyed by the radical political movements during the Cultural Revolution.

After retiring from Stony Brook, he returned as an honorary director of Tsinghua University, Beijing, where he is the Huang Jibei-Lu Kaiqun Professor at the Center for Advanced Study (CASTU). He is also a Distinguished Professor-at-Large at the Chinese University of Hong Kong.

Tu died in October 2003, and in December 2004 the then 82-year-old Yang caused a stir by marrying the then 28-year-old Weng Fan, calling Weng the "final blessing from God". Yang formally renounced his U.S. citizenship in late 2015.

Yang, C. N.[†] & Mills, R. (October, 1954). Conservation of Isotopic Spin and Isotopic Gauge Invariance*.

*Work performed under the auspices of the U. S. Atomic Energy Commission.

[†] On leave of absence from the Institute for Advanced Study, Princeton, New Jersey.

Phys. Rev., 96, 1, 191–5; https://journals.aps.org/pr/pdf/10.1103/PhysRev.96.191.

Brookhaven National Laboratory, Upton, New York.

Received June 28, 1954.

The *Yang–Mills theory*. Chen-Ning Yang and Robert Mills extended the concept of *gauge theory* for *abelian* groups, e.g. *quantum electrodynamics*, to *nonabelian* groups to provide an explanation for *strong interactions*.

[The *strong interaction*, also called the *strong force* or *strong nuclear force*, is a fundamental *interaction* that confines *quarks* into *protons*, neutrons, and other *hadron* particles. The *strong interaction* also binds *neutrons* and *protons* to create *atomic nuclei*, where it is called the *nuclear force*.

Most of the *mass* of a *proton* or *neutron* is the result of the *strong interaction* energy; the individual *quarks* provide only about 1% of the *mass* of a *proton*.

Before 1971, physicists were uncertain as to how the *atomic nucleus* was bound together. It was known that the *nucleus* was composed of *protons* and *neutrons* and that *protons* possessed positive *electric charge*, while *neutrons* were electrically neutral. By the understanding of physics at that time, positive charges would repel one another and the positively charged *protons* should cause the *nucleus* to fly apart. However, this was never observed. A stronger attractive force was postulated to explain how the *atomic nucleus* was bound despite the *protons'* mutual *electromagnetic* repulsion. This hypothesized force was called the *strong force*, which was believed to be a fundamental force that acted on the *protons* and *neutrons* that make up the *nucleus*.

In 1964, Murray Gell-Mann, and separately George Zweig, proposed that *baryons*, which include *protons* and *neutrons*, and *mesons* were composed of elementary particles. Zweig called the elementary particles "aces" while Gell-Mann called them "*quarks*"; the theory came to be called the *quark model*. [Gell-Mann. M. (February, 1964). A Schematic Model of Baryons and Mesons. See below.]

The *strong attraction* between *nucleons* was the side-effect of a more fundamental force that bound the *quarks* together into *protons* and *neutrons*. The theory of *quantum chromodynamics* explains that *quarks* carry what is called a *color charge*, although it has no relation to visible *color*. *Quarks* with unlike *color charge* attract one another as a result of the *strong interaction*, and the particle that mediates this was called the *gluon*.

Yang–Mills theory is a *quantum field theory* for *nuclear binding* devised by Chen Ning Yang and Robert Mills in 1953, as well as a generic term for the class of similar theories. The *Yang–Mills theory* is a *gauge theory* based on a special unitary group SU(n), or more generally any compact Lie group. A *Yang–Mills theory* seeks to describe the behavior of *elementary particles* using these non-abelian Lie groups and *is at the core of the unification of the electromagnetic force and weak forces* (i.e. U(1) × SU(2)) as well as *quantum chromodynamics*, the theory of the *strong force* (based on SU(3)). Thus, it forms the basis of the understanding of the *Standard Model* of particle physics.

Yang's core idea was to look for a *conserved quantity* in nuclear physics comparable to *electric charge* and use it to develop a corresponding *gauge theory* comparable to *electrodynamics*. He settled on *conservation of isospin*, a quantum number that distinguishes a *neutron* from a *proton*, but he made no progress on a theory. Taking a break from Princeton in the summer of 1953, Yang met a collaborator who could help: Robert Mills. As Mills himself describes:

> "During the academic year 1953–1954, Yang was a visitor to Brookhaven National Laboratory ... I was at Brookhaven also...and was assigned to the same office as Yang. Yang, who has demonstrated on a number of occasions his generosity to physicists beginning their careers, told me about his idea of generalizing *gauge invariance* and we discussed it at some length...I was able to contribute something to the discussions, especially with regard to the quantization procedures, and to a small degree in working out the formalism; however, the key ideas were Yang's."]

Abstract.

It is pointed out that the usual *principle of invariance under isotopic spin rotation* is not consistent with the concept of *localized fields*. The possibility is explored of having *invariance under local isotopic spin rotations*. This leads to formulating a *principle of isotopic gauge invariance* and the existence of a ***b*** *field which has the same relation to the isotopic spin that the electromagnetic field has to the electric charge*. The ***b*** *field* satisfies nonlinear differential equations. *The quanta of the **b** field are particles with spin unity, isotopic spin unity, and electric charge ± e or zero.*

INTRODUCTION

The *conservation of isotopic spin* is a much-discussed concept in recent years.

> [*Isospin*, also known as *isobaric spin* or *isotopic spin*, is a *quantum number* related to the *up- and down quark* content of the particle. *Isotopic spin symmetry* is a subset of the *flavor symmetry* seen more broadly in the interactions of *baryons* and *mesons*. The name of the concept contains the term spin because its quantum mechanical description is mathematically similar to that of *angular momentum* (in particular, in the way it *couples*; for example, a proton–neutron pair can be coupled either in a state of *total isospin* 1 or in one of 0). But unlike *angular momentum*, it is a dimensionless quantity and *is not actually any type of spin*.]

Historically an *isotopic spin parameter* was first introduced by Heisenberg[1] in 1932 to describe the two *charge states* (namely *neutron* and *proton*) of a *nucleon*.

> [1] Heisenberg, W. (January, 1932). Über den Bau der Atomkerne. I. (About the construction of atomic nuclei. I.); (March, 1932). Über den Bau der Atomkerne. II. (About the construction of atomic nuclei. II.); (September, 1933). Über den Bau der Atomkerne. III. See above.

The idea that the *neutron* and *proton* correspond to two *states* of the same particle was suggested at that time by the fact that their *masses* are nearly equal, and that the light stable even nuclei contain equal numbers of them. Then in 1937 Breit, Condon, and Present pointed out the approximate equality of p—p and n—p *interactions* in the ^{1}S *state*[2].

> [2] Breit, G., Condon, E. U., & Present, R. D. (November, 1936). Theory of Scattering of Protons by Protons. *Phys. Rev.*, 50, 825; https://journals.aps.org/pr/abstract/10.1103/PhysRev.50.825. Schwinger pointed out that the small difference may be attributed to magnetic interactions. [Schwinger, J. (April, 1950). On the Charge Independence of Nuclear Forces. *Phys. Rev.*, 78, 135; https://doi.org/10.1103/PhysRev.78.135.]

It seemed natural to assume that this equality holds also in the other states available to both the n—p and p—p systems. Under such an assumption one arrives at the concept of a *total isotopic spin*[3] which is conserved in nucleon-nucleon interactions.

> [3] The *total isotopic spin* T was first introduced by Wigner, E. (January, 1937). On the Consequences of the Symmetry of the Nuclear Hamiltonian on the Spectroscopy of Nuclei.

Phys. Rev. 51, 106 (1937), see above; Cassen, B. & Condon, E. U. (1936). On Nuclear Forces. *Phys. Rev.*, 50, 846; https://journals.aps.org/ pr/abstract/10.1103/PhysRev.50.846.

Experiments in recent years[4] on the *energy levels* of light nuclei strongly suggest that this assumption is indeed correct.

[4] Lauritsen, T. (1952). Energy Levels of Light Nuclei. *Ann. Rev. Nuclear Sci.*, 1, 67; https://doi.org/10.1146/annurev.ns.01.120152.000435; Inglis, D. R. (1953). The Energy Levels and the Structure of Light Nuclei. *Revs. Modern Phys.*, 25, 390; https://doi.org/ 10.1103/RevModPhys.25.390.

An implication of this is that all *strong interactions* such as the *pion-nucleon interaction*, must also satisfy the same conservation law. This and the knowledge that there are three charge states of the *pion*, and that *pions* can be coupled to the *nucleon field* singly, lead to the conclusion that *pions* have *isotopic spin unity*.

[A *pion* (or pi meson) is any of three subatomic particles: π^0, π^+, and π^-. Each *pion* consists of a *quark* and an *antiquark* and is therefore a *meson*. *Pions* are the lightest *mesons* and, more generally, the lightest *hadrons*. They are unstable, with the *charged pions* π^+ and π^- decaying after a mean lifetime of 26.033 nanoseconds (2.6033×10^{-8} seconds), and the *neutral pion* π^0 decaying after a much shorter lifetime of 85 attoseconds (8.5×10^{-17} seconds). *Charged pions* most often decay into *muons* and *muon neutrinos*, while *neutral pions* generally decay into *gamma rays*.

A *meson* is a type of hadronic subatomic particle composed of *an equal number of quarks and antiquarks, usually one of each, bound together by the strong interaction*. Because *mesons* are composed of *quark* subparticles, they have a meaningful physical size, a diameter of roughly one femtometre (10^{-15} m), which is about 0.6 times the size of a *proton* or *neutron*. *All mesons are unstable*, with the longest-lived lasting for only a few tenths of a nanosecond. Heavier *mesons* decay to lighter *mesons* and ultimately to stable *electrons, neutrinos* and *photons*.

The existence of *mesons* was predicted by Hideki Yukawa's 1935 *theory of mesons* that postulated the particle as mediating the nuclear force. [Yukawa, H. (1935). On the Interaction of Elementary Particles. *Proceedings of the Physico-Mathematical Society of Japan*, 17, 48.

In 1947, the *pion* (or *pi meson*) was discovered by members of C. F. Powell's group at the University of Bristol, in England. In the same year, Lattes, together with Powell and Occhialini, determined the new particle's mass. [Lattes, C. M. G.,

Muirhead, H., Occhialini, G. P. S. & Powell, C. F. (1947). Processes involving charged mesons. *Nature.* 159, 4047, 694–7; https://doi.org/10.1038/ 159694a0.]

In 1948, Lattes, Eugene Gardner, and their team first artificially produced *pions* at the University of California's cyclotron in Berkeley, California, by bombarding carbon atoms with high-speed alpha particles.]

A direct verification of this conclusion was found in the experiment of Hildebrand[5] which compares the differential *cross section* of the process $n + p \rightarrow \pi^0 + d$ with that of the previously measured process $p + p \rightarrow \pi^+ + d$.

[5] Hildebrand, R. H. (1953). Neutral Meson Production in *n−p* Collisions. *Phys. Rev.*, 89, 1090; https://doi.org/10.1103/PhysRev.89.1090.

The *conservation of isotopic spin* is identical with the requirement of *invariance of all interactions under isotopic spin rotation*. This means that when *electromagnetic interactions* can be neglected, as we shall hereafter assume to be the case, *the orientation of the isotopic spin is of no physical significance*. The differentiation between a *neutron* and a *proton* is then a purely arbitrary process. As usually conceived, however, this arbitrariness is subject to the following limitation: once one chooses what to call a *proton*, what a *neutron*, at one *space-time point*, one is then not free to make any choices at other *space-time points*.

It seems that this is not consistent with the localized field concept that underlies the usual physical theories. In the present paper we wish to explore the possibility of requiring all interactions to be invariant under independent rotations of the isotopic spin at all space-time points, so that the relative orientation of the isotopic spin at two space-time points becomes a physically meaningless quantity (the *electromagnetic field* being neglected).

We wish to point out that an entirely similar situation arises with respect to the ordinary *gauge invariance* of a *charged field* which is described by a *complex wave function* ψ.

[The term *gauge* refers to any specific mathematical formalism to regulate redundant degrees of freedom in the Lagrangian of a physical system. *Gauge fields are included in the Lagrangian to ensure its invariance under the local group transformations (called gauge invariance).*]

A change of *gauge*[6] means a change of *phase factor* $\psi \rightarrow \psi'$, $\psi' = (\exp i\alpha)\psi$, a change that is devoid of any physical consequences.

[6] Pauli, (May, 1941). Relativistic Field Theories of Elementary Particles. *Revs. Modern Phys.*, 13, 203; https://doi.org/10.1103/RevModPhys.13.203.

Since ψ may depend on x, y, z, and t, the relative *phase factor* of ψ at two different space-time points is therefore completely arbitrary. In other words, the arbitrariness in choosing the *phase factor* is *local* in character.

We define *isotopic gauge* as an arbitrary way of choosing the *orientation of the isotopic spin axes at all space-time points*, in analogy with the *electromagnetic gauge* which represents an arbitrary way of choosing the *complex phase factor of a charged field at all space-time points*. We then propose that all physical processes (not involving the *electromagnetic field*) *be invariant under an isotopic gauge transformation*, $\psi \rightarrow \psi'$, $\psi' = S^{-1}\psi$, where S represents a *space-time dependent isotopic spin rotation*.

To preserve *invariance*, one notices that in *electrodynamics* it is necessary to counteract the variation of α with x, y, z, and t by introducing the *electromagnetic field* A_μ which changes under a *gauge transformation* as

$$A_\mu' = A_\mu + 1/\varepsilon\, \partial\alpha/\partial x_\mu.$$

In an entirely similar manner, we introduce a B *field* in the case of the *isotopic gauge transformation* to counteract the dependence of S on x, y, z, and t. It will be seen that this natural generalization allows for very little arbitrariness. The *field equations* satisfied by the twelve independent components of the B *field*, which we shall call the **b** *field*, and their *interaction* with any *field* having an *isotopic spin* are essentially fixed, in much the same way that the *free electromagnetic field* and its *interaction* with *charged fields* are essentially determined by the requirement of *gauge invariance*.

In the following two sections we put down the mathematical formulation of the idea of *isotopic gauge invariance* discussed above. We then proceed to the *quantization* of the *field equations* for the **b** *field*. In the last section the properties of the *quanta* of the **b** *field* are discussed.

ISOTOPIC GAUGE TRANSFORMATION

Let ψ be a two-component *wave function* describing a *field* with *isotopic spin* ½. Under an *isotopic gauge transformation*, it transforms by

$$\psi = S\psi' \tag{1}$$

where S is a 2 x 2 *unitary matrix* with *determinant unity*. In accordance with the discussion in the previous section, we require, in analogy with the *electromagnetic case*, that all derivatives of ψ appear in the following combination:

$$(\partial_\mu - i\varepsilon\, B_\mu)\psi.$$

B_μ are 2 x 2 matrices such that[7] for $\psi = 1$, 2, and 3, B_μ is *Hermitian* and B_4 is *anti-Hermitian*.

> [A *Hermitian matrix* (or self-adjoint matrix) is a complex square matrix that is equal to its own conjugate transpose—that is, the element in the i-th row and j-th column is equal to the *complex conjugate* of the element in the j-th row and i-th column, for all indices i and j. The *complex conjugate* of a complex number is the number with an equal real part and an imaginary part equal in magnitude but opposite in sign.]

> [7] We use the conventions $\hbar = c = 1$, and $x_4 = it$. Bold-face type refers to vectors in isotopic space, not in space-time.

Invariance requires that

$$S(\partial_\mu - i\varepsilon\, B_\mu')\psi' = (\partial_\mu - i\varepsilon\, B_\mu)\psi. \tag{2}$$

Combining (1) and (2), we obtain the *isotopic gauge transformation* on B_μ:

$$B_\mu' = S^{-1}B_\mu S' + i/\varepsilon\; S^{-1}\partial S/\partial x_\mu. \tag{3}$$

The last term is similar to the *gradiant term* in the *gauge transformation* of *electromagnetic potentials*. *In analogy to the procedure of obtaining gauge invariant field strengths in the electromagnetic case*, we now define

$$F_{\mu\nu} = \partial B_\mu/\partial x_\nu - \partial B_\nu/\partial x_\mu + i\varepsilon\,(B_\mu B_\nu - B_\nu B_\mu). \tag{4}$$

One easily shows from (3) that

$$F_{\mu\nu}' = S^{-1}F_{\mu\nu}S \tag{5}$$

under an isotopic gauge transformation[†].

> [†] Note added in proof. — It may appear that B_μ could be introduced as an auxiliary quantity to accomplish *invariance*, but need not be regarded as a field variable by itself. It is to be emphasized that such a procedure violates the *principle of invariance*. Every quantity that is not a pure numeral (like 2, or M, or any definite representation of the γ matrices) should

206

be regarded as a dynamical variable, and should be varied in the Lagrangian to yield an equation of motion. Thus, the quantities B_μ must be regarded as independent fields.

Other simple functions of B than (4) do not lead to such a simple transformation property.

The above lines of thought can be applied to any *field* ψ with arbitrary *isotopic spin*. One need only use other *representations* S of rotations in three-dimensional space. *It is reasonable to assume that different fields with the same total isotopic spin, hence belonging to the same representation S, interact with the same matrix field B_μ.* (This is analogous to the fact that the *electromagnetic field* interacts in the same way with any *charged particle*, regardless of the nature of the particle. If different fields interact with different and independent B fields, there would be more conservation laws than simply the conservation of *total isotopic spin*.) To find a more explicit form for the *B fields* and to relate the B_μ's corresponding to different *representations* S, we proceed as follows.

Equation (3)

$$[B_\mu' = S^{-1}B_\mu S' + i/\varepsilon\, S^{-1}\, \partial S/\partial x_\mu. \tag{3}]$$

is valid for any S and its corresponding B_μ. Now the matrix $S^{-1}B_\mu S'$ appearing in (3) is a linear combination of the *isotopic spin "angular momentum" matrices* T^i (i = 1, 2, 3) corresponding to the *isotopic spin* of the ψ *field* we are considering. So B_μ itself must also contain a linear combination of the matrices T^i. But any part of B_μ in addition to this, B^*_μ say, is a scalar or tensor combination of the T's, and must transform by the homogeneous part of (3), $B^*_\mu' = S^{-1}B^*_\mu S'$. Such a field is extraneous; it was allowed by the very general form we assumed for the *B field*, but is irrelevant to the question of *isotopic gauge*. Thus, the relevant part of the *B field* is of the form

$$B_\mu = 2\mathbf{b}_\mu \cdot \mathbf{T}. \tag{6}$$

(Bold-face letters denote *three-component vectors* in isotopic space.) To relate the $\mathbf{b}_\mu$'s corresponding to different representations S we now consider the product representation $S = S^{(a)}S^{(b)}$. The *B field* for the combination transforms, according to (3), by

$$B_\mu' = \dots .$$

But the sum of $B_\mu^{(a)}$ and $B_\mu^{(b)}$, the *B fields* corresponding to $S^{(a)}$ and $S^{(b)}$, transforms in exactly the same way, so that

$$B_\mu = B_\mu^{(a)} + B_\mu^{(b)}$$

(plus possible terms which transform homogeneously, and hence are irrelevant and will not be included).

Decomposing $S^{(a)}S^{(b)}$ into irreducible *representations*, we see that the twelve-component field $\mathbf{b}_\mu$ in Eq. (6) is the same for all *representations*.

To obtain the *interaction* between any *field* ψ of arbitrary *isotopic spin* with the *b field* one therefore simply replaces the gradient of ψ by

$$(\partial_\mu - 2i\varepsilon\, \mathbf{b}_\mu \,.\, \mathbf{T})\psi, \tag{7}$$

where T^i ($i = 1, 2, 3$), as defined above, are the *isotopic spin* "angular momentum" matrices for the *field* ψ.

We remark that the nine components of $\mathbf{b}_\mu$, $\mu = 1, 2, 3$ are real and the three of $\mathbf{b}_4$ are pure imaginary. The *isotopic-gauge covariant* field quantities $F_{\mu\nu}$ are expressible in terms of $\mathbf{b}_\mu$:

$$F_{\mu\nu} = 2\, \mathbf{f}_{\mu\nu} \,.\, \mathbf{T}, \tag{8}$$

where

$$\mathbf{f}_{\mu\nu} = \partial\mathbf{b}_\mu/\partial x_\nu - \partial\mathbf{b}_\nu/\partial x_\mu - 2i\varepsilon\, \mathbf{b}_\mu \times \mathbf{b}_\nu. \tag{9}$$

$\mathbf{f}_{\mu\nu}$ transforms like a vector under an *isotopic gauge transformation*. Obviously, the same $\mathbf{f}_{\mu\nu}$ *interact* with all *fields* ψ irrespective of the *representation* S that it belongs to.

The corresponding transformation of $\mathbf{b}_\mu$ is cumbersome. One need, however, study only the infinitesimal *isotopic gauge transformations*,

$$S = 1 - 2i\mathbf{T} \,.\, \delta\omega.$$

Then

$$\mathbf{b}_\mu' = \mathbf{b}_\mu + 2\, \mathbf{b}_\mu \times \delta\omega + 1/\varepsilon\, \partial/\partial x_\mu\, \delta\omega. \tag{10}$$

FIELD EQUATIONS

To write down the *field equations* for the *b field* we clearly only want to use *isotopic gauge invariant* quantities. In analogy with the *electromagnetic* case, we therefore write down the following *Lagrangian density*[8]

$$- \tfrac{1}{4}\, \mathbf{f}_{\mu\nu} \,.\, \mathbf{f}_{\mu\nu}.$$

[8] Repeated indices are summed over, except where explicitly stated otherwise. Latin indices are summed from 1 to 3, Greek ones from 1 to 4.

Since the inclusion of a *field* with *isotopic spin* 2 is illustrative, and does not complicate matters very much, we shall use the following *total Lagrangian density*:

$$\mathscr{L} = -\tfrac{1}{4}\, \mathbf{f}_{\mu\nu} \cdot \mathbf{f}_{\mu\nu} - \psi^* \gamma_\mu (\partial_\mu - i\varepsilon\, \boldsymbol{\tau} \cdot \mathbf{b}_\mu)\psi - m\psi^*\psi \tag{11}$$

One obtains from this the following *equations of motion*:

$$\partial\mathbf{f}_{\mu\nu}/\partial x_\nu + 2\varepsilon\, (\mathbf{b}_\nu \times \mathbf{f}_{\mu\nu}) + \mathbf{J}_\mu = 0, \tag{12}$$
$$\gamma_\mu(\partial_\mu - i\varepsilon\, \boldsymbol{\tau} \cdot \mathbf{b}_\mu)\psi + m\psi = 0,$$

where

$$\mathbf{J}_\mu = i\varepsilon\, \psi^* \gamma_\mu\, \boldsymbol{\tau}\psi. \tag{13}$$

The *divergence* of $\mathbf{J}_\mu$ does not vanish. Instead, it can easily be shown from (13) that

$$\partial\mathbf{J}_\mu/\partial x_\mu = -2\varepsilon\mathbf{b}_\mu \times \mathbf{J}_\mu. \tag{14}$$

If we define, however,

$$\mathfrak{J}_\mu = \mathbf{J}_\mu + 2\varepsilon\mathbf{b}_\nu \times \mathbf{f}_{\mu\nu}, \tag{15}$$

then (12) leads to the *equation of continuity*,

$$\partial\mathfrak{J}_\mu/\partial x_\mu = 0. \tag{16}$$

$\mathfrak{J}_{1,2,3}$ and $\mathfrak{J}_4$ are respectively the *isotopic spin current density* and *isotopic spin density* of the system. The *equation of continuity* guarantees that the *total isotopic spin*

$$\mathbf{T} = \int \mathfrak{J}_4 \, d^3x$$

is independent of time and independent of a Lorentz transformation. It is important to notice that $\mathfrak{J}_\mu$ like $\mathbf{b}_\mu$ does not transform exactly like vectors under *isotopic space rotations*. But the *total isotopic spin*,

$$\mathbf{T} = -\int \partial\mathbf{f}_{4i}/\partial x_i \, d^3x,$$

is the integral of the *divergence* of $\mathbf{f}_{4i}$, *which transforms like a true vector under isotopic spin space rotations*. Hence, under a *general isotopic gauge transformation*, if $S \to S_0$ on an infinitely large sphere, $\mathbf{T}$ would transform like an *isotopic spin vector*.

Equation (15)
$$[\mathfrak{J}_\mu = \mathbf{J}_\mu + 2\varepsilon\mathbf{b}_\nu \times \mathbf{f}_{\mu\nu}, \tag{15}]$$

shows that the *isotopic spin* arises both from the *spin-½ field* ($\mathbf{J}_\mu$) and from the $\boldsymbol{b}_\mu$ *field* itself. Inasmuch as the *isotopic spin* is the source of the $\boldsymbol{b}$ *field*, this fact makes the *field equations* for the $\boldsymbol{b}$ *field* nonlinear, even in the absence of the *spin-½ field*. This is different from the case of the *electromagnetic field*, which is itself *chargeless*, and consequently satisfies linear equations in the absence of a *charged field*.

The Hamiltonian derived from (11)
$$[\mathscr{L} = -\tfrac{1}{4}\, \mathbf{f}_{\mu\nu} \cdot \mathbf{f}_{\mu\nu} - \psi^*\gamma_\mu(\partial_\mu - i\varepsilon\, \boldsymbol{\tau} \cdot \mathbf{b}_\mu)\psi - m\psi^*\psi \tag{11}]$$
is easily demonstrated to be positive definite in the absence of the field of *isotopic spin* ½. The demonstration is completely identical with the similar one in *electrodynamics*.

We must complete the set of *equations of motion* (12) and (13)
$$[\partial \mathbf{f}_{\mu\nu}/\partial x_\nu + 2\varepsilon\, (\mathbf{b}_\nu \times \mathbf{f}_{\mu\nu}) + \mathbf{J}_\mu = 0, \tag{12}$$
$$\gamma_\mu(\partial_\mu - i\varepsilon\, \boldsymbol{\tau} \cdot \mathbf{b}_\mu)\psi + m\psi = 0,$$
where
$$\mathbf{J}_\mu = i\varepsilon\, \psi^*\gamma_\mu\, \boldsymbol{\tau}\psi. \tag{13}]$$
by the supplementary condition,

$$\partial \mathbf{b}_\mu/\partial x_\mu = 0, \tag{17}$$

which serves to eliminate the *scalar* part of the field in $\mathbf{b}_\mu$. This clearly imposes a condition on the possible *isotopic gauge transformations*. That is, the infinitesimal *isotopic gauge transformation* $S = 1 - 2i\, \boldsymbol{\tau} \cdot \delta\omega$ must satisfy the following condition:

$$2\, \mathbf{b}_\mu \times \partial/\partial x\, \delta\omega + 1/\varepsilon\, \partial^2/\partial x_\mu^2\, \delta\omega = 0. \tag{18}$$

This is the analog of the equation $\partial^2\alpha/\partial x_\mu^2 = 0$ that must be satisfied by the *gauge transformation* $A_\mu' = A_\mu + 1/\varepsilon\, (\partial\alpha/\partial x_\mu)$ of the *electromagnetic field*.

QUANTIZATION

To quantize, it is not convenient to use the *isotopic gauge invariant Lagrangian density* (11). This is quite similar to the corresponding situation in *electrodynamics* and we adopt the customary procedure of using a *Lagrangian density* which is *not obviously gauge invariant*:

$$\mathscr{L} = \dots . \tag{19}$$

The *equations of motion* that result from this Lagrangian density can be easily shown to imply that

$$\dots$$
where

. . . .

Thus if, consistent with (17),

$$[\partial \mathbf{b}_\mu / \partial x_\mu = 0, \qquad\qquad\qquad (17)]$$

we put on one space-like surface $\mathbf{a} = 0$ together with $\partial \mathbf{a} / \partial t = 0$, it follows that $\mathbf{a} = 0$ at all times. Using this supplementary condition one can easily prove that the *field equations* resulting from the *Lagrangian densities* (19) and (11) are identical.

One can follow the canonical method of quantization with the *Lagrangian density* (19). Defining

$$\Pi_\mu = -\partial \mathbf{b}_\mu / \partial x_4 + 2\varepsilon \, (\mathbf{b}_\mu \times \mathbf{b}_4),$$

one obtains the *equal-time commutation rule*

$$\ldots ., \qquad\qquad\qquad (20)$$

where $b_\mu{}^i$, $i = 1, 2, 3$, are the three components of $\mathbf{b}_\mu$. The *relativistic invariance* of these *commutation rules* follows from the general proof for canonical methods of quantization given by Heisenberg and Pauli[9].

[9] Heisenberg W. & Pauli, W. (July, 1929). Zur Quantendynamik der Wellenfelder. (On the quantum dynamics of wave fields.) *Zeit. Phys.*, 56, 1-61; https://www.neo-classical-physics.info/uploads/3/4/3/6/34363841/heisenberg_and_pauli_-_qed_i.pdf; (January,1930). Zur Quantendynamik der Wellenfelder II. (On the quantum dynamics of wave fields II.) *Ibid.*, 59, 168-90; https://doi.org/10.1007/BF01341423; translation by D. H. Delphenich; https://neo-classical-physics.info/electromagnetism.html; also in Underwood T. G. (2023). *Quantum Electrodynamics - annotated sources.* Volume I; see Part II above.)

The Hamiltonian derived from (19) is identical with the one from (11), in virtue of the *supplementary condition*. Its *density* is

$$H = H_0 + H_{int}, \qquad\qquad\qquad (21)$$
$$H_0 = \ldots ,$$
$$H_{int} = \ldots .$$

The quantized form of the *supplementary condition* is the same as in *quantum electrodynamics*.

211

PROPERTIES OF THE b QUANTA

The *quanta* of the **b** field clearly have *spin unity* and *isotopic spin unity*. We know their *electric charge* too because all the *interactions* that we proposed must satisfy the law of *conservation of electric charge*, which is exact. The two *states* of the *nucleon*, namely *proton* and *neutron*, differ by *charge unity*. Since they can transform into each other through the emission or absorption of a **b** *quantum*, the latter must have three *charge states* with *charges* ± e and 0. Any measurement of *electric charges* of course involves the *electromagnetic field*, which necessarily introduces a *preferential direction* in *isotopic space* at all space-time points. Choosing the *isotopic gauge* such that this preferential direction is along the z axis in *isotopic space*, one sees that for the *nucleons*

$$Q = \text{electric charge} = e \ (\tfrac{1}{2} + \varepsilon^{-1} T^z),$$

and for the **b** *quanta*

$$Q = (e/\varepsilon) \ T^z.$$

The *interaction* (7)
$$[(\partial_\mu - 2i\varepsilon \ \mathbf{b}_\mu . \mathbf{T})\psi, \tag{7)]}$$
then fixes the *electric charge* up to an additive constant for all fields with any *isotopic spin*:

$$Q = e(\varepsilon^{-1} T^z + R). \tag{22}$$

The constants R for two *charge conjugate fields* must be equal but have opposite signs[10].

[10] See Gell-Mann, M. (1953). Isotopic Spin and New Unstable Particles. *Phys. Rev.*, 92, 833; https://doi.org/10.1103/PhysRev.92.833.

...

Fig. 1. Elementary vertices for b fields and nucleon fields. Dotted lines refer to b field, solid lines with arrow refer to nucleon field.

We next come to the question of the *mass* of the **b** *quantum*, to which we do not have a satisfactory answer. One may argue that without a *nucleon field* the Lagrangian would contain no quantity of the dimension of a *mass*, and that therefore the *mass* of the **b** *quantum* in such a case is zero. This argument is however subject to the criticism that, *like all field theories, the **b** field is beset with divergences, and dimensional arguments are not satisfactory*.

One may of course try to apply to the **b** field the *methods for handling infinities developed for quantum electrodynamics*. Dyson's approach[11] is best suited for the present case.

[11] Dyson, F. J. (1949). The Radiation Theories of Tomonaga, Schwinger, and Feynman. *Phys. Rev.* 75, 486; https://journals.aps.org/pr/abstract/10.1103/PhysRev.75.486; 1736; also in Underwood T. G. (2023). *Quantum Electrodynamics - annotated sources*. Volume II, pp. 441-69.

One first transforms into the *interaction representation* in which the state vector Ψ satisfies

$$i\partial\Psi/\partial t = H_{int}$$

where H_{int} was defined in Eq. (21). The *matrix elements* of the *scattering matrix* are then formulated in terms of contributions from *Feynman diagrams*. These diagrams have three elementary types of vertices illustrated in Fig. 1, instead of only one type as in *quantum electrodynamics*. The "*primitive divergences*" are still finite in number and are listed in Fig. 2.

…

Fig. 2. Primitive divergences.

Of these, the one labeled *a* is the one that effects the *propagation function* of the **b** *quantum*, and whose singularity determines the *mass* of the **b** *quantum*. In *electrodynamics*, by the requirement of *electric charge conservation*[12], it is argued that the *mass* of the *photon* vanishes.

[12] Schwinger, J. (1949). Quantum Electrodynamics. III. The Electromagnetic Properties of the Electron—Radiative Corrections to Scattering. *Phys. Rev.*, 76, 790; https://journals.aps.org/pr/abstract/10.1103/PhysRev.76.790; also in Underwood T. G. (2023). *Quantum Electrodynamics - annotated sources*. Volume II, pp. 569-85.

Corresponding arguments in the **b** *field* case do not exist[13] even though the *conservation of isotopic spin* still holds.

[13] In *electrodynamics* one can formally prove' that $G_{\mu\nu}k_\nu = 0$, where $G_{\mu\nu}$ is defined by Schwinger's Eq. (A12). ($G_{\mu\nu}A_\nu = 0$ is the current generated through virtual processes by the arbitrary external field A_ν.) No corresponding proof has been found for the present case. This is due to the fact that in *electrodynamics* the conservation of charge is a consequence of the *equation of motion* of the *electron field* alone, quite independently of the *electromagnetic field* itself. In the present case the **b** field carries an *isotopic spin* and destroys such general conservation laws.

We have therefore not been able to conclude anything about the *mass* of the **b** *quantum*.

A conclusion about the *mass* of the **b** *quantum* is of course very important in deciding whether the proposal of the existence of the **b** *field* is consistent with experimental information. For example, it is inconsistent with present experiments to have their *mass* less than that of the *pions*, because among other reasons they would then be created abundantly at high energies and the charged ones should live long enough to be seen. If they have a *mass* greater than that of the *pions*, on the other hand, they would have a short *lifetime* (say, less than 10^{-20} sec) for decay into *pions* and *photons* and would so far have escaped detection.

Sakata, S. (September, 1956). On a Composite Model for the New Particles*.

Progr. Theor. Phys., 16, 6, 686-8; https://doi.org/10.1143/PTP.16.686.

Institute for Theoretical Physics, Nagoya University, Nagoya

> * The content of this letter was read before the annual meeting of the Japanese Physical Society held in October 1955. A note on the same subject has also been published in *Bulletin de L' academie Polonaise des Sciences* (Cl, lll·vol. IV, No. 6, 1956)

September 3, 1956.

In 1956, Sakata proposed his *Sakata Model* which explains the physics behind the *Nakano-Nishijima-Gell-Mann (NNG) rule* by postulating that the fundamental building blocks of all strongly interacting particles are the *proton*, the *neutron* and the *lambda baryon*. The positively charged *pion* is made out of a *proton* and an *anti-neutron*, in a manner similar to the *Fermi-Yang composite Yukawa meson model*, while the positively charged *kaon* is composed of a *proton* and an *anti-lambda*. Aside from the integer *charges*, the *proton*, *neutron*, and *lambda* have similar properties as the *up quark*, *down quark*, and *strange quark* respectively.

Recently, *Nishijima-Gell-Mann's rule*[1] for the *systematization* of new particles has achieved a great success to account for various facts obtained from the experiments with cosmic rays and with high energy accelerators.

> [1] Nakano, T. & Nishijima, K. (November, 1953). Charge Independence for *V*-particle. *Prog. Theor. Phys.*, 10, 5, 581-2; https://doi.org/10.1143/PTP.10.581; Nishijima, K. (1954). *Prog. Theor. Phys.*, 12, 107; (March, 1955). Charge Independence Theory of *V* Particles. *Ibid.*, 13, 3, 285-304; https://doi.org/10.1143/PTP.13.285; Gell-Mann, M. (November, 1953). Isotopic Spin and New Unstable Particles. *Phys. Rev.*, 92, 3, 833; https://doi.org/10.1103/PhysRev.92.833.

Nevertheless, it would be desirable from the theoretical standpoint to find out a more profound meaning hidden behind this rule. The purpose of this work is concerned with this point.

It seems to me that the present state of the theory of new particles is very similar to that of the atomic nuclei 25 years ago. At that time, we had known a beautiful relation between the *spin* and the *mass* number of the atomic nuclei. Namely, the *spin* of the *nucleus* is always integer if the *mass number* is even, whereas the former is always half integer if the latter is odd. But unfortunately, we could not understand the profound meaning for this even-odd rule. This fact together with other mysterious properties of the *atomic nuclei*, for instance the *beta disintegration* in which the *conservation of energy* seemed to be invalid, led us to a very pessimistic view-point that the *quantum theory* would not be applicable in the domain of the *atomic nucleus*. However, the situation was entirely changed after the discovery of the *neutron*. Iwanenko and Heisenberg[2] proposed immediately a new model for the *atomic nuclei* in which *neutrons* and *protons* are considered to be their constituents.

[2] Iwanenko, D. (May, 1932). The Neutron Hypothesis. *Nature*, 129, 798; https://doi.org/ 10.1038/129798d0; Heisenberg, W. (January, 1932). Über den Bau der Atomkerne. I. (About the Structure of Atomic Nuclei. I.) *Zeit. Phys.*, 77, 1-11; https://doi.org/10.1007/ BF01342433.

By assuming that the *neutron* has the spin of one half, they explained the even-odd rule for the *spins* of *atomic nuclei* as the result of the addition law for the *angular momenta* of the constituents. Moreover, they could reduce all the mysterious properties of *atomic nuclei* to those of the *neutron* contained in them.

Supposing that the similar situation is realized at present, I proposed a compound hypothesis for new unstable particles to account for *Nishijima-Gell-Mann's rule. In our model, the new particles are considered to be composed of four kinds of fundamental particles in the true sense, that is, nucleon, anti-nucleon, Λ^0 and anti-Λ^0.* If we assume that Λ^0 has such intrinsic properties as were assigned by Nishijima and Gell-Mann, we can easily get their even-odd rule for the composite particles as the result of the addition laws for the ordinary *spin*, the *isotopic spin* and the *strangeness*. In the next table, the models and the properties of the new particles are shown together with those of the fundamental particles in the true sense.

…

So far as the internal structure is not concerned, our model for new particles is identical with that of Nishijima and Gell-Mann. However, it should be stressed that the curious properties of the new particles could be reduced to those of Λ^0, just like the mysterious properties of the *atomic nuclei* were reduced to those of *neutron*. Hence our theory contains less arbitrary elements than was the case for original one of Nishijima and Gell-Mann.

Though the rigorous treatment of our model is a very hard task[4], it is worthwhile to notice that most of the composite particles which seem to be stable against the *strong interaction* can be identified with the well-known new particles, and that there are possibilities of predicting some more new particles which have not been discovered up till now[5].

[4] Tanaka, S. (1956). *Prog. Theor. Phys.*, 16, 625. Maki, Z. (1956). *Prog. Theor. Phys.*, 16, 667.

[5] Matsumoto, K. (1956). *Prog. Theor. Phys.*, 16, 583.

Finally, it should be remarked that there are some other arguments in favor of the compound hypothesis for the elementary particles. In spite of the great success achieved by the advent of *Tomonaga-Schwinger's technique*, it has recently become clear that *we could not avoid the internal inconsistency of the quantum field theory*, so far as the *point model* for elementary particles was adopted. Moreover, in the case of π-*meson*, the cut-off prescription has recently been proved to be very powerful in order to account for the experimental results. These facts indicate strongly the necessity of substantial innovations in the model for the elementary particles, though some change has already been made by the discovery of the *renormalization* technique. Landau pointed out that the model for the *electron* would possibly be changed by the effect of the *gravitational field*. But in the case of π-*meson* we must look for another effect, because the cut-off radius is found to be as large as the order of the *nucleon Compton wave length* in contrast to $e^2/mc^2 \cdot e^{-137} \sim 10^{-58}$ cm which appeared in the quantum electrodynamics[6].

[6] Markov, M. A. (1953). *Uspekhi Fiz. Nauk*, 51, 317; L. Landau, L. et al. (1954). *DAN.*, 95, 497; 733; 1177; (1954). *Ibid.*, 96, 261; (1955). *Ibid.*, 102, 489; S. Kamefuchi, S. & Umezawa, H. (1956). *Prog. Theor. Phys.*, 15, 298; (1956). *Nuevo Cimento*, 3, 1060.

Tsung-Dao Lee (November 24, 1926-August 4, 2024).

Lee was a Chinese-American physicist, known for his work on parity violation, the Lee–Yang theorem, particle physics, relativistic heavy ion physics, non-topological *solitons*, and *soliton* stars. He was a university professor emeritus at Columbia University in New York City, where he taught from 1953 until his retirement in 2012.

In 1957, at the age of 30, Lee won the Nobel Prize in Physics with Chen Ning Yang for their work on the violation of the parity law in weak interactions. Lee was the third-youngest Nobel laureate in sciences after William L. Bragg (who won the prize at 25 with his father William H. Bragg in 1915) and Werner Heisenberg (who won in 1932 also at 30). Lee and Yang were the first Chinese laureates.

Lee was born in Shanghai, China, with his ancestral home in nearby Suzhou. His father Chun-kang Lee, one of the first graduates of the University of Nanking, was a chemical industrialist and merchant who was involved in China's early development of modern synthesized fertilizer. Lee has four brothers and one sister.

Lee received his secondary education in Shanghai (High School Affiliated to Soochow University) and Jiangxi (Jiangxi Joint High School). Due to the Second Sino-Japanese war, Lee's high school education was interrupted, thus he did not obtain his secondary diploma. Nevertheless, in 1943, Lee directly applied to and was admitted by the National Che Kiang University (now Zhejiang University). Initially, Lee registered as a student in the Department of Chemical Engineering. Very quickly, Lee's talent was discovered and his interest in physics grew rapidly. Several physics professors, including Shu Xingbei and Wang Ganchang, largely guided Lee, and he soon transferred into the Department of Physics of National Che Kiang University, where he studied in 1943–1944.

However, again disrupted by a further Japanese invasion, Lee continued at the National Southwestern Associated University in Kunming the next year, in 1945, where he studied with Professor Wu Ta-You.

In 1946, Lee went to the University of Chicago and was selected by Professor Enrico Fermi to become his PhD student. Lee received his PhD under Fermi in 1950 for his research work *Hydrogen Content of White Dwarf Stars*. Lee served as research associate and lecturer in physics at the University of California at Berkeley from 1950 to 1951.
In 1950, Lee married Jeannette Hui-Chun Chin. They have two sons: James Lee (born 1952) and Stephen Lee (born 1956).

In 1953, Lee joined Columbia University, where he remained until retirement. His first work at Columbia was on a solvable model of *quantum field theory* better known as the

Lee model. Soon, his focus turned to particle physics and the developing puzzle of K meson decays. Lee realized in early 1956 that the key to the puzzle was *parity non-conservation.* At Lee's suggestion, the first experimental test was on *hyperion decay* by the Steinberger group. At that time, the experimental result gave only an indication of a 2 standard deviation effect of possible *parity violation.* Encouraged by this feasibility study, Lee made a systematic study of possible *Time reversal* (T), *Parity* (P), *Charge Conjugation* (C), and *CP violations* in *weak interactions* with collaborators, including C. N. Yang. After the definitive experimental confirmation that showed that parity was not conserved, Lee and Yang were awarded the 1957 Nobel Prize in Physics.

In the early 1960s, Lee and collaborators initiated the important field of high-energy neutrino physics. Beginning in 1975, Lee and collaborators established the field of non-topological solitons, which led to his work on soliton stars and black holes throughout the 1980s and 1990s.

From 1997 to 2003, Lee was director of the RIKEN-BNL Research Center (now director emeritus), which together with other researchers from Columbia, completed a 1 teraflops supercomputer QCDSP for lattice quantum chromodynamics in 1998 and a 10 teraflops QCDOC machine in 2001.

Lee died at his home in San Francisco on August 4, 2024, at age 97.

Lee, T. D. & Yang, C. N. † (October, 1956). Question of Parity Conservation in Weak Interactions.*

Phys. Rev., 104, 254; https://journals.aps.org/pr/abstract/10.1103/PhysRev.104.254.

* Work supported in part by the U. S. Atomic Energy Commission.
† Permanent address: Institute for Advanced Study, Princeton, New Jersey.

———————————

Abstract.

The question of *parity conservation* in *β decays* and in *hyperon* and *meson decays* is examined. Possible experiments are suggested which might test *parity conservation* in these interactions.

[A *parity transformation* (also called *parity inversion*) is the flip in the sign of one spatial coordinate. In three dimensions, it can also refer to the simultaneous flip in the sign of all three spatial coordinates (a point reflection). It can also be thought of as a test for *chirality* of a physical phenomenon, in that a parity inversion transforms a phenomenon into its mirror image. All fundamental interactions of elementary particles, *with the exception of the weak interaction*, are symmetric under *parity*.

A *hyperon* is any *baryon* containing one or more *strange quarks*, but no *charm*, *bottom*, or *top quark*. Being *baryons*, all *hyperons* are *fermions*. That is, they have half-integer *spin* and obey Fermi–Dirac statistics. *Hyperons* all interact via the *strong nuclear force*, making them types of *hadron*. They are composed of three light *quarks*, at least one of which is a *strange quark*, which makes them *strange baryons*.

This form of matter may exist in a stable form within the core of some neutron stars. The first research into *hyperons* happened in the 1950s and spurred physicists on to the creation of an organized classification of particles. The term was coined by French physicist Louis Leprince-Ringuet in 1953.

Today, research in this area is carried out on data taken at many facilities around the world, including CERN, Fermilab, SLAC, JLAB, Brookhaven National Laboratory, KEK, GSI and others. Physics topics include searches for CP violation, measurements of spin, studies of excited states (commonly referred to as spectroscopy), and hunts for exotic forms such as *pentaquarks* and *dibaryons*.]

———————————

Recent experimental data indicate closely identical *masses*[1] and *lifetimes*[2] of the θ^+ ($\equiv K_{\pi 2}^+$) and the τ^+ ($\equiv K_{\pi 3}^+$) *mesons*.

[1] Whitehead, Stork, Perkins, Peterson, & Birge. (1956). *Bull. Am. Phys. Soc.*, Ser. II, 1, 184 (1956); Barkas, Heckman, & Smith. (1956). *Bull. Am. Phys. Soc.*, Ser. II, 1, 184.
[2] Harris, Orear, & Taylor. (1955). *Phys. Rev.*, 100, 932; Fitch, V. & Motley, K. (1956). *Phys. Rev.*, 101, 496; Alvarez, Crawford, Good, & Stevenson. (1956). *Phys. Rev.*, 101, 503.

On the other hand, analyses[3] of the decay products of τ^+ strongly suggest on the grounds of *angular momentum* and *parity conservation* that the τ^+ and θ^+ are not the same particle.

[3] Dalitz, R. (1953). *Phil. Mag.*, 44, 1068; Fabri, E. (1954). *Nuovo Cimento*, 11, 479. See Orear, Harris, & Taylor. (1956). *Phys. Rev.*, 102, 1676 for recent experimental results.

This poses a rather puzzling situation that has been extensively discussed[4].

[4] See, e.g., Report of the Sixth Annual Rochester Conference on High Energy Physics (Interscience Publishers, Inc., New York to be published).

One way out of the difficulty is to assume that *parity* is not strictly conserved, so that θ^+ and τ^+ are two different decay modes of the same particle, which necessarily has a single *mass value* and a single *lifetime. We wish to analyze this possibility in the present paper against the background of the existing experimental evidence of parity conservation.* It will become clear that existing experiments do indicate *parity conservation* in *strong* and *electromagnetic* interactions to a high degree of accuracy, but that for the *weak interaction* actions (i.e., *decay interactions* for the *mesons* and *hyperons*, and various *Fermi interactions*) *parity conservation* is so far only an extrapolated hypothesis unsupported by experimental evidence. (One might even say that the present θ - τ puzzle may be taken as an indication that *parity conservation* is violated in *weak interactions*. This argument is, however, not to be taken seriously because of the paucity of our present knowledge concerning the nature of the strange particles. It supplies rather an incentive for an examination of the question of *parity conservation*.) To decide unequivocally whether parity is conserved in *weak interactions*, one must perform an experiment to determine whether *weak interactions* differentiate the right from the left. Some such possible experiments will be discussed.

[The *weak interaction*, also called the *weak force*, is the mechanism of *interaction* between subatomic particles that is responsible for the *radioactive decay* of atoms: the *weak interaction* participates in nuclear fission and nuclear fusion. The effective range of the *weak force* is limited to subatomic distances and is less than the diameter of a *proton*.

In 1933, Enrico Fermi proposed the first theory of the *weak interaction*, known as *Fermi's interaction*. He suggested that *beta decay* could be explained by a four-*fermion* interaction, involving a contact force with no range. [Fermi, E. (March, 1934) Versuch einer Theorie der β-Strahlen. I. (Attempt at a theory of β rays. I.). See below.]

> [A *fermion* is a particle that follows *Fermi–Dirac statistics*. *Fermions* have a half-odd-integer spin (spin 1/2, spin 3/2, etc.) and obey the *Pauli exclusion principle*. These particles include all *quarks* and *leptons* and all composite particles made of an odd number of these, such as all *baryons* and many atoms and nuclei. Some *fermions* are elementary particles (such as *electrons*), and some are composite particles (such as *protons*). *Fermions* differ from *bosons*, which have integer *spin* and obey *Bose–Einstein statistics*.]

In the mid-1950s, Chen-Ning Yang and Tsung-Dao Lee first suggested that the handedness of the *spins* of particles in *weak interaction* might violate the *conservation law* or *symmetry*. [Lee, T. D. & Yang, C. N. (October, 1956). Question of Parity Conservation in Weak Interactions. See below.]

In the 1960s, Sheldon Glashow, Abdus Salam and Steven Weinberg unified the *electromagnetic force* and the *weak interaction* by showing them to be two aspects of a single force, now termed the *electroweak force*. [Glashow, S. L. (February, 1961). Partial-symmetries of weak interactions. See below.]

The *Standard Model* of particle physics provides a uniform framework for understanding *electromagnetic, weak,* and *strong interactions*. A *weak interaction* occurs when two particles (typically, but not necessarily, *half-integer spin fermions*) exchange *integer-spin*, force-carrying *bosons*. The *fermions* involved in such exchanges can be either *elementary* (e.g. *electrons* or *quarks*) or *composite* (e.g. *protons* or *neutrons*), although at the deepest levels, all *weak interactions* ultimately are between *elementary particles*.

In the *weak interaction, fermions* can exchange three types of force carriers, namely W+, W−, and Z *bosons*. The *masses* of these *bosons* are far greater than the *mass* of a *proton* or *neutron*, which is consistent with the short range of the *weak force*. In fact, the force is termed *weak* because its field strength over any set distance is typically several orders of magnitude less than that of the *electromagnetic force*, which itself is further orders of magnitude less than the *strong nuclear force*.

The existence of the W and Z *bosons* was not directly confirmed until 1983.

The *weak interaction* is the only fundamental *interaction* that breaks *parity symmetry*, and similarly, but far more rarely, the only *interaction* to break *charge–parity symmetry*.

The *weak interaction* is unique in that it allows *quarks* to swap their flavor for another. *Quarks*, which make up composite particles like *neutrons* and *protons*, come in six "*flavors*" – *up*, *down*, *charm*, *strange*, *top* and *bottom* – which give those composite particles their properties. The swapping of those properties is mediated by the force carrier *bosons*. For example, during *beta-minus decay*, a *down quark* within a *neutron* is changed into an *up quark*, thus converting the *neutron* to a *proton* and resulting in the emission of an *electron* and an *electron antineutrino*.

Most *fermions* decay by a *weak interaction* over time. Such decay makes radiocarbon dating possible, as carbon-14 decays through the *weak interaction* to nitrogen-14. It can also create *radioluminescence*, commonly used in tritium luminescence, and in the related field of betavoltaics (but not similar radium luminescence).

The *electroweak force* is believed to have separated into the *electromagnetic* and *weak forces* during the *quark* epoch of the early universe.]

PRESENT EXPERIMENTAL LIMIT ON PARITY NONCONSERVATION

If *parity* is not strictly conserved, all atomic and nuclear *states* become mixtures consisting mainly of the *state* they are usually assigned, together with small percentages of *states* possessing the opposite parity. The fractional weight of the latter will be called $\mathscr{F}^2$. It is a quantity that characterizes the *degree of violation of parity conservation*.

The existence of *parity selection rules* which work well in atomic and nuclear physics is a clear indication that the degree of mixing, $\mathscr{F}^2$ cannot be large. From such considerations one can impose the limit $\mathscr{F}^2 \leq (r/\lambda)^2$, which for atomic spectroscopy is, in most cases, $\sim 10^{-6}$. In general, a less accurate limit obtains for nuclear spectroscopy.

Parity non-conservation implies the existence of *interactions* which mix parities. The strength of such *interactions* compared to the usual *interactions* will in general be characterized by $\mathscr{F}$, so that the *mixing* will be of the order $\mathscr{F}^2$. The presence of such *interactions* would affect *angular distributions* in nuclear reactions. As we shall see, however, the accuracy of these experiments is not good. The limit on $\mathscr{F}^2$ obtained is not better than $\mathscr{F}^2 < 10^{-4}$.

To give an illustration, let us examine the *polarization experiments*, since they are closely analogous to some experiments to be discussed later. A proton beam polarized in a direction z perpendicular to its *momentum* was scattered by nuclei. The scattered *intensities* were compared[5] in two directions A and B related to each other by a reflection in the x-y plane, and were found to be identical to within ~ 1%.

[5] See, e.g., Chamberlain, Segre, Tripp, & Ypsilantis. (1954). *Phys. Rev.*, 93, 1430.

If the scattering originates from an ordinary parity-conserving plus a parity-non-conserving *interaction* (e.g., $\sigma \cdot r$), then the *scattering amplitudes* in the directions A and B are in the proportion $(1 + \mathscr{F})/(1 - \mathscr{F})$, where $\mathscr{F}$ represents the ratio of the strengths of the two kinds of *interactions* in the scattering. The experimental result therefore requires $\mathscr{F} < 10^{-2}$, or $\mathscr{F}^2 < 10^{-4}$.

The violation of *parity conservation* would lead to an *electric dipole moment* for all systems. The magnitude of the *moment* is

$$moment \sim e\, \mathscr{F} x \text{ (dimension of system)}. \tag{1}$$

The presence of such *electric dipole moments* would have interesting consequences. For example, if the *proton* has an *electric dipole moment* $\simeq e \times (10^{-16}$ cm), the *perturbation* caused by the presence of the neighboring *2p state* of the *hydrogen atom* would shift the energy of the *2s state* by about 1 Mc/sec. This would be inconsistent with the present theoretical interpretations of the *Lamb shift*. Another example is found in the *electron-neutron interaction*. An *electric dipole moment* for the *neutron* $\simeq e \times (10^{-18}$ cm) is the upper limit allowable by the present experiments.

By far the most accurate measurement of the *electric dipole moment* was made by Purcell, Ramsey, and Smith. They gave[6] an upper limit for the *electric dipole moment* of the *neutron* of $e \times (10^{-20}$ cm).

[6] Purcell, E. M. & Ramsey, N. F. (1950). *Phys. Rev.*, 78, 807; Smith el al. as quoted in Ramsey, N. F. (1956). *Molecular Beams*. Oxford University Press, London.

This value sets the upper limit for $\mathscr{F}^2$ as $\mathscr{F}^2 < 3 \times 10^{-12}$, which is also the most accurate verification of the *conservation of parity* in *strong* and *electromagnetic* interactions. We shall see, however, that even this high degree of accuracy is not sufficient to supply an experimental proof of *parity conservation* in the *weak interactions*. For such a proof an accuracy of $\mathscr{F}^2 < 10^{-24}$ is necessary.

QUESTION OF PARITY CONSERVATION IN β DECAY

At first sight it might appear that the numerous experiments related to *β decay* would provide a verification that the *weak β interaction* does conserve *parity*. We have examined this question in detail and found this to be not so. (See Appendix.) We start by writing down the five usual types of *couplings*. In addition to these we introduce the five types of *couplings* that conserve *angular momentum* but do not conserve *parity*. It is then apparent that the classification of *β decays* into allowed transitions, first forbidden, etc., proceeds exactly as usual. (The mixing of *parity* of the *nuclear states* would not measurably affect these selection rules. This phenomenon belongs to the discussions of the last section.) The following phenomena are then examined: allowed spectra, unique forbidden spectra, forbidden spectra with allowed shape, β-neutrino correlation, and β – γ correlation. It is found that these experiments have no bearing on the question of *parity conservation* of the *β-decay* interactions. *This comes about because in all of these phenomena no interference terms exist between the parity-conserving and parity-non-conserving interactions.* In other words, the calculations always result in terms proportional to $|C|^2$ plus terms proportional to $|C'|^2$. Here C and C' are, respectively, the *coupling constants* for the usual parity-conserving *interactions* (a sum of five terms) and the parity-non-conserving *interactions* (also a sum of five terms). Furthermore, it is well known[7]

[7] Yang, C. N. & Tiomno, J. (1950). *Phys. Rev.*, 79, 495.

that without measuring the *spin* of the *neutrino* we cannot distinguish the *couplings* C from the *couplings* C' (provided the *mass* of the *neutrino* is zero). The experimental results concerning the above-named phenomena, which constitute the bulk of our present knowledge about *β decay*, therefore cannot decide the degree of mixing of the C' type *interactions* with the usual type.

The reason for the absence of *interference terms* CC' is actually quite obvious. Such terms can only occur as a pseudoscalar formed out of the experimentally measured quantities. For example, if three *momenta* $\mathbf{p_1}$, $\mathbf{p_2}$, $\mathbf{p_3}$ are measured, the term CC' $\mathbf{p_1} \cdot (\mathbf{p_2} \times \mathbf{p_3})$ may occur. Or if a *momentum* $\mathbf{p}$ and a *spin* $\boldsymbol{\sigma}$ are measured, the term CC' $\mathbf{p} \cdot \boldsymbol{\sigma}$ may occur. In all the *β-decay* phenomena mentioned above, no such pseudoscalars can be formed out of the measured quantities.

POSSIBLE EXPERIMENTAL TESTS OF PARITY CONSERVATION IN β DECAYS

The above discussion also suggests the kind of experiments that could detect the possible *interference* between C and C' and consequently could establish whether *parity conservation* is violated in *β decay*. A relatively simple possibility is to measure the angular distribution of the *electrons* coming from *β decays* of oriented *nuclei*. If θ is the angle between the orientation of the parent *nucleus* and the *momentum* of the *electron*, an asymmetry of distribution between θ and $180° - \theta$ constitutes an unequivocal proof that *parity* is not conserved in β *decay*.

To be more specific, let us consider the allowed β *transition* of any oriented nucleus, say Co^{60}. The *angular distribution* of the β *radiation* is of the form (see Appendix):

$$I(\theta)d\theta = (\text{constant})\,(1 + \alpha \cos \theta)\sin \theta\, d\theta,$$

where α is proportional to the *interference term* CC'. If $\alpha \neq 0$, one would then have a positive proof of *parity non-conservation* in β *decay*. The quantity α can be obtained by measuring the fractional asymmetry between $\theta < 90°$ and $\theta > 90°$; i.e.,

$$\alpha = 2\,[\textstyle\int_0^{\pi/2} I(\theta)d\theta - \int_{\pi/2}^{\pi} I(\theta)d\theta]/\int_0^{\pi} I(\theta)d\,\theta.$$

It is noteworthy that in this case the presence of the *magnetic field* used for orienting the nuclei would automatically cause a spatial separation between the electrons emitted with $\theta < 90°$ and those with $\theta > 90°$. Thus, this experiment may prove to be quite feasible.

It appears at first sight that in the study of γ-*radiation* distribution from β-*decay* products of oriented nuclei one can form a pseudoscalar from the *spin* of the oriented nucleus and the γ-*ray momentum* $\mathbf{p}_\gamma$. Thus, it may seem to offer another possible experimental test of *parity conservation*. Unfortunately, *the nuclear levels have definite parities, and electromagnetic interactions conserve parity.* (Any small mixing of parities characterized by $\mathscr{F}^2 < 3 \times 10^{-15}$ would not affect the arguments here). Consequently, the γ rays carry away definite parities. Thus, the observed probability function must be an even function of $\mathbf{p}_\gamma$. This property eliminates the possibility of forming a pseudoscalar quantity. It is therefore not possible to use such experiments as a test of *parity conservation*.

In β-γ-γ' triple correlation experiments one can, by some rather similar but more complicated reasoning, prove that a measurement of the three *momenta* cannot supply any information on the question of *parity conservation* in β *decay*.

In β-γ correlation experiments the nature of the polarization of the γ ray could provide a test. To be more specific, let us consider the polarization state of γ rays emitted parallel to the β ray. If *parity conservation* holds for β decay, the γ ray will be unpolarized. On the other hand, if *parity conservation* is violated in β *decay*, the γ ray will in general be polarized. However, this polarization must be circular in nature and therefore may not lend itself to easy experimental detection. (The usual ways of measuring polarization through Compton effect, photoelectric effect, and photodisintegration of the deuteron are all incapable of detecting circular polarization. This is because circular polarization is specified by an axial vector parallel to the direction of propagation. From the observed momenta in these detection techniques such an axial vector cannot be formed). For other directions of γ -ray propagation, elliptical polarization will result if *parity* is not conserved. This effect will thus be more difficult to detect.

QUESTION OF PARITY CONSERVATION IN MESON AND HYPERON DECAYS

If the *weak. interactions*, such as the *β-decay interactions* or the *decay interactions* of *mesons* and *hyperons*, do not conserve *parity*, *parity mixing* will occur in all interactions by means of second-order processes. To examine this effect let us consider, for example, the decay of the Δ^0 [the *lambda baryon*]:

$$\Delta^0 \rightarrow \text{p} + \pi^-.$$

[In October 1950, the Λ^0 (or *lambda baryon*) was discovered during a study of cosmic-ray interactions by V. D. Hopper and S. Biswas of the University of Melbourne, as a neutral *V particle* with a *proton* as a decay product, thus correctly distinguishing it as a *baryon*, rather than a *meson*, i.e. different in kind from the *K meson* discovered in 1947 by Rochester and Butler.

> [In particle physics, V was a generic name for heavy, unstable subatomic particles that decay into a pair of particles, thereby producing a characteristic letter V in a bubble chamber or other particle detector.]

They were produced by *cosmic rays* and detected in photographic emulsions flown in a balloon at 70,000 feet (21,000 m). Though the particle was expected to live for $\sim 10^{-23}$ s, it actually survived for $\sim 10^{-10}$ s. The property that caused it to live so long was dubbed *strangeness* and led to the discovery of the *strange quark*. Furthermore, these discoveries led to a principle known as the *conservation of strangeness*, wherein lightweight particles do not decay as quickly if they exhibit *strangeness* (because *non-weak* methods of particle decay must preserve the *strangeness* of the decaying *baryon*).

A *baryon* is a type of composite subatomic particle, including the *proton* and the *neutron, that contains an odd number of valence quarks,* conventionally three. *Baryons* belong to the *hadron* family of particles; *hadrons* are composed of *quarks*. *Baryons* are also classified as *fermions* because they have half-integer spin.

The *lambda baryons* (Λ) are a family of subatomic *hadron* particles containing one *up quark*, one *down quark*, and a *third quark* from a *higher flavor generation*, in a combination where the *quantum wave function* changes *sign upon the flavor of any two quarks being swapped* (thus slightly different from a *neutral sigma baryon*, Σ^0). They are thus *baryons*, with *total isospin* of 0, and have either *neutral electric charge* or the *elementary charge* +1.

There are four types of *lambda baryons* represented by the symbols Λ^0, Λ^+_c, Λ^0_b, and Λ^+_t. In this notation, the superscript character indicates whether the particle is electrically *neutral* (^{0}) or carries a *positive charge* ($^+$). The subscript character, or its absence, indicates whether the third quark is a *strange quark* (Λ^0) (no subscript), a *charm quark* (Λ^+_c), a *bottom quark* (Λ^0_b), or a *top quark* (Λ^+_t). Physicists expect to not observe a *lambda baryon* with a *top quark*, because the *Standard Model* of particle physics predicts that the *mean lifetime* of *top quarks* is roughly 5×10^{-25} seconds; that is about 1/20 of the mean timescale for *strong interactions*, which indicates that the *top quark* would decay before a *lambda baryon* could form a *hadron*.]

The assumption that *parity* is not conserved in this decay implies that the [*lambda baryon*] Δ^0 exists virtually in *states* of opposite *parities*. It could therefore possess an *electric dipole moment* of a magnitude

$$\text{moment} \sim e\varsigma^2 \text{ x (dimension of } \Delta^0), \tag{3}$$

where ς is the *coupling strength* of the *decay interaction* of the Δ^0 ($\varsigma^2 \leq 10^{-12}$). The *electric dipole moment* of the Δ^0 is therefore $\leq$ e x 10^{-25} cm).

Clearly the *proton* would have an *electric dipole moment* of the same order of magnitude. The existence of such a small *electric dipole moment* is, as we have seen, completely consistent with the present experimental information. Another way of putting this is to observe that by comparing Eq. (3) with Eq. (1), one has

$$\mathscr{F} = \varsigma^2.$$

Since all the *weak interactions* including β *interactions* are characterized by *coupling strengths* $\varsigma^2 < 10^{-12}$, a *violation of parity* in *weak interactions* would introduce a *parity mixing* characterized by an $\mathscr{F}^2 < 10^{-24}$. This is outside the present limit of experimental knowledge, as we have discussed before.

If the *weak interactions* violate *parity conservation*, *parity* would be defined and measured in *strong* and *electromagnetic* interactions only, just as *strangeness* is. Furthermore, it is important to notice that with the conservation of *strangeness*, as with every conservation law, there is an element of arbitrariness introduced into the *parity* of all systems. The *parity* of all *strange* particles would be defined only up to a factor of $(-1)^S$, where S is the *strangeness*. The *parity* of the [*lambda baryon*] Δ^0 (relative to the nucleons) is therefore a matter of definition. But once this is defined, the parity of other *strange* particles would be measurable from the *strong interactions*.

POSSIBLE EXPERIMENTAL TESTS OF PARITY CONSERVATION IN MESON AND HYPERON DECAYS

To have a sensitive unequivocal test of whether *parity* is conserved in *weak interactions*, one must decide whether the *weak interactions* differentiate between the right and the left. This is possible only if one produces interference between states of opposite parities. The mere observation of two decay products of opposite parities originating from a "particle" cannot provide conclusive evidence that parity is not conserved. Such indeed is the state of affairs of the present $\theta - \tau$ puzzle.

As we have discussed before, these *interference terms* are possible only if the observed quantities can form a pseudoscalar such as $\mathbf{p}_1 \cdot (\mathbf{p}_2 \times \mathbf{p}_3)$. The observation of Δ^0 decays in association with their production does provide such a possible pseudoscalar and hence a possible test of whether *parity* is conserved in the Δ^0 *decay interaction*. Let us consider the experiment

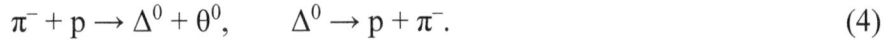

$$\pi^- + p \rightarrow \Delta^0 + \theta^0, \qquad \Delta^0 \rightarrow p + \pi^-. \tag{4}$$

Let $\mathbf{p}_{in}$, $\mathbf{p}_\Delta$, and $\mathbf{p}_{out}$ be, respectively, the *momenta* in the laboratory system of the incoming *pion* [*pi meson*], the Δ^0, and the decay *pion*. We define a parameter R as the projection of $\mathbf{p}_{out}$ in the direction of $\mathbf{p}_{in} \times \mathbf{p}_\Delta$. The value of R ranges from approximately -100 Mev/c to approximately $+100$ Mev/c. Switching from a right-handed convention for vector products (which we use) to a left-handed convention means a switch of the sign of R. *Parity conservation* in the *weak decay interaction* of Δ^0 can therefore be experimentally checked by investigating whether $+R$ and $-R$ have equal probabilities of occurrence.

To see more clearly the meaning of the parameter R, one transforms $\mathbf{p}_{out}$ ($\rightarrow \mathbf{p}'$) into the *center-of-mass system* of Δ^0. The new vector $\mathbf{p}'$ has a constant magnitude $\simeq 100$ Me v/c. The *frequency distribution*, of this vector $\mathbf{p}'$ can then be plotted on a spherical surface. Taking the z axis for this sphere to be in the direction of $\mathbf{p}_{in} \times \mathbf{p}_\Delta$, one can prove the following two *symmetries*:

(a) The frequency distribution on the sphere remains unchanged under a rotation through $180°$ around the z axis. This symmetry follows from *parity conservation* in the *strong reaction* producing the Δ^0. It does not depend on the nature of the *weak interaction*.

(b) If *parity* is conserved in the *decay interaction* of Δ^0, the frequency distribution on the sphere is unchanged under a reflection with respect to the production plane of Δ^0.

229

To prove statement (a), one need only consider the invariance of the production process under a reflection with respect to the production plane defined by $\mathbf{p}_{in}$ and $\mathbf{p}_\Delta$. This reflection is the resultant of an inversion and a rotation through $180°$ around the z axis (which is normal to the production plane). The state of polarization of Δ^0 is thus invariant under a $180°$ rotation around the z axis, leading to the stated symmetry[8].

[8] This proof for statement (a) is correct only if Δ^0 exists as a single particle with a definite parity in the *strong interactions*, (as discussed in the last section); i.e. if Δ^0 does not exist as two degenerate states Δ_1^0 and Δ_2^0 of opposite parity, as has been suggested [Lee, T. D. & Yang, C. N. (1956). *Phys. Rev.*, 102, 290]. [It is to be emphasized, that if *parity* is indeed not conserved in the *weak interactions*, there would be (at present) no necessity to introduce the complication of two degenerate states of opposite *parity* at all.] On the other hand, statement (b) is correct even if Δ^0 exists as two degenerate states Δ_1^0 and Δ_2^0 of opposite *parity*. To summarize, violation of the symmetry stated in (a) implies the existence of the *parity doublets* Δ_1^0 and Δ_2^0 with a *mass* difference less than their widths Violation of the symmetry stated in (b) implies the non-conservation of *parity* in a Δ decay. See also footnote 12 and Lee, T. D. & Yang, C. N. *Phys. Rev.* (to be published).

Statement (b) follows[8] directly from the assumption that the *weak interaction* as well as the *strong interaction* conserves *parity*. A reflection with respect to the production plane must then leave the whole process invariant.

The frequency distribution of R is just the projection of the distribution on the sphere onto the z axis. An asymmetry between $+ R$ and $- R$ therefore implies *parity non-conservation* in Δ^0 decay. However, if the *spin* of Δ^0 is unpolarized, no asymmetry[9] can obtain even if *parity* is not conserved in Δ^0 *decay*.

[9] Also the interference may accidentally be absent if the relative phase between the two parities in the decay product is $90°$. This, however, cannot be the case if time-reversal invariance is preserved in the decay process.

To obtain a polarized Δ^0 beam, the experiment is therefore best done at a definite non-forward angle of production of Δ^0 and at a definite incoming *energy*.

The above discussions apply also to any other *strange particle decay* if (1) the particle has a non-vanishing *spin* and (2) it decays into two particles at least one of which has a non-vanishing *spin*, or it decays into three or more particles. Thus, the above considerations can be applied also to the decays of $\Sigma^\pm$ and maybe also to $K_{\mu 2}^\pm$, $K_{\mu 3}^\pm$, and $K_{\pi 3}^\pm$ ($\equiv \tau^\pm$).

[The *sigma baryons*, $\Sigma^\pm$, are a family of subatomic *hadron* particles which have two *quarks* from the first *flavor* generation (*up and/or down quarks*), and a third *quark* from a higher flavor generation, in a combination where the *wavefunction* sign remains constant when any two *quark* flavors are swapped. They are thus *baryons*, with *total isospin* of 1, and can either be *neutral* or have an *elementary charge* of +2, +1, 0, or −1. They are closely related to the *lambda baryons*, which differ only in the wavefunction's behavior upon *flavor exchange*.

The third *quark* can hence be either a *strange* (symbols Σ^+, Σ^0, Σ^-), a *charm* (symbols Σ_c^{++}, Σ_c^+, Σ_c^0), a *bottom* (symbols Σ_b^+, Σ_b^0, Σ_b^-) or a *top* (symbols Σ_t^{++}, Σ_t^+, Σ_t^0) quark. However, the *top sigmas* are expected to never be observed, since the *Standard Model* predicts the mean lifetime of *top quarks* to be roughly 5×10^{-25} s. This is about 20 times shorter than the timescale for *strong interactions*, and therefore it does not form *hadrons*.]

In the decay processes

$$\pi \rightarrow \mu + \nu, \tag{5}$$
$$\mu \rightarrow e + \nu + \nu, \tag{6}$$

starting from a π *meson* at rest, one could study the distribution of the angle θ between the *μ-meson momentum* and the *electron momentum*, the latter being in the *center-of-mass system* of the μ meson. If *parity* is conserved in neither (5) nor (6), the distribution will not in general be identical for θ and $\pi - \theta$. To understand this, consider first the orientation of the *muon spin*. If (5) violates *parity conservation*, the *muon* would be in general polarized in its direction of motion, In the subsequent decay (6), the *angular distribution problem* with respect to θ is therefore closely similar to the *angular distribution problem* of β *rays* from oriented *nuclei*, which we have discussed before. (Entirely similar considerations can be applied to $\Xi \rightarrow \Delta^0 + \pi^-$ and $\Delta^0 \rightarrow p + \pi^-$.)

[Ξ (*Xi*) *baryons* or *cascade particles* are a family of subatomic *hadron* particles which may have an *electric charge* of +2e, +1e, 0, or −1e, where e is the *elementary charge*. They are historically called the *cascade particles* because of their *unstable state*; they are typically observed to decay rapidly into lighter particles, through a chain of decays (cascading decays).

Like all conventional *baryons*, *Xi baryons* contain three *quarks*. *Xi baryons*, in particular, contain either one *up* or one *down quark* and two other, more massive *quarks*. The two more massive *quarks* are any two of *strange*, *charm*, or *bottom* (doubles allowed). For notation, the assumption is that the two heavy quarks in the *Xi baryon* are both *strange*; subscripts "c" and "b" are added for each even heavier *charm* or *bottom quark* that replaces one of the two presumed *strange quarks*.

The first discovery of a *charged Xi baryon* was in cosmic ray experiments by the Manchester group in 1952. [Armenteros, R. et al. (Manchester group) (1952). The properties of charged V-particles. *Phil. Mag.*, 43, 341, 597; https://doi.org/10.1080/14786440608520216. The first discovery of the *neutral Xi baryon* was at the Lawrence Berkeley Laboratory in 1959. [Alvarez, L.W. et al. (1959). Neutral Cascade Hyperon Event. *Phys. Rev. Let.*, 2, 5, 215; https://doi.org/10.1103/PhysRevLett. 2.215.] It was later observed as a daughter product from the decay of the *omega baryon* (Ω^-) observed at Brookhaven National Laboratory in 1964.

The Xi^-_b particle, also known as the *cascade B* particle, was discovered by DØ and CDF experiments at Fermilab in 2007. It was the first known particle made of *quarks* from all three *quark generations* – namely, a *down quark*, a *strange quark*, and a *bottom quark*. It The DØ and CDF collaborations reported the consistent masses of the new state. The Particle Data Group world average *mass* is 5.7924 ± 0.0030 GeV/c^2.

In 2012, the CMS experiment at the Large Hadron Collider detected a Xi^{*0}_b *baryon* (reported mass 5945 ± 2.8 MeV/c2). (Here,"*" indicates a *baryon decuplet*.) The LHCb experiment at CERN discovered two new Xi *baryons* in 2014: Xi'^-_b and Xi^{*-}_b. And, in 2017, the LHCb researchers reported yet another Xi *baryon*: the double charmed Xi^{++}_{cc} *baryon*, consisting of two heavy *charm quarks* and one *up quark*. The *mass* of Xi^{++}_{cc} is about 3.8 times that of a *proton*.]

REMARKS

If parity conservation is violated in *hyperon decay*, the decay products will have mixed parities. This, however, does not affect the arguments of Adair[10] and of Treiman[11] concerning the relationship between the *spin* of the *hyperons* and the angular distribution of their decay products in certain special cases[12].

[10] Adair, R. K. (1955). *Phys. Rev.*, 100, 1540.
[11] Treiman, S. B. (1956). *Phys. Rev.*, 101, 1216.
[12] The existence of Δ_1^0 and Δ_2^0 of opposite parity may affect these relationships. This is similar to the violation of symmetry (*a*) discussed in footnote 8. See T. D. Lee, T. D. & Yang, C. N. *Phys. Rev.* (to be published).

One may question whether the other conservation laws of physics could also be violated in the *weak interactions*. Upon examining this question, one finds that the conservations of the *number of heavy particles*, of *electric charge*, of *energy*, and of *momentum* all appear to be inviolate in the *weak interactions*. The same cannot be said of the conservation of *angular momentum*, and of *parity*. Nor can it be said of the invariance under *time reversal*. It might appear at first sight that the equality of the *life times* of $\pi^\pm$ and of those of $\mu^\pm$ furnish proofs of the invariance under *charge conjugation* of the *weak interactions*. A closer examination of this problem reveals, however, that this is not so. In fact, the equality of the *life times* of a *charged particle* and its *charge conjugate* against decay through a *weak interaction* (to the lowest order of the strength of the *weak interaction*) can be shown to follow from the invariance under proper Lorentz transformations (i.e., Lorentz transformation with neither space nor time inversion). One has therefore at present no experimental proof of the invariance under *charge conjugation* of the *weak interactions*. In the present paper, only the question of parity non-conservation is discussed.

The *conservation of parity* is usually accepted without questions concerning its possible limit of validity being asked. There is actually no a priori reason why its violation is undesirable. As is well known, its violation implies the existence of a right-left asymmetry. We have seen in the above some possible experimental tests of this asymmetry. These experiments test whether the present elementary particles exhibit asymmetrical behavior with respect to the right and the left. If such asymmetry is indeed found, the question could still be raised whether there could not exist corresponding elementary particles exhibiting opposite asymmetry such that in the broader sense there will still be over-all right-left symmetry. If this is the case, it should be pointed out, there must exist two kinds of *protons* p_R and p_L, the right-handed one and the left-handed one. Furthermore, at the present time the *protons* in the laboratory must be predominantly of one kind in order to produce the supposedly observed asymmetry, and also to give rise to the observed Fermi-Dirac statistical character of the *proton*. This means that the free oscillation period between them must be longer than the age of the universe. They could therefore both be regarded as stable particles. Furthermore, the numbers of p_R and p_L must be separately conserved. However, the interaction between them is not necessarily weak. For example, p_R and p_L could interact with the same electromagnetic field and perhaps the same pion field. They could then be separately pair-produced, giving rise to interesting observational possibilities.

In such a picture the supposedly observed right-and left asymmetry is therefore ascribed not to a basic non-invariance under inversion, but to a cosmologically local preponderance of, say, p_R over p_L, a situation not unlike that of the preponderance of the positive *proton* over the negative. Speculations along these lines are extremely interesting, but are quite beyond the scope of this note.

APPENDIX

...

Chen Ning Yang – 1957 Nobel Lecture, December 11, 1957. *The law of parity conservation and other symmetry laws of physics.*

Chen Ning Yang – Nobel Lecture. NobelPrize.org. https://www.nobelprize.org/prizes/physics/1957/yang/lecture/.

[The Nobel Prize in Physics 1957 was awarded jointly to Chen Ning Yang and Tsung-Dao Lee "for their penetrating investigation of the so-called *parity laws* which has led to important discoveries regarding the elementary particles".

For a long time, physicists assumed that various symmetries characterized nature. In a kind of "mirror world" where right and left were reversed and *matter* was replaced by *antimatter*, the same physical laws would apply, they posited. The equality of these laws was questioned concerning the decay of certain elementary particles, however, and in 1956 Chen Ning Yang and Tsung Dao Lee formulated a theory that the *left-right symmetry law* is violated by the *weak interaction*. Measurements of electrons' direction of motion during a cobalt isotope's beta decay confirmed this. [Chen Ning Yang – Facts. NobelPrize.org. https://www.nobelprize.org/prizes/physics/1957/yang/facts/.]

In his Nobel Prize lecture Yang described the developments that led to the disproof of *parity conservation*.

It is a pleasure and a great privilege to have this opportunity to discuss with you the question of *parity conservation* and other *symmetry laws*. We shall be concerned first with the general aspects of the role of the *symmetry laws* in physics; second, with the development that led to the disproof of *parity conservation*; and last, with a discussion of some other symmetry laws which physicists have learned through experience, but which do not yet together form an integral and conceptually simple pattern. The interesting and very exciting developments since *parity conservation* was disproved, will be covered by Dr. Lee in his lecture.

I. The existence of *symmetry laws* is in full accordance with our daily experience. The simplest of these symmetries, the *isotropy* and *homogeneity* of space, are concepts that date back to the early history of human thought. The invariance of physical laws under a coordinate transformation of uniform velocity, also known as the *invariance under Galilean transformations*, is a more sophisticated *symmetry* that was early recognized, and formed one of the corner-stones of Newtonian mechanics. Consequences of these symmetry principles were greatly exploited by physicists of the past centuries and gave rise to many important results. A good example in this direction is the theorem that in an *isotropic* solid there are only two elastic constants.

Another type of consequences of the *symmetry laws* relates to the *conservation laws*. It is common knowledge today that in general a *symmetry principle* (or equivalently an *invariance principle*) generates a *conservation law*. For example, the invariance of physical laws under *space displacement* has as a consequence the *conservation of momentum*, the invariance under *space rotation* has as a consequence the *conservation of angular momentum*. While the importance of these conservation laws was fully understood, their close relationship with the symmetry laws seemed not to have been clearly recognized until the beginning of the twentieth century. ...

With the advent of *special and general relativity*, the *symmetry laws* gained new importance. Their connection with the dynamic laws of physics takes on a much more integrated and interdependent relationship than in classical mechanics, where logically the *symmetry laws* were only consequences of the dynamical laws that by chance possess the symmetries. Also, in the *relativity theories* the realm of the *symmetry laws* was greatly enriched to include *invariances* that were by no means apparent from daily experience. Their validity rather was deduced from, or was later confirmed by complicated experimentation. Let me emphasize that the conceptual simplicity and intrinsic beauty of the symmetries that so evolve from complex experiments are for the physicists great sources of encouragement. One learns to hope that Nature possesses an order that one may aspire to comprehend.

It was, however, not until the development of *quantum mechanics* that the use of the *symmetry principles* began to permeate into the very language of physics. The *quantum numbers* that designate the *states* of a system are often identical with those that represent the *symmetries* of the system. It in deed is scarcely possible to overemphasize the role played by the *symmetry principles* in *quantum mechanics*. To quote two examples: The general structure of the Periodic Table is essentially a direct consequence of the *isotropy* of *Coulomb's law*. The existence of the *antiparticles* - namely the *positron*, the *anti-proton*, and the *anti-neutron* - were theoretically anticipated as consequences of the symmetry of physical laws with respect to *Lorentz transformations*. In both cases Nature seems to take advantage of the simple mathematical representations of the *symmetry laws*. When one pauses to consider the elegance and the beautiful perfection of the mathematical reasoning involved and contrast it with the complex and far-reaching physical consequences, a deep sense of respect for the power of the *symmetry laws* never fails to develop.

One of the *symmetry principles*, the *symmetry between the left and the right*, is as old as human civilization. The question whether Nature exhibits such symmetry was debated at length by philosophers of the pasts. Of course, in daily life, left and right are quite distinct from each other. Our hearts, for example, are on our left sides. The language that people use both in the orient and the occident, carries even a connotation that right is good and left is evil. However, the laws of physics have always shown complete symmetry between the left and the right, the asymmetry in daily life being attributed to the accidental asymmetry of the environment, or initial conditions in organic life. To illustrate the point, we mention that if there existed a mirror image man with his heart on his right side, his internal organs reversed compared to ours, and in fact his body molecules, for example sugar molecules, the mirror image of ours, and if he ate the mirror image of the food that we eat, then according to the laws of physics, he should function as well as we do.

The law of *right-left symmetry* was used in classical physics, but was not of any great practical importance there. One reason for this derives from the fact that right-left symmetry is a discrete symmetry, unlike rotational symmetry which is continuous. Whereas the *continuous symmetries always lead to conservation laws* in classical mechanics, a discrete symmetry does not. With the introduction of *quantum mechanics*, however, this difference between the discrete and continuous symmetries disappears. The *law of right-left symmetry* then leads also to a *conservation law*: the *conservation of parity*.

The discovery of this conservation law dates back to 1924 when Laporte found that *energy levels* in complex atoms can be classified into « gestrichene » and « ungestrichene » types, or in more recent language, *even* and *odd* levels. In transitions between these levels during which one *photon* is emitted or absorbed, Laporte found that the level always changes from even to odd or vice versa. Anticipating later developments, we remark that the evenness or oddness of the levels was later referred to as the *parity* of the levels. Even levels are defined to have *parity* +1, odd levels *parity* −1. One also defines the *photon* emitted or absorbed in the usual atomic transitions to have odd *parity*. Laporte's rule can then be formulated as the statement that in an atomic transition with the emission of a *photon*, the *parity of the initial state is equal to the total parity of the final state*, i.e. the product of the parities of the *final atomic state* and the *photon* emitted. In other words, *parity is conserved*, or unchanged, in the transition.

In 1927 Wigner took the critical and profound step to prove that the empirical rule of Laporte is a consequence of the *reflection invariance*, or *right-left symmetry*, of the *electromagnetic forces* in the *atom*. This fundamental idea was rapidly absorbed into the language of physics. Since *right-left symmetry* was unquestioned also in other *interactions*, the idea was further taken over into new domains as the subject matter of physics extended into *nuclear reactions*, *meson-interactions*, and *strange-particle* physics. One became accustomed to the idea of *nuclear parities* as well as *atomic parities*, and one discusses and measures the intrinsic *parities* of the *mesons*. Throughout these developments the concept of *parity* and the *law of parity conservation* proved to be extremely fruitful, and the success had in turn been taken as a support for the validity of *right-left symmetry*.

II. Against such a background the so-called *ϑ-τ puzzle* developed in the last few years. …

…

Now to return to the *ϑ-τ puzzle*. In 1953, Dalitz and Fabri pointed out that in the decay of the ϑ and τ *mesons*

$$\vartheta \rightarrow \pi + \pi$$
$$\tau \rightarrow \pi + \pi + \pi$$

some information about the *spins* and *parities* of the τ and ϑ *mesons* can be obtained. The argument is very roughly as follows. It has previously been determined that the *parity* of a π *meson* is odd (i.e. = −1). Let us first neglect the effects due to the relative motion of the π *mesons*. To conserve *parity* in the decays, the ϑ *meson* must have the *total parity*, or in other words, the product *parity*, of two π *mesons*, which is even (i.e. = +1). Similarly, the τ *meson* must have the *total parity* of three π *mesons*, which is odd. Actually because of the relative motion of the π *mesons* the argument was not as simple and unambiguous as we just discussed. To render the argument conclusive and definitive it was necessary to study experimentally the *momentum* and *angular distribution* of the π *mesons*. Such studies were made in many laboratories, and by the spring of 1956 the accumulated experimental data seemed to unambiguously indicate, along the lines of reasoning discussed above, that ϑ and τ do not have the same *parity*, and consequently are not the same particle. This conclusion, however, was in marked contradiction with other experimental results which also became definite at about the same time. The contradiction was known as the ϑ -τ puzzle and was widely discussed. …

…

That direction turned out to lie in the *faultiness of the law of parity conservation for the weak interactions*. But to uproot an accepted concept one must first demonstrate why the previous evidence in its favor were insufficient. Dr. Lee and I examined this question in detail, and in May 1956 we came to the following conclusions: (A) Past experiments on the *weak interactions* had actually no bearing on the question of *parity conservation*. (B) In the *strong interactions*, … , there were indeed many experiments that established *parity conservation* to a high degree of accuracy, but not to a sufficiently high degree to be able to reveal the effects of a lack of *parity conservation* in the *weak interactions*. The fact that *parity conservation* in the *weak interactions* was believed for so long without experimental support was very startling. But what was more startling was the prospect that a *space-time symmetry law* which the physicists have learned so well may be violated. This prospect did not appeal to us. Rather we were, so to speak, driven to it through frustration with the various other efforts at understanding the puzzle that had been made. …

237

Tsung-Dao Lee – 1957 Nobel Lecture, December 11, 1957. *Weak Interactions and Non-conservation of Parity.*

Tsung-Dao Lee – Nobel Lecture. NobelPrize.org. https://www.nobelprize.org/prizes/physics/1957/lee/lecture/.

In his Nobel Prize lecture Lee reviewed the current knowledge about elementary particles and their *interactions*.

In the previous talk Professor Yang has outlined to you the position of our understanding concerning the various *symmetry principles* in physics prior to the end of last year. Since then, in the short period of one year, the proper roles of these principles in various physical processes have been greatly clarified. This remarkably rapid development is made possible only through the efforts and ingenuity of many physicists in various laboratories all over the world. To have a proper perspective and understanding of these new experimental results it may be desirable to review very briefly our knowledge about elementary particles and their interactions.

...

Yoichiro Nambu (January 18, 1921–July 5, 2015).

Nambu was a Japanese-American physicist and professor at the University of Chicago.

Known for his contributions to the field of theoretical physics, he was awarded half of the 2008 Nobel Prize in Physics "for the discovery [in 1960] of the mechanism of *spontaneous broken symmetry* in subatomic physics". The other half was split equally between Makoto Kobayashi and Toshihide Maskawa "for the discovery of the origin of the *broken symmetry* which predicts the existence of at least three families of *quarks* in nature".

Nambu was born on 18 January 1921 in Tokyo, Empire of Japan (Now Japan). After graduating from the then-Fukui Secondary High School in Fukui City, he enrolled in the Imperial University of Tokyo (Now University of Tokyo) and studied physics. He received his Bachelor of Science in 1942 and Doctorate of Science in 1952. In 1949, he was appointed to associate professor at Osaka City University and promoted to professorship the next year at the age of 29.

In 1952, Nambu was invited by the Institute for Advanced Study in Princeton, New Jersey, United States, to study. He moved to the University of Chicago in 1954 and was promoted to professor in 1958.

Nambu's work on *spontaneous symmetry breaking* was groundbreaking. In 1960/1 he explored models of *symmetry breaking* in particle physics, drawing inspiration from the *theory of superconductivity*. [Nambu, Y. (April, 1960). Axial Vector Current Conservation in Weak Interactions; *Phys. Rev. Lett.*, 4, 7, 380-2; https://journals.aps.org/prl/abstract/ 10.1103/PhysRevLett.4.380. See below; Nambu, Y. & Jona-Lasinio, G. (April 1961). Dynamical Model of Elementary Particles Based on an Analogy with Superconductivity. I. *Phys. Rev.*, 122, 1, 345–58; https://doi.org/10.1103/physrev.122.345; (October 1961). Dynamical Model of Elementary Particles Based on an Analogy with Superconductivity. II. *Idem.* 124, 1, 246–54; https://doi.org/10.1103/physrev.124.246.]

Specifically, Nambu considered *chiral symmetry breaking*, which involves the approximate symmetries (*flavor symmetries*) observed in particle physics. His models predicted that the *masses* of spin-$\frac{1}{2}$ particles (like *protons* and *neutrons*) could arise from this *symmetry breaking*. However, there was a challenge: his model also predicted massless spin-0 particles, which contradicted experimental evidence. These massless spin-0 particles are indeed the *pions*, which play a crucial role in our understanding of *strong interactions*. Despite this initial obstacle, Nambu's work laid the foundation for further developments in particle physics. [*Spontaneous symmetry breaking* was later related to the *electroweak interaction* and *Higgs mechanism*.]

From 1974 to 1977, Nambu was also served as the Chairman of the Department of Physics, and then he became as an American citizen from 1970 until his death in 2015.

Nambu proposed the "*color charge*" of *quantum chromodynamics*, having discovered that the dual resonance model could be explained as a quantum mechanical theory of *strings*. He was accounted as one of the founders of *string theory*. [Nambu, Y. (1970). Quark model and the factorization of the Veneziano amplitude. In R. Chand (ed.), *Symmetries and quark models* Singapore: World Scientific, pp. 269–277).]

After more than fifty years as a professor, he was Henry Pratt Judson Distinguished Service Professor emeritus at the University of Chicago's Department of Physics and Enrico Fermi Institute.

The *Nambu–Goto action* in *string theory* is named after Nambu and Tetsuo Goto. Also, massless *bosons* arising in *field theories* with *spontaneous symmetry breaking* are sometimes referred to as *Nambu–Goldstone bosons*.

[The *Nambu–Goto action* is the simplest invariant action in *bosonic string theory*, and is also used in other theories that investigate string-like objects (for example, cosmic strings). It is the starting point of the analysis of zero-thickness (infinitely thin) string behavior, using the principles of Lagrangian mechanics. Just as the action for a free point particle is proportional to its proper time — i.e., the "length" of its world-line — a *relativistic string's action* is proportional to the area of the sheet which the string traces as it travels through spacetime.]

Nambu died of heart failure at the hospital in Osaka on 5 July 2015, at the age of 94. The announcement of his death was delayed until 17 July, just 12 days after his death. Nambu was survived by his wife, Chieko, and his son, John.

Nambu, Y. (April, 1960). Axial Vector Current Conservation in Weak Interactions*.

Phys. Rev. Lett., 4, 7, 380-2; https://journals.aps.org/prl/abstract/10.1103/PhysRevLett.4.380.

Enrico Fermi Institute for Nuclear Studies and Department of Physics University of Chicago, Chicago, Illinois.

* This work was supported by the U. S. Atomic Energy Commission.

Received February 23, 1960.

In analogy to the conserved *vector current interaction* in the *beta decay* suggested by Feynman and Gell-Mann, some speculations have been made about a possibly conserved *axial vector current*. We would like to suggest that there may not be a strict *pseudovector current conservation*, but that we may have an approximate conservation which becomes rigorous in the limit $q^2 \gg m_\pi^2$, m_π being the *pion* mass and q^2 a *massless, pseudo scalar*, and *charged quantum* bridging the *nucleon* and *lepton currents*. We are tempted to extend this approximate conservation of the *axial vector current* (and naturally also the *vector current*) to the *strangeness-non conserving beta decays*.

In analogy to the conserved *vector current interaction* in the *beta decay* suggested by Feynman and Gell-Mann, some speculations have been made about a possibly conserved *axial vector current*[1-3].

[1] Taylor, J. C. (1958). *Phys. Rev.*, 110, 1216.
[2] Polkinghorne, J. C. (1958). *Nuovo Cimento*, 8, 179 and 781.
[3] Goldberger, M. L. & S. B. Treiman, S. B. (1958) *Phys. Rev.*, 110, 1478.

One can formally construct an *axial vector nucleon current*, which satisfies a *continuity equation*,

$$\Gamma_\mu^A (p', p) = i\gamma_5\gamma_\mu - 2M\gamma_5 q_\mu/q^2, \qquad q = p' - p, \qquad (1)$$

where p and p' are the initial and final *nucleon momenta*. Such an attempt has some appeal in view of the apparently modest *renormalization* effect on the *axial vector beta decay constant* ($g_A/g_V \approx 1.25$), although the second appealing point[1], namely, the possible forbidding of $\pi \to e+ \nu$, has now lost its relevance.

The expression (1), unfortunately, easily ruled out experimentally, can be as was pointed out by Goldberger and Treiman[3], since it introduces a large admixture of *pseudoscalar interaction.*

On the other hand, Eq. (1) arouses theoretical curiosity as to the origin of the second term if it really exists; according to our conventional *field theory*, we would have to interpret the denominator q^2 as implying a *massless, pseudo scalar*, and *charged quantum* bridging the *nucleon* and *lepton currents.*

[A *lepton* is an elementary particle of half-integer *spin* (*spin* ½) that does not undergo *strong interactions*. Two main classes of *leptons* exist: *charged leptons* (also known as the *electron-like leptons* or *muons*), including the *electron, muon,* and *tauon,* and *neutral leptons,* better known as *neutrinos. Charged leptons* can combine with other particles to form various composite particles such as atoms and *positronium,* while *neutrinos* rarely interact with anything, and are consequently rarely observed. The best known of all *leptons* is the *electron.*

There are six types of *leptons,* known as flavors, grouped in three generations. The *first-generation leptons,* also called *electronic leptons,* comprise the *electron* (e^-) and the *electron neutrino* (ν_e); the second are the *muonic leptons,* comprising the *muon* (μ^-) and the *muon neutrino* ($\nu\mu$); and the third are the *tauonic leptons,* comprising the *tau* (τ^-) and the *tau neutrino* ($\nu\tau$). *Electrons* have the least mass of all the *charged leptons*. The heavier *muons* and *taus* will rapidly change into *electrons* and *neutrinos* through a process of particle decay: the transformation from a higher mass state to a lower mass state. Thus, *electrons* are stable and the most common charged *lepton* in the universe, whereas *muons* and *taus* can only be produced in high-energy collisions (such as those involving cosmic rays and those carried out in particle accelerators).

Leptons have various intrinsic properties, including *electric charge, spin,* and *mass.* Unlike *quarks,* however, *leptons* are not subject to the *strong interaction,* but they are subject to the other three fundamental interactions: *gravitation,* the *weak interaction,* and to *electromagnetism,* of which the latter is proportional to charge, and is thus zero for the electrically neutral *neutrinos.*

For every *lepton flavor,* there is a corresponding type of *antiparticle,* known as an *antilepton,* that differs from the *lepton* only in that some of its properties have equal magnitude but opposite sign. According to certain theories, *neutrinos* may be their own antiparticle. It is not currently known whether this is the case.

The first *charged lepton*, the *electron*, was theorized in the mid-19th century by several scientists and was discovered in 1897 by J. J. Thomson. The next *lepton* to be observed was the *muon*, discovered by Carl D. Anderson in 1936, which was classified as a meson at the time. After investigation, it was realized that the *muon* did not have the expected properties of a meson, but rather behaved like an *electron*, only with higher mass. It took until 1947 for the concept of "*leptons*" as a family of particles to be proposed. The term *lepton* was first used by physicist Léon Rosenfeld in 1948.

The first *neutrino*, the *electron neutrino*, was proposed by Wolfgang Pauli in 1930 to explain certain characteristics of *beta decay*. It was first observed in the Cowan–Reines *neutrino* experiment conducted by Clyde Cowan and Frederick Reines in 1956. The *muon neutrino* was discovered in 1962 by Leon M. Lederman, Melvin Schwartz, and Jack Steinberger, and the *tau* discovered between 1974 and 1977 by Martin Lewis Perl and his colleagues from the Stanford Linear Accelerator Center and Lawrence Berkeley National Laboratory. The *tau neutrino* remained elusive until July 2000, when the DONUT collaboration from Fermilab announced its discovery.

Leptons are an important part of the *Standard Model*. *Electrons* are one of the components of atoms, alongside *protons* and *neutrons*. Exotic atoms with *muons* and *taus* instead of *electrons* can also be synthesized, as well as *lepton–antilepton* particles such as *positronium*.]

We would like to suggest that there may not be a strict *pseudovector current conservation*, but that we may have an approximate conservation which becomes rigorous in the limit $q^2 \gg m_\pi^2$, m_π being the *pion* [*pi meson*] mass.

[A *meson* is a type of hadronic subatomic particle composed of *an equal number of quarks and antiquarks, usually one of each, bound together by the strong interaction.* Because *mesons* are composed of *quark* subparticles, they have a meaningful physical size, a diameter of roughly one femtometre (10^{-15} m), which is about 0.6 times the size of a *proton* or *neutron*. *All mesons are unstable*, with the longest-lived lasting for only a few tenths of a nanosecond. Heavier *mesons* decay to lighter *mesons* and ultimately to stable *electrons*, *neutrinos* and *photons*.

The existence of *mesons* was predicted by Hideki Yukawa's 1935 *theory of mesons* that postulated the particle as mediating the nuclear force. [Yukawa, H. (1935). On the Interaction of Elementary Particles. *Proceedings of the Physico-Mathematical Society of Japan*, 17, 48.

In 1947, the *pion* (or *pi meson*) was discovered by members of C. F. Powell's group at the University of Bristol, in England. In the same year, Lattes, together with Powell and Occhialini, determined the new particle's mass. [Lattes, C. M. G., Muirhead, H., Occhialini, G. P. S. & Powell, C. F. (1947). Processes involving charged mesons. *Nature*. 159, 4047, 694–7; https://doi.org/10.1038/159694a0.]

In 1948, Lattes, Eugene Gardner, and their team first artificially produced *pions* at the University of California's cyclotron in Berkeley, California, by bombarding carbon atoms with high-speed alpha particles.]

Specifically, we propose that the *axial vector* part of the *nucleon beta decay vertex* has the following form and properties:

$$g_A \Gamma_\mu^A (p', p) = g_V [\dots],$$
$$F_1(0) = g_A/g_V \approx F_2(0),$$
$$F_1(q^2) \sim F_1(q^2) \text{ for } q^2 \gg m_\pi^2. \tag{2}$$

The *pion* is then the analog of the massless *quantum* mentioned above. This is consistent with the *dispersion relations* expected for Γ_μ^A. Namely, F_1 and F_2 should have in general the form

$$F_i(q^2) = \dots \qquad (i = 1, 2) \tag{3}$$

where $m_0 = 3m_\pi$ unless there are new particles of low mass. Thus, the F's will be slowly varying for $|q^2| \ll m_0^2$. The conditions in Eq. (2) imply that $F_1/F_2 = 1$ for all q^2. If $m_\pi = 0$ and $F_1/F_2 = 1$, then we restore exact current conservation[2], and we also expect $F_1(0) = g_A/g_V = 1$.

If we adopt Eq. (2), the second term of Γ_μ^A immediately gives a relation between g_A, the *pion decay* (pseudovector) constant g_π, and the *pion-nucleon* (pseudoscalar) *coupling* G_π:

$$2Mg_A \approx 2Mg_V F_2(-m_\pi^2) = \sqrt{2} \ G_\pi g_\pi. \tag{4}$$

With $g_A = 1.25$ $g_V = 1.75 \times 10^{-49}$ erg cm³, [4] $G_z^2/4\pi = 13.5$, this gives a $\pi - \mu$ *decay life* of 2.7×10^{-8} sec as compared with the observed value 2.5×10^{-8} sec.

[4] Alikhanov, A. I. (1959). Ninth Annual International Conference on High-Energy Physics, Kiev, (unpublished).

Goldberger and Treiman[5] have arrived at the same relation Eq. (4) (in the limit of their self-energy integral $J \to \infty$) from an entirely different approach.

[5] Goldberger, M. L. & S. B. Treiman, S. B. (1958). *Phys. Rev.*, 110, 1178; Goldberger, M. L. (1959). *Revs. Modern Phys.*, 31, 797.

In our opinion, this is not a coincidence, as will be explained elsewhere.

We are tempted to extend this approximate conservation of the *axial vector* (and naturally also the *vector current*) to the *strangeness-non conserving beta decays*. We take, for example, the ΛN *axial vector* in the form

$$\Gamma_\mu^A (p_{N'}, p_\Lambda) \approx \dots , \tag{5}$$

and attribute the second term to the pseudoscalar K *meson*[6].

[6] It is also possible to associate a scalar K *meson* with the ΛN vector current conservation, while leaving the axial vector unaccounted for.

The degree of accuracy of the relation (5) will be poorer than in the previous case in view of the Λ - N mass difference (which destroys vector conservation) and the large *K-meson* mass. At any rate, we obtain an analog of Eq. (4):

$$(M_\Lambda + M_N) g_A' \approx G_K g_{K'} \tag{6}$$

which relates the Λ beta decay axial vector coupling g_A', the ΛNK coupling G_K, and the K_μ decay coupling g_K.

With the observed $K_{\mu2}$ lifetime 2.1×10^{-8} sec. and a tentative value $G_K^2/4\pi = \frac{1}{4} G\pi^2/4\pi$, we get

$$g_A'/g_A = 1/10. \tag{7}$$

This is not inconsistent with the observed beta decay of Λ which seems an order of magnitude less than predicted from a universal coupling scheme $g_V' = g_A' = g_V$. [7]

[7] Again Eq. (5) and the subsequent conclusions are essentially the same as those of Albright, C. H. (1959). *Phys. Rev.*, 114, 1648, and Sakita, B. (1959). *Phys. Rev.*, 114, 1650, which are based on the Goldberger - Treiman method. For the Λ-decay case below, see L. Tenaglia, L. (1959). *Nuovo Cimento*, 14, 499.

We can still go further, though the argument becomes more arbitrary. Let us assume that a fundamental weak coupling (N-NN-A) gives rise to an effective V-A interaction (or at least part of it) of the form

$$\dots . \tag{8}$$

Here $\Gamma_\mu{}^V = i\gamma_\mu$ which is approximately conserved itself, and $\Gamma_\mu{}^A$ stands for Eq. (2) or (5). We see easily that Eq. (8) contains information about the $\Lambda - N + \pi$ decay matrix element:

$$\cdots . \tag{9}$$

Combined with the assumption of $\Delta T = \frac{1}{2}$ selection rule, this gives a lifetime of 2.5×10^{-10} sec. for $g'' = g_V$ as compared with the observed value 2.8×10^{-10} sec.

It is possible to apply this kind of consideration to other *hyperons*. Moreover, if the Feynman —Gell-Mann coupling scheme such as ($\pi\pi ev$) is formally extended to (Kπev), etc. as has been tried by some people, all the observed decay processes may be covered. Here we would like to point out that if all *baryons* should satisfy Eqs. (4) and (6), the ratios g_A/G_π and g_A'/G_K must be approximately common constants.

[A *baryon* is a type of composite subatomic particle, including the *proton* and the *neutron, that contains an odd number of valence quarks*, conventionally three. *Baryons* belong to the *hadron* family of particles; *hadrons* are composed of *quarks*. *Baryons* are also classified as *fermions* because they have half-integer spin.

A *hyperon* is any *baryon* containing one or more *strange quarks*, but no *charm*, *bottom*, or *top quark*. Being *baryons*, all *hyperons* are *fermions*. That is, they have half-integer *spin* and obey Fermi–Dirac statistics. *Hyperons* all interact via the *strong nuclear force*, making them types of *hadron*. They are composed of three light *quarks*, at least one of which is a *strange quark*, which makes them *strange baryons*.

This form of matter may exist in a stable form within the core of some neutron stars. The first research into *hyperons* happened in the 1950s and spurred physicists on to the creation of an organized classification of particles. The term was coined by French physicist Louis Leprince-Ringuet in 1953.]

Our final remark concerns the theoretical basis for the assumptions made here. If the *baryons* are derived from some fundamental field ψ which possesses an invariance under a transformation of the type $\psi \rightarrow \exp(i\vec{\alpha} \cdot \vec{\tau}\gamma_5)$,[8] then there will be a *conservation of the pseudovector charge-current.*

[8] F. Gursey (private communication) has recently obtained similar results on the π decay based on this γ_5 invariance. We do not here specify the interaction of the ψ field, which may be of the nonlinear Heisenberg type, or due to an intermediate *boson* (different from π or K).

A finite observed *mass* can be compatible with the conservation if the particle is coupled with a *boson* as was noted in Eq. (1).

This situation may be understood by making an analogy to the *theory of superconductivity* originated by Bardeen, Cooper, and Schrieffer[9], and refined by Bogoliubov[10].

[9] Bardeen, J., Cooper, L. N. & Schrieffer, J. R. (1957). *Phys. Rev.*, 106, 162.

[10] Bogoliubov, N., Tolmachev, V. V. & Shirkov, D. V. (1958). *A New Method in the Theory of Superconductivity*. (Academy of Sciences of USSR, Moscow).

There *gauge invariance*, the *energy gap*, and the collective *excitations* are logically related to each other as was shown by the author[11].

[11] Nambu, Y. (1960). Quasi-Particles and Gauge Invariance in the Theory of Superconductivity. *Phys. Rev.*, 117, 648; https://doi.org/10.1103/PhysRev.117.648.

[*Abstract*: Ideas and techniques known in *quantum electrodynamics* have been applied to the Bardeen-Cooper-Schrieffer *theory of superconductivity*. In an approximation which corresponds to a generalization of the Hartree-Fock fields, one can write down an integral equation defining the *self-energy* of an *electron* in an *electron gas* with *phonon* and *Coulomb interaction*. The form of the equation implies the existence of a particular solution which does not follow from perturbation theory, and which leads to the *energy gap equation* and the quasi-particle picture analogous to Bogoliubov's.

The *gauge invariance*, to the first order in the external *electromagnetic field*, can be maintained in the quasi-particle picture by taking into account a certain class of corrections to the *charge-current operator* due to the *phonon* and *Coulomb interaction*. In fact, generalized forms of the *Ward identity* are obtained between certain vertex parts and the self-energy. The *Meissner effect* calculation is thus rendered strictly *gauge invariant*, but essentially keeping the BCS result unaltered for transverse fields.

It is shown also that the integral equation for vertex parts allows homogeneous solutions which describe collective excitations of quasi-particle pairs, and the nature and effects of such collective states are discussed.]

In the present case we have only to replace them by γ_5 *invariance*, *baryon mass*, and the *mesons*.

In fact, the mathematical method used in superconductivity particles may be taken over to study the *self-energy problem* of *elementary particles*. It is interesting that pseudoscalar *mesons* automatically emerge in this theory as *bound states* of *baryon* pairs. The nonzero *meson masses* and *baryon mass splitting* would indicate that the γ_5 *invariance* of the bare *baryon field* is not rigorous, possibly because of a small bare *mass* of the order of the *pion mass*.

The above-mentioned model of *elementary particles* will be studied in a separate paper.

Sakurai, J. J. (September, 1960). Theory of strong interactions.*

Ann. Phys., 11, 1, 1-48; https://www.sciencedirect.com/science/article/abs/pii/0003491660901263.

* This work was supported by the U. S. Atomic Energy Commission.

Abstract

All the *symmetry* models of *strong interactions* which have been proposed up to the present are devoid of deep physical foundations. It is suggested that, instead of postulating artificial "higher" *symmetries* which must be broken anyway within the realm of *strong interactions*, we take the *existing exact* symmetries of *strong interactions* more seriously than before and exploit them to the utmost limit. A new theory of *strong interactions* is proposed on this basis.

[The *strong interaction*, also called the *strong force* or *strong nuclear force*, is a fundamental *interaction* that confines *quarks* into *protons*, neutrons, and other *hadron* particles. The *strong interaction* also binds *neutrons* and *protons* to create *atomic nuclei*, where it is called the *nuclear force*.

Most of the *mass* of a *proton* or *neutron* is the result of the *strong interaction* energy; the individual *quarks* provide only about 1% of the *mass* of a *proton*.

Before 1971, physicists were uncertain as to how the *atomic nucleus* was bound together. It was known that the *nucleus* was composed of *protons* and *neutrons* and that *protons* possessed positive *electric charge*, while *neutrons* were electrically neutral. By the understanding of physics at that time, positive charges would repel one another and the positively charged *protons* should cause the *nucleus* to fly apart. However, this was never observed. A stronger attractive force was postulated to explain how the *atomic nucleus* was bound despite the *protons'* mutual *electromagnetic* repulsion. This hypothesized force was called the *strong force*, which was believed to be a fundamental force that acted on the *protons* and *neutrons* that make up the *nucleus*.

In 1964, Murray Gell-Mann, and separately George Zweig, proposed that *baryons*, which include *protons* and *neutrons*, and *mesons* were composed of elementary particles. Zweig called the elementary particles "aces" while Gell-Mann called them "*quarks*"; the theory came to be called the *quark model*. [Gell-Mann. M. (February, 1964). A Schematic Model of Baryons and Mesons. See below.]

249

The *strong attraction* between *nucleons* was the side-effect of a more fundamental force that bound the *quarks* together into *protons* and *neutrons*. The theory of *quantum chromodynamics* explains that *quarks* carry what is called a *color charge*, although it has no relation to visible *color*. *Quarks* with unlike *color charge* attract one another as a result of the *strong interaction*, and the particle that mediates this was called the *gluon*.]

Following Yang and Mills we require that the *gauge transformations* that are associated with the three "internal" conservation laws—*baryon conservation, hypercharge conservation,* and *isospin conservation*—be "consistent with the local field concept that underlies the usual physical theories." In analogy with *electromagnetism* there emerge three kinds of couplings such that in each case a *massive vector field* is coupled linearly to the conserved *current* in question. Each of the three fundamental *couplings* is characterized by a single universal constant. Since, as Pais has shown, there are no other internal *symmetries* that are exact, and since any successful theory must be simple, there are no other fundamental *strong couplings. Parity conservation in strong interactions follows as the direct consequence of parity conservation of the three fundamental vector couplings.* The three *vector couplings* give rise to corresponding *current-current* interactions. Yukawa-type *couplings* of *pions* and *K particles* to *baryons* are "phenomenological," and may arise, for instance, out of *four-baryon current-current interactions* along the lines suggested by Fermi and Yang. All the successful features of Chew-Low type *meson* theories and of relativistic dispersion relations can, in principle, be in accordance with the theory whereas *none of the predictions based on relativistic Yukawa-type Lagrangians are meaningful unless ωM is considerably less than unity.*

Simple and direct experimental tests of the theory should be looked for in those phenomena in which phenomenological *Yukawa-type couplings* are likely to play unimportant roles. The fundamental *isospin current coupling* in the static limit gives rise to a short-range repulsion (attraction) between two particles whenever the *isospins* are parallel (antiparallel). Thus, the low-energy *s-wave πN interaction* should be repulsive in the $T = 32$ state and attractive in the $T = 12$ state in agreement with observation. In *$\pi \Sigma$ s-wave scattering* the $T = 0$ state is strongly attractive, and there definitely exists the possibility of an *s-wave resonance* at energies of the order of the $K^- p$ threshold, while the $T = 1$ *$\pi \Sigma$ phase shift* is likely to remain small; using the *K matrix formalism* of Dalitz and Tuan, we might be able to compare the "ideal" *phase shifts* derived in this manner with the "actual" *phase shifts* deduced from $K^- p$ reactions. It is expected that the *two-pion system* exhibits a resonant behavior in the $T = 1$ *(p-wave)* state in agreement with the conjecture of Frazer and Fulco based on the electromagnetic structure of the *nucleon*. The *three-pion system* is expected to exhibit two $T = 0, J = 1$ *resonances*. It is conjectured that the two $T = 12$ and one $T = 32$ *"higher resonances"* in the *πN interactions* may be due to the two $T = 0$

3π resonances and the one $T = 1$ *2π resonance* predicted by the theory. Multiple *pion* production is expected at all energies to be more frequent than that predicted on the basis of statistical considerations. The fundamental *hypercharge current coupling* gives rise to a short-range repulsion (attraction) between two *charge-doublet particles* when their *hypercharges* are like (opposite). If the *isospin current coupling* is effectively weaker than the *hypercharge current coupling*, the *KN "potential"* should be repulsive and the KN *"potential"* should be attractive, and the *charge exchange scattering* of K^+ and K^- should be relatively rare, at least in *s states*. All these features seem to be in agreement with current experiments. Conditions for the validity of Pais' *doublet* approximation are discussed. The theory offers a possible explanation for the long-standing problem as to why associated production *cross sections* are small and K^- *cross sections* are large. The empirical fact that the ratio of (KK2N) to $(K\Lambda N) + (K\Sigma N)$ in *NN* collisions seems to be about twenty to thirty times larger than simple statistical considerations indicate is not surprising. The fundamental *baryonic current coupling* gives rise to a short-range repulsion for *baryon-baryon interactions* and an attraction for *baryon-antibaryon interactions*. There should be effects similar to those expected from *"repulsive cores"* for all *angular momentum* and *parity states* in both the $T = 1$ and $T = 0$, *NN interactions* at short distances though the $T = 1$ *state* may be more repulsive. A simple Thomas-type calculation gives rise to a *spin-orbit force* of the right sign with not unreasonable order of magnitude. The *ΛN* and *ΣN interactions* at short distances should be somewhat less repulsive than the *NN interactions*. *Annihilation cross sections* in NN *collisions* are expected to be large even in Bev regions in contrast to the predictions of Ball and Chew. The observed large *pion* multiplicity in NN *annihilations* is not mysterious. It is possible to invent a reasonable mechanism which makes the reaction p +p→ π++ π− very rare, as recently observed. Fermi-Landau-Heisenberg type theories of high energy collisions are not expected to hold in *relativistic NN* collisions; instead, the theory offers a theoretical justification for the *"two-fire-ball model"* of *high-energy jets* previously proposed on purely phenomenological grounds.

Because of the strong *short-range attraction* between a *baryon* and *antibaryon* there exists a mechanism for a *baryon-antibaryon pair* to form a *meson*. The dynamical basis of the Fermi-Yang-Sakata-Okun model as well as that of the Goldhaber-Christy model follows naturally from the theory; all the *ad hoc* assumptions that must be made in order that the compound models work at all can be explained from first principles. It is suggested that one should not ask which *elementary particles* are "more elementary than others," and which compound model is right, but rather characterize each particle only by its internal properties such as *total hypercharge* and *mean-square baryonic radius*. Although the fundamental *couplings* of the theory are highly symmetric and universal, it is possible for the three *couplings alone* to account for the observed *mass spectrum*. The theory can

251

explain, in a trivial manner, why there are no "elementary" particles with *baryon number* greater than unity provided that the *baryonic current coupling* is sufficiently strong. The question of whether or not an $|S| = 2$ *meson* exists is a dynamical one (not a group-theoretic one) that depends on the strength of the *hypercharge current coupling*. A possible reason for the nonexistence of a $\pi^{0'}$ (*charge-singlet, nonstrange boson*) is given. The theory realizes Pais' principles of economy of constants and of a hierarchy of *interactions* in a natural and elegant manner.

It is conjectured that there exists a deep connection between the *law of conservation of fermions* and the universal V-A *weak coupling*. In the absence of *strong* and *electromagnetic interactions, baryonic charge, hypercharge,* and *electric charge* all disappear, and only the sign of γ_5 can distinguish a *fermion* from an *antifermion*, the *fermionic charge* being diagonalized by γ_5; hence $1 + \gamma_5$ appears naturally in *weak interactions. Parity conservation* in *strong interactions, parity conservation* in *electromagnetic interactions,* parity non-conservation in *weak interactions* can all be understood from the *single common* principle of *generalized gauge invariance.* It appears that in the future ultimate theory of *elementary particles* all *elementary particle interactions* will be manifestations of the *five fundamental vector-type couplings* corresponding to the *five conservation laws* of "internal attributes"— *baryonic charge, hypercharge, isospin, electric charge,* and *fermionic charge.* Gravity and cosmology are briefly discussed; it is estimated that the *Compton wavelength* of the *graviton* is of the order of 10^8 light years.

It is suggested that every conceivable experimental attempt be made to detect directly quantum manifestations of the *vector fields* introduced in the theory, especially by studying Q values of *pions* in various combinations in NN *annihilations* and in *multiple pion production.*

Jeffrey Goldstone (born September 3, 1933).

Goldstone is a British theoretical physicist and an emeritus physics faculty member at the MIT Center for Theoretical Physics.

He worked at the University of Cambridge until 1977. He is noted for *Goldstone's theorem* and the discovery of the Nambu–Goldstone *boson*.

Born in Manchester, he was educated at Manchester Grammar School and Trinity College, Cambridge, (B.A. 1954, Ph.D. 1958). He worked on the theory of nuclear matter under the guidance of Hans Bethe and developed modifications of Feynman diagrams for non-relativistic many-*fermion* systems, which are currently referred to as *Goldstone diagrams*. In 1957, he proved the linked-cluster theorem, showing that only connected diagrams contribute to the calculation.

Goldstone was a research fellow of Trinity College, Cambridge, from 1956 to 1960 and held visiting research posts at Copenhagen, CERN and Harvard. During this time, his research focus shifted to particle physics and he investigated the nature of *relativistic field theories* with *spontaneously broken symmetries*. In Goldstone, J. (January, 1961). Field theories with "Superconductor" solutions, he examined the conditions for the existence of non-perturbative type "superconductor" solutions of field theories, resulting in what is known as *Goldstone's theorem*, which states that if there is an exact *continuous symmetry* of the Hamiltonian or Lagrangian defining the system, and this is not a symmetry of the vacuum state (i.e. there is *broken symmetry*), then there must be at least one *spin-zero massless particle* called a *Goldstone boson*.

From 1962 to 1976, Goldstone was a faculty member at Cambridge. In the early 1970s, with Peter Goddard, Claudio Rebbi and Charles Thorn, he worked out the light-cone quantization theory of *relativistic strings*.

> [The author studied Mathematics and Physics, in the Physics Department at Cambridge University between 1962-65, but does not remember Goldstone.]

He moved to the USA in 1977 as Professor of Physics at MIT, where he has been the Cecil and Ida Green Professor of Physics since 1983 and was Director of the MIT Center for Theoretical Physics from 1983-89.

Goldstone published research on *solitons* in *quantum field theory* with Roman Jackiw and Frank Wilczek, and on the *quantum strong law of large numbers* with Edward Farhi and Samuel Gutmann.

Goldstone, J. (January, 1961). Field theories with "Superconductor" solutions.

Nuovo Cimento, 19, 1, 154–64; https://doi.org/10.1007/BF02812722; also at https://sci-hub.st/10.1007/BF02812722.

CERN, Geneva.

Received September 8, 1960.

Abstract

The conditions for the existence of non-perturbative type *"superconductor"* solutions of *field theories* are examined. A *non-covariant canonical transformation method* is used to find such solutions for a theory of a *fermion* interacting with a *pseudoscalar boson*. A *covariant renormalizable method* using Feynman integrals is then given. A *"superconductor"* solution is found whenever in the normal perturbative-type solution the *boson* mass squared is negative and the *coupling constants* satisfy certain inequalities. The *symmetry properties* of such solutions are examined with the aid of a simple model of self-interacting *boson fields*. The solutions have lower *symmetry* than the Lagrangian, and contain *mass zero bosons*.

[*Goldstone's theorem* in *relativistic quantum field theory* states that if there is an exact *continuous symmetry* of the Hamiltonian or Lagrangian defining the system, and this is not a symmetry of the vacuum state (i.e. there is *broken symmetry*), then there must be at least one *spin-zero massless particle* called a *Goldstone boson*. In the *quantum theory of many-body systems Goldstone bosons* are collective excitations such as *spin waves*. An important exception to Goldstone's theorem is provided in *gauge theories* with the *Higgs mechanism*, whereby the *Goldstone bosons* gain *mass* and become *Higgs bosons*.]

---

1. - Introduction.

This paper reports some work on the possible existence of *field theories* with solutions analogous to the *Bardeen model of a superconductor*. This possibility has been discussed by Nambu[1] in a report which presents the general ideas of the theory which will not be repeated here.

[1] Nambu, Y. *Enrico Fermi Institute for Nuclear Studies, Chicago, Report 60-21.*

The present work merely considers models and has no direct physical applications but the nature of these theories seems worthwhile exploring. The models considered here all have a *boson field* in them from the beginning.

> [A *boson* is a subatomic particle whose *spin* quantum number has an integer value (0, 1, 2, ...). *Bosons* form one of the two fundamental classes of subatomic particle, the other being *fermions*, which have odd half-integer spin (1/2, 3/2, 5/2, ...). Every observed subatomic particle is either a *boson* or a *fermion*.
>
> Some *bosons* are elementary particles occupying a special role in particle physics, distinct from the role of *fermions* (which are sometimes described as the constituents of "ordinary matter"). Certain elementary *bosons* (e.g. *gluons*) act as force carriers, which give rise to forces between other particles, while one (the *Higgs boson*) contributes to the phenomenon of *mass*. Other *bosons*, such as *mesons*, are composite particles made up of smaller constituents.
>
> Paul Dirac coined the name *boson* to commemorate the contribution of Satyendra Nath Bose, an Indian physicist.]

It would be more desirable to construct *bosons* out of *fermions* and this type of theory does contain that possibility[1].

> [A *fermion* is a particle that follows *Fermi–Dirac statistics*. *Fermions* have a half-odd-integer spin (spin 1/2, spin 3/2, etc.) and obey the *Pauli exclusion principle*. These particles include all *quarks* and *leptons* and all composite particles made of an odd number of these, such as all *baryons* and many atoms and nuclei. *Fermions* differ from *bosons*, which obey *Bose–Einstein statistics*.
>
> Some *fermions* are elementary particles (such as *electrons*), and some are composite particles (such as *protons*). Particles with integer *spin* are *bosons*; particles with half-integer *spin* are *fermions*.
>
> In addition to the *spin* characteristic, *fermions* have another specific property: they possess conserved *baryon* or *lepton* quantum numbers. Therefore, what is usually referred to as the *spin-statistics relation* is, in fact, a *spin statistics-quantum number relation*.
>
> As a consequence of the *Pauli exclusion principle*, only one *fermion* can occupy a particular *quantum state* at a given time. Suppose multiple *fermions* have the same spatial probability distribution. *Then, at least one property of each fermion, such as its spin, must be different*. *Fermions* are usually associated with *matter*, whereas

255

bosons are generally *force carrier* particles. However, in the current state of particle physics, the distinction between the two concepts is unclear. Weakly interacting fermions can also display bosonic behavior under extreme conditions. For example, at low temperatures, *fermions* show superfluidity for uncharged particles and superconductivity for charged particles. Composite *fermions*, such as *protons* and *neutrons*, are the key building blocks of everyday matter.

Paul Dirac coined the name *fermion* from the surname of Italian physicist Enrico Fermi.]

The theories of this paper have the *dubious advantage of being renormalizable*, which at least allows one to find simple conditions in finite terms for the existence of *"superconducting" solutions*. It also appears that in fact many features of these solutions can be found in very simple models with only *boson fields*, in which the analogy to the Bardeen theory has almost disappeared. In all these theories the relation between the *boson field* and the actual particles is more indirect than in the usual perturbation type solutions of field theory.

2. - Non-covariant theory.

The first model has a single *fermion* interacting with a single *pseudoscalar boson field* with the Lagrangian

$$\mathscr{L} = \psi^- \left(i\gamma^\mu (\partial\psi/\partial x^\mu - m\psi) + \frac{1}{2} \left(\partial\psi/\partial x^\mu \, \partial\varphi/\partial x_\mu - \mu_0^2\varphi^2 \right) - g_0\psi^-\gamma^5\psi\varphi - 1/24 \, \lambda_0\varphi^4.$$

(The last term is necessary to obtain finite results, as in *perturbation theory*.) The new solutions can be found by a non-covariant calculation which perhaps may show more clearly what is happening than the *covariant theory* which follows.
…

3. - Covariant theory.

A first approach to a *covariant theory* can be made by calculating the *fermion* Green's function in a self-consistent field approximation. In *perturbation theory* the term represented by Fig. 4 vanishes by *reflection invariance*.
…

4. - Symmetry properties and a simple model.

It is now necessary to discuss the principal peculiar feature of this type of solution. The original Lagrangian had a *reflection symmetry*. From this it follows that $F(\chi)$ must be an even function. Thus, $\chi = \chi_1$ is one solution of $F'(\chi_0) = 0$, $\chi_1 = -\chi_1$ is another. By choosing

256

one solution, the *reflection symmetry* is effectively destroyed. It is possible to make a very simple model which shows this kind of behavior, and also demonstrates that so long as there is a *boson field* in the theory to start with, the essential features of the abnormal solutions have very little to do with *fermion pairs*.

Consider the theory of a single *neutral pseudoscalar boson* interacting with itself,

$$\mathscr{L} = \tfrac{1}{2} \left(\partial\varphi/\partial x_\mu \, \partial\varphi/\partial x^\mu - \mu_0^2\varphi^2 \right) - 1/24 \, \lambda_0\varphi^4.$$

Normally this theory is quantized by letting each mode of oscillation of the classical field correspond to a quantum oscillator whose quantum number gives the number of particles. When $\mu_0^2 < 0$, this approach will not work. However, if $\lambda_0 > 0$, the function

$$\mu_0^2/2 \; \varphi^2 + \lambda_0/24 \; \varphi^4$$

is as shown in Fig. 7. ...

The classical equations

$$(\Box^2 - 2\mu_0^2)\varphi + \lambda_0/6 \; \varphi^2 = 0,$$

now have solutions $\varphi = \pm \sqrt{(-6\mu_0^2/\lambda_0)}$ corresponding to the minima of this curve. Infinitesimal oscillations round one of these minima obey the equation

$$(\Box^2 - 2\mu_0^2) \, \delta\varphi = 0,$$

These can now be quantized to represent particles of *mass* $\sqrt{(-2\mu_0^2}$. This is simply done by making the transformation $\varphi = \varphi' + \chi$

$$\chi^2 = -6\mu_0^2/\lambda_0.$$

Then

$$\mathscr{L} = \tfrac{1}{2} \left(\partial\varphi'/\partial x_\mu \, \partial\varphi'/\partial x^\mu + 2\mu_0^2\varphi'^2 \right) - \lambda_0/24 \; \varphi'^4 - \lambda_0\chi /6 \; \varphi'^3 + 3/2 \; \mu_0^2/\lambda_0.$$

This new Lagrangian can be treated by the canonical methods.

In any state with a finite number of particles, the expectation value of φ is infinitesimally different from the vacuum expectation value. Thus, the eigenstates corresponding to oscillations round $\varphi = \chi$ are all orthogonal to the usual states corresponding to oscillations round $\varphi = 0$, and also to the *eigenstates* round $\varphi = -\chi$. This means that the theory has two vacuum states, with a complete set of particle states built on each vacuum, but that there is a *super-selection rule* between these two sets so that it is only necessary to consider one

257

of them. The *symmetry* $\varphi \rightarrow -\varphi$ has disappeared. Of course, it can be restored by introducing linear combinations of states in the two sets but because of the *super-selection rule* this is a highly artificial procedure.

Now consider the case when the *symmetry group* of the Lagrangian is *continuous* instead of discrete. A simple example is that of a complex *boson* field, $\varphi = (\varphi_1 + i\varphi_2)/\sqrt{2}$

$$\mathcal{L} = \partial\varphi^*/\partial x_\mu \, \partial\varphi/\partial x^\mu - \mu_0^2 \varphi^*\varphi - \lambda_0/6 \, (\varphi^*\varphi)^2.$$

The *symmetry* is $\varphi \rightarrow \exp[ix]\varphi$. The canonical transformation is $\varphi = \varphi' + \chi$

$$|\chi^2| = -3\mu_0^2/\lambda_0.$$

The *phase* of χ is not determined. Fixing it destroys the *symmetry*. With χ real the new Lagrangian is

$$\mathcal{L} = \tfrac{1}{2}\,(\partial\varphi_1'/\partial x_\mu \, \partial\varphi_1'/\partial x^\mu + 2\mu_0^2\varphi_1'^2) + \tfrac{1}{2}\,\partial\varphi_2'/\partial x_\mu \, \partial\varphi_2'/\partial x^\mu - \lambda_0\chi/6 \, \varphi_1'(\varphi_1'^2 + \varphi_2'^2)$$
$$- \lambda_0/24 \, (\varphi_1'^2 + \varphi_2'^2)^2.$$

The particle corresponding to the φ_2' field has *zero mass*. This is true even when the *interaction* is included, and is the new way the original *symmetry* expresses itself.

A simple picture can be made for this theory by thinking of the two-dimensional vector φ at each point of space. In the vacuum state the vectors have magnitude χ and are all lined up (apart from the *quantum fluctuations*). The massive particles φ_1' correspond to oscillations in the direction of χ. The massless particles φ_2' correspond to "*spin-wave*" *excitations* in which only the direction of φ makes infinitesimal oscillations. The *mass* must be zero, because when all the $\varphi(x)$ rotate in phase there is no gain in *energy* because of the *symmetry*.

This time there are infinitely many vacuum states. A state can be specified by giving the *phase* of χ and then the numbers of particles in the two different oscillation modes. There is now a *super-selection rule* on the *phase* of χ. States with a definite *charge* can only be constructed artificially by superposing states with different *phases*.

5. - Conclusion.

This result is completely general. Whenever the original Lagrangian has a *continuous symmetry group*, the new solutions have a reduced *symmetry* and contain *massless bosons*. One consequence is that *this kind of theory cannot be applied to a vector particle without losing Lorentz invariance.* A method of losing *symmetry* is of course highly desirable in elementary particle theory but these theories will not do this without introducing nonexistent *massless bosons* (unless *discrete symmetry groups* can be used). Skyrme[3] has hoped that one set of fields could have *excitations* both of the usual type and of the "*spin-wave*" type, thus for example obtaining the π-*mesons* as collective oscillations of the four *K-meson* fields, but this does not seem possible in this type of theory.

[3] Skyrme, T. H. R. (1959). *Proc. Roy. Soc.*, A, 252, 236.

Thus, *if any use is to be made of these solutions something more complicated than the simple models considered in this paper will be necessary.*

Sheldon Lee Glashow (born December 5, 1932).

Glashow is a Nobel Prize-winning American theoretical physicist. He is the Metcalf Professor of Mathematics and Physics at Boston University and Eugene Higgins Professor of Physics, emeritus, at Harvard University.

Glashow was born on December 5, 1932, in New York City, to Jewish immigrants from Russia, Bella (née Rubin) and Lewis Gluchovsky, a plumber. He graduated from Bronx High School of Science in 1950. Glashow was in the same graduating class as Steven Weinberg, whose own research, independent of Glashow's, would result in Glashow, Weinberg, and Abdus Salam sharing the 1979 Nobel Prize in Physics.

Glashow received a Bachelor of Arts degree from Cornell University in 1954 and a PhD degree in physics from Harvard University in 1959 under Nobel-laureate physicist Julian Schwinger. Afterwards, Glashow became a NSF fellow at NORDITA and met Murray Gell-Mann, who convinced him to become a research fellow at the California Institute of Technology. Glashow then became an assistant professor at Stanford University before joining the University of California, Berkeley where he was an associate professor from 1962 to 1966. He joined the Harvard physics department as a professor in 1966, and was named Eugene Higgins Professor of Physics in 1979; he became emeritus in 2000.

In 1961, Glashow extended *electroweak unification models* due to Schwinger by including a short-range *neutral current*, the Z_0. The resulting *symmetry structure* that Glashow proposed, $SU(2) \times U(1)$, forms the basis of the accepted theory of the *electroweak interactions*. [Glashow, S. L. (February, 1961). Partial-symmetries of weak interactions. Nuclear Physics. See below.] For this discovery, Glashow along with Steven Weinberg and Abdus Salam, was awarded the 1979 Nobel Prize in Physics.

In collaboration with James Bjorken, Glashow was the first to predict a fourth *quark*, the charm quark, in 1964. [Bjørken, B.J. & Glashow, S.L. (August, 1964). Elementary particles and SU(4). *Physics Letters*, 11, 3, 255-7; https://doi.org/10.1016/0031-9163(64)90433-0.] This was at a time when 4 *leptons* had been discovered but only 3 *quarks* proposed. The development of their work in 1970, the Glashow–Iliopoulos–Maiani (GIM) mechanism showed that the two *quark* pairs: (d.s), (u,c), would largely cancel out *flavor changing neutral currents*, which had been observed experimentally at far lower levels than theoretically predicted on the basis of 3 *quarks* only. The prediction of the *charm quark* also removed a technical disaster for any *quantum field theory* with unequal numbers of *quarks* and *leptons* — an anomaly — where classical field theory *symmetries* fail to carry over into the *quantum theory*.

In 1972, Glashow married to Joan Shirley Alexander. They have four children. Lynn Margulis was Joan's sister, making Carl Sagan his former brother-in-law. Daniel Kleitman, who was another doctoral student of Julian Schwinger, is also his brother-in-law, through Joan's other sister, Sharon.

In 1973, Glashow and Howard Georgi proposed the first *grand unified theory*. They discovered how to fit the *gauge forces* in the *Standard Model* into an SU(5) Lie group group, and the *quarks* and *leptons* into two simple representations. Their theory qualitatively predicted the general pattern of *coupling constant* running, with plausible assumptions, it gave rough *mass ratio values* between third generation *leptons* and *quarks*, and it was the first indication that the *law of Baryon number* is inexact, that the *proton* is unstable. This work was the foundation for all future unifying work. [Georgi, H. & Glashow, S. L. (1974). Unity of All Elementary Particle Forces. *Phys. Rev. Lett.*, 32, 438; https://journals.aps.org/prl/abstract/10.1103/PhysRevLett.32.438.]

Glashow is a skeptic of superstring theory due to its lack of experimentally testable predictions. He had campaigned to keep string theorists out of the Harvard physics department, though the campaign failed. About ten minutes into "String's the Thing", the second episode of The Elegant Universe TV series, he describes *superstring theory* as a discipline distinct from physics, saying "...you may call it a tumor, if you will...".

Glashow, S. L. (February, 1961). Partial-symmetries of weak interactions[†].

Nuclear Physics, 22, 4, 579–88; https://doi.org/10.1016/0029-5582(61)90469-2.

[†] The article is hidden behind pay walls but can be purchased from the publisher.

Glashow combined the *electromagnetic* and *weak interactions* and extended *electroweak unification models* due to Schwinger by including a short-range *neutral current*, the Z_0. The resulting *symmetry structure* that Glashow proposed, SU(2) × U(1), forms the basis of the accepted *theory of the electroweak interactions*. The *W and Z bosons* were predicted in detail by Sheldon Glashow, Mohammad Abdus Salam, and Steven Weinberg. For this discovery, Glashow along with Steven Weinberg and Abdus Salam, was awarded the 1979 Nobel Prize in Physics.

———————————

Abstract.

Weak and electromagnetic interactions of the *leptons* were examined under the hypothesis that the *weak interactions* are mediated by *vector bosons*. With only an *isotopic* triplet of *leptons* coupled to a *triplet* of *vector bosons* (two charged decay intermediaries and the photon), the theory possesses no *partial symmetries*. Such symmetries may be established if additional vector *bosons* or additional *leptons* are introduced. Since the latter possibility yields a theory disagreeing with experiment, *the simplest partially symmetric model reproducing the observed electromagnetic and weak interactions of leptons requires the existence of at least four vector-boson fields* (including the *photon*). Corresponding partially conserved quantities *suggest leptonic analogs to the conserved quantities associated with strong interactions: strangeness and isobaric spin.*

[A *lepton* is an elementary particle of half-integer *spin* (*spin* ½) that does not undergo *strong interactions*. Two main classes of *leptons* exist: *charged leptons* (also known as the *electron-like leptons* or *muons*), including the *electron, muon,* and *tauon,* and *neutral leptons*, better known as *neutrinos. Charged leptons* can combine with other particles to form various composite particles such as atoms and *positronium*, while *neutrinos* rarely interact with anything, and are consequently rarely observed. The best known of all *leptons* is the *electron*.

There are six types of *leptons*, known as flavors, grouped in three generations. The *first-generation leptons*, also called *electronic leptons*, comprise the *electron* (e⁻) and the *electron neutrino* (v_e); the second are the *muonic leptons*, comprising the *muon* (μ⁻) and the *muon neutrino* ($v\mu$); and the third are the *tauonic leptons*, comprising the *tau* (τ⁻) and the *tau neutrino* (vτ). *Electrons* have the least mass of all the *charged leptons*. The heavier *muons* and *taus* will rapidly change into *electrons* and *neutrinos* through a process of particle decay: the transformation from a higher mass state to a lower mass state. Thus, *electrons* are stable and the most common charged *lepton* in the universe, whereas *muons* and *taus* can only be produced in high-energy collisions (such as those involving cosmic rays and those carried out in particle accelerators).

Leptons have various intrinsic properties, including *electric charge*, *spin*, and *mass*. Unlike *quarks*, however, *leptons* are not subject to the *strong interaction*, but they are subject to the other three fundamental interactions: *gravitation*, the *weak interaction*, and to *electromagnetism*, of which the latter is proportional to charge, and is thus zero for the electrically neutral *neutrinos*.

For every *lepton flavor*, there is a corresponding type of *antiparticle*, known as an *antilepton*, that differs from the *lepton* only in that some of its properties have equal magnitude but opposite sign. According to certain theories, *neutrinos* may be their own antiparticle. It is not currently known whether this is the case.

The first *charged lepton*, the *electron*, was theorized in the mid-19th century by several scientists and was discovered in 1897 by J. J. Thomson. The next *lepton* to be observed was the *muon*, discovered by Carl D. Anderson in 1936, which was classified as a meson at the time. After investigation, it was realized that the *muon* did not have the expected properties of a meson, but rather behaved like an *electron*, only with higher mass. It took until 1947 for the concept of "*leptons*" as a family of particles to be proposed. The term *lepton* was first used by physicist Léon Rosenfeld in 1948.

The first *neutrino*, the *electron neutrino*, was proposed by Wolfgang Pauli in 1930 to explain certain characteristics of *beta decay*. It was first observed in the Cowan–Reines *neutrino* experiment conducted by Clyde Cowan and Frederick Reines in 1956. The *muon neutrino* was discovered in 1962 by Leon M. Lederman, Melvin Schwartz, and Jack Steinberger, and the *tau* discovered between 1974 and 1977 by Martin Lewis Perl and his colleagues from the Stanford Linear Accelerator Center and Lawrence Berkeley National Laboratory. The *tau neutrino* remained elusive until July 2000, when the DONUT collaboration from Fermilab announced its discovery.

Leptons are an important part of the *Standard Model*. *Electrons* are one of the components of atoms, alongside *protons* and *neutrons*. Exotic atoms with *muons* and *taus* instead of *electrons* can also be synthesized, as well as *lepton–antilepton* particles such as *positronium*.

A *boson* is a subatomic particle whose *spin* quantum number has an integer value (0, 1, 2, ...). *Bosons* form one of the two fundamental classes of subatomic particle, the other being *fermions*, which have odd half-integer spin (1/2, 3/2, 5/2, ...). Every observed subatomic particle is either a *boson* or a *fermion*.

Some *bosons* are elementary particles occupying a special role in particle physics, distinct from the role of *fermions* (which are sometimes described as the constituents of "ordinary matter"). Certain elementary *bosons* (e.g. *gluons*) act as force carriers, which give rise to forces between other particles, while one (the *Higgs boson*) contributes to the phenomenon of *mass*. Other *bosons*, such as *mesons*, are composite particles made up of smaller constituents.

Paul Dirac coined the name *boson* to commemorate the contribution of Satyendra Nath Bose, an Indian physicist.

A *vector boson* is a boson whose spin equals one. *Vector bosons* that are also elementary particles are *gauge bosons*, the force carriers of fundamental interactions. Some composite particles are *vector bosons*, for instance any *vector meson* (*quark* and *antiquark*). During the 1970s and 1980s, intermediate *vector bosons* (the *W* and *Z bosons*, which mediate the *weak interaction*) drew much attention in particle physics.]

Murray Gell-Mann (September 15, 1929–May 24, 2019).

Gell-Mann was an American theoretical physicist who played a preeminent role in the development of the *theory of elementary particles*. Gell-Mann introduced the concept of *quarks* as the fundamental building blocks of the *strongly interacting particles*, and the *renormalization group* as a foundational element of *quantum field theory* and statistical mechanics. He played key roles in developing the concept of *chirality* in the *theory of the weak interactions* and *spontaneous chiral symmetry breaking* in the *strong interactions*, which controls the physics of the light *mesons*. In the 1970s he was a co-inventor of *quantum chromodynamics* (QCD) which explains the *confinement* of *quarks* in *mesons* and *baryons* and forms a large part of the *Standard Model* of elementary particles and forces.

Gell-Mann received the 1969 Nobel Prize in Physics "for his contributions and discoveries concerning the classification of elementary particles and their interactions".

Gell-Mann was born in Lower Manhattan to a family of Jewish immigrants from the Austro-Hungarian Empire, specifically from Czernowitz in present-day Ukraine. His parents were Pauline (née Reichstein) and Arthur Isidore Gelman, who taught English as a second language. Propelled by an intense boyhood curiosity and love for nature and mathematics, he graduated valedictorian from the Columbia Grammar & Preparatory School aged 14 and subsequently entered Yale College as a member of Jonathan Edwards College. Gell-Mann graduated from Yale with a bachelor's degree in physics in 1948 and intended to pursue graduate studies in physics. He sought to remain in the Ivy League for his graduate education and applied to Princeton University as well as Harvard University. He was rejected by Princeton and accepted by Harvard, but the latter institution was unable to offer him needed financial assistance.

He was accepted by the Massachusetts Institute of Technology (MIT) and received a letter from Victor Weisskopf urging him to attend MIT and become Weisskopf's research assistant. This would provide Gell-Mann with the financial assistance he required. He received his Ph.D. in physics from MIT in 1951 after completing a doctoral dissertation, titled "*Coupling strength and nuclear reactions*", under the supervision of Weisskopf.

Subsequently, Gell-Mann was a postdoctoral fellow at the Institute for Advanced Study at Princeton in 1951, and a visiting research professor at the University of Illinois at Urbana–Champaign from 1952 to 1953. He was a visiting associate professor at Columbia University and an associate professor at the University of Chicago in 1954–1955, before moving to the California Institute of Technology, where he taught from 1955 until he retired in 1993.

Gell-Mann was the Robert Andrews Millikan Professor of Theoretical Physics Emeritus at California Institute of Technology, where he joined the faculty in 1955, as well as a university professor in the physics and astronomy department of the University of New Mexico in Albuquerque, New Mexico, and the Presidential Professor of Physics and Medicine at the University of Southern California.

Gell-Mann married J. Margaret Dow in 1955; they had a daughter and a son. Margaret died in 1981, and in 1992 he married Marcia Southwick, whose son became his stepson.

In 1958, Gell-Mann in collaboration with Richard Feynman, in parallel with the independent team of E. C. George Sudarshan and Robert Marshak, discovered the *chiral* structures of the *weak interaction* of physics and developed the *V-A theory* (*vector minus axial vector theory*). This work followed the experimental discovery of the *violation of parity* by Chien-Shiung Wu, as suggested theoretically by Chen-Ning Yang and Tsung-Dao Lee.

Gell-Mann's work in the 1950s involved recently discovered cosmic ray particles that came to be called *kaons* and *hyperons*. Classifying these particles led him to propose that a *quantum number*, called *strangeness*, would be conserved by the *strong* and the *electromagnetic interactions*, but not by the *weak interaction*. [Gell-Mann, M. (1956). The Interpretation of the New Particles as Displaced Charge Multiplets. *Il Nuovo Cimento*, 4, (supplement 2), 848–66; https://doi.org/10.1007/BF02748000. S2CID 121017243.]

Another of Gell-Mann's ideas is the *Gell-Mann–Okubo formula*, which was, initially, a formula based on empirical results, but was later explained by his *quark model*. Gell-Mann and Abraham Pais were involved in explaining this puzzling aspect of the *neutral kaon mixing*.

Gell-Mann's fortunate encounter with mathematician Richard Earl Block at Caltech, in the fall of 1960, "enlightened" him to introduce in 1961 a novel *classification scheme* for *hadrons*. [Gell-Mann, M. (1961). The Eightfold Way: *A Theory of Strong Interaction Symmetry*. California Institute of Technology, Synchrotron Laboratory Report No. CTSL-20. See below.]

A similar scheme had been independently proposed by Yuval Ne'eman, and has come to be explained by the *quark model*. [Ne'eman, Y. (August 1961). Derivation of Strong Interactions from a Gauge Invariance. Nuclear Physics, 26, 2, 222–9; https://doi.org/10.1016/0029-5582(61)90134-1]

Gell-Mann referred to the scheme as *the eightfold way*, because of the *octets* of particles in the classification (the term is a reference to the Eightfold Path of Buddhism).

Gell-Mann, along with Maurice Lévy, developed the sigma model of *pions*, which describes low-energy *pion interactions*. [Gell-Mann, M. & Lévy, M. (1960). The axial vector current in beta decay. *Il Nuovo Cimento*, 16, 4, 705–726; https://doi.org/ 10.1007/BF02859738]

In 1964, Gell-Mann and, independently, George Zweig went on to postulate the existence of *quarks*, particles which make up the *hadrons* of this scheme. The name "*quark*" was coined by Gell-Mann, and is a reference to the novel "*Finnegans Wake*", by James Joyce ("Three quarks for Muster Mark!" book 2, episode 4). Zweig had referred to the particles as "aces", but Gell-Mann's name caught on. *Quarks*, *antiquarks*, and *gluons* were soon established as the underlying elementary objects in the study of the structure of *hadrons*.

In the 1960s, he introduced *current algebra* as a method of systematically exploiting *symmetries* to extract predictions from *quark models*, in the absence of reliable dynamical theory. This method led to model-independent sum rules confirmed by experiment, and provided starting points underpinning the development of the *Standard Model*.

In 1972 Gell-Mann, while on sabbatical leave to CERN, together with Harald Fritzsch, Heinrich Leutwyler and William A. Bardeen, considered a *Yang-Mills theory* of "quark color", and coined the term *quantum chromodynamics* (QCD) as the *gauge theory* of the *strong interaction. The quark model is a part of QCD, and it has been robust enough to accommodate in a natural fashion the discovery of new "flavors" of quarks, which has superseded the eightfold way scheme.*

Gell-Mann was responsible, with Pierre Ramond and Richard Slansky, and independently of Peter Minkowski, Rabindra Mohapatra, Goran Senjanović, Sheldon Glashow, and Tsutomu Yanagida, for the proposal of the *seesaw theory of neutrino masses*. This produces *masses* at the large scale in any theory with a *right-handed neutrino*. He is also known to have played a role in *keeping string theory alive* through the 1970s and early 1980s, supporting that line of research at a time when it was a topic of niche interest.

In 1984 Gell-Mann was one of several co-founders of the Santa Fe Institute—a non-profit theoretical research institute in Santa Fe, New Mexico intended to study various aspects of a complex system and disseminate the notion of a separate interdisciplinary study of complexity theory.

Gell-Mann died on May 24, 2019, at his home in Santa Fe, New Mexico.

Gell-Mann, M. (March, 1961). The Eightfold Way: A Theory of Strong Interaction Symmetry*.

Pasadena, CA: California Inst. of Tech., Synchrotron Laboratory; https://www.osti.gov/biblio/4008239.

* Research supported in part by the U.S. Atomic Energy Commission and the Alfred P. Sloan Foundation.

Report No. CTSL-20, March 15, 1961. (Preliminary version circulated Jan. 20, 1961.)

Gell-Mann introduces his formulation of a particle classification system for *hadrons* known as *the Eightfold Way* – or, in more technical terms, SU(3) *flavor symmetry*, streamlining its structure.

Abstract

A new model of the higher symmetry of elementary particles is introduced in which the eight known *baryons* are treated as a *super-multiplet*, degenerate in the limit of *unitary symmetry* but split into isotopic *spin multiplets* by a *symmetry-breaking* term.

[A *multiplet* is the *state space* for 'internal' degrees of freedom of a particle, that is, degrees of freedom associated to a particle itself, as opposed to 'external' degrees of freedom such as the particle's position in space. Examples of such degrees of freedom are the *spin state* of a particle in *quantum mechanics*, or the *color, isospin* and *hypercharge state* of particles in the *Standard Model* of particle physics. Formally, this *state space* is described by a *vector space* which carries the *action* of a group of *continuous symmetries*. See above, Wigner, E. (January, 1937). On the Consequences of the Symmetry of the Nuclear Hamiltonian on the Spectroscopy of Nuclei.]

The *symmetry violation* is ascribed phenomenologically to the *mass* differences. The *baryons* correspond to an eight-dimensional irreducible representation of the *unitary group*. The *pion* and *K meson* fit into a similar set of eight particles along with a predicted *pseudoscalar meson* χ^0 having I = 0. A ninth *vector meson* coupled to the *baryon* current can be accommodated naturally in the scheme. It is predicted that the eight *baryons* should all have the same *spin* and *parity* and that *pseudoscalar* and *vector mesons* should form octets with possible additional singlets. The mathematics of the *unitary group* is described by considering three fictitious *leptons*, ν, e^-, and μ^-, which may throw light on the structure of *weak interactions*. (D. L.C.)

CONTENTS

I   Introduction   p. 3
II   The "Leptons" as a Model for Unitary Symmetry   p. 7
III   Mathematical Description of the Baryons   p. 13
IV   Pseudoscalar Mesons   p. 17
V   Vector Mesons   p. 22
VI   Weak Interactions   p. 28
VII   Properties of the New Mesons   p. 30
VIII   Violations of Unitary Symmetry   p. 35
IX   Acknowledgments   p. 38

We attempt once more, as in the *global symmetry* scheme, to treat the eight known *baryons* as a super-multiplet, degenerate in the limit of a certain symmetry but split into isotopic *spin multiplets* by a symmetry-breaking term.

[*Baryon* [*classification*]

Protons and neutrons are *baryons*, a type of composite subatomic particle *that contains an odd number of valence quarks and antiquarks*, conventionally three. *Baryons* belong to the *hadron* family of particles; *hadrons* are composed of quarks. Because *quarks* have a *spin* ½ , the difference in *quark* number results in *baryons* being *fermions* because they have half-integer *spin*.

Each *baryon* has a corresponding antiparticle (*antibaryon*) where their corresponding *antiquarks* replace *quarks*. For example, a *proton* is made of two *up quarks* and one *down quark*; and its corresponding antiparticle, the *antiproton*, is made of two *up antiquarks* and one *down antiquark*.

Baryons participate in the residual *strong force*, which is mediated by particles known as *mesons*. The most familiar *baryons* are *protons* and *neutrons*, both of which contain three *quarks*, and for this reason they are sometimes called *triquarks*. These particles make up most of the mass of the visible matter in the universe and compose the nucleus of every atom (*electrons*, the other major component of the atom, are members of a different family of particles called *leptons*; *leptons* do not interact via the *strong force*). Exotic *baryons* containing five quarks, called *pentaquarks*, have also been discovered and studied.

Baryons are strongly interacting *fermions*; that is, they are acted on by the *strong nuclear force* and are described by *Fermi–Dirac statistics*, which apply to all particles obeying the Pauli exclusion principle. This is in contrast to the *bosons*, which do not obey the exclusion principle.

The name "*baryon*", introduced by Abraham Pais, comes from the Greek word for "heavy" (βαρύς, barýs), because, at the time of their naming, most known elementary particles had lower masses than the baryons.]

[*Omega (Ω) baryon* [*composite particle*]

In 1964, the *omega baryon* was discovered at Brookhaven National Laboratory. The first *omega baryon* discovered was the Ω^-, made of three *strange quarks*. [V. E. Barnes; et al. (1964). Observation of a Hyperon with Strangeness Minus Three. Phys. Rev. Let., 12, 8, 204; https://doi.org/10.1103/ PhysRevLett.12.204.]

The *omega baryons* are a family of subatomic *hadron* (a *baryon*) particles that are either neutral or have a +2, +1 or −1 elementary *charge*. They are *baryons containing no up or down quarks*. Since *omega baryons* do not have any *up or down quarks*, they all have *isospin* 0. *Omega baryons* containing *top quarks* are not expected to be observed. This is because the *Standard Model* predicts the mean lifetime of *top quarks* to be roughly 5×10^{-25} s, which is about a twentieth of the timescale for *strong interactions*, and therefore that they do not form *hadrons*.

Besides the Ω^-, a charmed omega particle (Ω^0_c) was discovered in 1985, in which a *strange quark* is replaced by a *charm quark*. The $\Omega-$ decays only via the *weak interaction* and has therefore a relatively long lifetime. *Spin* and *parity* values for unobserved baryons are predicted by the *quark model*.

The existence of the *omega baryon*, and its mass, and decay products had been predicted in 1961 by the American physicist Murray Gell-Mann and, independently, by the Israeli physicist Yuval Ne'eman. In 1961, Gell-Mann introduced his formulation of a particle classification system for *hadrons* known as *the Eightfold Way* – or, in more technical terms, SU(3) *flavor symmetry*, streamlining its structure. A similar scheme had been independently proposed by Yuval Ne'eman in the same year. [Ne'eman, Y. (August, 1961). Derivation of Strong Interactions from a Gauge Invariance. *Nuclear Physics*, 26, 2; https://doi.org/10.1016/0029-5582(61)90134-1.]]

Here we do not try to describe the *symmetry violation* in detail, but we ascribe it phenomenologically to the *mass differences* themselves, supposing that there is some analogy to the μ-e *mass difference*.

The *symmetry* is called *unitary symmetry* and corresponds to the "*unitary group*" in three dimensions in the same way that *charge independence* corresponds to the "unitary group" in two dimensions. The eight infinitesimal generators of the group form a simple *Lie algebra*, just like the three components of *isotopic spin*. In this important sense, *unitary symmetry is the simplest generalization of charge independence*.

The *baryons* then correspond naturally to an eight-dimensional irreducible representation of the group; when the *mass differences* are turned on, the familiar *multiplets* appear. The *pion* and *K meson* fit into a similar set of eight particles, along with a predicted pseudoscalar meson χ^0 having I = 0.

[*Pion (or pi meson)* [*composite particle*]

In 1947, the *pion* (or *pi meson*) was discovered by members of C. F. Powell's group at the University of Bristol, in England, including César Lattes, Giuseppe Occhialini and Hugh Muirhead. In the same year, Lattes, together with Powell and Occhialini, determined the new particle's mass. [Lattes, C. M. G., Muirhead, H., Occhialini, G. P. S. & Powell, C. F. (1947). Processes involving charged mesons. *Nature*. 159, 4047, 694–7; https://doi.org/10.1038/ 159694a0.]

Cecil Frank Powell, (December 5,1903 –August 9, 1969) was a British physicist, who headed the team that developed the photographic method of studying nuclear processes. The Nobel Prize in Physics 1950 was awarded to Powell "for his development of the photographic method of studying nuclear processes and his discoveries regarding mesons made with this method".

Since the advent of particle accelerators had not yet come, high-energy subatomic particles were only obtainable from atmospheric cosmic rays. Photographic emulsions based on the gelatin-silver process were placed for long periods of time in sites located at high-altitude mountains, first at Pic du Midi de Bigorre in the Pyrenees, and later at Chacaltaya in the Andes Mountains, where the plates were struck by cosmic rays. After development, the photographic plates were inspected under a microscope by a team of about a dozen women.

In 1948, Lattes, Eugene Gardner, and their team first artificially produced *pions* at the University of California's cyclotron in Berkeley, California, by bombarding carbon atoms with high-speed alpha particles.

Meson [*classification*]

A *meson* is a type of hadronic subatomic particle composed of *an equal number of quarks and antiquarks, usually one of each, bound together by the strong interaction*. Because *quarks* have a *spin ½* , the difference in quark number between *mesons* and *baryons* results in *mesons* being *bosons*, whereas *baryons,* the other members of the *hadron* family, composed of *odd numbers of valence quarks* (at least three), are *fermions*.

Some experiments show evidence of exotic *mesons*, which do not have the conventional *valence quark* content of two *quark*s (one *quark* and one *antiquark*), but four or more.

Because *mesons* are composed of *quark* subparticles, they have a meaningful physical size, a diameter of roughly one femtometre (10^{-15} m), which is about 0.6 times the size of a *proton* or *neutron*. *All mesons are unstable*, with the longest-lived lasting for only a few tenths of a nanosecond. Heavier *mesons* decay to lighter *mesons* and ultimately to stable *electrons*, *neutrinos* and *photons*.

Each type of *meson* has a corresponding antiparticle (*antimeson*) in which *quarks* are replaced by their corresponding *antiquarks* and vice versa.

Because *mesons* are composed of *quarks*, they participate in both the *weak interaction* and *strong interaction*. *Mesons* with net *electric charge* also participate in the *electromagnetic interaction*. *Mesons* are classified according to their *quark content*, *total angular momentum*, *parity* and various other properties, such as *C-parity* and *G-parity*. Although no *meson* is stable, those of lower mass are nonetheless more stable than the more massive, and hence are easier to observe and study in particle accelerators or in cosmic ray experiments. The lightest group of *mesons* is less massive than the lightest group of *baryons*, meaning that they are more easily produced in experiments, and thus exhibit certain higher-energy phenomena more readily than do *baryons*. But *mesons* can be quite massive: for example, the J/Psi meson (J/ψ) containing the charm quark, first seen 1974, is about three times as massive as a proton, and the upsilon meson (Υ) containing the bottom quark, first seen in 1977, is about ten times as massive as a proton.

The existence of *mesons* was predicted by Hideki Yukawa's 1935 *theory of mesons* that postulated the particle as mediating the nuclear force. [Yukawa, H. (1935). On the Interaction of Elementary Particles. *Proceedings of the Physico-Mathematical Society of Japan*, 17, 48.]

Kaon (or *K meson*) [*composite particle*]

In the same year, 1947, the *kaon* (or *K meson*), the first *strange* particle, was co-discovered by George Dixon Rochester, and Clifford Charles Butler, two British physicists at the University of Manchester. They published two cloud chamber photographs of cosmic ray-induced events, one showing what appeared to be a neutral particle decaying into two charged *pions*, and one which appeared to be a charged particle decaying into a charged *pion* and something neutral.

A *kaon*, also called a *K meson,* is any of a group of four *mesons* distinguished by a *quantum number* called *strangeness.* In the *quark model* they are understood to be bound states of a *strange quark* (or *antiquark*) and an *up or down antiquark* (or *quark*).

The four *kaons* are:

K^-, *negatively charged* (containing a *strange quark* and an *up antiquark*) has *mass* 493.677 ± 0.013 MeV and *mean lifetime* $(1.2380 \pm 0.0020) \times 10^{-8}$ s;

K^+ (*antiparticle* of above) *positively charged* (containing an *up quark* and a *strange antiquark*) must (by CPT invariance) have *mass* and *lifetime* equal to that of K^-. Experimentally, the *mass* difference is 0.032 ± 0.090 MeV, consistent with zero; the difference in lifetimes is $(0.11 \pm 0.09) \times 10^{-8}$ s, also consistent with zero;

K^0, *neutrally charged* (containing a *down quark* and a *strange antiquark*) has *mass* 497.648 ± 0.022 MeV. It has mean squared *charge radius* of -0.076 ± 0.01 fm^2;

K^0, *neutrally charged* (*antiparticle* of above) (containing a *strange quark* and a *down antiquark*) has the same *mass*.

As the *quark model* shows, assignments that the *kaons* form two doublets of *isospin*; that is, they belong to the fundamental representation of SU(2) called the 2. One doublet of *strangeness* +1 contains the K^+ and the K^0. The *antiparticles* form the other doublet (of *strangeness* −1).

Kaons have proved to be a copious source of information on the nature of fundamental interactions since their discovery in cosmic rays in 1947. They were essential in establishing the foundations of the *Standard Model* of particle physics, such as the quark model of *hadrons* and the theory of *quark* mixing.

Kaons have played a distinguished role in our understanding of fundamental conservation laws: *CP violation*, a phenomenon generating the observed *matter–*

antimatter asymmetry of the universe, was discovered in the *kaon* system in 1964. [Christenson, J. H., Cronin, J. W., Fitch, V. L. & Turlay, R. (July 1964). Evidence for the 2π Decay of the K20 Meson. *Phys. Rev. Let.*, 13, 4, 138–40; https://doi.org/10.1103/PhysRevLett.13.13.] This was acknowledged by the award of the 1980 Nobel Prize in Physics jointly to James Watson Cronin and Val Logsdon Fitch "for the discovery of violations of fundamental symmetry principles in the decay of neutral K-mesons".

Direct *CP violation* was discovered in the *kaon* decays in the early 2000s by the NA48 experiment at CERN and the KTeV experiment at Fermilab. As the *quark model* shows, assignments that *the kaons form two doublets of isospin*; that is, they belong to the fundamental representation of SU(2) called the **2**. One doublet of strangeness +1 contains the K^+ and the K^0. The antiparticles form the other doublet (of strangeness −1).]

The pattern of *Yukawa couplings* of π, K and χ is then nearly determined, in the limit of *unitary symmetry*.

The most attractive feature of the scheme is that it permits the description of eight *vector mesons* by a unified theory of the Yang-Mills type (with a *mass* term).

[*Vector meson [classification]*

A *vector meson* is a *meson* with *total spin* 1 and odd parity. *Vector mesons* have been seen in experiments since the 1960s, and are well known for their spectroscopic pattern of masses.

The *vector mesons* contrast with the *pseudovector mesons*, which also have a *total spin* 1 but instead have *even parity*. The *vector* and *pseudovector mesons* are also dissimilar in that the spectroscopy of *vector mesons* tends to show nearly pure states of constituent *quark flavors*, whereas *pseudovector mesons* and *scalar mesons* tend to be expressed as composites of mixed states.]

Like Sakurai, we have a *triplet* ρ of *vector mesons* coupled to the isotopic *spin* current and a *singlet vector meson* ω^0 coupled to the *hypercharge current*. We also have a pair of *doublets* M and M⁻, *strange vector mesons* coupled to *strangeness-changing currents* that are conserved when the *mass differences* are turned off. There is only one *coupling constant*, in the symmetric limit, for the system of eight *vector mesons*. There is some experimental, evidence for the existence of ω^0 and M, while ρ is presumably the famous I = 1, J = 1, *π-π resonance*.

A ninth *vector meson* coupled to the *baryon current* can be accommodated naturally in the scheme.

The most important prediction is the qualitative one that the eight baryons should all have the same spin and parity and that the *pseudoscalar* and *vector mesons* should form "*octets*", with possible additional "*singlets*".

If the *symmetry* is not too badly broken in the case of the *renormalized coupling constants* of the eight *vector mesons*, then numerous detailed predictions can be made of experimental results.

> [*Renormalization* is a collection of techniques in *quantum field theory, statistical field theory*, and the *theory of self-similar geometric structures*, that are used to treat infinities arising in calculated quantities *by altering values of these quantities to compensate for effects of their self-interactions*. But even if no infinities arose in *loop diagrams* in *quantum field theory*, it could be shown that it would be necessary to *renormalize* the *mass* and *fields* appearing in the original Lagrangian.
>
> For example, an *electron theory* may begin by postulating an *electron* with an initial *mass* and *charge*. In *quantum field theory* a cloud of virtual particles, such as *photons, positrons*, and others surround and interact with the initial *electron*. Accounting for the interactions of the surrounding particles (e.g. collisions at different energies) shows that the electron-system behaves as if it had a different *mass* and *charge* than initially postulated. *Renormalization*, in this example, mathematically replaces the initially postulated *mass* and *charge* of an *electron* with *the experimentally observed mass and charge*. ???!]

The mathematics of the *unitary group* is described by considering three fictitious "*leptons*", ν, e^-, and μ^-, which may or may not have something to do with real *leptons*. If there is a connection, then it may throw light on the structure of the *weak interactions*.

I Introduction

It has seemed likely for many years that the *strongly interacting particles*, grouped as they are into isotopic *multiplets*, would show traces of a higher *symmetry* that is somehow broken. Under the higher *symmetry*, the eight familiar *baryons* would be degenerate and form a *supermultiplet*. As the higher *symmetry* is broken, the Ξ, Δ, Σ, and N would split apart, leaving inviolate only the conservation of *isotopic spin*, of *strangeness*, and of *baryons*. Of these three, the first is partially broken by *electromagnetism* and the second is broken by the *weak interactions*. Only the conservation of *baryons* and of *electric charge* are absolute.

An attempt[1,2] to incorporate these ideas in a concrete model was the scheme of "*global symmetry*", in which the higher *symmetry* was valid for the interactions of the π *meson*, but broken by those of the *K*.

[1] Gell-Mann, M. (June, 1957). Model of the Strong Couplings. *Phys. Rev.*, 106, 6, 1296; https://journals.aps.org/pr/abstract/10.1103/PhysRev.106.1296.

[2] Schwinger, J. (November, 1957). A theory of the fundamental interactions. *Ann. Phys.*, 2, 5, 407-34; https://www.sciencedirect.com/science/article/abs/pii/0003491657900155.

The *mass differences* of the *baryons* were thus attributed to the *K couplings*, the *symmetry* of which vas unspecified, and the strength of which was supposed to be significantly less than that of the π *couplings*.

The *theory of global symmetry* has not had great success in predicting experimental results. Also, it has a number of defects. The peculiar distribution of isotopic *multiplets* among the observed *mesons* and *baryons* is left unexplained. The arbitrary *K couplings* (which are not really particularly weak) bring in several adjustable constants. Furthermore, as admitted in Reference 1 and reemphasized recently by Sakurai[34]

[3] Sakurai, J. J. (September, 1960). Theory of strong interactions.
Ann. Phys., 11, 1, 1-48. See above.

[4] Sakurai, J. J. "*Vector Theory of Strong Interactions*", unpublished.

in his remarkable articles predicting *vector mesons*, the *global model* makes no direct connection between physical *couplings* and the *currents* of the conserved *symmetry operators*.

In place of *global symmetry*, we introduce here a new model of the *higher symmetry* of *elementary particles* which has none of these faults and a number of virtues.

We note that the *isotopic spin group* is the same as the group of all unitary 2 x 2 matrices with unit determinant. Each of these matrices can be written as exp(iA), where A is a hermitian 2 x 2 matrix. Since there are three independent hermitian 2 x 2 matrices (say those of Pauli), there are three components of the *isotopic spin*.

Our *higher symmetry group* is the simplest generalization of *isotopic spin*, namely the group of all unitary 3x 3 matrices with unit determinant. *There are eight independent traceless 3 x 3 matrices* and consequently the new "*unitary spin*" has eight components. The first three are just the components of the *isotopic spin*, the eighth is proportional to the *hypercharge* Y (which is + 1 for N and K, − 1 for Ξ and K⁻, 0 for Λ, Σ, π, etc.), and the remaining four are *strangeness-changing operators*.

Just as *isotopic spin* possesses a three-dimensional representation (*spin* 1), so the *"unitary spin"* group has an eight-dimensional irreducible *representation*, which we shall call simply 8. *In our theory, the baryon supermultiplet corresponds to this representation.* When the *symmetry* is reduced, then I and Y are still conserved but the four other components of *unitary spin* are not; the *supermultiplet* then breaks up into Ξ, Σ, Λ, and N. Thus, the distribution of *multiplets* and the nature of *strangeness* or *hypercharge* are to some extent explained.

The *pseudoscalar mesons* are also assigned to the *representation* 8. When the *symmetry* is reduced, they become the *multiplets* K, K⁻, π, and χ, where χ is a *neutral isotopic singlet meson* the existence of which we predict. Whether the *pseudoscalar mesons* are regarded as fundamental or as bound *states*, their *Yukawa couplings* in the limit of *"unitary"* symmetry are describable in terms of only two *coupling* parameters.

The *vector mesons* are introduced in a very natural way, by an extension of the *gauge principle* of Yang and Mills[5].

[5] Yang, C. N. & Mills, R. (October, 1954). Conservation of Isotopic Spin and Isotopic Gauge Invariance. *Phys. Rev.*, 96, 1, 191–5. See above. Also, Shaw, R. unpublished.

Here too we have a *supermultiplet* of eight *mesons*, corresponding to the *representation* 8. In the limit of *unitary symmetry* and with the *mass* of these *vector mesons* "turned off", we have a completely *gauge-invariant* and *minimal* theory, just like *electromagnetism*. When the *mass* is turned on, the *gauge invariance* is reduced (the *gauge function* may no longer be space-time-dependent) but *the conservation of unitary spin remains exact.* The sources of the *vector mesons* are the *conserved currents* of the eight components of the unitary spin.[6]

[6] After the circulation of the preliminary version of this work (January, 1961) the author has learned of a similar theory put forward independently and simultaneously by Ne'eman, Y. (*Nuclear Physics*, to be published). Earlier uses of the 3-dimensional *unitary group* in connection with the Sakata model are reported by Y. Ohnuki at the 1960 Rochester Conference on High Energy Physics. Salam, A. & Ward, J. (*Nuovo Cimento*, to be published) have considered related questions. The author would like to thank Dr. Ne'eman and Professor Salam for communicating their results to him.

When the *symmetry* is reduced, the eight *vector mesons* break up into a *triplet* ρ (coupled to the still-conserved *isotopic spin current*), a *singlet* ω (coupled to the still-conserved *hypercharge current*), and a pair of *doublets* M and M⁻ (coupled to a *strangeness- changing current* that is no longer conserved). The particles ρ and ω were both discussed by Sakurai. The ρ *meson* is presumably identical to the I = 1, J = 1, π-π *resonance* postulated by Frazer and Fulco[7] in order to explain the *isovector electromagnetic* form factors of the *nucleon*.

[7] Frazer, W. R. & Fulco, J. R. (1960). *Phys. Rev.*, 117, 1609. See also Bowcock, J., Cottingham, W. N. & Lurie, D. (1960). *Phys. Rev. Lett.*, 5, 386.

The ω *meson* is no doubt the same as the I = 1, J = 0 particle or 3π *resonance* predicted by Nambu[8] and later by Chew[9] and others in order to explain the *isoscalar* form factors of the *nucleon*.

[8] Nambu, Y. (1957). *Phys. Rev.*, 106, 1366.
[9] Chew, G. F. (1960). *Phys. Rev. Lett.*, 4, 142.

The *strange meson* M may be the same as the K* particle observed by Alston et al.[10].

[10] Alston, M. et al., to be published.

Thus, we predict that the eight *baryons* have the same *spin* and *parity*, that K is *pseudoscalar* and that χ exists, that ρ and ω exist with the properties assigned to them by Sakurai, and that M exists. But besides these qualitative predictions *there are also the many symmetry rules associated with the unitary spin*. All of these are broken, though, *by whatever destroys the unitary symmetry*, and it is a delicate matter to find ways in which these effects of a broken symmetry can be explored experimentally.

Besides the eight *vector mesons* coupled to the *unitary spin*, there can be a ninth, which is invariant under *unitary spin* and is thus not degenerate with the other eight, even in the limit of *unitary symmetry*. We call this *meson* B^0. Presumably it exists too and is coupled to the *baryon current*. It is the *meson* predicted by Teller[11]

[11] Teller, E. (1956). *Proceedings of the Rochester Conference, 1956*.

and later by Sakurai[3] and explains most of the hard-core repulsion between *nucleons* and the attraction between *nucleons* and *antinucleons* at short distances.

We begin our exposition of the "*eightfold way*" in the next Section by discussing *unitary symmetry* using fictitious "*leptons*" which may have nothing to do with real *leptons* but help to fix the physical ideas in a rather graphic way. If there is a parallel between these "*leptons*" and the real ones, that would throw some light on the *weak interactions*, as discussed briefly in Section VI.

Section III is devoted to the 8 *representation* and the *baryons* and Section IV to the *pseudoscalar mesons*. In Section V we present the theory of the *vector mesons*.

The physical properties to be expected of the predicted *mesons* are discussed in Section VII, along with a number of experiments that bear on those properties.

In Section VIII we take up the vexed question of the *broken symmetry*, how badly it is broken, and how we might succeed in testing it.

II The "Leptons" as a Model for Unitary Symmetry

For the sake of a simple exposition, we begin our discussion of *unitary symmetry* with "*leptons*", although our theory really concerns the *baryons* and *mesons* and the *strong interactions*.

[*Lepton* [*classification*]

A *lepton* is an elementary particle of half-integer *spin* (*spin* ½) that does not undergo *strong interactions*. Two main classes of *leptons* exist: *charged leptons* (also known as the *electron-like leptons* or *muons*), including the *electron*, *muon*, and *tauon*, and *neutral leptons*, better known as *neutrinos*. *Charged leptons* can combine with other particles to form various composite particles such as atoms and *positronium*, while *neutrinos* rarely interact with anything, and are consequently rarely observed. The best known of all *leptons* is the *electron*.

There are six types of *leptons*, known as flavors, grouped in three generations. The *first-generation leptons*, also called electronic leptons, comprise the electron (e^-) and the electron neutrino (ν_e); the second are the *muonic leptons*, comprising the *muon* (μ^-) and the *muon neutrino* (ν_μ); and the third are the *tauonic leptons*, comprising the *tau* (τ^-) and the *tau neutrino* (ν_τ). *Electrons* have the least mass of all the *charged leptons*. The heavier *muons* and *taus* will rapidly change into *electrons* and *neutrinos* through a process of particle decay: the transformation from a higher mass state to a lower mass state. Thus, *electrons* are stable and the most common charged *lepton* in the universe, whereas *muons* and *taus* can only be produced in high-energy collisions (such as those involving cosmic rays and those carried out in particle accelerators).

Leptons have various intrinsic properties, including *electric charge*, *spin*, and *mass*. Unlike *quarks*, however, *leptons* are not subject to the *strong interaction*, but they are subject to the other three fundamental interactions: *gravitation*, the *weak interaction*, and to *electromagnetism*, of which the latter is proportional to charge, and is thus zero for the electrically neutral *neutrinos*.

For every *lepton flavor*, there is a corresponding type of *antiparticle*, known as an *antilepton*, that differs from the *lepton* only in that some of its properties have equal magnitude but opposite sign. According to certain theories, *neutrinos* may be their own antiparticle. It is not currently known whether this is the case.

The first *charged lepton*, the *electron*, was theorized in the mid-19th century by several scientists and was discovered in 1897 by J. J. Thomson. The next *lepton* to be observed was the *muon*, discovered by Carl D. Anderson in 1936, which was classified as a meson at the time. After investigation, it was realized that the *muon* did not have the expected properties of a meson, but rather behaved like an *electron*, only with higher mass. It took until 1947 for the concept of "*leptons*" as a family of particles to be proposed. The term *lepton* was first used by physicist Léon Rosenfeld in 1948.

The first *neutrino*, the *electron neutrino*, was proposed by Wolfgang Pauli in 1930 to explain certain characteristics of *beta decay*. It was first observed in the Cowan–Reines *neutrino* experiment conducted by Clyde Cowan and Frederick Reines in 1956. The *muon neutrino* was discovered in 1962 by Leon M. Lederman, Melvin Schwartz, and Jack Steinberger, and the *tau* discovered between 1974 and 1977 by Martin Lewis Perl and his colleagues from the Stanford Linear Accelerator Center and Lawrence Berkeley National Laboratory. The *tau neutrino* remained elusive until July 2000, when the DONUT collaboration from Fermilab announced its discovery.

Leptons are an important part of the *Standard Model*. Electrons are one of the components of atoms, alongside protons and neutrons. Exotic atoms with muons and *taus* instead of *electrons* can also be synthesized, as well as *lepton–antilepton* particles such as *positronium*.]

The particles we consider here for mathematical purposes do not necessarily have anything to do with real *leptons*, but there are some suggestive parallels. We consider three *leptons*, ν, e^-, and μ^-, and their *antiparticles*. The *neutrino* is treated on the same footing as the other two, although experience suggests that if it is treated as a four-component Dirac field, only two of the components have physical interaction. (Furthermore, there may exist two *neutrinos*, one coupled to the *electron* and the other to the *muon*.)

As far as we know, the *electrical* and *weak interactions* are absolutely symmetrical between e^- and μ^-, which are distinguished, however from ν. The *charged particles* e^- and μ⁻ are separated by the mysterious difference in their *masses*. We shall not necessarily attribute this difference to any *interaction*, nor shall we explain it in any way. (If one insists on connecting it to an *interaction*, one might have to consider a *coupling* that becomes important only at exceedingly high energies and is, for the time being, only of academic interest.) We do, however, guess that the μ-e *mass splitting* is related to the equally mysterious mechanism that breaks the *unitary symmetry* of the *baryons* and *mesons* and splits the *super-multiplets* into *isotopic multiplets*. For practical purposes, we shall put all of these *splittings* into the mechanical *masses* of the particles involved.

It is well known that in present *quantum electrodynamics*, no one has succeeded in explaining the e-ν *mass difference* as an *electromagnetic* effect. Without prejudice to the question of its physical origin, we shall proceed with our discussion as if that *mass difference* were "turned on" along with the *charge* of the *electron*.

If we now "turn off" the μ-e *mass difference, electromagnetism*, and the weak *interactions* we are left with a physically vacuous theory of three exactly similar Dirac particles with no *rest mass* and no known *couplings*. This empty model is ideal for our mathematical purposes, however, and is physically motivated by the analogy with the *strongly interacting particles*, because it is at the corresponding stage of total *unitary symmetry* that we shall introduce the basic *baryon mass* and the *strong interactions* of *baryons* and *mesons*.

The *symmetric model* is, of course, invariant under all *unitary transformations* on the three *states*, ν, e⁻, and μ⁻.

Let us first suppose for simplicity that we had only two particles ν and e⁻. ...
...

III Mathematical Description of the Baryons

In the case of *isotopic spin* I, we know that the various possible *charge multiplets* correspond to "irreducible *representations*" of the simple 2 x 2 matrix algebra described above for (ν, e⁻). ...
...

In this *representation*, the 8 x 8 matrices $F_i{}^{jk}$ of the eight components F_i of the *unitary spin* are given by the relation

$$F_i{}^{jk} = {}^{-i}f_{ijk} \tag{3.4}$$

analogous to Eq. (3.2). ...

IV Pseudoscalar Mesons

We have supposed that the *baryon fields* N_j transform like an *octet* 8 under F, so that the matrices of F for the *baryon fields* are given by Eq. (3.4). We now demand that all *mesons* transform under in such a way as to have *F-invariant strong couplings*. If the 8 *mesons* π_i are to have *Yukawa couplings*, they must be coupled to $N^-\Theta_i N$ for some matrices Θ_i, and we must investigate how such bilinear forms transform under F. ...

...

In general, we may write the *Yukawa coupling* (whether fundamental or phenomenological, depending on whether the π_i are *elementary* or not) in the form

$$L_{int} = 2ig_0\, N^-\, \gamma_5\, [\alpha D_i + (1-\alpha)\, F_i]\, N\, \pi_i. \tag{4.7}$$

...

V Vector Mesons

The possible transformation properties of the *vector mesons* under F are the same as those we have already examined in the *pseudoscalar* case. Again, it seems that for low *mass states* we can safely ignore the *representations* ... . We are left with 1 and the two cases of 8.

A *vector meson* transforming according to 1 would have $\Theta = 0$, $I = 0$, $Y = 0$ and would be coupled to the *total baryon current* $iN^-\gamma_\mu N$, which is exactly conserved. Such a *meson* may well exist and be of great importance. The possibility of its existence has been envisaged for a long time.

We recall that the *conservation of baryons* is associated with the invariance of the theory under infinitesimal transformations,

  ...

...

VI Weak Interactions

So far, the role of the *leptons* in *unitary symmetry* has been purely symbolic. Although we introduced a mathematical F *spin* for v, e^-, and μ^-, that *spin* is not coupled to the eight *vector mesons* that take up the F *spin gauge* for *baryons* and *mesons*. If we take it seriously at all, we should probably regard it as a different *spin*, but one with the same mathematical properties.

Let us make another point, which may seem irrelevant but possibly is not. The *photon* and the *charge operator* to which it is *coupled* have not so far been explicitly included in our scheme. They must be put in as an afterthought, along with the corresponding *gauge transformation*, which was the model for the more peculiar *gauge transformations* we have treated. If the *weak interactions* are carried[17] by *vector bosons* X_α and generated by a *gauge transformtion*[18,19] of their own, then these *bosons* and *gauges* have been ignored as well.

[17] Feynman, R. P. & Gell-Mann, M. (1958). *Phys. Rev.*, 109, 193.

[18] Bludman, S. (1958). *Nuovo Cimento*, 9, 433.

[19] Gell-Mann, M. & Levy, M. (1960). *Nuovo Cimento*, 16, 705.

Such considerations might cause us, if we are in a highly speculative frame of mind, to wonder about the possibility that each kind of *interaction* has its own type of *gauge* and its own set of *vector particles* and that the algebraic properties of these *gauge transformations* conflict with one another.

When we draw a parallel between the "*F spin*" of *leptons* and the *F spin* of *baryons* and *mesons*, and when we discuss the *weak interactions* at all, we are exploring phenomena that transcend the scheme we are using. *Everything we say in this Section must be regarded as highly tentative and useful only in laying the groundwork for a possible future theory. The same is true of any physical interpretation of the mathematics in Sections II and III.*

We shall restrict our discussion to *charge-exchange weak currents* and then only to the vector part. A complete discussion of the *axial vector weak currents* may involve more complicated concepts and even new *mesons* (scalar and/or axial vector) lying very high in *energy*, ...

...

VII Properties of the New Mesons

The theory we have sketched is fairly solid only in the realm of the *strong interactions*, and we shall restrict our discussion of predictions to the interactions among *baryons* and *mesons*. We predict the existence of 8 *baryons* with equal *spin* and *parity* following the pattern of N, Λ, Σ, and Ξ. Likewise, given the π and its *coupling constant*, we predict a *pseudoscalar* K and a new particle, the χ^0, both coupled (in the absence of *mass differences*) as in Eq. (4.7), and we predict *pion couplings* to *hyperons* as in the same equation.

Now in the limit of *unitary symmetry* an enormous number of *selection* and *intensity* rules apply. For example, for the reactions *PS meson + baryon → PS meson + baryon*, there are only 7 independent *amplitudes*. Likewise, *baryon-baryon forces* are highly *symmetric*. *However, the apparent smallness of $g_1^2/4\pi$ for NKΛ and NKΣ compared to NπN indicates that unitary symmetry is badly broken, assuming that it is valid at all. We must thus rely* principally on qualitative predictions for tests of the theory; in Section VIII we take up the question of how quantitative testing may be possible.

The most clear-cut new prediction for the *pseudoscalar mesons* is the existence of χ^0, which should decay into 2γ like the π_0, unless it is heavy enough to yield $\pi^+ + \pi^- + \gamma$ with appreciable probability. (In the latter case, we must have $(\pi^+\pi^-)$ in an odd state.) $\chi^0 \rightarrow 3\pi$ is forbidden by *conservation of I and C*. For a sufficiently heavy χ^0, the decay $\chi^0 \rightarrow 4\pi$ is possible, but hampered by centrifugal barriers.

Now we turn to the *vector mesons*, with coupling pattern as given in Table IV. We predict, like Sakurai, the ρ *meson*, presumably identical with the *resonance* of Frazer and Fulco, and the ω *meson*, coupled to the *hypercharge*. In addition, we predict, the *strange vector meson* M, which may be the same as the K* of Alston et al.

Some of these are unstable with respect to the *strong interactions* and their physical *coupling constants* to the decay products are given by the *decay widths*. ...
...

VIII Violations of Unitary Symmetry

We have mentioned that within the *unitary scheme* there is no way that the *coupling constants* of K to both NΛ and NΣ much smaller than 15, except through large violations of the *symmetry*. Yet *experiments on photoproduction of K particles seem to point to such a situation.* Even if *unitary symmetry* exists as an underlying pattern, whatever mechanism is responsible for the *mass differences* apparently produces a wide spread among the *renormalized coupling constants* as well. It is true that the binding of Λ particles in *hypernuclei* indicates a $\pi\Lambda\Sigma$ *coupling* of the same order of magnitude as the πNN *coupling*, but the anomalously small *renormalized constants* of the K *meson* indicate that a quantitative check of *unitary symmetry* will be very difficult.

What about the *vector mesons*? Let us discuss first the ρ and ω fields, which are coupled to *conserved currents*. ...
...

We have seen that the prospect is rather gloomy for a quantitative test of *unitary symmetry*, or indeed of any proposed higher *symmetry* that is broken by *mass differences* or *strong interactions*. The best hope seems to lie in the possibility of direct study of the ratios of bare constants in experiments involving very high energies and momentum transfers, much larger than all masses[24].

[24] Gell-Mann, M. & Zachariasen, F. "*Broken Symmetries and Bare Coupling Constants*", to be published.

However, the theoretical work on this subject is restricted to *renormalizable* theories. At present, *theories of the Yang-Mills type with a mass do not seem to be renormalizable*[25], and no one knows how to improve the situation.

[25] Kamefuchi & Umezawa, to be published. Salam and Kumar, to be published.

It is in any case an important challenge to theoreticians to construct a satisfactory theory of *vector mesons*. It may be useful to remark that the difficulty in Yang-Mills theories is caused by the *mass*. It is also the *mass* which spoils the *gauge invariance* of the first kind. Likewise, as in the μ-e case, it may be the *mass* that produces the violation of *symmetry*. Similarly, the *nucleon* and *pion masses* break the conservation of any *axial vector current* in the theory of *weak interactions*. It may be that a new approach to the *rest masses* of elementary particles can solve many of our present theoretical problems.

IX Acknowledgments

The author takes great pleasure in thanking Dr. S. L. Glashow and Professor R. P. Feynman for their enthusiastic help and encouragement and for numerous ideas, although they bear none of the blame for any errors or defects in the theory. Conversations with Professor R. Block about Lie algebras have been very enlightening,

Glashow, S. L. & Gell-Mann, M. (September, 1961). Gauge Theories of Vector Particles.*

Ann. Phys. (N.Y.), 15, 437-60; https://doi.org/10.1016/0003-4916(61)90193-2; also at https://www.semanticscholar.org/paper/Gauge-Theories-of-Vector-Particles-Glashow-Gell-Mann/c0184d2962be36beb622736b14e0484651ec0ba0.

* Research supported in part by the U.S. Atomic Energy Commission and the Alfred P. Sloan Foundation.

California Institute of Technology, Synchrotron Laboratory Pasadena, California.

Report No. CTSL-28, April 24, 1961.

Abstract

The possibility of generalizing the Yang-Mills trick is examined. Thus, *we seek theories of vector bosons invariant under continuous groups of coordinate-dependent linear transformations*. All such theories may be expressed as super-positions of certain "simple" theories; we show that each "simple" theory is associated with a simple *Lie algebra*. We may introduce *mass* terms for the *vector bosons* at the price of destroying the *gauge-invariance* for *coordinate-dependent gauge functions*.

The theories corresponding to three particular simple Lie algebras—*those which admit precisely two commuting quantum numbers*—are examined in some detail as examples. *One of them might play a role in the physics of the strong interactions if there is an underlying super-symmetry, transcending charge independence, that is badly broken.*

The intermediate *vector boson theory of weak interactions* is discussed also. The so-called *"schizon"* model cannot be made to conform to the requirements of *partial gauge-invariance*. It is possible, however, to find a formal theory of four intermediate *bosons* that is *partially gauge-invariant* and gives an approximate $| \Delta I | = 1/2$ rule.

Contents

I	Introduction	p. 4
II	The One-Parameter Gauge Theory	p. 7
III	The Three-Parameter Gauge Theory of Yang and Mills	p. 13
IV	Generalizations	p. 15

286

V	On Simple Lie Algebras	p. 22
VI	Examples of Simple Lie Algebras	p. 29
VII	Possible Applications to Physics	p. 33

I. Introduction

The *electromagnetic interaction* of elementary particles is remarkably simple. It is of universal strength and form and is associated with a *principle of gauge invariance*. In fact, *starting with the idea of invariance under gauge transformations with coordinate - dependent gauge functions, one can deduce the existence of a massless vector field coupled to a conserved current.* If all *charged fields* are subjected to the same *gauge transformation*, then the *electric charges* of all particles are the same. The fact that the *weak interactions* are *vectorial* in character (apart from *non-conservation of parity*) and nearly universal in strength has suggested to many physicists that they may be mediated by *vector fields*[1,2]

[1] Feynman, R. P. & Gell-Mann, M. (January, 1958). Theory of the Fermi Interaction. *Phys. Rev.*, 109, 193 (1958); https://journals.aps.org/pr/abstract/10.1103/PhysRev.109.193.

[2] Schwinger, J. (November, 1957). A theory of the fundamental interactions. *Ann. Phys.*, 2, 407; https://www.sciencedirect.com/science/article/abs/pii/0003491657900155.

and that *there may be a useful parallel between them and electromagnetism, perhaps even extending to the notion of gauge invariance*[3,4,5,6].

[3] Bludman, S. A. (1958). On the universal fermi interaction. *Nuovo Cim.*, 9, 433–45; https://doi.org/10.1007/BF02725099.

[4] Glashow, S. L. (February-May, 1959). The renormalizability of vector meson interactions. *Nuclear Physics*, 10, 107; https://www.sciencedirect.com/science/article/abs/pii/0029558259901968.

[5] Salam, A. & Ward, J. C. (February, 1959). Weak and Electromagnetic interactions. *Nuovo Cim.*, 11, 568; https://doi.org/10.1007/BF02726525.

[6] Gell-Mann, M., & Lévy, M. (May, 1960). The axial vector current in beta decay. Nuovo Cim., 16, 705–26; https://doi.org/10.1007/BF02859738.

The *strong interactions*, too, seem to exhibit some degree of universality. Moreover, the approximate *conservation laws of isotopic spin* and *of strangeness*, as well as the exact law of *conservation of baryons*, present an *analogy with the conservation of charge* and suggest that some principles of *gauge invariance* may be at work. Until recently, it seemed that the *strong couplings* were not *vectorial*, but there is mounting evidence that there are objects (like the I = 1, J = 1, $\pi\pi$ resonance) that can be interpreted as *vector mesons* and that may play a very significant role in the *strong interactions*[7,8].

[7] Sakurai, J. (1960). Theory of strong interactions. *Ann. Phys.*, 11, 1-48. See above.

[8] Gell-Mann, M. (1961). The Eightfold Way: A Theory of Strong Interaction Symmetry. *California Institute of Technology, Synchrotron Laboratory Report No. CTSL-20.* See above. See also Y. Ne'eman, *Nuclear Physics*, (to be published).

There are two great difficulties in the way of constructing *theories of weak and strong interactions* by analogy with *electrodynamics*. One is that *some of the relevant currents are not conserved*. The *isotopic spin and strangeness currents* that may enter into a *vectorial theory of the strong couplings* fail to be conserved on account of *electromagnetic and weak interactions*, while the *conservation of the weak current* is broken not only by *electromagnetism* but, in the case of the *axial vector* and *strangeness*-changing parts, by *masses* and perhaps by *strong interactions* as well.

The other difficulty is that whereas *photons are massless (as the quanta must be in a theory that is fully gauge invariant with a coordinate-dependent gauge function)* the *vector particles* that *mediate* the *strong and weak interactions* must be *massive* if they exist at all. Thus, the notion has arisen[3,4,5,6,7,8] of a theory that is *partially gauge-invariant*. In each case we have a Lagrangian like the electromagnetic one, *fully invariant under coordinate-dependent gauge transformations*, plus other terms. The remaining terms are of two kinds: a) *those which break the full gauge invariance*, while leaving intact the *conservation law* and the *invariance under constant gauge transformations*; b) those which *destroy the gauge invariance altogether*, along with the *conservation law*. In the case where the conservation law is exact (conservation of baryons) the terms of type b) are, of course, absent.

Now the idea of *partial gauge-invariance* poses a number of questions, to which we shall return briefly in Section VII. For the moment, let us concentrate on the straightforward part of the problem, *the construction of the fully gauge-invariant part of the theory. The coupling of a vector meson field to a single quantity like baryon number follows exactly the pattern of electromagnetic coupling to the charge, as long as the complete gauge-invariance is maintained.* But, when we go over to the case of three non-commuting quantities like the components of the *isotopic spin current*, the situation becomes different and a more sophisticated theory becomes necessary. The intermediary *vector meson field* now carries *isotopic spin* 1 and its own *isotopic spin current* contributes a source term. Thus, the *theory of the vector meson field* becomes *non-linear*. The problem of constructing the theory in question has been solved by Yang and Mills[9] and by Shaw[10].

[9] Yang, C. N. & Mills, R. (October, 1954). Conservation of Isotopic Spin and Isotopic Gauge Invariance. (See above.)

[10] Shaw, R., unpublished.

In the next two sections, we review the simple case of *charge* or *baryon number* and the more complicated case of *isotopic spin*. Then, in Section IV, we go on to the main point of this article — the description of all possible straightforward generalizations of the Yang-Mills trick. *We are interested in such generalizations because we do not know, for either the strong or the weak interactions, exactly how many intermediate vector fields may be involved* (if any). To give just one example, it has been suggested[11,12,13] that there may he four such (*hermitian*) fields for the *weak interactions* -- the so-called *schizon model*, set up to give $|\Delta I| = 1/2$ and $\Delta S = 0, \pm 1$ *for the nonleptonic weak interactions of baryons and mesons*.

[11] Lee, T. D. & Yang, C. N. (August, 1960). Implications of the Intermediate Boson Basis of the Weak Interactions: Existence of a Quartet of Intermediate Bosons and Their Dual Isotopic Spin Transformation Properties. *Phys. Rev.*, 119, 1410; https://journals.aps.org/pr/abstract/10.1103/PhysRev.119.1410.

[12] Treiman, S. B. (March, 1959). Weak global symmetry. *Nuovo Cim.*, 15, 916–24; https://doi.org/10.1007/BF02860196.

[13] Gell-Mann, M. (1959). *Bull. Am. Phys. Soc.*, 4, 256 (T).

We shall show in Section VII that the ideas of *partial gauge-invariance* lead to severe restrictions on *four-field models*; in fact, the restrictions are so strong as to make it impossible to construct the *schizon model* according to the gauge principles of this article. The *classification of generalized Yang-Mills theories* discussed in Section IV is described further in Section V; some examples are given in Section VI; and some possible *physical applications* are touched on briefly in Section VII.

II. The One-Parameter Gauge Theory

The *gauge* formalism of *electromagnetism* is, of course, well-known. The generalization from *charge* to *baryon number* was discussed by Yang and Lee[14];

[14] Lee, T. D. & Yang, C. N. (June, 1955). Conservation of Heavy Particles and Generalized Gauge Transformations. *Phys. Rev.*, 98, 1501; https://journals.aps.org/pr/abstract/10.1103/PhysRev.98.1501.

it is clear from their work that the generalization contradicts experiment unless either the *coupling constant is ridiculously small* or the *gauge invariance is broken*, say by a *mass term* for the *vector field*. Let us review the method.

We start with an additive quantity like *charge* or *baryon number*; call it Q. Let the fields $\psi_a(x)$ destroy particles of *charge* Q_a and create their antiparticles. We then discuss invariance under the infinitesimal *gauge transformations*

$$\dots . \tag{2.1}$$

$\dots$

We have discussed several ways in which the *strong interactions* may constitute a *partially gauge invariant theory*, and have sketched a *gauge-invariant theory* of the *weak interactions*. In general, the *"weak"* and *"strong"* *gauge symmetries* will not be mutually compatible. There will be conflicts with the *electromagnetic symmetry*, conflicts that must be resolved in favor of electromagnetism, since its *gauge invariance* is exact. *We have not attempted here to describe the three kinds of interactions together, but only to speculate about what the symmetry of each might look like* in an ideal limit where *symmetry-breaking* effects disappear.

Schwinger, J. (January, 1962). Gauge Invariance and Mass.

Phys. Rev., 125, 397; https://doi.org/10.1103/PhysRev.125.397; also at https://harvest. aps.org/v2/journals/articles/10.1103/PhysRev.125.397/fulltext.

Harvard University, Cambridge, Massachusetts, and University of California, Los Angeles, California.

Received July 20, 1961.

Abstract

It is argued that the *gauge invariance* of a *vector field* does not necessarily imply zero *mass* for an associated particle if the *current vector coupling* is sufficiently strong. This situation may permit a deeper understanding of nucleonic *charge conservation* as a manifestation of a *gauge invariance*, without the obvious conflict with experience that a massless particle entails.

Does the requirement of *gauge invariance* for a *vector field* coupled to a dynamical *current* imply the existence of a corresponding particle with *zero mass*? Although the answer to this question is invariably given in the affirmative[1], the author has become convinced that there is no such necessary implication, once the assumption of *weak coupling* is removed.

  [1] For example, Schwinger, J. (February, 1949). Quantum Electrodynamics. II. Vacuum Polarization and Self-Energy. *Phys. Rev.*, 75, 4, 651-79; https://doi.org/10.1103/ PHYSREV.75.651. [Also in Underwood, T. G. (2023). *Quantum Electrodynamics – annotated sources*, pp. 492-4].

Thus, the path to an understanding of *nucleonic (baryonic) charge conservation* as an aspect of a *gauge invariance*, function in strict analogy with *electric charge*[2], may be open for the first time.

  [2] Lee, T. D. & Yang, C. N. (June, 1955). Conservation of Heavy Particles and Generalized Gauge Transformations. *Phys. Rev.*, 98, 5, 1501; https://doi.org/10.1103/ PhysRev.98.1501.

One potential source of error should be recognized at the outset. A *gauge-invariant* system is not the continuous limit of one that fails to admit such an arbitrary group. The discontinuous change of invariance properties produces a corresponding discontinuity of the dynamical degrees of freedom and of the *operator commutation relations*. No reliable conclusions about the *mass spectrum* of a *gauge invariant system* can be drawn from the properties of an apparently neighboring system, with a smaller *invariance group*. Indeed, if one considers a *vector field* coupled to a *divergenceless current*, where *gauge invariance* is destroyed by a so-called *mass term* with parameter m_c, it is easily shown[3] that the *mass spectrum* must extend below m_0.

[3] Johnson, K. (1961). *Nuclear Phys.*, 25, 435.

The lowest *mass value* will therefore become arbitrarily small as m_0 approaches zero. Nevertheless, if m_0 is exactly zero the *commutation relations*, or equivalent properties, upon which this conclusion is based become entirely different and the argument fails.

If *invariance* under arbitrary *gauge transformations* is asserted, one should distinguish sharply between *numerical gauge functions* and *operator gauge functions*, for the various *operator gauges* are not on the same *quantum* footing. In each coordinate frame there is a unique *operator gauge*, characterized by *three-dimensional transversality (radiation gauge)*, for which one has the standard operator construction in a *vector space* of positive norm, with a physical probability interpretation. When the theory is formulated with the aid of *vacuum expectation values* of time-ordered operator products, the *Green's functions*, the freedom of formal *gauge transformation* can be restored[4]. The

[4] Schwinger, J. (August, 1959). Euclidean Quantum Electrodynamics. *Phys. Rev.*, 115, 721; https://doi.org/10.1103/PhysRev.115.721.

Green's functions of other *gauges* have more complicated operator realizations, however, and will generally lack the positiveness properties of the *radiation gauge*.
Let us consider the simplest *Green's function* associated with the field $A_\mu(x)$, which can be derived from the unordered product

$$<A_\mu(x)A_\nu(x')> = \int (dp)/(2\pi)^3 \, e^{ip(x-x')} \, dm^2 \, \eta_+(p) \, \delta(p^2 + m^2) \, A_{\mu\nu}(p),$$

where the factor $\mu_+(p)\delta(p^2 + m^2)$ enforces the spectral restriction to *states* with *mass* $m \geq 0$ and positive *energy*. The requirement of non-negativeness for the matrix $A_{\mu\nu}(p)$ is satisfied by the structure associated with the *radiation gauge*, in virtue of the *gauge-dependent asymmetry between space and time* (the time axis is specified by the unit vector n_μ):

$$A_{\mu\nu}{}^B(p) = B(m^2) \, [g_{\mu\nu} - \{(p_\mu n_\nu + p_\nu n_\mu)(np) + p_\mu p_\nu\}/\{p^2 + (np)^2\}]$$

Here $B(m^2)$ is a real non-negative number. It obeys the sum rule

$$1 = \int_0^\infty dm^2 \, B(m^2),$$

which is a full expression of all the fundamental *equal time commutation relations*.

The *field equations* supply the analogous construction for the *vacuum expectation value* of *current* products $(j_\mu(x)j_\nu(x'))$, in terms of the non-negative matrix

$$j_{\mu\nu}(p) = m^2 B(m^2) \, (p_\mu p_\nu - g_{\mu\nu}p^2).$$

The factor m^2 has the decisive consequence that $m = 0$ is not contained in the *current vector's* spectrum of *vacuum fluctuations*. The latter determines $B(m^2)$ for $m > 0$, but leaves unspecified a possible delta function contribution at $m = 0$,

$$B(m^2) = B_0 \, \delta(m^2) + B_1(m^2).$$

The non-negative constant B_0 is then fixed by the sum rule,

$$1 = B_0 + \int_0^\infty dm^2 \, B_1(m^2).$$

We have now recognized that the vacuum fluctuations of the vector A_μ are composed of two parts. One, with $m > 0$, is directly related to corresponding *current fluctuations*, while the other part, with $m = 0$, can be associated with a pure *radiation field*, which is transverse in both three- and four-dimensional senses and has no accompanying *current*. Imagine that the *current vector* contains a variable numerical factor. If this is set equal to zero, we have $B_1(m^2) = 0$ and $B_0 = 1$ or, just the *radiation field*. For a sufficiently small nonzero value of the parameter, B_0 will be slightly less than unity, which may be the situation for the *electromagnetic field*. Or it may be that the electrodynamic coupling is quite considerable and gives rise to a small value of B_0, which has the appearance of a fairly weak *coupling*. Can we increase further the magnitude of the variable parameter until $\int_0^\infty dm^2 \, B_1(m^2)$ attains its limiting value of unity, at which point $B_0 = 0$, and $m = 0$ disappears from the spectrum of A_μ? The general requirement of *gauge invariance* no longer seems to dispose of this essentially dynamical question.

Would the absence of a massless particle imply the existence of a stable unit *spin* particle of nonzero mass? Not necessarily, since the vacuum fluctuation spectrum of A_μ becomes identical with that of j_μ, which is governed by all of the dynamical properties of the fields that contribute to this *current*. For the particularly interesting situation of a *vector field* that is coupled to the *current* of *nucleonic charge*, the relevant *spectrum*, in the approximate *strong-interaction* framework, is that of the *states* with $N = V - T = 0$, $R_T = -1$, $J = 1$, and odd parity. This is a continuum, beginning at three *pion masses*[5].

^{5} The very short range of the resulting nuclear interaction together with the qualitative inference that like *nucleonic charges* are thereby repelled suggests that the *vector field* which defines *nucleonic charge* is also the ultimate instrument of nuclear stability.

It is entirely possible, of course, that $B(m^2)$ shows a more or less pronounced maximum which could be characterized approximately as an unstable particle[6].

^{6} *Note added in proof.* Experimental evidence for an unstable particle of this type has recently been announced by B. C. Maglic, L. W. Alvarez, A. H. Rosenfeld, and M. L. Stevenson, in (September, 1961). Evidence for a *T*=0 Three-Pion Resonance. *Phys. Rev. Lett.*, 7, 178; https://doi.org/10.1103/PhysRevLett.7.178.

But the essential point is embodied in the view that the observed physical world is the outcome of the dynamical play among underlying primary fields, and the relationship between these fundamental fields and the phenomenological particles can be comparatively remote, in contrast to the immediate correlation that is commonly assumed.

Goldstone, J.[†], Salam, A. [§] & Weinberg, S. [§,#] (August, 1962). Broken Symmetries*.

Phys. Rev. 127, 965; https://doi.org/10.1103/PhysRev.127.965; also at https://sci-hub.ru/10.1103/physrev.127.965.

* This research was supported in part by the U. S. Air Force under a contract monitored by the Air Force Office of Scientific Research of the Air Development Command and the Office of Naval Research.

† Trinity College, Cambridge University, Cambridge, England.
§ Imperial College of Science and Technology, London, England.
# Alfred P. Sloan Foundation Fellow; Permanent address: University of California, Berkeley, California.

Received March 16, 1962.

Abstract

Some proofs are presented of Goldstone's conjecture, that if there is continuous *symmetry transformation* under which the Lagrangian is invariant, then *either the vacuum state is also invariant under the transformation, or there must exist spinless particles* of *zero mass*.

I. INTRODUCTION

In the past few years several authors have developed an idea which might offer hope of understanding the *broken symmetries* that seem to be characteristic of *elementary particle* physics. Perhaps the fundamental Lagrangian is invariant under all *symmetries*, but the *vacuum state*[1] is not.

[1] Nambu, Y. & Iona-Lasinio, G, (1961). *Phys. Rev.*, 122, 345; W. Heisenberg, W. (1959). *Z. Naturforsch.*, 14, 441.

It would then be impossible to prove the usual sort of symmetry relations among *S-matrix* elements, but enough symmetry might remain (perhaps at high energy) to be interesting.

But whenever this idea has been applied to specific models, there has appeared an intractable difficulty. For example, Nambu suggested that the Lagrangian might be invariant under a *continuous chirality transformation* $\psi \rightarrow \exp(i\vartheta \cdot \tau\gamma_5)\psi$ even if the *fermion* physical *mass* M were nonzero. But then there would be a conserved *current* J_λ, with matrix element

$$\langle p' \mid J_\lambda \mid p \rangle = f(q^2) \ldots \ldots ,$$

where $q = p - p'$. The *pole* at $q^2 = 0$ can only arise from a *spinless particle of mass zero*, which almost certainly does not exist. Of course, the *pole* would not occur if $f(0) = 0$, which might be the case if we do not insist on identifying J_λ with the *axial vector current* of β decay. But Nambu showed that this unwanted *massless "pion"* also appears as a solution of the approximate Bethe Salpeter equation[1].

Goldstone[2] has examined another model, in which the manifestation of *"broken"* symmetry was the nonzero *vacuum expectation value* of a *boson field*. (This was suggested as an explanation of the $\Delta I = \frac{1}{2}$ rule by Salam and Ward[3].)

[2] Goldstone, J. (January, 1961). Field theories with "Superconductor" solutions. See above.

[3] Salam, A. & Ward, J. C. (1960). *Phys. Rev. Lett.*, 5, 512.

Here again there appeared a *spinless particle of zero mass*. Goldstone was led to conjecture that this will always happen whenever a *continuous symmetry group* leaves the Lagrangian but not the vacuum *invariant*.

We will present here three proofs of this result. The first uses *perturbation theory*; the other two are much more general.

II. PERTURBATION THEORY

We will consider a *multiplet* of n spinless fields ϕ_i which interact among themselves and perhaps also with other fields. The Lagrangian is assumed to be *invariant* under a set of infinitesimal transformations:

$$\delta^\alpha \phi_i = \varepsilon T_{ij}{}^\alpha \phi_j. \tag{1}$$

If the *vacuum state* were also *invariant* under these transformations, the *vacuum expectation values* of the ϕ_i would be subject to a set of linear relations,

$$T_{ij}{}^\alpha \langle \phi_j \rangle_0 = 0. \tag{2}$$

(Usually, there would be enough such relations to imply that all ϕ_i have zero *vacuum expectation value*. This is the case in the example to be discussed at the end of this section, where the ϕ_i transform as the representation of the orthogonal group, so that the T span the space of all antisymmetric matrices.)

We are going to examine the possibility that the *vacuum state* is not *invariant* under these transformations; in particular we will consider the consequences that ensue if

$$T_{ij}{}^{\alpha}\langle\phi_j\rangle_0 \neq 0. \tag{3}$$

for some 0. and some i. ...

...

We see that for zero *momentum* the inverse of the propagator becomes singular and so some elements of the propagator become infinite. This does not prove that there is a *pole* at *zero mass*, but we certainly expect the propagator to be infinite at $P^2 = 0$ only if the theory involves particles of *zero mass*. The fields with nonvanishing matrix element between the *vacuum* and *states* of *zero mass* are

$$\chi^{\alpha}\phi_i \equiv T_{ik}{}^{\alpha}\eta_k\chi_i. \tag{19}$$

Clearly, none of this trouble would occur if it were not for our assumption (3) that $T_{ij}{}^{\alpha}\langle\phi_j\rangle_0 \neq 0$. It is. the *broken symmetry*, and not merely the *nonzero vacuum expectation value* η, that necessitates massless *bosons*[4].

> [4] It is clear from (19) that the maximum number of *zero-mass fields* is L, the number of Lie generators. There may in special cases be fewer than L *zero-mass fields* if not all *fields* χ^{α} given by (19) are linearly independent. This happens for example when $T_{ik}{}^{\alpha}$ correspond to the "tensor" representations of simple Lie groups. For this case $T_{ik}{}^{\alpha}$ are antisymmetric for all three indices. Therefore $\eta_{\alpha}\chi^{\alpha} = \eta_{\alpha}T_{ik}{}^{\alpha}\eta_k\chi_i = 0$, and only (L—1) of the fields χ^{α} are linearly independent. These results are unaltered even if we allow in the theory more than one set of *scalar fields* ϕ_i with *nonzero vacuum expectation values*. To take a concrete case, the *spurion theory* proposed by Salam and Ward (reference 3) to explain the $\Delta I = \frac{1}{2}$ rule rests on assuming $\langle K_1{}^0\rangle \neq 0$. This would mean that the three companion fields to $K_1{}^0$, i.e., K^-, K^+, and $K_1{}^0$, must possess zero *masses*.

...

III. GENERAL PROOFS.

If the Lagrangian is invariant under an n-dimensional set of infinitesimal transformations which transform a general *field* ϕ_a according to

$$\delta\phi_a = \varepsilon T_{ab}{}^{\alpha}\phi_b. \tag{33}$$

then there will exist a, set of *conserved currents*

$$J^{\mu\alpha} = i\ \partial L/\partial(\partial_\mu\phi_\alpha)\ T_{ab}{}^\alpha\phi_b, \tag{34}$$
$$\partial_\mu J^{\mu\alpha} = 0. \tag{35}$$

The usual proof of the *conservation equations* (35) makes use only of the *invariance* of the Lagrangian, and hence should not be affected by the *non-invariance* of the *vacuum*. Also, from the canonical commutation relations we always expect that

$$[Q^\alpha, \phi_a] = T_{ab}{}^\alpha\phi_b, \tag{36}$$
where
$$Q^\alpha = \int d^3x\ J^{0\alpha}(x). \tag{37}$$

We will begin by assuming again that there exists a set of spinless *fields* ϕ_i transforming according to Eq. (1),
$$[\delta^\alpha\phi_i = \varepsilon T_{ij}{}^\alpha\phi_j. \tag{1)]}$$
i.e.,
$$[Q^\alpha, \phi_i] = T_{ij}{}^\alpha\phi_j. \tag{38}$$
...

... The role of the *massless particles* is apparently just to give meaning to the various possible *vacua*.

IV. PROSPECTS FOR THE UNSYMMETRIC VACUUM

The general proofs of the last section rest entirely on the assumption that there exists a *conserved current*, and that the integral of its time-component satisfies (38). This follows formally from the *invariance* of the Lagrangian, but in a *quantum* field theory the non-commutativity of the factors in the *current*, and the possible *nonconvergence* of the integral of its time component, make our arguments essentially non-rigorous.

Therefore, it seems reasonable to defer belief in the necessity of *massless bosons* in a theory with unsymmetric *vacua* until such a *bete noire* is found in an actual calculation based on such a theory. We have already shown in Sec. II that the *massless bosons* do appear when we perform calculations using *perturbation theory*, provided that the *symmetry* of the theory is broken only by the choice of the *vacuum expectation value* η of the *boson field*.

But this is not the most general possibility. The original work of Nambu[2] indicates that the choice of a *fermion mass* can also break a *symmetry*. In this theory the *fermion mass* is

$$- ip^\mu\gamma_\mu = m_1 + i\gamma_5 m_2,$$

where (m_1, m_2) transform under *chirality transformations* like the components of a 2-vector. If m_1 and m_2 are not zero they must satisfy a condition of form

$$F(m_1^2 + m_2^2) = 0.$$

Any particular choice of direction for the vector (m_1, m_2) breaks the *chirality invariance*. (It should be noted that Nambu's choice $m_2 = 0$ is purely arbitrary and not dictated by *parity conservation*. For a general *mass* we must simply define the matrix associated with *parity transformations* to be

$$[(m_1 + i\gamma_5 m_2)/(m_1^2 + m_2^2)^{1/2}] \beta,$$

rather than just β.)

In Nambu's theory there is no "bare" *spinless boson*, but it is possible to construct a two-vector

$$\phi_1 = \bar{\psi}\psi, \qquad \phi_2 = i\bar{\psi}\gamma_5\psi.$$

With Nambu's definition of *parity* (i.e., $m_2 = 0$) the *vacuum expectation value* of ϕ_2 but not of ϕ_1 vanishes, so the vector $\langle\phi\rangle_0$ points in the 1-direction. An infinitesimal *chirality transformation* would rotate $\langle\phi\rangle_0$ towards the 2-axis, so we are led to conjecture that the propagator of ϕ_2 has a *zero-mass pole*. In fact, just such a *pole* was found by Nambu in an approximate treatment of the *bound-state* problem. However, to show that the *pole* remains at *zero mass* when more complicated diagrams are considered would require a more thorough understanding of the treatment of *bound states* in *perturbation theory*. We are attempting this at present.

In a more complicated situation, we could have an *invariance* broken both by the choice of a *vacuum expectation value* of a "bare" field and also simultaneously by the choice of a *mass*. For example, if we specialize the model discussed in Sec. II to the case of *chirality invariance*, we must take

$$M = 0, \qquad O_1 = 1, \qquad O_2 = i\gamma_5,$$

so that

$$L = \dots.$$

In this case our conjecture would be that:

(1) If part of the loss of *symmetry* is due to the choice of a two-vector $\langle\phi\rangle_0$, then there must appear a *zero-mass pole* in the part of ϕ perpendicular to this *vacuum expectation value*.

(2) If part of the loss of *symmetry* is due to the choice of a *non-zero Fermion mass*

$m_1 + i\gamma_5 m_2$ then there must appear a *zero-mass pole* in the propagator of

$$\phi' = - m_2\bar{\psi}\psi + m_1\bar{\psi}\gamma_5\psi.$$

[Presumably this is the same *pole* as for (I). *Parity conservation* would require (m_1, m_2) to be in the direction of $\langle\phi\rangle_0$]

(3) If part of the loss of *symmetry* is due to the choice of a *non-invariant boson mass* (i.e., if the residue of the *pole* at *mass* m in the propagator of ϕ_i and ϕ_j is a matrix which is not just a constant times δ_{ij}), then there must appear a *two-bosom pole* at *zero mass* in the propagator of ϕ^2.

These "conjectures" can be taken as proved if we accept the arguments of Sec. III. We believe that we will also soon be able to prove these conjectures, in general, within the framework of *perturbation theory*.

If this is so, then there seem only three roads open to an understanding of *broken symmetries* based on the *non-invariance of the vacuum*:

(A) The particle interpretation of such theories might be revised (as in the Gupta-Bleuler method) so that *the massless particles are not physically present in final states* if they are absent in *initial states*. How ever, all our attempts in this direction have failed.

(B) The *massless particles might really exist*. The argument against this based on the *Eotvos experiment* might not apply *if the particles carry quantum numbers*, since then the *scattering cross section* of two macroscopic bodies *due to exchange of the massless bosons* would be proportional only to the numbers of atoms in each body and *not (as for Coulomb forces or gravitation) to the squares of the numbers of atoms*. But the *couplings* of these *massless particles* would presumably be quite strong, and would have shown up in exotic decay modes.

(C) Goldstone has already remarked that nothing seems to go wrong *if it is just discrete symmetries that fail to leave the vacuum invariant*. A more appealing possibility is that the "*ur symmetry*" broken by the vacuum involves an inextricable combination of *gauge* and *space-time transformations*.

Note added in proof. Recently, one of us (S. W., *Proceedings of the 1962 Geneva Conference on High Energy Nuclear Physics*) has developed a method of rewriting any Lagrangian in order to introduce *fields* for bound as well as "elementary" particles. This allows the proof of Sec. II to be extended to the case where the *field* with *non-vanishing vacuum expectation value* is any scalar function of the *elementary particle fields*, hence completing our argument.

Philip Warren Anderson (December 13, 1923–March 29, 2020).

Anderson was an American theoretical physicist and Nobel laureate. Anderson made contributions to the theories of localization, *antiferromagnetism*, *symmetry breaking* (including a paper in 1962 discussing *symmetry breaking* in particle physics, leading to the development of the *Standard Model* around 10 years later), and high-temperature superconductivity, and to the philosophy of science through his writings on emergent phenomena. Anderson is also responsible for naming the field of physics that is now known as condensed matter physics.

The Nobel Prize in Physics 1977 was awarded jointly to Anderson, together with Sir Nevill Francis Mott and John Hasbrouck Van Vleck "for their fundamental theoretical investigations of the electronic structure of magnetic and disordered systems".

Anderson was born in Indianapolis, Indiana, and grew up in Urbana, Illinois. His father, Harry Warren Anderson, was a professor of plant pathology at the University of Illinois at Urbana-Champaign; his maternal grandfather was a mathematician at Wabash College, where Anderson's father studied; and his maternal uncle was a Rhodes Scholar who became a professor of English, also at Wabash College. He graduated from University Laboratory High School in Urbana in 1940. Under the encouragement of a math teacher by the name of Miles Hartley, Anderson enrolled at Harvard University to study under a fully-funded scholarship. He concentrated in "Electronic Physics" and completed his B.S. in 1943, after which he was drafted into the war effort and built antennas at the Naval Research Laboratory until the end of the Second World War in 1945.

As an undergraduate, his close associates included particle-nuclear physicist H. Pierre Noyes, philosopher and historian of science Thomas Kuhn and molecular physicist Henry Silsbee. Anderson married Joyce Gothwaite in 1947 and they had a daughter, Susan.

After the war, Anderson returned to Harvard to pursue graduate studies in physics under the mentorship of John Hasbrouck van Vleck; he received his Ph.D. in 1949 after completing a doctoral dissertation titled "*The theory of pressure broadening of spectral lines in the microwave and infrared regions*".

From 1949 to 1984, Anderson was employed by Bell Laboratories in New Jersey, where he worked on a wide variety of problems in condensed matter physics. During this period he developed what is now called Anderson localization (the idea that extended states can be localized by the presence of disorder in a system) and Anderson's theorem (concerning impurity scattering in superconductors); invented the Anderson Hamiltonian, which describes the site-wise interaction of electrons in a transition metal; proposed *symmetry breaking* within particle physics (this played a role in the development of the *Standard Model* and the development of the theory behind the *Higgs mechanism*, which in turn generates *mass* in some elementary particles); created the pseudospin approach to the BCS theory of superconductivity; made seminal studies of non-s-wave pairing (both symmetry-breaking and microscopic mechanism) in the superfluidity of helium-3, and helped found the area of spin-glasses.

Anderson spent a year as lecturer at Cambridge University in 1961–1962, and recalled that having Brian Josephson in a class was "a disconcerting experience for a lecturer, I can assure you, because everything had to be right or he would come up and explain it to me after class".

From 1967 to 1975, Anderson was a professor of theoretical physics at Cambridge. In 1977 Anderson was awarded the Nobel Prize in Physics for his investigations into the electronic structure of magnetic and disordered systems, which allowed for the development of electronic switching and memory devices in computers. Co-researchers Sir Nevill Francis Mott and John van Vleck shared the award with him. He retired from Bell Labs in 1984 and was Joseph Henry Professor Emeritus of Physics at Princeton University.

Anderson also made conceptual contributions to the philosophy of science through his explication of emergent phenomena, which became an inspiration for the science of complex systems. In 1972, he wrote an article called "More is Different" in which he emphasized the limitations of reductionism and the existence of hierarchical levels of science, each of which requires its own fundamental principles for advancement.

In 1984, he participated in the founding workshops of the Santa Fe Institute, a multidisciplinary research institute dedicated to the science of complex systems. Anderson also co-chaired the institute's 1987 conference on economics with Kenneth Arrow and W. Brian Arthur, and participated in its 2007 workshop on models of emergent behavior in complex systems.

In 1987, Anderson testified to the US Congress, "against the construction of the Superconducting Super Collider (SSC), a 40 TeV proton-proton collider in Texas that would have been the biggest experiment in particle physics. Anderson's opposition to the SSC did not directly lead to its cancellation in 1993—spiraling costs were the main factor—but he was perhaps its most high-profile opponent." He was, "skeptical of the supposed boost it would provide to science in the US and the claim that the spin-offs would provide great return on investment".

Anderson died in Princeton, New Jersey, on March 29, 2020, at the age of 96.

Anderson, P. W. (April, 1963). Plasmons, Gauge Invariance, and Mass.

Phys. Rev., 130, 1, 439; https://journals.aps.org/pr/abstract/10.1103/PhysRev.130.439; also at https://web.archive.org/web/20160307022433/https://www.physics.rutgers.edu/grad/601/Anderson_Plasmons.pdf.

Bell Telephone Laboratories, Murray Hill, New Jersey.

Received November 8, 1962.

Abstract

Schwinger has pointed out that the *Yang-Mills vector boson* implied by associating a generalized *gauge transformation* with a conservation law (of *baryonic charge*, for instance) does not necessarily have zero *mass*, if a certain criterion on the *vacuum fluctuations* of the generalized *current* is satisfied. We show that the theory of *plasma oscillations* is a simple *nonrelativistic* example exhibiting all of the features of Schwinger's idea. It is also shown that Schwinger's criterion that the *vector field* $m \neq 0$ implies that the *matter* spectrum before including the *Yang-Mills interaction* contains $m = 0$, but that the example of *superconductivity* illustrates that the physical spectrum need not. Some comments on the relationship between these ideas and the *zero-mass* difficulty in theories with *broken symmetries* are given.

Recently, Schwinger[1] has given an argument strongly suggesting that associating a *gauge transformation* with a *local conservation law* does not necessarily require the existence of a *zero-mass vector boson*.

[1] Schwinger, J. (January, 1962). Gauge Invariance and Mass. *Phys. Rev.*, 125, 397. See above.

For instance, it had previously seemed impossible to describe the *conservation of baryons* in such a manner because of the absence of a *zero-mass boson* and of the accompanying long-range forces[2].

[2] Lee, T. D. & Yang, C. N. (June, 1955). Conservation of Heavy Particles and Generalized Gauge Transformations. *Phys. Rev.*, 98, 5, 1501; https://doi.org/10.1103/PhysRev.98.1501.

The problem of the *mass of the bosons* represents the major stumbling block in Sakurai's attempt to treat the dynamics of *strongly interacting particles* in terms of the *Yang-Mills gauge fields* which seem to be required to accompany the known *conserved currents* of *baryon number* and *hypercharge*[3]. (We use the term "Yang Mills" in Sakurai's sense, to denote any generalized *gauge field* accompanying a *local conservation law*).

[3] Sakurai, J. J. (September, 1960). Theory of strong interactions. *Ann. Phys.*, 11, 1, 1-48. See above.

The purpose of this article is to point out that the familiar *plasmon theory* of the *free-electron gas* exemplifies Schwinger's theory in a very straightforward manner.

[A *plasmon* is a quantum of *plasma oscillation*. Just as light (an optical oscillation) consists of *photons*, the *plasma oscillation* consists of *plasmons*. The plasmon can be considered as a quasiparticle since it arises from the quantization of *plasma oscillations*, just like *phonons* are quantizations of *mechanical vibrations*. Thus, *plasmons* are collective (a discrete number) oscillations of the *free electron gas density*. For example, at optical frequencies, *plasmons* can couple with a *photon* to create another quasiparticle called a *plasmon polariton*.

Plasmons can be described in the classical picture as *an oscillation of electron density* with respect to the fixed positive ions in a metal. Since *plasmons* are the quantization of classical *plasma oscillations*, most of their properties can be derived directly from Maxwell's equations.

The *plasmon* was initially proposed in 1952 by David Pines and David Bohm and was shown to arise from a Hamiltonian for the long-range *electron-electron* correlations.]

In the *plasma, transverse electromagnetic waves* do not propagate below the "*plasma frequency*", which is usually thought of as the frequency of long wavelength longitudinal oscillation of the *electron gas*. At and above this frequency, three modes exist, in close analogy (except for problems of *Galilean invariance* implied by the inequivalent dispersion of longitudinal and transverse modes) with the *massive vector boson* mentioned by Schwinger. *The plasma frequency is equivalent to the mass*, while the finite *density* of *electrons* leading to divergent "*vacuum*" *current* fluctuations resembles the strong *renormalized coupling* of Schwinger's theory. In spite of the absence of low-frequency *photons, gauge invariance* and *particle conservation* are clearly satisfied in the plasma.

In fact, one can draw a direct parallel between the *dielectric constant treatment* of plasmon theory[4] and Schwinger's argument.

304

Nozieres, P. & Pines, D. (February, 1958). Electron Interaction in Solids. General Formulation. *Phys. Rev.*, 109, 3, 741; https://doi.org/10.1103/PhysRev.109.741.

Schwinger comments that the *commutation relations* for the *gauge field* A give us one sum rule for the *vacuum fluctuations* of A, while those for the *matter field* give a completely independent value for the fluctuations of *matter current* j. Since j is the source for A and the two are connected by *field equations*, the two sum rules are normally incompatible unless there is a contribution to the A rule from a free, homogeneous, weakly interacting, massless solution of the *field equations*. If, however, the *source term* is large enough, there can be no such contribution and the massless solutions cannot exist.

The usual *theory of the plasmon* does not treat the *electromagnetic field* quantum-mechanically or discuss *vacuum fluctuations*; yet there is a close relationship between the two arguments, and we, therefore, show that the quantum nature of the *gauge field* is irrelevant. Our argument is as follows: The equation for the *electromagnetic field* is

$$p^2 A_\mu = (k^2 - \omega^2) \, A_\mu \, (\mathbf{k},\omega) = 4\pi j_\mu \, (\mathbf{k},\omega).$$

A given distribution of current j_μ will, therefore, lead to a response A_μ given by

$$A_\mu = 4\pi/(k^2 - \omega^2) \, j_\mu = 4\pi/p^2 \, j_\mu. \tag{1}$$

(1) is merely the statement that only the *electromagnetic current* can be a source of the field; it is required for *general gauge invariance* and *charge conservation* according to the usual arguments.

The dynamics of the *matter system* — of the plasma in that case, of the *vacuum* in the *elementary particle problem* — determine a second *response function*, the response of the *current* to a given *electromagnetic* or *Yang-Mills field*. Let us call this *response function*

$$j_\mu = - K_{\mu\nu}(\mathbf{k},\omega) \, A_\mu(\mathbf{k},\omega). \tag{2}$$

By well-known arguments of *gauge invariance*, $K_{\mu\nu}$ must have a certain form: Schwinger points out that in the *relativistic* case it must be proportional to $p_\mu p_\nu - g_{\mu\nu}p^2$, and equivalent arguments give one the same form in superconductivity[5].

[5] Schafroth, M. R. (1951). *Helv. Phys. Acta*, 24, 645.

It will be convenient to consider, for simplicity, only the *gauge*

$$p_\mu A_\mu = 0. \tag{3}$$

305

Then the response is diagonal: $K_{\mu\nu} = -g_{\mu\nu}K$. For a *plasma* with n carriers of *charge* e and *mass* M it is simply (in the limit $p \to 0$)

$$K = ne^2/M. \tag{4}$$

In an insulator the response is not *relativistically* invariant. If the insulator has *magnetic polarizability* α_m and *electric* α_e, the response equations may be written, in the *gauge* (3),

$$j_\mu = \alpha_e p^2 A_\mu \quad \text{(longitudinal and time components)},$$
$$j = -\alpha_m p^2 A \quad \text{(transverse components)}.$$

In a truly *relativistic* situation such as our normal picture of a *vacuum*, we expect

$$j_\mu = \alpha p^2 A_\mu \tag{5}$$

to describe normal polarizable behavior.

Since we cannot turn off the *interactions*, we do not actually observe the responses (1), (2), or (5). If we insert a test particle, its field A_μ^e induces a *current* j_μ which in turn acts as the source for an internal *field* A_μ^i:

$$j_\mu = -K(A_\mu^i + A_\mu^e), \quad A_\mu^i = +4\pi j_\mu/p^2,$$

or, the *total field* is modified to

$$A_\mu = [p^2/(p^2 + 4\pi K)A_\mu^e. \tag{6}$$

The *pole* at which A propagates freely occurs at a *mass (frequency)*

$$m^2 = -p^2 = 4\pi K, \tag{7}$$

which in a *conductor* is

$$m^2 = \omega^2 - k^2 = \omega_p^2. \tag{8}$$

ω_p is the usual *plasma frequency* $(4\pi ne^2/M)^{1/2}$.

It is not necessary here to go in detail into the relationship between longitudinal and transverse behavior of the *plasmon*. In the limit $p \to 0$ both waves propagate according to (8). The *longitudinal plasmon* is generally thought of as entirely an attribute of the *plasma*, while the *transverse* ones are considered to result from modification of the *propagation of real photons by the medium*. This is reasonable in the classical case because the *longitudinal plasmon* disappears at a certain cutoff *energy* and has a different *dispersion*

306

law; but in a *Lorentz-covariant theory of the vacuum* it would be indistinguishable from the third component of a massive *vector boson* of which the *transverse photons* are the two transverse components.

How, then, if we were confined to the *plasma* as we are to the vacuum and could only measure *renormalized* quantities, might we try to determine whether, before turning on the effects of *electromagnetic interaction*, A had been a massless *gauge field* and K had been finite? *As far as we can see, this is not possible*; it is, nonetheless, interesting to see what the criterion is in terms of the actual *current response function* to a perturbation in the Lagrangian

$$\delta L = j_\mu \delta A_\mu. \tag{9}$$

This will turn out to be identical to Schwinger's criterion. The original *"bare" response function* was K:

$$j_\mu = -K_{\mu\nu}\delta A_\mu.$$

Taking into account the *interaction*, however, we must correct for the *induced fields* and *currents*, and we get

$$j_\mu = -K' \, \delta A_\mu^e = -K[p^2/(p^2 + 4\pi K)]\delta A_\mu^e \rightarrow (p^2/4\pi) \, \delta A_\mu^e, \quad p^2 \rightarrow 0. \tag{10}$$

Thus, the new response to an applied *perturbing field* (9) is very like that of an ordinary *polarizable medium*. The only difference from an ordinary polarizable *"vacuum"* with bare response (5) is that in that case as $p \rightarrow 0$

$$K' \rightarrow -[\alpha/(1 + 4\pi\alpha)] \, p^2, \tag{11}$$

so that the coefficient of $p^2/4\pi$ is less than unity.

This criterion is precisely the same as Schwinger's criterion

$$\int B_1(m^2) \, dm^2 = 1,$$

where $B_1(m^2)$ is the *weight function* for the *current vacuum fluctuations*. This can be shown by a simple dispersion argument. Schwinger expresses the unordered product expectation value of the *current* as

$$<j_\mu(x)j_\nu(x')> = \int dm^2 \, m^2 B_1(m^2) \int dp/(2\pi)^3 \, e^{ip(x-x')} \, \eta_+(p) \, \delta(p^2 + m^2)(p_\mu p_\nu - g_{\mu\nu}p^2).$$

The Fourier transform of the corresponding retarded *Green's function* is our *response function*:

$$K'(p) = \int dm^2\, m^2 B_1(m^2)/(p^2 - m^2)\, [p_\mu p_\nu - g_{\mu\nu}p^2],$$

and

$$\lim_{p\to 0} K'(p) = (p_\mu p_\nu - g_{\mu\nu}p^2)\int dm^2\, B_1(m^2).$$

Thus, (aside from a factor 4π which Schwinger has not used in his field equation) his criterion is also that the *polarizability* α', here expressed in terms of a *dispersion integral*, have its maximum possible value, 1.

The *polarizability of the vacuum* is not generally considered to be observable[6] except in its p dependence (terms of order p^4 or higher in K).

[6] We follow here, as elsewhere, the viewpoint of W. Thirring, W. (1958). *Principles of Quantum Electrodynamics*. Academic Press Inc., New York, Chap. 14.

In fact, we can remove (11) entirely by the conventional *renormalization* of the *field* and *charge*

$$A_\tau = AZ^{-1/2}, \quad e_\tau = eZ^{1/2}, \quad j_\tau = jZ^{1/2}.$$

Z, here, can be shown to be precisely

$$Z = 1 - 4\pi\alpha' = 1 - \int_0^\infty dm^2\, B_1(m^2).$$

Thus, the *renormalization* procedure is possible for any merely *polarizable "vacuum"*, but not for the special case of the *conducting "plasma" type of vacuum*. In this case, no net true *charge* remains localized in the region of the dressed particle; all of the *charge* is carried "*at infinity*" corresponding to the fact, well known in the theory of metals, that all the *charge* carried by a quasi-particle in a *plasma* is actually on the surface. Nonetheless, conservation of particles, if not of bare *charge*, is strictly maintained. Note that the situation does not resemble the case of *infinite" charge renormalization* because the infinity in the *vacuum polarizability* need only occur at $p^2 = 0$.

Either in the case of the *polarizable vacuum* or of the "*conducting*" one, no low-energy experiment, and even possibly no high-energy one, seems capable of directly testing the value of the *vacuum polarizability* prior to *renormalization*. Thus, we conclude that the *plasmon* is a physical example demonstrating Schwinger's contention that *under some circumstances the Yang-Mills type of vector boson need not have zero mass*. In addition, aside from the short range of forces and the finite *mass*, which we might interpret without resorting to Yang-Mills, it is not obvious how to characterize such a case mathematically in terms of observable, *renormalized* quantities.

We can, on the other hand, try to turn the problem around and see what other conclusions we can draw about possible *Yang-Mills models* of *strong interactions* from the solid-state analogs. What properties of the vacuum are needed for it to have the analog of a *conducting* response to the *Yang-Mills field*?

Certainly, the fact that the *polarizability* of the "*matter*" system, without taking into account the *interaction* with the *gauge field*, is infinite need not bother us, since that is unobservable. In physical *conductors* we can see it, but only because we can get outside them and apply to them true *electromagnetic fields*, not only internal test *charges*.

More serious is the implication —obviously physically from the fact that a has a pole at $p^2 = 0$— that the "*matter*" spectrum, at least for the "*undressed*" matter system, must extend all the way to $m^2 = 0$. In the normal *plasma*, even the final spectrum extends to zero frequency, the *coupling* rather than the *spectrum* being affected by the screening. Is this necessarily always the case? The answer is no, obviously, since the *superconducting electron gas* has no *zero-mass* excitations whatever. In that case, the *fermion mass* is finite because of the *energy* gap, while the *boson* which appears as a result of the theorem of Goldstone[7,8] and has zero *unrenormalized mass* is converted into a finite-mass *plasmon* by interaction with the appropriate *gauge field*, which is the *electromagnetic field*.

[7] Goldstone, J. (January, 1961). Field theories with "Superconductor" solutions. *Nuovo Cimento*, 19, 154–64. See above.

[8] Goldstone, J., Salam, A. & Weinberg, S. (August, 1962). Broken Symmetries. *Phys. Rev.* 127, 965. See above.

The same is true of the *charged Bose gas*.

It is likely, then, considering the *superconducting* analog, that the way is now open for a degenerate-vacuum theory of the Nambu type[9] without any difficulties involving either *zero-mass Yang-Mills gauge bosons* or *zero-mass Goldstone bosons*.

[9] Nambu, Y. & Jona-Lasinio, G. (April, 1961). Dynamical Model of Elementary Particles Based on an Analogy with Superconductivity. I. *Phys. Rev.*, 122, 1, 345; https://doi.org/ 10.1103/PhysRev.122.345.

These two types of *bosons* seem capable of "canceling each other out" and leaving *finite mass bosons* only. It is not at all clear that the way for a *Sakurai theory*[3] is equally uncluttered. The only mechanism suggested by the present work (of course, we have not discussed *non-Abelian gauge groups*) for giving the *gauge field* mass is the degenerate vacuum type of theory, in which the original *symmetry* is not manifest in the observable domain. Therefore, it needs to be demonstrated that the necessary *conservation laws* can be maintained.

I should like to close with one final remark on the *Goldstone theorem*. This theorem was initially conjectured, one presumes, because of the solid-state analogs, via the work of Nambu[10] and of Anderson[11].

[10] Nambu, Y. (February, 1960). Quasi-Particles and Gauge Invariance in the Theory of Superconductivity. *Phys. Rev.*, 117, 3, 648; https://doi.org/10.1103/PhysRev.117.648.

[11] Anderson, P. W. (May, 1958). Coherent Excited States in the Theory of Superconductivity: Gauge Invariance and the Meissner Effect. *Phys. Rev.* 110, 4, 827; https://doi.org/10.1103/PhysRev.110.827.

The theorem states, essentially, that *if the Lagrangian possesses a continuous symmetry group under which the ground or vacuum state is not invariant, that state is, therefore, degenerate with other ground states*. This implies a *zero-mass boson*. Thus, the solid crystal violates *translational* and *rotational invariance*, and possesses *phonons*; liquid helium violates (in a certain sense only, of course) *gauge invariance*, and possesses a *longitudinal phonon*; *ferro-magnetism* violates *spin rotation symmetry*, and possesses *spin waves*; *superconductivity* violates *gauge invariance*, and would have a *zero-mass collective mode* in the absence of long-range *Coulomb forces*.

It is noteworthy that in most of these cases, upon closer examination, *the Goldstone bosons do indeed become tangled up with Yang-Mills gauge bosons* and, thus, do not in any true sense really have *zero mass*. *Superconductivity* is a familiar example, but a similar phenomenon happens with *phonons*; when the *phonon frequency* is as low as the *gravitational plasma frequency*, $(4\pi G_p)^{1/2}$ (wavelength $\sim 10^4$ km in normal matter) there is a *phonon-graviton interaction*: in that case, because of the peculiar sign of the *gravitational interaction*, leading to instability rather than finite *mass*[12].

[12] Jeans, J. H. (1903). *Phil. Trans. Roy. Soc. London*, 101, 157.

Utiyama and Feynman[13] have pointed out that *gravity* is also a *Yang-Mills field*.

[13] Utiyama, R. (1956). *Phys. Rev.*, 101, 1597; R. P. Feynman (unpublished).

It is an amusing observation that the *three phonons* plus two *gravitons* are just enough components to make up the appropriate tensor particle which would be required for a *finite-mass graviton*.

Spin waves also are known to interact strongly with *magnetostatic* forces at very long wavelengths[14], for rather more obscure and less satisfactory reasons.

[14] Walker, L. R. (1957). *Phys. Rev.* 105, 390.

We conclude, then, that the Goldstone zero-mass difficulty is not a serious one, because we can probably cancel it off against an equal Yang-Mills zero-mass problem. What is not clear yet, on the other hand, is whether it is possible to describe a truly strong *conservation law* such as that of *baryons* with a *gauge group* and a *Yang-Mills field* having finite *mass*.

Gell-Mann. M. (February, 1964). A Schematic Model of Baryons and Mesons.*

Physics Letters, 8, 3, 214–5; https://doi.org/10.1016/S0031-9163(64)92001-3; also at https://www.nssp.uni-saarland.de/lehre/Vorlesung/Kernphysik_SS19/History/Papers/Gell-Mann.pdf.

* Work supported in part by the U. S. Atomic Energy Commission.

California Institute of Technology, Pasadena, California.

Received January 4, 1964

In 1964, Murray Gell-Mann, and separately George Zweig, proposed that *baryons*, which include *protons* and *neutrons*, and *mesons* were composed of *elementary particles*. [Zweig, G. (February 21, 1964). An SU(3) Model for Strong Interaction Symmetry and its Breaking: II. *CERN Document Server*. CERN-TH-412; doi:10.17181/CERN-TH-412.] Zweig called the *elementary particles* "aces" while Gell-Mann called them "*quarks*"; the theory came to be called the *quark model*. The bootstrap model for *strongly interacting particles* described in terms of the *broken eightfold way* is discussed to determine algebraic properties of the interactions with scattering amplitudes on the *mass shell*. A mathematical model based on *field theory* is described.

If we assume that the *strong interactions* of *baryons* and *mesons* are correctly described in terms of the *broken "eightfold way"*[1,2,3] we are tempted to look for some fundamental explanation of the situation.

[1] Gell-Mann, M. (1961). *California Institute of Technology Synchrotron Laboratory, Report CTSL-20.*
[2] Ne'eman, Y. (1961). *Nuclear Phys.*, 26, 222.
[3] Gell-Mann, M. (1962). *Phys. Rev.*, 125, 1067.

A highly promised approach is the purely dynamical "bootstrap" model for all the strongly interacting particles within which one may try to derive *isotopic spin* and *strangeness conservation* and *broken eightfold symmetry* from self-consistency alone[4].

[4] E.g.: Capps, R. H. (1963). *Phys. Rev. Lett.*, 10, 312; R. E. Cutkosky, R. E., Kalckar, J. & Tarjanne, P. (1962). *Physics Letters*, 1, 93; Abers, E., Zachariasen, F. & Zemaeh, A. C. (1963). *Phys. Rev.*, 132, 1831; Glashow, s. (1963). *Phys. Rev.* 130, 2132; R. E. Cutkosky, R. E. & P. Tarjanne, P. (1963). *Phys. Rev.*, 132, 1354.

Of course, with only *strong interactions*, the orientation of the asymmetry in the *unitary space* cannot be specified; one hopes that in some way the selection of specific components of the *F-spin* by *electromagnetism* and the *weak interactions* determines the choice of *isotopic spin* and *hypercharge directions*.

[The *strong interaction*, also called the *strong force* or *strong nuclear force*, is a fundamental *interaction* that confines *quarks* into *protons*, neutrons, and other *hadron* particles. The *strong interaction* also binds *neutrons* and *protons* to create *atomic nuclei*, where it is called the *nuclear force*.

Most of the *mass* of a *proton* or *neutron* is the result of the *strong interaction* energy; the individual *quarks* provide only about 1% of the *mass* of a *proton*.

Before 1971, physicists were uncertain as to how the *atomic nucleus* was bound together. It was known that the *nucleus* was composed of *protons* and *neutrons* and that *protons* possessed positive *electric charge*, while *neutrons* were electrically neutral. By the understanding of physics at that time, positive charges would repel one another and the positively charged *protons* should cause the *nucleus* to fly apart. However, this was never observed. A stronger attractive force was postulated to explain how the *atomic nucleus* was bound despite the *protons'* mutual *electromagnetic* repulsion. This hypothesized force was called the *strong force*, which was believed to be a fundamental force that acted on the *protons* and *neutrons* that make up the *nucleus*.

The *strong attraction* between *nucleons* was the side-effect of a more fundamental force that bound the *quarks* together into *protons* and *neutrons*. The theory of *quantum chromodynamics* explains that *quarks* carry what is called a *color charge*, although it has no relation to visible *color*. *Quarks* with unlike *color charge* attract one another as a result of the *strong interaction*, and the particle that mediates this was called the *gluon*.]

Even if we consider the *scattering amplitudes* of *strongly interacting* particles on the *mass shell* only and treat the matrix elements of the *weak, electromagnetic,* and *gravitational interactions* by means of *dispersion theory*, there are still meaningful and important questions regarding the algebraic properties of these *interactions* that have so far been discussed only by abstracting the properties from a formal *field theory model* based on fundamental entities[5] from which the *baryons* and *mesons* are built up.

[5] Tarjanne, P. & Teplitz, V. L. (1963). *Phys. Rev. Lett.*, 11, 447.

If these entities were *octets*, we might expect the underlying *symmetry group* to be SU(8) instead of SU(3); it is therefore tempting to try to use *unitary triplets* as fundamental objects. A *unitary triplet* t consists of an *isotopic singlet* s of *electric charge* z (in units of e) and an *isotopic doublet* (u, d) with *charges* z + 1 and z respectively. The *anti-triplet* has, of course, the opposite signs of the *charges*. *Complete symmetry among the members of the triplet gives the exact eightfold way*, while a *mass difference*, for example, between the *isotopic doublet* and *singlet* gives the first-order violation.

For any value of z and of *triplet spin*, we can construct *baryon octets* from a basic neutral *baryon singlet* b by taking combinations (btt⁻), (bttt⁻t⁻), etc. [**].

[**] This is similar to the treatment in ref. 1. See also ref. 5.

From (btt⁻), we get the *representations* **1** and **8**, while from (bttt⁻t⁻) we get **1**, **8**, **10**, **10**, and **27**. In a similar way, *meson singlets* and *octets* can be made out of (tt⁻), (ttt⁻t⁻), etc. The *quantum number* $n_t - n_{t^-}$ would be zero for all known *baryons* and *mesons*. The most interesting example of such a model is one in which the *triplet* has spin ½ and z = − 1, so that the four particles d⁻, s⁻, u° and b° exhibit a parallel with the *leptons*.

A simpler and more elegant scheme can be constructed if we allow non-integral values for the *charges*. We can dispense entirely with the basic *baryon* b if we assign to the *triplet* t the following properties: *spin* ½, z = − 1/3 and *baryon number* 1/3. We then refer to the members $u^{2/3}$, $d^{-1/3}$, and $s^{-1/3}$ of the *triplet* as "*quarks*"[6], q and the members of the *anti-triplet* as *anti-quarks* q⁻.

[6] James Joyce, (1939). *Finnegan's Wake*. Viking Press, New York, p. 383.

[Quarks and antiquarks [*classification*]

A *quark* is a type of elementary particle and a fundamental constituent of matter. *Quarks* combine to form composite particles called *hadrons*, the most stable of which are *protons* and *neutrons*, the components of atomic nuclei. All commonly observable matter is composed of *up quarks*, *down quarks* and *electrons*. Owing to a phenomenon known as *color confinement*, *quarks* are never found in isolation; they can be found only within *hadrons*, which include *baryons* (such as protons and neutrons) and *mesons*, or in *quark–gluon* plasmas. For this reason, much of what is known about *quarks* has been drawn from observations of *hadrons*.

Quarks have various intrinsic properties, including *electric charge*, *mass*, *color charge*, and *spin*. They are the only elementary particles in the *Standard Model* of particle physics to experience all four fundamental interactions, also known as

314

fundamental forces (*electromagnetism, gravitation, strong interaction,* and *weak interaction*), as well as the only known particles whose *electric charges* are not integer multiples of the elementary *charge*.

There are six types, known as *flavors*, of *quarks*: *up, down, charm, strange, top,* and *bottom. Up and down quarks* have the lowest *masses* of all *quarks*. The heavier *quarks* rapidly change into *up and down quarks* through a process of *particle decay*: the transformation from a higher *mass state* to a lower *mass state*. Because of this, *up and down quarks* are generally stable and the most common in the universe, whereas *strange, charm, bottom,* and *top quarks* can only be produced in high energy collisions (such as those involving cosmic rays and in particle accelerators). For every *quark flavor* there is a corresponding type of *antiparticle*, known as an *antiquark*, that differs from the *quark* only in that some of its properties (such as the *electric charge*) have equal magnitude but opposite sign.

Quarks were introduced as parts of an ordering scheme for *hadrons*, and there was little evidence for their physical existence until deep inelastic scattering experiments at the Stanford Linear Accelerator Center in 1968. Accelerator program experiments have provided evidence for all six *flavors*. The *top quark*, first observed at Fermilab in 1995, was the last to be discovered.

The name "*quark*" was coined by Gell-Mann, and is a reference to the novel Finnegans Wake, by James Joyce ("*Three quarks for Muster Mark!*" book 2, episode 4). Zweig had referred to the particles as "aces", but Gell-Mann's name caught on. *Quarks, antiquarks,* and *gluons* were soon established as the underlying elementary objects in the study of the structure of *hadrons*. The 1969 Nobel Prize in Physics was awarded to Gell-Mann "for his contributions and discoveries concerning the classification of elementary particles and their interactions".]

Baryons can now be constructed from *quarks* by using the combinations (qqq), (qqqqq⁻), etc., while *mesons* are made out of (qq⁻), (qqq⁻q⁻), etc. It is assuming that the lowest *baryon* configuration (qqq) gives just the *representations* **1**, **8**, and **10** that have been observed, while the lowest *meson* configuration (qq⁻) similarly gives just **1** and **8**.

A formal mathematical model based on *field theory* can be built up for the *quarks* exactly as for p, n, Λ in the old Sakata model, for example[3] with all *strong interactions* ascribed to a neutral *vector meson field* interacting symmetrically with the three particles. Within such a framework, the *electromagnetic current* (in units of e) is just

$$i\{2/3 \; u^- \gamma_\alpha u - 1/3 \; d^- \gamma_\alpha d - 1/3 \; s^- \gamma_\alpha s\}$$

or $\mathscr{F}_{3\alpha} + \mathscr{F}_{8\alpha}/\sqrt{3}$ in the notation of ref. 3. For the *weak current*, we can take over from the Sakata model the form suggested by Gell-Mann and Levy[7],

[7] Gell-Mann, M. & Levy, M. (1960). *Nuovo Cimento*, 16, 705.

namely i p⁻ $\gamma_\alpha(1 + \gamma_5)$ (n cos θ + Δ sin θ), which gives in the *quark scheme* the expression ***

i u⁻ $\gamma_\alpha(1 + \gamma_5)$ (d cos θ + s sin θ)

or, in the notation of ref. 3,

$$[\mathscr{F}_{1\alpha} + \mathscr{F}_{1\alpha}{}^5 + i\,(\mathscr{F}_{2\alpha} + \mathscr{F}_{2\alpha}{}^5)]\cos\theta + [\mathscr{F}_{4\alpha} + \mathscr{F}_{4\alpha}{}^5 + i\,(\mathscr{F}_{5\alpha} + \mathscr{F}_{5\alpha}{}^5)]\sin\theta.$$

*** The parallel with i v⁻$_e$ $\gamma_\alpha(1 + \gamma_5)$ e and i v⁻$_\mu$ $\gamma_\alpha(1 + \gamma_5)$ μ is obvious. Likewise, in the model with d⁻, s⁻, u°, and b° discussed above, we would take the *weak current* to be i(b⁻° cos θ + u⁻° sin θ) $\gamma_\alpha(1 + \gamma_5)$ s⁻ + i(u⁻° cos θ – b⁻° sin θ) $\gamma_\alpha(1 + \gamma_5)$ d⁻. The part with $\Delta(n_t - n_{\bar{t}}) = 0$ is just i u⁻° $\gamma_\alpha(1 + \gamma_5)$ (d⁻ cos θ – s⁻° sin θ).

We thus obtain all the features of Cabibbo's picture[8]

[8] Cabibbo, N. (1963). *Phys. Rev. Lett.*, 10, 531.

of the *weak current*, namely the rules $|\Delta I| = 1$, $\Delta Y = 0$ and $|\Delta I| = \frac{1}{2}$, $|\Delta Y/\Delta Q = +1$, the conserved $\Delta Y = 0$ *current* with coefficient cos θ, the *vector current* in general as a component of the *current* of the F-spin, and the *axial vector current* transforming under SU(3) as the same component of another *octet*. Furthermore, we have[3] the equal-time commutation rules for the fourth components of the *currents*:

$$[\mathscr{F}_{j4}(x) \pm \mathscr{F}_{j4}{}^5(x),\ \mathscr{F}_{k4}(x') \pm \mathscr{F}_{k4}{}^5(x')] = -\,2f_{jkl}\,[\mathscr{F}_{k4}(x) \pm \mathscr{F}_{k4}{}^5(x)]\,\delta(x - x'),$$
$$[\mathscr{F}_{j4}(x) \pm \mathscr{F}_{j4}{}^5(x),\ \mathscr{F}_{k4}(x') \pm \mathscr{F}_{k4}{}^5(x')] = 0,$$

i = 1, ... 8, yielding the group SU(3) × SU(3). We can also look at the behavior of the *energy density* $\theta_{44}(x)$ (in the gravitational interaction) under equal-time commutation with the *operators* $\mathscr{F}_{j4}(x') \pm \mathscr{F}_{j4}{}^5(x')$. That part which is non-invariant under the group will transform like particular *representations* of SU(3) × SU(3), for example like (3, 3⁻) and (3⁻, 3) if it comes just from the *masses* of the *quarks*.

All these relations can now be abstracted from the field theory model and used in a *dispersion theory* treatment. The *scattering amplitudes* for *strongly interacting particles* on the *mass shell* are assumed known; there is then a system of linear *dispersion relations* for the matrix elements of the *weak currents* (and also the *electromagnetic* and *gravitational interactions*) to lowest order in these *interactions*. These *dispersion relations*, unsubtracted and supplemented by the non-linear *commutation rules* abstracted from the *field theory*, may be powerful enough to determine all the matrix elements of the *weak currents*, including the effective strengths of the *axial vector current* matrix elements compared with those of the *vector current*.

It is fun to speculate about the way *quarks* would behave if they were *physical particles* of finite *mass* (instead of purely mathematical entities as they would be in the limit of infinite mass). Since *charge* and *baryon number* are exactly conserved, one of the *quarks* (presumably $u^{2/3}$ or $d^{-1/3}$) would be absolutely stable[*] while the other member of the *doublet* would go into the first member very slowly by β-*decay* or K-*capture*.

> [*] There is the alternative possibility that the quarks are unstable under decay into *baryon* plus *anti-di-quark* or *anti-baryon* plus *quadri-quark*. In any case, some particle of fractional *charge* would have to be absolutely stable.

The *isotopic singlet quark* would presumably decay into the *doublet* by *weak interactions*, much as Δ goes into N. Ordinary *matter* near the earth's surface would be contaminated by stable *quarks* as a result of high energy *cosmic ray* events throughout the earth's history, but the contamination is estimated to be so small that it would never have been detected. A search for stable *quarks* of charge − 1/3 or + 2/3 and/or stable *di-quarks* of charge − 2/3 or + 1/3 or + 4/3 at the highest energy accelerators would help to reassure us of the non-existence of real *quarks*.

These ideas were developed during a visit to Columbia University in March 1963; the author would like to thank Professor Robert Serber for stimulating them.

Murray Gell-Mann – 1969 Nobel Prize in Physics. *Presentation Speech.*

[The 1969 Nobel Prize in Physics was awarded to Murray Gell-Mann "for his contributions and discoveries concerning the classification of *elementary particles* and their *interactions*".

During the 1950s and 1960s, new accelerators and apparatuses helped identify many new *elementary particles*. In theoretical works from the same period, Murray Gell-Mann classified *particles* and their *interactions*. He proposed that observed particles are in fact composite, that is, comprised of smaller building blocks called *quarks*. According to this theory, as-yet-undiscovered particles should exist. When these were later found in experiments, the theory was accepted. [Murray Gell-Mann – Facts. NobelPrize.org. https://www.nobelprize. org/prizes/physics/1969/gell-mann/facts/.]

Gellman's Nobel Prize lecture was not submitted by Gellman and consequently was not published. The Award ceremony speech for his prize is provided in its place.

———————

Presentation Speech by Professor Ivar Waller, member of the Nobel Committee for Physics

"Your Majesty, Your Royal Highnesses, Ladies and Gentlemen.

Elementary particle physics which is now so vigorous was still in its infancy when Murray Gell-Mann in 1953 published the first of the papers which have been honored with this years' Nobel Prize in physics.

The physicists were, however, already then acquainted with a rather large number of particles which apparently were indivisible and therefore elementary building stones of all matter. The *elementary particle* known for the longest time was the *electron*.

New particles were added when the *atomic nuclei* were explored. It was found that the *atomic nuclei* consist of positively charged *protons* and electrically neutral *neutrons*. These particles are held together in the atomic nuclei by enormously *strong forces* called *nuclear forces* which do not distinguish between *protons* and *neutrons*. *This symmetry of the nuclear forces was expressed by saying that the nuclear forces are charge-independent.* A *proton* and a *neutron* have further very nearly the same *mass*. They form a *doublet* of particles and have been given the common name of *nucleons*.

An increase already expected and desired occurred in the family of *elementary particles* at the end of the 1940's, when new particles called *pi-mesons* were discovered. They were named *mesons* because they have a *mass between the electron and the nucleon masses*. The *pi-mesons* had been predicted by the Japanese physicist Yukawa. They form a *triplet* of particles having nearly the *same mass but different charges* which are + 1, 0 and –1 in units of the *proton charge*. Their *interaction* with the *nucleons* is strong and *charge*-independent. Their most important task is to be an intermediary agent for the *strong interactions* between the *nucleons*.

A very remarkable discovery which marked a new area in particle physics was made by the British physicists Rochester and Butler about the same time. [Rochester, G. D. & Butler, C. C. (1947). Evidence for the Existence of New Unstable Elementary Particles. *Nature*. 160, 4077, 855–7; https://doi.org/10.1038/160855a0. Discovery of *kaon* (*K meson*), the first *strange* particle.] They found new unstable particles which did not fit in with the theoretical ideas developed so far. Some of the new particles are heavier than the *nucleons* and were grouped together with them under the common name of *baryons*. The others were lighter than the *nucleons* but heavier than the *electrons* and were called *K-mesons*. The new particles were copiously produced when high-energy *pi-mesons* collide with *nucleons* and were therefore assumed to interact strongly with other particles. But they had such a long lifetime that some law must exist which prevent the strong forces to act when they disintegrate into other particles. Gell-Mann discovered this law after some preliminary results had been found by Pais.

It had been assumed earlier that the new *baryons* form *doublets* like the *nucleons* and that the *K-mesons* form *triplets* like the *pi-mesons*. Gell-Mann made the fundamental new assumption that *the new baryons instead form a singlet, a triplet and a doublet, the latter being different from the nucleon doublet*, and that *the new mesons form two kinds of doublets, one consisting of the antiparticles of the other*. Gell-Mann assumed further that *the principle of charge-independence was generally valid for strong interactions*. He could thereby explain the mysterious properties of the new particles. He introduced a new fundamental characteristic of a *multiplet* called its *hypercharge*. This is defined as twice the mean value of the *charges* in the *multiplet*. Gell-Mann's proposed the new rule: *Elementary particles can be transformed in others by the strong and the electromagnetic interactions only if the total hypercharge is conserved*. This rule reminds of the *law of conservation of the electric charge*. It should be remarked that Gell-Mann initially used instead of the *hypercharge* a closely related number called the *strangeness*.

This discovery by Gell-Mann was admirable considering in particular the very meagre experimental material available to him. In the predicted *baryon multiplets* there occurred

319

empty places. Gell-Mann could on this ground predict two new *baryons*. One of them was soon discovered but the other not until six years later.

This classification of the elementary particles and their interaction discovered by Gell-Mann has turned out to be applicable to all strongly interacting particles found later and these are practically all particles discovered after 1953. His discovery is therefore fundamental in *elementary particle* physics.

It should be added that two Japanese physicists, Nakano and Nishijima, published a similar classification some months later than Gell-Mann.

Many theoretical physicists tried during the following years to find *new symmetries* which should give relations between the particle *multiplets*. Initiated by Sakata a series of papers were published in particular by Japanese physicists. They indicated that a certain kind of *symmetry* could be of interest. *Gell-Mann showed in a new fundamentally important paper of 1961 that this symmetry which had since long been studied in pure mathematics could be used for the classification of all strongly interacting particles.* [Gell-Mann, M. (March, 1961). The Eightfold Way: A Theory of Strong Interaction Symmetry (Report). See above.] Assuming the validity of the new *symmetry* which includes the *symmetry* corresponding to *charge-independence*, Gell-Mann found that *his earlier multiplets could be brought together into larger groups called supermultiplets each containing all baryons or all mesons which have the same spin and the same parity*, i. e. have the same measure for their rotation around their axes and are transformed in the same way by reflections. Gell-Mann called this classification "*The Eightfold Way*". The nucleons were found to belong to a *supermultiplet* of eight particles i.e. an *octet*. For the *mesons* an *octet* was proposed where the pi- and K-*mesons* filled seven places. Because one place was empty a new *meson* was predicted. Its existence had been suspected already by some of the Japanese physicists mentioned above. It was soon discovered which meant that Gell-Mann's theory was strongly supported. Still more famous is Gell-Mann's prediction in 1962 of a new *baryon* called *omega minus*.

A similar classification was proposed by Y. Néeman somewhat later than Gell-Mann.

Gell-Mann has also found that "*The Eightfold Way*" can be described very simply *by assuming that all particles which interact strongly with each other are composed of only three kinds of particles which he called quarks and of the corresponding antiparticles.* [Gell-Mann. M. (February, 1964). A Schematic Model of Baryons and Mesons. See above.] The *quarks* are peculiar in particular because their *charges are fractions of the proton charge* which according to all experience up to now is the indivisible *elementary charge*. It has not yet been possible to find individual *quarks* although they have been eagerly looked for. Gell-Mann's idea is none the less of great heuristic value.

And interesting application of *"The Eightfold Way"* is the so-called *current algebra* which was founded by Gell-Mann. It has e.g. made evident that there are important connections between the different kinds of *elementary particle interactions*.

Gell-Mann has given many fundamental contributions to the theory of *elementary particles* besides those which have been mentioned here. He has during more than a decade been considered as the leading scientist in this field.

Professor Gell-Mann. You have given fundamental contributions to our knowledge of *mesons* and *baryons* and their interactions. You have developed new algebraic methods which have led to a far-reaching classification of these particles according to their *symmetry properties*. The methods introduced by you are among the most powerful tools for further research in particle physics.

On behalf of the Royal Swedish Academy of Science, I congratulate you on your successful work and ask you to receive your Nobel Prize from the hands of His Majesty the King."

[From (1972). *Nobel Lectures, Physics 1963-1970*, Elsevier Publishing Company, Amsterdam.]

James Watson Cronin (September 29, 1931–August 25, 2016).

Cronin was an American particle physicist. He and co-researcher Val Logsdon Fitch were awarded the 1980 Nobel Prize in Physics for a 1964 experiment that proved that certain subatomic reactions do not adhere to fundamental *symmetry principles*. Specifically, they proved, by examining the decay of *kaons*, that a reaction run in reverse does not merely retrace the path of the original reaction, which showed that *the interactions of subatomic particles are not invariant under time reversal*. Thus, the phenomenon of *CP violation* was discovered.

> [*CP violation* is a violation of *CP-symmetry* (or *charge conjugation parity symmetry*): the combination of *C-symmetry* (*charge conjugation symmetry*) and *P-symmetry* (*parity symmetry*). *CP-symmetry* states that the laws of physics should be the same if a particle is interchanged with its *antiparticle* (*C-symmetry*) while its *spatial coordinates* are inverted ("mirror" or *P-symmetry*).
>
> *A parity transformation* (also called *parity inversion)* is the flip in the sign of one spatial coordinate. *Charge conjugation* is a transformation that switches all particles with their corresponding *antiparticles*, thus changing the sign of all *charges*: not only *electric charge* but also the *charges* relevant to other forces.]

James Cronin was born in Chicago on September 29, 1931. His father, James Farley Cronin, was a graduate student of classical languages at the University of Chicago. After his father had obtained his doctorate, the family first moved to Alabama, and later in 1939 to Dallas, Texas, where his father became a professor of Latin and Greek at Southern Methodist University. After high school Cronin stayed in Dallas and obtained an undergraduate degree at SMU in physics and mathematics in 1951. He is of Irish descent, with his Irish ancestors immigrating from County Cork, Ireland.

For graduate school Cronin moved back to Illinois to attend the University of Chicago. His teachers there included Nobel Prize laureates Enrico Fermi, Maria Mayer, Murray Gell-Mann and Subrahmanyan Chandrasekhar.

After obtaining his doctorate in 1955, Cronin joined the group of Rodney L. Cool and Oreste Piccioni at Brookhaven National Laboratory, where the new Cosmotron particle accelerator had just been completed. There he started to study *parity violation* in the decay of *hyperon* particles.

> [*Hyperon* [*classification*]
> A *hyperon* is any *baryon* containing one or more *strange quarks*, but no *charm*, *bottom*, or *top quark*. Being *baryons*, all *hyperons* are *fermions*. That is, they have

half-integer *spin* and obey Fermi–Dirac statistics. *Hyperons* all interact via the *strong nuclear force*, making them types of *hadron*. They are composed of three light *quarks*, at least one of which is a *strange quark*, which makes them *strange baryons*.

This form of matter may exist in a stable form within the core of some neutron stars. The first research into *hyperons* happened in the 1950s and spurred physicists on to the creation of an organized classification of particles. The term was coined by French physicist Louis Leprince-Ringuet in 1953.

Today, research in this area is carried out on data taken at many facilities around the world, including CERN, Fermilab, SLAC, JLAB, Brookhaven National Laboratory, KEK, GSI and others. Physics topics include searches for *CP violation*, measurements of *spin*, studies of *excited states* (commonly referred to as spectroscopy), and hunts for exotic forms such as *pentaquarks* and *dibaryons*.]

During that time, he also met Val Fitch, who brought him to Princeton University in Fall 1958. After the Cosmotron underwent magnet failure, Cronin and the Brookhaven group moved to Bevatron at the University of California, Berkeley during the first half of 1958. Cronin and Fitch studied the decays of neutral K *mesons*, in which they discovered *CP violation* in 1964. This discovery earned the duo the 1980 Nobel Prize in Physics.

After the discovery, Cronin spent a year in France at the Centre d'Études Nucléaires at Saclay. After returning to Princeton, he continued studying the neutral *CP violating decay modes* of the long-lived neutral K *meson*. In 1971, he moved back to the University of Chicago to become a full professor. This was attractive for him because of a new 400 GeV particle accelerator being built at nearby Fermilab.

When he moved to Chicago, he began a long series of experiments on particle production at high transverse momentum. Following these experiments Cronin took a sabbatical at CERN in 1982–83, where he performed an experiment to measure of the lifetime of the neutral *pion*. He then switched to the study of *cosmic rays*. The first was a series of measurements looking for point sources of *cosmic rays*. No sources were found.

In 1998 he joined the faculty at the University of Utah on a half-time basis to work on ultra-high-energy *cosmic ray* physics and to jumpstart the Pierre Auger Observatory project. His appointment was to last five years, but he left after a year to continue gathering international support for the Observatory with Alan Watson and Murat Boratav.

Cronin was Professor Emeritus at the University of Chicago.

Val Logsdon Fitch (March 10, 1923–February 5, 2015).

Fitch was an American nuclear physicist who, with co-researcher James Cronin, was awarded the 1980 Nobel Prize in Physics for a 1964 experiment using the Alternating Gradient Synchrotron at Brookhaven National Laboratory that proved that certain subatomic reactions do not adhere to fundamental *symmetry principles*. Specifically, they proved, by examining the decay of K-*mesons*, that a reaction run in reverse does not retrace the path of the original reaction, which showed that the reactions of subatomic particles are not indifferent to time. Thus, the phenomenon of *CP violation* was discovered. This demolished the faith that physicists had that natural laws were governed by *symmetry*.

Fitch was born on a cattle ranch near Merriman, Nebraska, on March 10, 1923, the youngest of three children of Fred Fitch, a cattle rancher, and his wife Frances née Logsdon, a school teacher. He had an older brother and sister. The family farm was about 4 square miles (10 km2) in size. The ranch was small; his father specialized in raising breeding stock. Soon after his birth, his father was badly injured in a horse-riding accident and could no longer work on his ranch, so the family moved to the nearby town of Gordon, Nebraska, where his father entered the insurance business. Here he attended school, graduating from Gordon High School in 1940 as valedictorian.

Fitch attended Chadron State College for three years, then transferred to Northwestern University. This was during WWII; his studies were interrupted by being drafted into the US Army in 1943. After completing basic training, he was sent to Carnegie Institute of Technology for training under the Army Specialized Training Program. Under this program, some 200,000 soldiers attended colleges for intensive courses. Fitch was in the program for less than a year before the manpower requirements of the war became too great, and the Army terminated the program. Most of the soldiers in the ASTP were posted to combat units, but Fitch was one of a hundred or so ASTP soldiers who joined the Special Engineer Detachment (SED), which provided much-needed technicians to the Manhattan Project.

The Army sent Fitch to the Manhattan Project's Los Alamos Laboratory in New Mexico. By mid-1944, about a third of the technicians at Los Alamos were from the SED. There he met many of the greats of physics including Niels Bohr, James Chadwick, Enrico Fermi, Isidor Isaac Rabi, Bruno Rossi, Emilio Segrè, Edward Teller and Richard C. Tolman, in some cases attending physics courses taught by them. He worked in the group headed by Ernest Titterton, a member of the British Mission, and became well-acquainted with the techniques of experimental physics. He participated in the drop testing of mock atomic bombs that was conducted at Wendover Army Air Field and the Naval Auxiliary Air Station Salton Sea, and worked at the Trinity site, where he witnessed the Trinity nuclear test on July 16, 1945. He was discharged from the Army in 1946. He continued to work at Los Alamos as a civilian for another year to earn money. He briefly returned to Los Alamos in summer 1948.

His wartime experiences led Fitch to decide to become a physicist. Robert Bacher, the head of the physics division at Los Alamos, offered him a graduate assistantship at Cornell University, but first he needed to complete his undergraduate degree. Rather than return to Northwestern or Carnegie Mellon, he elected to enter McGill University, which Titterton had recommended. Fitch graduated from McGill with a bachelor's degree in electrical engineering in 1948. On the advice of Jerry Kellogg, who had been a student of Rabi's at Columbia University, and was a division head at the Los Alamos, Fitch decided to pursue his doctoral studies at Columbia. Kellogg wrote him a letter of introduction to Rabi. James Rainwater became his academic supervisor. Rainwater gave him a paper by John Wheeler concerning *mu-mesic atoms*, atoms in which an *electron* is replaced by a *muon*. These had never been observed; they were completely theoretical and there was no evidence that they existed, but it made a good thesis topic.

Fitch designed and built an experiment to measure the gamma rays emitted from mumesic atoms. As it turned out, this was a good time to search for them. Columbia had recently commissioned a cyclotron at the Nevis Laboratories that could produce muons; Robert Hofstadter had developed the thallium-activated sodium iodide gamma ray detector; and wartime advances in electronics yielded advances in components such as new phototubes needed to bring it all together. Initially nothing was found, but Rainwater suggested expanding the search beyond the energy range predicted by Wheeler on the basis of the then-accepted size of the radius of the atomic nucleus as around $1.4 \times 10{-}15$ m. When this was done, they found what they had been looking for, discovering in the process that the nucleus was closer to $1.2 \times 10{-}15$ m. He completed his PhD in 1954, writing his thesis on "Studies of X-rays from mu-mesonic atoms".

In 1949, Fitch married Elise Cunningham, a secretary who worked in the laboratory at Columbia. They had two sons. Elise died in 1972, and in 1976 he married Daisy Harper Sharp, thereby acquiring two stepdaughters and a stepson.

After obtaining his doctorate, Fitch's interest shifted to *strange particles* and K *mesons*. In 1954, he joined the physics faculty at Princeton University, where he spent the rest of his career until his retirement in 2005.

Fitch conducted much of his research at the Brookhaven National Laboratory, where he became acquainted with James Cronin. The two of them played bridge at nights while they waited for the Cosmotron to become available. Cronin had built a new kind of detector, a *spark chamber spectrometer*, and Fitch realized that it would be perfect for experiments with K *mesons* (also known as *kaons*), which Yale University physicist Robert Adair had suggested had interesting properties worth investigating. They could decay into either *matter* or *antimatter*.

Along with two colleagues, James Christenson and René Turlay, they set up their experiment on the Alternating Gradient Synchrotron at Brookhaven. They discovered an unexpected result. The decay of neutral K *mesons* did not respect CP *symmetry*. K *mesons* that decayed into *positrons* did so faster than those that decayed into *electrons*. The importance of this result was not immediately appreciated; but as evidence of the Big Bang accumulated, Andrei Sakharov realized in 1967 that it explained why the universe is largely made of *matter* and not *antimatter*. Put simply, they had found "the answer to the physicist's 'Why do we exist?'" For this discovery, Fitch and Cronin received the 1980 Nobel Prize in Physics.

He died at his home in Princeton, New Jersey, at the age of 91 on February 5, 2015.

James W. Cronin and Val L. Fitch - the 1980 Nobel Prize in Physics.

14 October 1980

Demonstration in 1964 using *neutral K-mesons* of the violation of *all three symmetry principles* (1) that the laws of Nature are exactly alike for both *antimatter* and ordinary *matter*; (2) that the fundamental laws have exact *mirror symmetry*; and (3) that the fundamental laws have exact *time reflection symmetry* – symmetry under motion reversal. [Christenson, J. H., Cronin, J. W., Fitch, V. L. & Turlay, R. (July, 1964). Evidence for the 2π Decay of the K20 Meson. *Phys. Rev. Lett.*, 13, 4, 138–40; https://doi.org/10.1103/PhysRevLett.13.138.]

[The *Standard Model* of particle physics has three related *natural near-symmetries*. These state that the universe in which we live should be indistinguishable from one where a certain type of change is introduced.

Charge conjugation symmetry (*C-symmetry*), a universe where every *particle* is replaced with its *antiparticle*

Parity symmetry (*P-symmetry*), a universe where everything is *mirrored* along the three physical axes, known as *chirality*.

Time reversal symmetry (*T-symmetry*), a universe where the direction of time is reversed. *T-symmetry* is counterintuitive (the future and the past are not symmetrical) but explained by the fact that the *Standard Model* describes *local properties*, not *global* ones like entropy.

These symmetries are *near-symmetries* because each is broken in the present-day universe. However, the *Standard Model* predicts that the combination of the three (that is, the simultaneous application of all three transformations) must be a *symmetry*, called *CPT symmetry*. *CP violation*, the violation of the combination of *C- and P-symmetry*, is necessary for the presence of significant amounts of *baryonic* matter in the universe.]

The Royal Swedish Academy of Sciences has decided to award the 1980 Nobel Prize in Physics to Professor James W. Cronin, University of Chicago, USA and Professor Val L. Fitch, Princeton University, USA, *"for the discovery of violations of fundamental symmetry principles in the decay of neutral K-mesons"*.

Symmetries play a great role in many sciences and also in many other areas. This year's Nobel Prize in Physics is awarded for an unexpected discovery in an experiment devoted to a critical scrutiny of the validity of three related *symmetry principles*. These are of importance to the formulation of fundamental laws of Nature.

The Experiment

The discovery was made at Brookhaven National Laboratory by a research group led by James Cronin and Val Fitch who also initiated the search. Using the *proton* accelerator AGS a beam of neutral elementary particles was produced. Their radioactive decay in flight was recorded and measured with great precision. The specially designed detector arrangement was large and complicated. All the difficulties encountered in the analyses of the data were overcome in a skillful and convincing way. The type of neutral *K-mesons* which Cronin and Fitch chose to study are remarkable since they can be regarded to consist of one-half ordinary *matter* and the other half *antimatter*.

Three symmetry principles

One of the three *symmetry principles* says that the laws of Nature are exactly alike for both *antimatter* and ordinary *matter*. The neutral K-*mesons* are the most suitable test bodies for a critical and sensitive test of the validity of this principle, which was shown by Cronin and Fitch. The other two *symmetry principles* state that the fundamental laws have exact *mirror symmetry* and *time reflection symmetry* – by the latter is understood *symmetry* under motion reversal.

The situation before the prize-winning discovery

Complete *symmetry* is valid for the laws which describe *electric* and *magnetic* phenomena, which encompass most things in our daily lives. This is true of all three *symmetries*. They are also respected by *gravitation* and by *strong interactions* (= a force between elementary-particles). On the other hand, there is a maximal *lack of left-right symmetry*, i.e. *mirror symmetry* in one type of physical processes – the *radioactive decays*. It was understood by T.D. Lee and C.N. Yang, Nobel Prize-winners in 1957, that the *violation of the symmetry* was deeply rooted in the very *law of weak interactions*, which cause the radioactive decays and related processes. The almost self-evident statement which ceased to be valid in 1957, says that the *mirror image* of a physical process is always a possible physical process.

However, already in 1957, the conclusion could not be avoided that Nature makes an absolute distinction between *left and right*. Nor did the radioactive processes show *complete symmetry* between *matter* and *antimatter*. *One lack of symmetry was cancelled by the other in a complete and elegant way.* Thus, the *mirror image* of a physical process in our world is always a possible process in the *anti-world* and vice versa. If the universe consists also of *antimatter*, possible inhabitants on another planet could not then by themselves determine if they consist of the one or the other type of atoms.

Symmetry by time reversal

The conclusion in 1957 that *the two symmetry violations cancelled each other* was highly satisfactory since it allowed the third *symmetry principle* to keep its validity. This principle says that the fundamental *laws do not change when all motions are reversed.* Such *symmetry by time reversal* is in fact valid for all processes governed by *electromagnetic forces.* It is therefore, a cornerstone in physics and chemistry. The *symmetry* is also valid for processes controlled by *gravitational* and *strong forces. Due to the mutual cancellation of the symmetry violations in weak processes* we could consider *time reversal symmetry to continue to be generally valid.* This became a new cornerstone.

Consequences of the prize-winning discovery

The result of the prize-winning work showed for the first, time that the left-right asymmetry is not always completely compensated by transforming from matter to antimatter. This result has been verified in several similar experiments in other laboratories and by other research groups. This led to a situation in which the new cornerstone was overthrown. All attempts have been unsuccessful to avoid such a radically new conclusion as that which says that *perfect symmetry by time reversal is not always true.* The new knowledge permits us to make a distinction between *matter* and *antimatter* in an absolute and not only relative way. The *left and right* directions could then also be given absolute meaning, thus losing the arbitrariness of definition.

The search for the deeper causes of the *symmetry violations* discovered in the experiment by Cronin and Fitch is actively pursued at present. The progress in elementary particle physics during recent years has created new interesting possibilities.

The new truth reached by the discovery has recently also been incorporated as an important ingredient in cosmological speculations The aim has been to try to understand how a universe, originally very hot and symmetric, could avoid that *matter* and *antimatter* almost immediately annihilated each other. In other words, efforts have been made to describe how the matter we are made of was once created in a Big Bang and how it could survive the birth pains.

329

The discovery emphasizes, once again, that even almost self-evident principles in science cannot be regarded fully valid until they have been critically examined in precise experiments. [NobelPrize.org. https://www.nobelprize.org/prizes/physics/1980/press-release/.]

François Englert (born November 6, 1932).

Englert is a Belgian theoretical physicist and 2013 Nobel Prize laureate.

Englert is professor emeritus at the Université Libre de Bruxelles (ULB), where he is a member of the Service de Physique Théorique. He is also a Sackler Professor by Special Appointment in the School of Physics and Astronomy at Tel Aviv University and a member of the Institute for Quantum Studies at Chapman University in California.

Englert has made contributions in statistical physics, quantum field theory, cosmology, string theory and supergravity.

Englert was awarded the 2013 Nobel Prize in Physics, together with Peter Higgs for the discovery of the *Brout–Englert–Higgs* (BEH) *mechanism*.

François Englert is a Holocaust survivor. He was born in a Belgian Jewish family. During the German occupation of Belgium in World War II, he had to conceal his Jewish identity and live in orphanages and children's homes in the towns of Dinant, Lustin, Stoumont and, finally, Annevoie-Rouillon. These towns were eventually liberated by the US Army.

He graduated as an electromechanical engineer in 1955 from the Free University of Brussels (ULB) where he received his PhD in physical sciences in 1959. From 1959 until 1961, he worked at Cornell University, first as a research associate of Robert Brout and then as assistant professor. He then returned to the ULB, where he became a university professor and was joined there by Robert Brout who, in 1980, with Englert co-headed the theoretical physics group. In 1998 Englert became professor emeritus. In 1984 Englert was first appointed as a Sackler Professor by Special Appointment in the School of Physics and Astronomy at Tel-Aviv University. Englert joined Chapman University's Institute for Quantum Studies in 2011, where he serves as a distinguished visiting professor.

Brout and Englert showed in 1964 that *gauge vector fields*, abelian and non-abelian, could acquire *mass* if empty space were endowed with a particular type of structure that one encounters in material systems. [Englert, F. & Brout, R. (August, 1964). Broken Symmetry and the Mass of Gauge Vector Mesons. See below.]

Focusing on the failure of the *Goldstone theorem* for *gauge fields*, later that year, Higgs reached essentially the same result. [Higgs, P. W. (September, 1964). Broken symmetries, massless particles and gauge fields. *Physics Letters.*, 12, 2, 132–3; https://doi.org/10.1016/0031-9163(64)91136-9; (October, 1964). Broken Symmetries and the Masses of Gauge Bosons. See below.]

A third paper on the subject was published later in the same year by Gerald Guralnik, C. R. Hagen, and Tom Kibble. [Guralnik, G., Hagen, C. & Kibble, T. (November, 1964). Global Conservation Laws and Massless Particles. *Phys. Rev. Lett.*, 13, 20, 585; doi:10.1103/PhysRevLett.13.585].

The three papers written on this *boson* discovery by Englert and Brout, Higgs, and Guralnik, Hagen, and Kibble were each recognized as milestone papers by *Physical Review Letters* 50th anniversary celebration. While each of these famous papers took similar approaches, the contributions and differences between the 1964 *Physical Review Letters* symmetry breaking papers is noteworthy. *The mechanism had been proposed in 1962 by Philip Anderson although he did not include a crucial relativistic model.*

To illustrate the structure, consider a *ferromagnet* which is composed of atoms each equipped with a tiny magnet. When these magnets are lined up, the inside of the *ferromagnet* bears a strong analogy to the way empty space can be structured. *Gauge vector fields* that are sensitive to this structure of empty space can only propagate over a finite distance. Thus, *they mediate short range interactions and acquire mass.* Those fields that are not sensitive to the structure propagate unhindered. They remain *massless* and are responsible for the *long-range interactions.* In this way, the mechanism accommodates within a single unified theory both short and long-range interactions.

Brout and Englert, Higgs, and Gerald Guralnik, C. R. Hagen, and Tom Kibble introduced as agent of the *vacuum structure* a *scalar field* (subsequently referred to as the *Higgs field*) which many physicists view as the agent responsible for the *masses* of fundamental particles. Brout and Englert also showed that the mechanism may remain valid if the *scalar field* is replaced by a more structured agent such as a *fermion condensate.* Their approach led them to conjecture that the theory is *renormalizable.* The eventual proof of *renormalizability,* is due to Gerardus 't Hooft and Martinus Veltman who were awarded the 1999 Nobel Prize for this work. The *Brout–Englert–Higgs* (BEH) *mechanism* is the building stone of the *electroweak theory* of *elementary particles* and laid the foundation of a unified view of the basic laws of nature.

Englert, F. & Brout, R. (August, 1964). Broken Symmetry and the Mass of Gauge Vector Mesons*.

Phys. Rev. Lett., 13, 9, 321–3; https://journals.aps.org/prl/pdf/10.1103/PhysRevLett.13.321.

Faculte des Sciences, Universite Libre de Bruxelles, Bruxelles, Belgium.

* This work has been supported in part by the U. S. Air Force under grant No. AFEOAR 63-51 and monitored by the European Office of Aerospace Research.

Received June 26, 1964.

Brout and Englert showed that *gauge vector fields*, abelian and non-abelian, could acquire *mass* if empty space were endowed with a particular type of structure that one encounters in material systems. Other physicists, Peter Higgs and Gerald Guralnik, C. R. Hagen and Tom Kibble had reached similar conclusions at about the same time. The *Brout–Englert–Higgs* (BEH) *mechanism* is believed to give rise to the *masses* of all the elementary particles in the *Standard Model*. This includes the *masses* of the W and Z *bosons*, and the *masses* of the *fermions*, i.e. the *quarks* and *leptons*.

It is of interest to inquire whether *gauge vector mesons* acquire *mass* through *interaction*[1]; by a *gauge vector meson* we mean a *Yang-Mills field*[2] associated with the extension of a Lie group from *global* to *local symmetry*.

[1] Schwinger, J. (January, 1962). Gauge Invariance and Mass. *Phys. Rev.*, 125, 397
[2] Yang, C. N. & Mills, R. (October, 1954). Conservation of Isotopic Spin and Isotopic Gauge Invariance. *Phys. Rev.*, 96, 1, 191–5. See above.

The importance of this problem resides in the possibility that *strong-interaction* physics originates from massive *gauge fields* related to a system of *conserved currents*[3].

[3] Sakurai, J. J. (September, 1960). Theory of strong interactions. *Ann. Phys.*, 11, 1, 1-48. See above.

In this note, we shall show that in certain cases *vector mesons* do indeed acquire *mass* when the vacuum is degenerate with respect to a compact Lie group.

Theories with degenerate vacuum (*broken symmetry*) have been the subject of intensive study since their inception by Nambu[4-6].

[4] Nambu, Y. (April, 1960). Axial Vector Current Conservation in Weak Interactions. *Phys. Rev. Lett.*, 4, 7, 380-2. See above.

[5] Nambu, Y. & Jona-Lasinio, G. (April, 1961). Dynamical Model of Elementary Particles Based on an Analogy with Superconductivity. I. *Phys. Rev.*, 122, 1, 345; https://doi.org/ 10.1103/PhysRev.122.345.

[6] "*Broken symmetry*" has been extensively discussed by various authors in the *Proceedings of the Seminar on Unified Theories of Elementary Particles,* University of Rochester, Rochester, New York, 1963 (unpublished).

A characteristic feature of such theories is the possible existence of zero-mass *bosons* which tend to restore the *symmetry*[7,8].

[7] Goldstone, J., Salam, A. & Weinberg, S. (August, 1962). Broken Symmetries. *Phys. Rev.* 127, 965. See above.

[8] Bludman, S. A. & Klein, A. (1963). *Phys. Rev.*, 131, 2364.

We shall show that it is precisely these singularities which maintain the *gauge invariance* of the theory, despite the fact that the *vector meson* acquires *mass*.

We shall first treat the case where the original fields are a set of *bosons* ψ_A which transform as a basis for a representation of a compact Lie group. This example should be considered as a rather general phenomenological model. As such, we shall not study the particular mechanism by which the *symmetry* is broken but simply assume that such a mechanism exists. A calculation performed in lowest order *perturbation theory* indicates that those *vector mesons* which are coupled to *currents* that "rotate" the original vacuum are the ones which acquire *mass* [see Eq. (6)].

We shall then examine a particular model based on *chirality invariance* which may have a more fundamental significance. Here we begin with a *chirality-invariant Lagrangian* and introduce both *vector* and *pseudovector gauge fields*, thereby guaranteeing invariance under both *local phase* and *local γ_5-phase* transformations. In this model the *gauge fields* themselves may break the γ_5 *invariance* leading to a *mass* for the original *Fermi field*. We shall show in this case that the *pseudovector field* acquires *mass*.

In the last paragraph we sketch a simple argument which renders these results reasonable.

(1)   Lest the simplicity of the argument be shrouded in a cloud of indices, we first consider a one-parameter *Abelian group*, representing, for example, the *phase transformation* of a *charged boson*; we then present the generalization to an arbitrary *compact Lie group*.

The *interaction* between the ψ and the A_μ *fields* is

$$H_{int} = ieA_\mu\varphi^*\overset{\leftrightarrow}{\partial}_\mu\varphi - e^2\varphi^*\varphi A_\mu A_\mu, \tag{1}$$

where $\varphi = (\varphi_1 + i\varphi_2)/\sqrt{2}$. We shall break the *symmetry* by fixing $\langle\varphi\rangle \neq 0$ in the *vacuum*, with the *phase* chosen for convenience such that $\langle\varphi\rangle = \langle\varphi^*\rangle = \langle\varphi_1\rangle/\sqrt{2}$.

We shall assume that the application of the theorem of Goldstone, Salam, and Weinberg[7] is straightforward and thus that the *propagator* of the *field* φ_2, which is "orthogonal" to φ_1 has a *pole* at $q = 0$ which is not isolated.

We calculate the *vacuum polarization loop* $II_{\mu\nu}$ for the *field* A_μ in lowest order *perturbation theory* about the self-consistent *vacuum*. We take into consideration only the *broken-symmetry* diagrams (Fig. I).

. . .

Fig. 1. Broken-symmetry diagram leading to a mass for the gauge field. ...

The conventional terms do not lead to a *mass* in this approximation if *gauge invariance* is carefully maintained. One evaluates directly

$$II_{\mu\nu}(q) = (2\pi)^4 ie^2[g_{\mu\nu}\langle\varphi_1\rangle^2 - (q_\mu q_\nu/q^2)\,\langle\varphi_1\rangle^2]. \tag{2}$$

Here we have used for the *propagator* of φ_2 the value $[i/(2\pi)^4]/q^2$; the fact that the *renormalization* constant is 1 is consistent with our approximation[9].

[9] Klein, A., reference 6.

We then note that Eq. (2) both maintains *gauge invariance* ($II_{\mu\nu}q_\nu = 0$) and causes the A_μ *field* to acquire a *mass*

$$\mu^2 = e2\langle\varphi_1\rangle^2 \tag{3}$$

We have not yet constructed a proof in arbitrary order; however, the similar appearance of higher order graphs leads one to surmise the general truth of the theorem.

Consider now, in general, a set of *boson-field operators* φ_A (which we may always choose to be Hermitian) and the associated *Yang-Mills field* $A_{a,\mu}$. The Lagrangian is *invariant* under the *transformation*[10]

[10] Utiyama, R. (1956). *Phys. Rev.*, 101, 1597.

$$\delta\varphi_A = \Sigma \dots \tag{4}$$
$$\delta A_{\alpha,\mu} = \Sigma \dots$$

where c_{abc} are the structure constants of a compact Lie group and $T_{a,AB}$ the *antisymmetric generators* of the group in the *representation* defined by the φ_B.

$\dots$

(2) Consider the *interaction Hamiltonian*

$$H_{int} = -\eta\bar\psi\gamma_\mu\gamma_5\psi B_\mu - \epsilon\bar\psi\gamma_\mu\psi A_\mu, \tag{7}$$

where A_μ and B_μ are *vector* and *pseudovector gauge fields*. The *vector field* causes attraction whereas the *pseudovector* leads to repulsion between *particle* and *antiparticle*. For a suitable choice of ϵ and η there exists, as in Johnson's model[11], a *broken-symmetry* solution corresponding to an arbitrary *mass* m for the ψ *field* fixing the scale of the problem.

[11] Johnson, K. A., reference 6.

Thus, the *fermion propagator* S(p) is

$$S^{-1}(p) = \gamma p - \Sigma(p) = \dots , \tag{8}$$

with

$\dots$

and

$\dots$.

We define the *gauge-invariant current* J_μ^5 by using Johnson's method[12]:

[12] Johnson, K. A., reference 6.

$$J_\mu^5 = \dots ,$$
$$\psi'(x) = \dots . \tag{9}$$

This gives for the *polarization tensor* of the *pseudovector field*

$$\Pi_{\mu v}^5(q) = \dots . \tag{10}$$

where the *vertex function* $\Gamma_{v5} = \gamma_v\gamma_5 + \Delta_{v5}$ satisfies the *Ward identity*[5]

$$\dots , \tag{11}$$

which for low q reads

$$qv\Gamma_{v5} = \ldots .\qquad(12)$$

The singularity in the longitudinal Γ_{v5} *vertex* due to the *broken-symmetry* term $2\Sigma_1\gamma_5$ in the *Ward identity* leads to a nonvanishing *gauge invariant* $\Pi_{\mu v}{}^5(q)$ in the limit $q\to0$, while the usual spurious "*photon mass*" drops because of the second term in (10). The *mass* of the *pseudo vector field* is roughly π^2m^2 as can be checked by inserting into (10) the lowest approximation for Γ_{v5} consistent with the *Ward identity*.

> [A *Ward–Takahashi identity* is an identity between correlation functions that follows from the *global* or *gauge symmetries* of the theory, and which remains valid after *renormalization*. The *Ward–Takahashi identity* of *quantum electrodynamics* was originally used by John Clive Ward and Yasushi Takahashi to relate the *wave function renormalization* of the *electron* to its *vertex renormalization factor*, guaranteeing the cancellation of the ultraviolet divergence to all orders of perturbation theory. Later uses include the extension of the proof of *Goldstone's theorem* to all orders of perturbation theory.]

Thus, in this case the general feature of the phenomenological *boson* system survives. We would like to emphasize that here the *symmetry* is broken through the *gauge fields* themselves. One might hope that such a feature is quite general and is possibly instrumental in the realization of Sakurai's program[3].

(3) We present below a simple argument which indicates why the *gauge vector field* need not have zero *mass* in the presence of *broken symmetry*. Let us recall that these fields were introduced in the first place in order to extend the *symmetry group* to transformations which were different at various *space-time points*. Thus, one expects that when the *group transformations* become homogeneous in *space-time*, that is $q\to0$, no dynamical manifestation of these *fields* should appear. This means that it should cost no *energy* to create a *Yang-Mills quantum* at $q=0$ and thus the *mass* is zero. However, if we break *gauge invariance of the first kind* and still maintain *gauge invariance of the second kind* this reasoning is obviously incorrect. Indeed, in Fig. 1, one sees that the A_μ *propagator* connects to intermediate *states*, which are "*rotated*" *vacua*. This is seen most clearly by writing $\langle\varphi_1\rangle = \langle[Q\varphi_2]\rangle$ where Q is the *group generator*. This effect cannot vanish in the limit $q=0$.

Peter Ware Higgs (29 May 1929 – 8 April 2024).

Higgs was a British theoretical physicist, professor at the University of Edinburgh, and 2013 Nobel laureate in Physics for his work on the mass of subatomic particles.

Higgs was born in the Elswick district of Newcastle upon Tyne, England, to Thomas Ware Higgs (1898–1962) and his wife Gertrude Maude née Coghill (1895–1969). His father worked as a sound engineer for the BBC, and as a result of childhood asthma, together with the family moving around because of his father's job and later World War II, Higgs missed some early schooling and was taught at home. When his father relocated to Bedford, Higgs stayed behind in Bristol with his mother, and was largely raised there. He attended Cotham Grammar School in Bristol from 1941 to 1946, where he was inspired by the work of one of the school's alumni, Paul Dirac, a founder of the field of quantum mechanics.

In 1946, at the age of 17, Higgs moved to City of London School, where he specialized in mathematics, then in 1947 to King's College London, where he graduated with a first-class honors degree in physics in 1950 and achieved a master's degree in 1952. He was awarded a Research Fellowship from the Royal Commission for the Exhibition of 1851, and performed his doctoral research in molecular physics under the supervision of Charles Coulson and Christopher Longuet-Higgins. He was awarded a PhD degree in 1954 with a thesis entitled "Some problems in the theory of molecular vibrations".

After finishing his doctorate, Higgs was appointed a Senior Research Fellow at the University of Edinburgh (1954–56). He then held various posts at Imperial College London, and University College London (where he also became a temporary lecturer in mathematics). He returned to the University of Edinburgh in 1960 to take up the post of Lecturer at the Tait Institute of Mathematical Physics, allowing him to settle in the city he had enjoyed while hitchhiking to the Western Highlands as a student in 1949.

In 1963, Higgs married Jody Williamson, an American lecturer in linguistics at Edinburgh and a fellow activist with the Campaign for Nuclear Disarmament (CND). Their first son was. They had two sons: Christopher, born in August 1965, and Jonny, a jazz musician. Higgs and Williamson separated in 1972 but remained friends until she died in 2008.

At Edinburgh, Higgs first became interested in *mass*, developing the idea that particles – massless when the universe began – acquired *mass* a fraction of a second later as a result of interacting with a theoretical field (subsequently referred to as the *Higgs field*). Higgs postulated that this field permeates space, giving *mass* to all *elementary subatomic particles* interacting with it.

The original basis of Higgs's work came from the Japanese-born theorist and Nobel Prize laureate Yoichiro Nambu from the University of Chicago. Nambu had proposed a theory known as *spontaneous symmetry breaking* based on what was known to happen in superconductivity in condensed matter, which incorrectly predicted *massless* particles.

Higgs reportedly developed the fundamentals of his theory after returning to his Edinburgh New Town apartment from a failed weekend camping trip to the Highlands. He stated that there was no "eureka moment" in the development of the theory. He wrote a short paper exploiting a loophole in *Goldstone's theorem* (massless Goldstone particles need not occur when *local symmetry* is spontaneously broken in a *relativistic theory*) and published it in *Physics Letters*, a European physics journal edited at CERN, in Switzerland, in 1964.

Higgs wrote a second paper describing a theoretical model (subsequently known as the *Higgs mechanism*), but the paper was rejected (the editors of *Physics Letters* judged it "of no obvious relevance to physics"). Higgs wrote an extra paragraph and sent his paper to *Physical Review Letters*, another leading physics journal, which published it later in 1964. [Higgs, P. W. (October, 1964). Broken Symmetries and the Masses of Gauge Bosons. See below.] This paper predicted a new massive *spin-zero boson* (later referred to as the *Higgs boson*).

Robert Brout and François Englert [Englert, F. & Brout, R. (August, 1964). Broken Symmetry and the Mass of Gauge Vector Mesons. See above.] and Gerald Guralnik, C. R. Hagen and Tom Kibble [Guralnik, G., Hagen, C. & Kibble, T. (November, 1964). Global Conservation Laws and Massless Particles. See below.] had reached similar conclusions at about the same time. In the published version, Higgs quotes Brout and Englert, and the third paper quotes the previous ones.

Higgs was the author of one of the three milestone papers published in *Physical Review Letters* that proposed *that spontaneous symmetry breaking in electroweak theory could explain the origin of mass of elementary particles in general and of the W and Z bosons in particular*. The *Brout–Englert–Higgs* [BEH] *mechanism* (also referred to as the *Higgs mechanism*) predicted the existence of a new particle, which became known as the *Higgs boson*, the detection of which became one of the great goals of physics.

The *Brout–Englert–Higgs* [BEH] *mechanism* is generally accepted as an important ingredient in the *Standard Model* of particle physics, without which certain particles would have no *mass*. It postulates the existence of the *Higgs field*, which confers *mass* on *quarks* and *leptons*; this causes only a tiny portion of the *masses* of other subatomic particles, such as *protons* and *neutrons*. In these, *gluons* that bind *quarks* together confer most of the particle *mass*.

Higgs was promoted to a personal chair of Theoretical Physics in 1980. On his retirement in 1996, he became an emeritus professor.

On 4 July 2012, CERN announced the ATLAS and Compact Muon Solenoid (CMS) experiments had seen strong indications for the presence of a new particle, which could be the Higgs boson, in the mass region around 126 gigaelectronvolts (GeV). Speaking at the seminar in Geneva, Higgs commented "It's really an incredible thing that it's happened in my lifetime." Ironically, this probable confirmation of the Higgs boson was made at the same place where the editor of *Physics Letters* rejected Higgs's paper.

On 8 October 2013, it was announced that Higgs and François Englert would share the 2013 Nobel Prize in Physics "for the theoretical discovery of a mechanism that contributes to our understanding of the origin of mass of subatomic particles, and which recently was confirmed through the discovery of the predicted fundamental particle, by the ATLAS and CMS experiments at CERN's Large Hadron Collider".

Higgs died after a short illness at home in Edinburgh on April 8, 2024, at the age of 94.

Higgs, P. W. (October, 1964). Broken Symmetries and the Masses of Gauge Bosons.

Phys. Rev. Lett., 13, 16, 508–9; https://journals.aps.org/prl/pdf/10.1103/PhysRevLett.13.508

Tait Institute of Mathematical Physics, University of Edinburgh, Edinburgh, Scotland

Received August 31, 1964.

In a recent note it was shown that the *Goldstone theorem*, that Lorentz-covariant field theories in which spontaneous breakdown of symmetry under an internal Lie group occurs contain zero-mass particles, *fails if and only if the conserved currents associated with the internal group are coupled to gauge fields*. The purpose of the present note is to report that, *as a consequence of this coupling, the spin-one quanta of some of the gauge fields acquire mass*; the longitudinal degrees of freedom of these particles (which would be absent if their *mass* were zero) go over into the *Goldstone bosons when the coupling tends to zero. The model is discussed mainly in classical terms*; nothing is proved about the quantized theory. It should be understood, therefore, that *the conclusions which are presented concerning the masses of particles are conjectures based on the quantization of linearized classical field equations*.

In a recent note[1] it was shown that the *Goldstone theorem*[2], that *Lorentz-covariant field theories* in which *spontaneous breakdown* of *symmetry* under an internal Lie group occurs contain *zero-mass particles, fails if and only if the conserved currents associated with the internal group are coupled to gauge fields*.

[1] Higgs, P. W., to be published.
[2] Goldstone, J. (January, 1961). Field theories with "Superconductor" solutions. *Nuovo Cimento*, 19, 154, see above; Goldstone, J., Salam, A. & Weinberg, S. (August, 1962). Broken Symmetries. *Phys. Rev.* 127, 965, see above.

The purpose of the present note is to report that, *as a consequence of this coupling, the spin-one quanta of some of the gauge fields acquire mass*; the longitudinal degrees of freedom of these particles (which would be absent if their *mass* were zero) go over into the Goldstone bosons *when the coupling tends to zero*. This phenomenon is just the *relativistic analog of the plasmon phenomenon* to which Anderson[3] has drawn attention: that the *scalar zero-mass excitations of a superconducting neutral Fermi gas become longitudinal plasmon modes of finite mass when the gas is charged*.

341

[3] Anderson, P. W. (1963). Plasmons, Gauge Invariance, and Mass. *Phys. Rev.*, 130, 439; https://journals.aps.org/pr/abstract/10.1103/PhysRev.130.439.

The simplest theory which exhibits this behavior is a *gauge-invariant* version of a model used by Goldstone[2] himself: Two real[4] *scalar fields* φ_1, φ_2, and a real *vector field* A_μ *interact* through the *Lagrangian density*

$$L = \tfrac{1}{2}\,(\nabla\varphi_1)^2 - \tfrac{1}{2}\,(\nabla\varphi_2)^2 - V(\varphi_1{}^2 + \varphi_2{}^2) - \tfrac{1}{4}\,F_{\mu\nu}\,F^{\mu\nu}, \qquad (1)$$

where

$$\nabla_\mu\varphi_1 = \partial_\mu\varphi_1 - eA_\mu\varphi_2,$$
$$\nabla_\mu\varphi_2 = \partial_\mu\varphi_2 + eA_\mu\varphi_1,$$
$$F_{\mu\nu} = \partial_\mu A_\nu - \partial_\nu A_\mu,$$

e is a dimensionless *coupling constant*, and the metric is taken as $-+++$. L is *invariant* under *simultaneous gauge transformations of the first kind* on $\varphi_1 \pm i\,\varphi_2$, and *of the second kind* on A_μ.

[4] *In the present note the model is discussed mainly in classical terms*; nothing is proved about the quantized theory. It should be understood, therefore, that *the conclusions which are presented concerning the masses of particles are conjectures based on the quantization of linearized classical field equations.* However, essentially the same conclusions have been reached independently by F. Englert and R. Brout, *Phys. Rev. Letters*, 13, 321 (1964): These authors discuss the same model *quantum mechanically* in lowest order perturbation theory about the self-consistent vacuum.

Let us suppose that $V'(\varphi_0{}^2) = 0$, $V''(\varphi_0{}^2) > 0$; then *spontaneous breakdown of U(1) symmetry* occurs. Consider the equations [derived from (1) by treating $\Delta\varphi_1$, $\Delta\varphi_2$, A_μ as small quantities] governing the propagation of small oscillations about the "*vacuum*" solution $\varphi_1(x) = 0$, $\varphi_2(x) = \varphi_0$:

$$\partial^\mu\,\{\partial_\mu\,(\Delta\varphi_1) - e\varphi_0\,A_\mu\} = 0, \qquad (2a)$$
$$\{\partial^2 - 4\,\varphi_0{}^2\,V''(\varphi_0{}^2)\}(\Delta\varphi_2) = 0, \qquad (2b)$$
$$\partial F^{\mu\nu} = e\varphi_0\,\{\partial^\mu\,(\Delta\varphi_1) - e\varphi_0\,A_\mu\}. \qquad (2c)$$

Equation (2b) describes waves whose *quanta* have (bare) *mass* $2\varphi_0\{V''(\varphi_0{}^2)\}^{1/2}$; Eqs. (2a) and (2c) may be transformed, by the introduction of new variables

$$B_\mu = A_\mu - (e\varphi_0)^{-1}\{\partial_\mu\,(\Delta\varphi_1), \qquad (3)$$
$$G_{\mu\nu} = \partial_\mu B_\nu - \partial_\nu B_\mu = F_{\mu\nu},$$

into the form

$$\partial_\mu B^\mu = 0, \qquad \partial_\nu G^{\mu\nu} + e^2\, \varphi_0{}^2\, B^\mu = 0. \tag{4}$$

Equation (4) describes *vector waves* whose *quanta* have (bare) *mass* $e\varphi_0$. In the absence of the *gauge field coupling* ($e = 0$) the situation is quite different: Equations (2a) and (2c) describe *zero-mass scalar* and *vector bosons*, respectively. In passing, we note that the right-hand side of (2c) is just the linear approximation to the *conserved current*: It is linear in the *vector potential*, *gauge invariance* being maintained by the presence of the gradient term[5].

[5] In the *theory of superconductivity* such a term arises from collective excitations of the Fermi gas.

When one considers theoretical models in which *spontaneous breakdown* of *symmetry* under a *semi-simple group* occurs, one encounters a variety of possible situations corresponding to the various distinct irreducible representations to which the *scalar fields* may belong; the *gauge field* always belongs to the *adjoint representation*[6].

[6] See, for example, Glashow, S. L. & Gell-Mann, M. (September, 1961). Gauge Theories of Vector Particles. *Ann. Phys. (N.Y.)*, 15, 437-60. See above.

The model of the most immediate interest is that in which the scalar fields form an octet under SU(3): Here one finds the possibility of two non-vanishing *vacuum expectation values*, which may be chosen to be the two $Y = 0$, $I_3 = 0$ members of the *octet*[7].

[7] These are just the parameters which, if the *scalar octet* interacts with *baryons* and *mesons*, lead to the *Gell-Mann-Okubo* and *electromagnetic mass splittings*: See Coleman, S. & Glashow, S. L. (May, 1964). Departures from the Eightfold Way: Theory of Strong Interaction Symmetry Breakdown. *Phys. Rev.*, 134, B671; https://journals.aps.org/pr/abstract/10.1103/PhysRev.134.B671.

There are two massive *scalar bosons* with just these *quantum numbers*; the remaining six components of the *scalar octet* combine with the corresponding components of the *gauge-field octet* to describe massive *vector bosons*. There are two $I = \frac{1}{2}$ *vector doublets*, degenerate in *mass* between $Y = \pm 1$ but with an *electromagnetic mass splitting* between $I_3 = +\frac{1}{2}$, and the $I_3 = +1$ components of a $Y = 0$, $I = 1$ *triplet* whose *mass* is entirely *electromagnetic*. The two $Y = 0$, $I = 0$ *gauge fields* remain massless: This is associated with the residual *unbroken symmetry* under the *Abelian group* generated by Y and I_3. It may be expected that when a further mechanism (presumably related to the *weak interactions*) is introduced in order to break Y *conservation*, one of these *gauge fields* will acquire *mass*, leaving the *photon* as the only massless *vector particle*. A detailed discussion of these questions will be presented elsewhere.

It is worth noting that an essential feature of the type of theory which has been described in this note is the prediction of incomplete *multiplets* of *scalar* and *vector bosons*[8].

[8] Tentative proposals that *incomplete SU(3) octets* of *scalar particles* exist have been made by a number of people. Such a role, as an isolated $Y = \pm 1$, $I = \frac{1}{2}$ *state*, was proposed for the *K meson* (725 MeV) by Nambu, Y. & Sakurai, J. J. (1963). κ Meson and the Strangeness-Changing Currents of Unitary Symmetry. *Phys. Rev. Letters*, 11, 42. More recently the possibility that the σ *meson* (385 MeV) may be the $Y = I = 0$ member of an incomplete *octet* has been considered by Brown, L. M. (1964). Apparent Regularity in the Masses of the Mesons. *Phys. Rev. Lett.*, 13, 42; https://journals.aps.org/prl/abstract/10.1103/PhysRevLett.13.42.

It is to be expected that this feature will appear also in theories in which the *symmetry-breaking scalar fields* are not elementary dynamic variables but bilinear combinations of *Fermi fields*[9].

[9] In the *theory of superconductivity* the *scalar fields* are associated with *fermion pairs*; the doubly charged excitation responsible for the quantization of *magnetic flux* is then the surviving member of a *U(1) doublet*.

Gerald Stanford Guralnik (September 17, 1936–April 26, 2014).

Guralnik was the Chancellor's Professor of Physics at Brown University. In 1964 he co-discovered the *Brout–Englert–Higgs* [BEH] *mechanism* and *Higgs boson* with C. R. Hagen and Tom Kibble. [Guralnik, G., Hagen, C. & Kibble, T. (November, 1964). Global Conservation Laws and Massless Particles. See below.] While widely considered to have authored the most complete of the early papers on the theory, Guralnik, Hagen and Kibble were controversially not included in the 2013 Nobel Prize in Physics.

Guralnik received his BS degree from the Massachusetts Institute of Technology in 1958 and his PhD degree from Harvard University in 1964. He went to Imperial College London as a postdoctoral fellow supported by the National Science Foundation and then became a postdoctoral fellow at the University of Rochester. In the fall of 1967 Guralnik went to Brown University and frequently visited Imperial College and Los Alamos National Laboratory where he was a staff member from 1985 to 1987. While at Los Alamos, he did extensive work on the development and application of computational methods for lattice *quantum chromodynamics*.

Guralnik died of a heart attack at age 77 in 2014.

Guralnik, G.[†], Hagen, C.[§] & Kibble, T. (November, 1964). Global Conservation Laws and Massless Particles*.

Phys. Rev. Lett., 13, 20, 585; https://doi.org/10.1103/PhysRevLett.13.585; https://journals.aps.org/prl/pdf/10.1103/PhysRevLett.13.585.

Department of Physics, Imperial College, London, England.

† National Science Foundation Postdoctoral Fellow.
§ On leave of absence from the University of Rochester, Rochester, New York.

* The research reported in this document has been sponsored in whole, or in part, by the Air Force Office of Scientific Research under Grant No. AF EOAR 64-46 through the European Office of Aerospace Research (OAR), U. S. Air Force.

Received October 12, 1964.

A third paper on the subject was written later in the same year by Gerald Guralnik, C. R. Hagen, and Tom Kibble.

In all of the fairly numerous attempts to date to formulate a consistent *field theory* possessing a *broken symmetry*, Goldstone's remarkable theorem[1] has played an important role.

[1] Goldstone, J. (January, 1961). Field theories with "Superconductor" solutions. *Nuovo Cimento*, 19, 154, see above; Goldstone, J., Salam, A. & Weinberg, S. (August, 1962). Broken Symmetries. *Phys. Rev.* 127, 965, see above; Bludman, S. A. & Klein, A. (1963). *Phys. Rev.*, 131, 2364.

This theorem, briefly stated, asserts that if there exists a conserved operator Q_i such that

$$[Q_i, A_j (x)] = \Sigma_k t_{ijk} A_k(x),$$

and if it is possible consistently to take $\Sigma_k t_{ijk} x < 0 \mid A_k \mid 0 > \neq 0$, then $A_j(x)$ has a *zero-mass particle* in its spectrum. It has more recently been observed that *the assumed Lorentz invariance essential to the proof* [2] may allow one the hope of avoiding such massless particles through the introduction of *vector gauge fields* and the consequent breakdown of manifest *covariance*[3].

[2] Gilbert, W. (June, 1964). Broken Symmetries and Massless Particles. *Phys. Rev. Lett.*, 12, 713; https://doi.org/10.1103/PhysRevLett.12.713.

[3] Higgs, P. W. (September, 1964). Broken symmetries, massless particles and gauge fields. *Physics Letters.*, 12, 2, 132–3; https://doi.org/10.1016/0031-9163(64)91136-9.

This, of course, represents a departure from the assumptions of the theorem, and a limitation on its applicability which in no way reflects on the general validity of the proof.

In this note we shall show, within the framework of a simple soluble *field theory*, that it is possible consistently to break a *symmetry* (in the sense that Σ_k t_{ijk} x $< 0 \mid A_k \mid 0 > \neq 0$ without requiring that A(x) excite a *zero-mass particle*. While this result might suggest a general procedure for the elimination of unwanted *massless bosons*, it will be seen that this has been accomplished by giving up the *global conservation law* usually implied by *invariance* under a *local gauge group*. The consequent time dependence of the *generators* Q_i destroys the usual global operator rules of *quantum field theory* (while leaving the local algebra unchanged), in such a, way as to preclude the possibility of applying the *Goldstone theorem*. It is clear that such a modification of the basic operator relations is a far more drastic step than that taken in the usual *broken-symmetry theories* in which a degenerate *vacuum* is the sole symmetry-breaking agent, and the operator algebra possesses the full *symmetry*. However, since *superconductivity* appears to display a similar behavior, the possibility of breaking such *global conservation laws* must not be lightly discarded.

Normally, the time independence of

$$Q_i = \int d^3x \, j_i^0(\vec{x}, t)$$

is asserted to be a consequence of the *local conservation law* $\partial_\mu j^\mu = 0$. However, the relation

$$\partial_\mu < 0 \mid [j_i^\mu(x), A_j(x')] \mid 0 > = 0$$

implies that

$$\int d^3x < 0 \mid [j_i^0(x), A_j(x')] \mid 0 > = \text{const.}$$

only if the contributions from spatial infinity vanish. This, of course, is always the case in a fully causal theory whose *commutators* vanish outside the light cone. If, however, the theory is not manifestly *covariant* (e.g., *radiation-gauge electrodynamics*), *causality* is a requirement which must be imposed with caution. Since Q, consequently may not be time independent, it will not necessarily generate *local gauge transformations* upon $A_j(x')$ for $x^0 \neq x'^0$ despite the existence of the *differential conservation laws* $\partial_\mu j^\mu = 0$.

The phenomenon described here has previously been observed by Zumino[4] in the *radiation-gauge* formulation of two-dimensional *electrodynamics* where the usual *electric charge* cannot be conserved.

[4] Zumino, B. (1964). *Phys. Letters*, 10, 224.

The same effect is not present in the *Lorentz gauge* where *zero-mass excitations* which preserve *charge conservation* are found to occur. (These correspond to *gauge* parts rather than physical particles.) We shall, however, allow the possibility of the breakdown of such *global conservation laws*, and seek solutions of our model consistent only with the *differential conservation laws*.

We consider, as our example, a theory which was partially solved by Englert and Brout[5], and bears some resemblance to the classical theory of Higgs[6],

[5] Englert, F. & Brout, R. (August, 1964). Broken Symmetry and the Mass of Gauge Vector Mesons. *Phys. Rev. Lett.*, 13, 9, 321–3. See above.
[6] Higgs, P. W. to be published.

Our starting point is the ordinary *electrodynamics* of massless *spin-zero* particles, characterized by the Lagrangian

$$\mathscr{L} = -\tfrac{1}{2} F^{\mu\nu}(\partial_\mu A_\nu - \partial_\nu A_\mu) + \tfrac{1}{4} F^{\mu\nu} F_{\mu\nu} + \varphi^\mu \partial_\mu \varphi + \tfrac{1}{2} \varphi^\mu \varphi_\mu + ie_0 \varphi^\mu q \varphi A_\mu,$$

where φ is a two-component Hermitian field, and q is the Pauli matrix σ_2. The *broken-symmetry* condition

$$ie_0 q < 0 \mid \varphi \mid 0 > = \eta \equiv \binom{\eta_1}{\eta_2}$$

will be imposed by approximating $ie_0 \varphi^\mu q \varphi A_\mu$ in the Lagrangian by $\varphi^\mu \eta A_\mu$. The resulting *equations of motion*,

$$F^{\mu\nu} = \partial^\mu A^\nu - \partial^\nu A^\mu,$$
$$\partial_\mu F^{\mu\nu} = \varphi^\mu \eta,$$
$$\varphi^\mu = -\partial^\mu \varphi - \eta A^\mu,$$
$$\partial_\mu \varphi^\mu = 0,$$

are essentially those of the *Brout-Englert model*, and can be solved in either the *radiation*[7] or *Lorentz gauge*.

[7] This is an extension of a model considered in more detail in another context by Boulware D. G. & Gilbert, W. (1962). Phys. Rev., 126, 1563.

The Lorentz-gauge formulation, however, suffers from the fact that the usual canonical quantization is inconsistent with the field equations. (The quantization of A_μ leads to an

348

indefinite metric for one component of φ.) Since we choose to view the theory as being imbedded as a linear approximation in the full theory of *electrodynamics*, these equations will have significance only in the *radiation gauge*.

With no loss of generality, we can take $\eta_2 = 0$, and find

$$(-\partial^2 + \eta_1{}^2)\, \varphi_1 = 0,$$
$$-\partial^2 \varphi_2 = 0,$$
$$(-\partial^2 + \eta_1{}^2)\, A_k{}^T = 0,$$

where the superscript T denotes the transverse part. The two degrees of freedom of $A_k{}^T$ combine with φ_1 to form the three components of a *massive vector field*. While one sees by inspection that there is a *massless particle* in the theory, it is easily seen that it is completely decoupled from the other (massive) excitations, and *has nothing to do with the Goldstone theorem.*

It is now straightforward to demonstrate the failure of the *conservation law of electric charge*. If there exists a *conserved charge* Q, then the relation expressing Q as the generator of rotations in *charge space* is

$$[Q, \varphi(x)] = e_0\, q\, \varphi(x).$$

Our *broken symmetry requirement* is then

$$<0 \,|\, [Q, \varphi_1(x)] \,|\, 0> = -i\,\eta$$

or, in terms of the soluble model considered here,

$$\int d^3x'\, \eta_1 <0 \,|\, \varphi_1{}^0(x'),\, \varphi_1(x)] \,|\, 0> = -i\,\eta_1.$$

From the result

$$<0 \,|\, \varphi_1{}^0(x'),\, \varphi_1(x)] \,|\, 0> = \partial_0\, \Delta(+)(x' - x;\, \eta_1{}^2),$$

one is led to the consistency condition

$$\eta_1 \exp\left[-i\eta_1\,(x_0' - x_0)\right) = \eta_1,$$

which is clearly incompatible with a nontrivial η_1. Thus, *we have a direct demonstration of the failure of Q to perform its usual function as a conserved generator of rotations in charge space.* It is well to mention here that this result not only does not contradict, but is actually required by, the *field equations*, which imply

349

$$(\partial_0{}^2 + \eta_1{}^2)\, Q = 0.$$

It is also remarkable that if A_μ is given any bare *mass*, the entire theory becomes manifestly *covariant*, and Q is consequently conserved. *Goldstone's theorem can therefore assert the existence of a massless particle.* One indeed finds that in that case φ_1 has only *zero mass* excitations.

In summary then, *we have established that it may be possible consistently to break a symmetry by requiring that the vacuum expectation value of a field operator be non-vanishing without generating zero-mass particles.* If the theory lacks manifest *covariance*, it may happen that what should be the generators of the theory fail to be time-independent, despite the existence of a *local conservation law*. Thus, *the absence of massless bosons is a consequence of the inapplicability of Goldstone's theorem rather than a contradiction of* it. Preliminary investigations indicate that *superconductivity* displays an analogous behavior.

François Englert – 2013 Nobel Lecture, December 8, 2013. *The BEH Mechanism and its Scalar Boson.*

François Englert – Nobel Lecture. NobelPrize.org. <https://www.nobelprize.org/prizes/physics/2013/englert/lecture/.

[The Nobel Prize in Physics 2013 was awarded jointly to François Englert and Peter W. Higgs "for the theoretical discovery of a mechanism that contributes to our understanding of the origin of mass of subatomic particles, and which recently was confirmed through the discovery of the predicted fundamental particle, by the ATLAS and CMS experiments at CERN's Large Hadron Collider".

According to modern physics, matter consists of a set of particles that act as building blocks. Between these particles lie forces that are mediated by another set of particles. A fundamental property of the majority of particles is that they have a *mass*. Independently of one another, in 1964 both Peter Higgs and the team of François Englert and Robert Brout proposed a theory about the existence of a particle that explains why other particles have a *mass*. In 2012, two experiments conducted at the CERN laboratory confirmed the existence of the *Higgs particle*. [François Englert – Facts. NobelPrize.org. https://www.nobelprize.org/prizes/physics/2013/englert/facts/.]

[***Press release***: François Englert and Peter W. Higgs are jointly awarded the Nobel Prize in Physics 2013 for the theory of how particles acquire *mass*. In 1964, they proposed the theory independently of each other (Englert together with his now deceased colleague Robert Brout). In 2012, their ideas were confirmed by the discovery of a so-called *Higgs particle* at the CERN laboratory outside Geneva in Switzerland.

The awarded theory is a central part of the *Standard Model* of particle physics that describes how the world is constructed. According to the *Standard Model*, everything, from flowers and people to stars and planets, consists of just a few building blocks: matter particles. These particles are governed by forces mediated by force particles that make sure everything works as it should.

The entire *Standard Model* also rests on the existence of a special kind of particle: the *Higgs particle*. This particle originates from an invisible field that fills up all space. Even when the universe seems empty this field is there. Without it, we would not exist, because it is from contact with the field that particles acquire *mass*. The theory proposed by Englert and Higgs describes this process.

On 4 July 2012, at the CERN laboratory for particle physics, the theory was confirmed by the discovery of a *Higgs particle*. CERN's particle collider, LHC (Large Hadron Collider), is probably the largest and the most complex machine ever constructed by humans. Two research groups of some 3,000 scientists each, ATLAS and CMS, managed to extract the *Higgs particle* from billions of particle collisions in the LHC.

Even though it is a great achievement to have found the *Higgs particle* — the missing piece in the *Standard Model* puzzle — the *Standard Model* is not the final piece in the cosmic puzzle. One of the reasons for this is that the *Standard Model* treats certain particles, *neutrinos*, as being virtually massless, whereas recent studies show that they actually do have *mass*. Another reason is that the model only describes visible matter, which only accounts for one fifth of all matter in the cosmos. To find the mysterious *dark matter* is one of the objectives as scientists continue the chase of unknown particles at CERN. [Press release. NobelPrize.org. https://www.nobelprize.org/prizes/physics/2013/press-release/.]

In his Nobel Prize lecture François Englert provided a brief history of the theory.

---

1 INTRODUCTION: SHORT- AND LONG-RANGE INTERACTIONS

Physics, as it is conceived today, attempts to interpret diverse phenomena as particular manifestations of testable general laws. Since its inception in the Renaissance, mainly through Galileo's revolutionary concepts, this has been an extraordinarily successful adventure—to the point where after impressive developments in the first half of the twentieth century, one might have even conceived that all phenomena, from the atomic scale to the edge of the visible universe, are governed solely by two fundamental laws, and two known laws. Namely classical general relativity, Einstein's generalization of Newtonian gravity, and quantum electrodynamics, the quantum version of Maxwell's electromagnetic theory.

Gravitational and *electromagnetic interactions* are *long range interactions*, meaning they act on objects no matter how far they are separated from each other. The progress in the understanding of such physics applicable to large scales is certainly attributable to the fact they can be perceived without the mediation of highly sophisticated technical devices. But the discovery of subatomic structures had revealed the existence of other *fundamental interactions* that are short range, that is, negligible at larger distance scales. In the early 1960s, there was no consistent theoretical interpretation of *short-range fundamental interactions*, nor of the *"weak interactions"* responsible for radioactive decay, nor of the *"strong interactions"* responsible for the formation of nuclear structures.

Robert Brout and I[1], and independently Peter Higgs[2], constructed a mechanism to describe *short range fundamental interactions*.

[1] Englert, F. & Brout, R. (August, 1964). Broken Symmetry and the Mass of Gauge Vector Mesons. *Phys. Rev. Lett.*, 13, 9, 321–3. See above.

[2] Higgs, P. W. (October, 1964). Broken Symmetries and the Masses of Gauge Bosons. *Phys. Rev. Let.*, 13, 16, 508–9. See above.

Robert Brout passed away in 2011 and left me alone to tell our story. I will explain how we were led to propose the mechanism, and how it allows for consistent fundamental theories of *short-range interactions* and for building elementary particle *masses*. It became a cornerstone of the *Standard Model* and was recently confirmed by the magnificent discovery at CERN of its predicted *scalar boson*.

We became convinced that a consistent formulation of *short-range interactions* would require a common origin for both *short-* and *long-range interactions*.

While both classical general relativity and quantum electrodynamics describe *long range interactions* and are both built upon very large symmetries, labelled *"local symmetries"*, they have very different structures: *in contradistinction to general relativity, long range quantum electrodynamics is fully consistent at the quantum level* and was experimentally verified at that level, in particular by the successful inclusion of chemistry in the realm of known physics. As a valid theory of *short-range interactions* clearly required *quantum* consistency, we were naturally driven to take, as a model of the corresponding *long-range interactions*, the generalization of *quantum electrodynamics*, known as *Yang Mills theory*.

The *quantum* constituents of *electromagnetic waves* are *"photons"*, massless neutral particles travelling with the velocity of light. *Their massless character implies that the corresponding waves are polarized only in directions perpendicular to their propagation.* These features are apparently protected by *local symmetry*, as the latter does not survive the explicit inclusion of a *mass* term in the theory. *Yang-Mills theory* is built upon similar *local symmetries*, enlarged to include several massless interacting quantum constituents, neutral and charged ones.

[*Yang–Mills theory* is a *quantum field theory* for *nuclear binding* devised by Chen Ning Yang and Robert Mills in 1953, as well as a generic term for the class of similar theories. The *Yang–Mills theory* is a *gauge theory* based on a special unitary group SU(n), or more generally any compact Lie group. A *Yang–Mills theory* seeks

353

to describe the behavior of *elementary particles* using these non-abelian Lie groups and *is at the core of the unification of the electromagnetic force and weak forces* (i.e. U(1) × SU(2)) as well as *quantum chromodynamics*, the theory of the *strong force* (based on SU(3)). Thus, it forms the basis of the understanding of the *Standard Model* of particle physics.

Yang's core idea was to look for a *conserved quantity* in nuclear physics comparable to *electric charge* and use it to develop a corresponding *gauge theory* comparable to *electrodynamics*. He settled on *conservation of isospin*, a quantum number that distinguishes a *neutron* from a *proton*.]

These massless objects are labelled *gauge vector bosons* (or often simply *gauge bosons*).

To transmute long range *interactions* into short range ones in the context of *Yang-Mills theory* it would suffice to give these generalized *photons* a *mass*, a feature that, as we just indicated, is apparently forbidden by *local symmetries*. Momentarily leaving aside this feature, let us first recall why massive particles transmit in general *short-range interactions*.

Figure 1 is a *Feynman diagram* whose intuitive appearance hides a precise mathematical content. Viewing time as running from bottom to top, *it describes the scattering of two electrons* resulting from the *exchange of a massive particle* labeled Z of *mass* m_Z. Classically such process could not occur, as the presence of the Z particle would violate *energy conservation*. Quantum mechanically it is allowed if the violation takes place within a time span of the order $\hbar/mc^2$. This process then describes in lowest order *perturbation theory* a *short-range interaction* cut-off at a range $\sim \hbar/mc$.

...

Figure 1. Massive particle mediating short-range interactions.

As *local symmetries* apparently prevent the introduction of massive *gauge bosons* in the theory, we turn our attention to a class of theories where the *state* of a system is asymmetric with respect to the *symmetry principles* that govern its dynamics. This is often the case in the statistical physics of *phase transitions*[3].

[3] Landau, L. D. (1937). On the theory of phase transitions I. *Phys. Z. Sowjet.*, 11, 26; [JETP 7 (1937) 19].

This is not surprising, since more often than not *energetic considerations* dictate that the *ground state* or low-lying *excited states* of a many-body system become ordered. A collective variable such as *magnetization* picks up *expectation value*, which defines an order parameter that otherwise would vanish by virtue of the *symmetry* encoded in the formulation of the theory (*isotropy* in the aforementioned example). This is an example of *Spontaneous Symmetry Breaking* (SSB) which frequently occurs in the statistical theory of second order *phase transitions*. Could *mass* of *gauge bosons* arise through a similar SSB? This question arises naturally from the seminal work of Yoichiro Nambu, who showed that SSB could be transferred from the statistical theory of *phase transitions* to the realm of *relativistic quantum field theory*[4,5,6], the mathematical framework designed to analyze the world of *elementary particles* [???].

[4] Nambu, Y. (1960). Quasi-particles and gauge invariance in the theory of superconductivity. *Phys. Rev.*, 117, 648.

[5] Nambu, Y. (1960). Axial vector current conservation in weak interactions. *Phys. Rev. Lett.*, 4, 380.

[6] Nambu, Y. & Jona-Lasinio, G. (April, 1961). Dynamical Model of Elementary Particles Based on an Analogy with Superconductivity. I. *Phys. Rev.*, 122, 1, 345; https://doi.org/ 10.1103/PhysRev.122.345.; (1961) Dynamical Model of Elementary Particles Based on an Analogy with Superconductivity. II. *Ibid.*, 124, 1, 246.

This raises a deeper question: could *Spontaneous Symmetry Breaking* (SSB) be the agent of the transmutation of *long-range interactions* mediated by massless *gauge fields* to *short range interactions* mediated by massive ones, without impairing the validity of the quantum behavior that characterize the simplest *Yang-Mills theory*, namely *quantum electrodynamics*?

As we shall see, the answer is yes to both questions provided that the notion of SSB is traded for a more subtle one: the *BEH mechanism*[1,2]. To prepare for the discussion of the mechanism, I will first review how SSB can be transferred from the *theory of phase transitions* to *relativistic quantum field theory*.

2 SPONTANEOUS SYMMETRY BREAKING

2.1 Spontaneous symmetry breaking in phase transitions

Consider a *condensed matter system*, whose dynamics is invariant under a continuous *symmetry*. As the temperature is lowered below a critical one, the *symmetry* may be reduced by the appearance of an ordered *phase*. The breakdown of the original *symmetry* is always a discontinuous event at the phase transition point, but the order parameters may set in continuously as a function of temperature. In the latter case the *phase transition* is second order. *Symmetry breaking* by a second order *phase transition* occurs in particular in *ferromagnetism*, *superfluidity* and *superconductivity*.

I will first discuss the *ferromagnetic phase transition*, which illustrates three general features of the SSB which set in at the transition point in the low temperature *phase*: *ground state degeneracy*, the appearance of a *"massless mode"* when the dynamics is *invariant* under a *continuous symmetry*, and the occurrence of a *"massive mode"* characterizing the rigidity of the order parameter.

In the absence of external *magnetic fields* and of surface effects, a *ferromagnetic* substance below the Curie point displays a *global orientation* of the *magnetization*, while the dynamics of the system is clearly *rotation invariant*; namely, the Hamiltonian of the system is invariant under the *full rotation group*. This is SSB [*Spontaneous Symmetry Breaking*].

A *ferromagnetic* system is composed of microscopic atomic magnets (in simplified models such as the *Heisenberg Model* these are *spin 1/2 objects*) whose interactions tend to orient neighboring ones parallel to each other. No global orientation appears at high temperature where the disordering thermal motion dominates. Below a critical "Curie temperature" *energy* considerations dominate and the system picks up a global *magnetization*. The parallel orientation of neighboring magnets propagates, ending up in a macroscopic *magnetization*.

> [*Ferromagnetism* is the result of the *"exchange process"* resulting from the *quantum entanglement* of the *spins* of neighboring *electrons*. See Heisenberg, W. (September, 1928). Zur Theory of Ferromagnetismus. (On the theory of ferromagnetism.) *Zeit. Phys.*, 49, 619–36; translation in Underwood, T. G. (2024), *Electricity & Magnetism*, pp. 370-9.]

This selects a direction, which for an infinite isolated *ferromagnet* is arbitrary. It is easily proven that for an infinite system any pair of possible orientations defines orthogonal *ground states* and any local excitations on top of these *ground states* are also orthogonal to each other. Thus, the full Hilbert space of the system becomes split into an infinity of disjointed Hilbert spaces. This is *ground state degeneracy* (Figure 2).

…

Figure 2. Classical representation of a ferromagnet ground state.

The effective *thermodynamical potential* V, whose minimum yields *magnetization* in the absence of an external magnetic field, is depicted in Figure 3.

…

Figure 3. Effective thermodynamical potential of a ferromagnet above and below the Curie point.

Above the Curie point T_C the *magnetization* $\vec{M}$ vanishes. Below the Curie point the potential develops in a plane $V M_z$ a double minimum which generate a valley in the M_x, M_y directions. Each point of the valley defines one of the degenerate *ground states* with the same $|\vec{M}|$.

At a given minimum, say, $\vec{M} = M^z \vec{I_z}$, the curvature of the *effective potential* measures the *inverse susceptibility* which determines the *energy* for infinite wavelength fluctuations. This is the analogue of *mass* in *relativistic* particle physics. The *inverse susceptibility* is zero in directions transverse to the order parameter and positive in the longitudinal direction. One thus obtains from the *transverse susceptibility* a "massless" transverse mode characteristic of *broken continuous symmetry*: these are the "*spin-waves*" whose quantum constituents are interacting *bosons* called "*magnons*". The *longitudinal susceptibility* yields a (possibly unstable) *"massive" longitudinal mode* which corresponds to fluctuations of the order parameter. In contradistinction to the *massless mode* which exists only in continuous *Spontaneous Symmetry Breaking* (SSB) for which there is a valley, the *massive mode* is present in any SSB, continuous or discrete, and measures the *rigidity* of the ordered structure.

The structure of Figure 3 is common to many second order *phase transitions* and leads to similar consequences. However, in *superconductivity* a new phenomenon occurs. The quantum *phase symmetry* is broken by a condensation of *electron pairs* bounded by an attractive force due to *phonon exchange* in the vicinity of the *Fermi surface*. The condensation leads to an *energy gap* at the *Fermi surface*. For neutral superconductors, this gap would host a *massless mode* and one would recover the general features of SSB. But the presence of the *long-range Coulomb interactions* modifies the picture. The *massless mode* disappears: it is absorbed in *electron density oscillations*, namely in the *"massive" plasma mode*. As will be apparent later, this is a precursor of the *BEH mechanism*[7,4,8].

[7] Anderson, P. W. (1958). Random-phase approximation in the theory of superconductivity. *Phys. Rev.*, 112, 1900.

[8] Anderson, P. W. (1963). Plasmons, gauge invariance, and mass. *Phys. Rev.*, 130, 1, 439. See above.

2.2 Spontaneous symmetry breaking in field theory

Spontaneous symmetry breaking was introduced in *relativistic quantum field theory* by Nambu in analogy with the *Bardeen–Cooper–Schrieffer* (BCS) *theory of superconductivity*[4]. The problem studied by Nambu[5] and Nambu & Jona-Lasinio[6] is the *spontaneous breaking of the U(1) symmetry of massless fermions* resulting from the arbitrary relative *(chiral) phase* between their decoupled right and left constituent *neutrinos*. *Chiral invariant interactions* cannot generate a *fermion mass* in perturbation theory but may do so from a (non-perturbative) *fermion condensate*: *the condensate breaks the chiral symmetry spontaneously*. Nambu[5] showed that such *spontaneous symmetry breaking* is accompanied by a massless *pseudoscalar*. This is interpreted as the *chiral* limit of the (tiny on the *hadron* scale) *pion mass*. Such an interpretation of the *pion* constituted a breakthrough in our understanding of *strong interaction* physics.

[A *pion* or *pi meson*, is any of three subatomic particles: π^0, π^+, and π^-. Each pion consists of a *quark* and an *antiquark* and is therefore a *meson*. *Pions* are the lightest *mesons* and, more generally, the lightest *hadrons*. They are unstable, with the *charged pions*, π^+ and π^-, decaying after a mean lifetime of 26.033 nanoseconds (2.6033×10^{-8} seconds), and the *neutral pion* π^0 decaying after a much shorter lifetime of 85 attoseconds (8.5×10^{-17} seconds). *Charged pions* most often decay into *muons* and *muon neutrinos*, while *neutral pions* generally decay into *gamma rays*.

The exchange of virtual *pions*, along with *vector, rho* and *omega mesons*, provides an explanation for the residual *strong force* between *nucleons*. *Pions* are not produced in radioactive decay, but commonly are in high-energy collisions between *hadrons*. *Pions* also result from some *matter–antimatter annihilation* events. All types of *pions* are also produced in natural processes when high-energy cosmic-ray *protons* and other hadronic cosmic-ray components interact with matter in Earth's atmosphere. In 2013, the detection of characteristic *gamma rays* originating from the decay of *neutral pions* in two supernova remnants has shown that *pions* are produced copiously after supernovas, most probably in conjunction with production of high-energy protons that are detected on Earth as cosmic rays.]

The *massless pseudoscalar* is the field-theoretic counterpart of the *"massless" spin-wave mode* in *ferromagnetism*. In the model of reference[6] it is shown that *Spontaneous Symmetry Breaking* (SSB) also generates a *massive scalar boson* which is the counterpart of the *"massive mode"* measuring in *phase transitions* the *rigidity* of the order parameter in the *spontaneously broken phase*.

The significance of the *massless boson* and of the *massive scalar boson* occurring in SSB is well illustrated in a simple model devised by Jeffrey Goldstone[9].

[9] Goldstone, J. (January, 1961). Field theories with "Superconductor" solutions. *Nuovo Cimento*, 19, 154–64. See above.

The potential $V(\phi_1, \phi_2)$ depicted in Figure 4 has a *rotational symmetry* in the plane of the real fields (ϕ_1, ϕ_2), or equivalently is *invariant under the U(1) phase* of the complex field $\phi = (\phi_1 + i\phi_2)/\sqrt{2}$.

...

Figure 4. Spontaneous symmetry breaking in the Goldstone mode.

This *symmetry* is *spontaneously broken* by the *expectation value* $\langle\phi\rangle$ of the ϕ-*field* acquired at a minimum of the *potential* in some direction of the (ϕ_1, ϕ_2) plane, say $\langle\phi_1\rangle$. Writing $\phi = \langle\phi\rangle + \varphi$

$$\phi_1 = \langle\phi_1\rangle + \varphi_1, \tag{2.1}$$
$$\phi_2 = \varphi_2. \tag{2.2}$$

For small φ_1 and φ_2 we may identify the *quantum fluctuation* φ_1 climbing the potential as the *massive mode* measuring the *rigidity* of the SSB *ground state* selected by $\langle\phi_1\rangle$, and the *quantum fluctuation* φ_2 in the orthogonal valley direction as the *massless mode* characteristic of a continuous SSB.

Their significance is illustrated in Figure 5 and Figure 6 depicting respectively classical φ_2 and φ_1 *wave modes*, on the classical background $\langle\phi_1\rangle$. The corresponding *massless* and *massive bosons* are the *quantum constituents* of these waves.

Figure 5(a) represents schematically a *lowest energy state* (a "*vacuum*") of the system: a constant non-zero value of the field $\phi_1 = \langle\phi_1\rangle$ pervades *space-time*. Figure 5(b) depicts the excitation resulting from the rotation of half the fields in the (ϕ_1, ϕ_2) *plane*.

...

Figure 5. Massless Nambu-Goldstone mode φ_2.

This costs only an *energy* localized near the surface separating the rotated fields from the chosen vacuum. *Spontaneous Symmetry Breaking* (SSB) *indeed implies that rotating all the fields would cost no energy at all*: one would merely trade the initial chosen vacuum for an equivalent one with the same energy. This is the characteristic *vacuum degeneracy* of SSB. Figure 5(c) mimics a wave of φ_2. Comparing 5(c) with 5(b), we see that as the *wavelength* of the wave increases indefinitely, its *energy* tends to zero, and may be viewed as generating in that limit a motion along the valley of Figure 4. *Quantum excitations* carried by the wave reach thus zero energy at zero momentum and the mass m_{φ_2} is zero. Figure 5 can easily be generalized to more complex spontaneous *symmetry breaking* of *continuous symmetries*. *Massless bosons* are thus a general feature of such SSB already revealed by Nambu's discovery of the *massless pion* resulting from *spontaneous chiral symmetry breaking*[5]. They will be labelled *massless Nambu-Goldstone (NG) bosons*. Formal proofs corroborating the above simple analysis can be found in the literature[10].

[10] Goldstone, J., Salam, A. & Weinberg, S. (August, 1962). Broken Symmetries. *Phys. Rev.* 127, 965.

...

Figure 6. Massive scalar mode φ_1.

Figure 6 depicts similarly a *classical wave* corresponding to a stretching of the vacuum fields. These excitations in the φ_1 direction describe fluctuations of the order parameter $\langle \varphi_1 \rangle$. They are volume effects and their energy does not vanish when the wavelength becomes increasingly large. They correspond in Figure 4 to a climbing of the potential. The *quantum excitations* φ_1 are thus now *massive*. These considerations can be again extended to more general SSB (even to discrete ones) to account for order parameter fluctuations. *Lorentz invariance imposes that such massive excitations are necessarily scalar particles.* They were also already present in reference[6] and will be denoted in general as *massive scalar bosons*. To summarize, φ_2 describes *massless bosons*, φ_1 *massive* ones, and the *"order parameter"* $\langle \phi_1 \rangle$ may be viewed as a condensate of φ_1 *bosons*.

3 THE BEH MECHANISM [Brout Englert Higgs mechanism, also known as the Higgs mechanism].

The above considerations are restricted to *spontaneous symmetry breaking of global continuous symmetries*. *Global* means that the *symmetry operations* are independent of the space-time point x. For instance in the *Goldstone model*, the global rotations of the fields in Figure 5 (a) in the (ϕ_1, ϕ_2) plane by angles independent of the space-time point x are *symmetries* of the theory (they describes motion in the valley of Figure 4: these rotations cost no energy and simply span the *degenerate vacua*. *We will now discuss the fate of Spontaneous Symmetry Breaking (SSB) when the global symmetry is extended to a local one.*

3.1 The fate of the Nambu-Goldstone boson and vector boson masses

We extend the U(1) symmetry of the Goldstone model from global to local. Thus the rotation angle in the (ϕ_1, ϕ_2) plane in Figure 5, or equivalently the rotation in the valley of Figure 4, can now be chosen independently at each space-time point (x) with no cost of energy and no physical effect. To allow such feature, one has to invent a new field whose transformation would cancel the energy that such motion would generate in its absence. This is a *"gauge vector field"* A_μ. It has to be a *vector field* to compensate energy in all space directions and it has to transform in a definite way under a rotation in the (ϕ_1, ϕ_2) plane: this is called a *gauge transformation* and results in a large arbitrariness in the choice of the A_μ field corresponding to arbitrary "internal" rotations at different points of space.

The consequence of this *gauge symmetry* is that the *waves are polarized in directions perpendicular to their direction of propagation* and that their quantum constituents have to be introduced as *massless* objects.

Local U(1) symmetry is the simplest *gauge field theory* and is the *symmetry group* of *quantum electrodynamics*. In the local generalization (the *gauging*) of the *Goldstone model*, the introduction of the *potential* of Figure 4 will deeply affect the "*electromagnetic potential*" A_μ.

As in the Goldstone model of Section 1.2, the *SSB Yang-Mills phase* is realized by a non-vanishing *expectation value* for $\phi = (\phi_1 + i\phi_2)/\sqrt{2}$, which we choose to be in the ϕ_1-direction. Thus

$$\phi = \langle\phi\rangle + \varphi, \tag{3.1}$$

with $\phi_1 = \langle\phi_1\rangle + \varphi_1$ and $\phi_2 = \varphi_2$. As previously φ_2 and φ_1 appear to describe a NG *massless boson* and a *massive scalar boson*.

However, a glance on Figure 5 depicting the *NG [Nambu-Goldstone] mode* immediately shows that Figure 5 (b) and Figure 5 (c) differ from Figure 5 (a) only by *local rotations* and hence in the *local Goldstone model* they are just *symmetry* (or equivalently *gauge*) *transformations*. They cost no energy and therefore the NG *boson* has disappeared: the corresponding fluctuations in the valley are redundant (*gauge transformed*) descriptions of the same *gauge invariant vacuum*. It is easy to see that this argument remains valid for any *local symmetry* and hence *Nambu-Goldstone bosons* do not survive the *gauging* of a *global SSB* to a *local symmetry*. The *vacuum* is no longer degenerate and strictly speaking there is no *spontaneous symmetry breaking* of a *local symmetry*. The reason why the *phase* with non-vanishing *scalar expectation value* is often labelled *Spontaneous Symmetry Breaking* (SSB) is that one uses *perturbation theory* to select at *zero-gauge field coupling* a *scalar field* configuration from *global SSB*; but this preferred choice is only a convenient one.

The disappearance of the NG boson is thus an immediate consequence of local symmetry. The above argument[11] was formalized much later[12] but formal proofs not directly based on the *gauge invariance* of the *vacuum* were already presented in 1964 [13,14].

[11] Englert, F. "Broken symmetry and Yang-Mills theory" in (2005). *50 years of Yang-Mills Theory*, ed. by G.'t Hooft, World Scientific; hep-th/0406162; pp. 65–95.

[12] Elitzur, S. (1975). Impossibility of spontaneously breaking local symmetries. *Phys. Rev.*, D12, 3978.

[13] Higgs, P. W. (September, 1964). Broken symmetries, massless particles and gauge fields. *Physics Letters.*, 12, 2, 132–3; https://doi.org/10.1016/0031-9163(64)91136-9.

[14] Guralnik, G., Hagen, C. & Kibble, T. (November, 1964). Global Conservation Laws and Massless Particles. *Phys. Rev. Lett.*, 13, 20, 585. See above.

One may now understand in qualitative terms the consequence of the disappearance of the *NG boson*. Clearly, one does not expect that the degrees of freedom carried by the *NG* ϕ_2 *field* could vanish. As the *NG boson* disappears because of its *coupling* to the *gauge field*, one expects that these degrees of freedom should be transferred to it. This can only occur by adding to the *transverse polarization* of the *gauge field* a *longitudinal* one. But such polarization is forbidden as mentioned earlier, for a *massless field. Therefore, the coupling of the would-be NG boson to the gauge field must render the latter massive! This is the essence of the BEH mechanism.*

These qualitative considerations can be made quantitative[1] by considering the *Feynman graphs* (time runs horizontally) describing the propagation of the A_μ *gauge field* in the *vacuum* with non-vanishing *scalar field expectation value*, say $\langle \phi_1 \rangle \neq 0$. This propagation is depicted in lowest order in Figure 7 (time runs horizontally) and the interaction of A_μ with the *condensate* $\langle \phi_1 \rangle$ amounts to a "*polarization*" of the *vacuum*.

...

Figure 7. Interaction of the gauge field with the condensate.

The first graph shows the *local interaction* of the *gauge field* with the *condensate* while the second one gives a *non-local interaction* due to the propagation of a *NG boson*. Here e is the *coupling* of the *gauge vector* to *matter*, q_μ is a *four-momentum* (q_0 = *energy*; $\vec{q}$ = *momentum*), $q^2 = q_0^2 - \vec{q}^2$ and $g_{\mu\nu}$ has only non-zero values if $\mu \neq \nu$: 1,−1,−1,−1. The two graphs add up to

$$\Pi_{\mu\nu} = (g_{\mu\nu} - q_\mu q_\nu / q^2)\, \Pi(q^2), \qquad (3.2)$$
where
$$\Pi(q^2) = e^2 \langle \phi_1 \rangle^2. \qquad (3.3)$$

The second factor of Equation (3.2) does not vanish when $q^2 = 0$. In *field theory* this means that the *gauge field* has acquired a *mass*

$$(M_V^2) = e^2 \langle \phi_1 \rangle^2. \qquad (3.4)$$

The first factor describes the projection at $q^2 = m_V^2$ of $g_{\mu\nu}$ on a three-dimensional space of *polarizations*, which, as explained in qualitative terms above, is required for a *massive vector*. Its *transversality* (i.e. its vanishing under multiplication by q_μ) is characteristic of a "*Ward Identity*" which expresses the fact that the *local gauge symmetry* has not been broken

362

and is identical to the analogous factor in *quantum electrodynamics*, an important fact that will be commented on in the following section.

The generalization of these results to *more complicated symmetries* yields (for *real fields*) a *mass matrix*

$$(M_V^2)^{ab} = -e^2 \langle \varphi^B \rangle \, T^{aBC} T^{bCA} \langle \varphi^A \rangle, \tag{3.5}$$

where T^{aBC} is a *real anti-symmetric generator* coupled to a *gauge field* A_μ and $\langle \phi^A \rangle$ designates a *non-vanishing expectation value*.

In these cases, some gauge fields may remain massless. Consider for instance instead of the *invariance* of the *Goldstone model* on a *circle* in the *plane* (ϕ_1, ϕ_2), an *invariance* on a *sphere* in a *3-dimensional space* (ϕ_1, ϕ_2, ϕ_3) broken by $\langle \phi_1 \rangle \neq 0$. There are now three *gauge fields* associated to the *rotations on the sphere*, and while A_μ^2 and A_μ^3 acquire *mass* A_μ^1 remains *massless*. This can be understood in the following way: *rotation generators* around the directions 2 and 3 would move $\langle \phi_1 \rangle$ if the *symmetry* were *global* and would thus give rise, as in Figure 5, to *NG bosons*; their degrees of freedom are transferred in *local symmetries* to the *massive gauge vector fields* $A\mu^2$ and $A\mu^3$, providing their third degree of *polarization*. The *expectation value* $\langle \phi_1 \rangle$ is not affected by *rotation generators* around the direction 1 and does not generate *NG bosons* in the *global symmetry* case and hence the corresponding A_μ^1 remains *massless*.

Thus, the *BEH mechanism* can unify *long* and *short-range interactions* in the same theory by leaving unbroken a subgroup of *symmetry transformation* (e.g. *rotation* around the direction 1) whose corresponding *gauge fields* remain *massless*.

3.2 The fate of the massive scalar boson

A glance at Figure 5 shows that the stretching of (classical) *scalar fields* is independent of *local rotations* of the *ϕ-field* in the (ϕ_1, ϕ_2) *plane*. This translates the fact that the *modulus* of the *φ-field* is *gauge invariant*. Hence the *scalar bosons* survive the *gauging* and their classical analysis is identical to the one given for the *Goldstone model* in Section 1.2. The *coupling* of the *scalar boson* φ_1 to the *massive gauge bosons* follows from the Figure 7, by viewing the *Feynman diagrams* with time going from top to bottom and using Equation (3.1). One gets the two vertices of Figure 8, where the heavy wiggly lines on the right-hand side represent the *massive gauge propagators*. The *vertex couplings* follow from Equation (3.4).

$$[(M_V^2) = e^2 \langle \varphi_1 \rangle^2. \tag{3.4}]$$

363

3.3 Fermion masses

Let us couple the *Yang-Mills fields* to *massless fermions* in a way that respects *Yang-Mills symmetry*. This *coupling* preserves the *chiral symmetry* of the *massless fermions* and *fermion mass* requires *Spontaneous Symmetry Breaking* (SSB). In the Nambu theory of *spontaneous breaking* of *chiral symmetry*, this gives rise to *NG bosons* which are eaten up here by *massive gauge fields*. This can be done by suitable *couplings* of the *scalar fields* whose *expectation value* breaks the *symmetry*. *Mass generation* for *fermions* is depicted in Figure 9.

...

Figure 9. *Mass generation* $m_f = \lambda f \langle \phi \rangle$ from a *coupling* λ_f of *fermions* to the *scalar field* ϕ.

3.4 Why is the mechanism needed?

Equation (3.2) expresses the fact that the *mass generation* from the *BEH mechanism* does not destroy *local symmetry*, in contradistinction to a *mass term* introduced by hand *ab initio*. This equation remains valid at higher orders in *perturbation theory* and has the same form as the *polarization* in *quantum electrodynamics*. As in the latter case, it implies that in *covariant gauges*, the *gauge vector boson propagator* tames the *quantum fluctuations*, and therefore suggests that the theory is *renormalizable*[15].

[15] Englert, F., Brout, R. & Thiry, M. (1966). Vector mesons in presence of broken symmetry. *Nuovo Cimento*, 43A, 244; Englert, F. (1997). *Proceedings of the 1997 Solvay Conference, Fundamental Problems in Elementary Particle Physics*, Interscience Publishers J. Wiley and Sons, p. 18.

However, it is a highly non-trivial matter to prove that it does not introduce contributions from unphysical particles, and it is therefore a very difficult problem to prove *quantum consistency* to all orders. That this is indeed the case has been proven by 't Hooft and Veltman[16]. (See also Ref. 17).

[16] 't Hooft, G. (1972). Renormalizable Lagrangians for massive Yang-Mills fields. *Nucl. Phys.*, B35, 167; 't Hooft, G. & Veltman, M. (1972). Regularization and renormalization of gauge fields, *Nucl. Phys.*, B44, 189.
[17] Lee, B. W. & Zinn-Justin, J. (1972). Spontaneously broken gauge symmetries. *Phys. Rev.*, D5, 3121; 3137; 3155.

The quantum consistency of the BEH mechanism is the basic reason for its success. Precision experiments can be predicted and were indeed verified. *Quantum consistency* played a critical role in the analysis of the production of the *scalar boson* at the LHC and of its decay products, leading to the confirmation of the detailed validity of the mechanism.

3.5 Dynamical symmetry breaking

The *symmetry breaking* giving *mass* to *gauge vector bosons* may also arise from a *fermion condensate*. This is labelled *dynamical symmetry breaking*. If a *spontaneously broken global symmetry* is extended to a *local one* by introducing *gauge fields*, the *massless NG bosons* disappear as previously from the physical spectrum and their absorption by *gauge fields* renders these *massive*. In contradistinction with breaking by *scalar field condensate*, it is very difficult in this way to give *mass* in a *renormalizable theory* simultaneously to both *gauge vector fields* and *fermions*. [N.B.]

3.6 The electroweak theory and the Standard Model

The most impressive success of the *BEH mechanism* is the *electroweak theory* for *weak* and *electromagnetic interactions*[18] applied to all particles of the *Standard Model*.

> [18] Glashow, S. L. (February, 1961). Partial-symmetries of weak interactions. *Nuclear Physics*, 22, 4, 579–88, see above; Weinberg, S. (November, 1967). Model of Leptons. *Phys. Rev. Let.*, 19, 21, 1264–6S. Weinberg, A model of leptons, Phys. Rev. Lett. 19 (1967) 1264, see below; Salam, A. "The BEH Mechanism and its Scalar Boson." *Proceedings of the 8th Nobel Symposium, Elementary Particle Physics*, ed. by N. Svartholm, (Almqvist and Wiksell, Stockholm), p. 367.

These encompass all known particles. These are a) the *fermions* which are listed in Figure 10, [e, μ, τ, p, n, and their antiparticles]; b) γ and W$^+$, W$^-$, Z, the *gauge vector bosons* transmitting the *electromagnetic* and the *weak interactions*, c) eight "*gluons*," the *gauge vectors bosons* of the "color group" SU(3) mediating the *strong interactions*, and d) last but not least, one *massive scalar boson* [also known as the *Higgs boson*] which was recently discovered and identified as the *scalar* predicted by the *BEH mechanism*.

…

Figure 10. Fermion constituents of the Standard Model.

The first row in Figure 10 contains the basic constituents of the atom, namely the *electron*, the three *up and down colored quarks* building the *proton* and *neutron* bonded by the *gluons*, to which is added the *electron neutrino*. The second[19] and third row[20] [μ and τ, and their antiparticles] were completed as predictions in the 1970s and verified afterwards.

[19] Glashow, S. L., Iliopoulos, J. & Maiani, L. (October, 1970). Weak Interactions with Lepton-Hadron Symmetry. *Phys. Rev. D*, 2, 7, 1285; https://doi.org/10.1103/ PhysRevD.2.1285.

[20] Kobayashi, M. & Maskawa, T. (February, 1973). CP-Violation in the Renormalizable Theory of Weak Interaction. *Progress of Theoretical Physics*, 49, 2, 652–7; https://doi.org/10.1143/PTP.49.652. See below.

Color was also introduced in the 60s. The particles in the first and the second row are called *leptons*. To all *fermions* of the table, one must of course also add their *antiparticles*.

All the *fermions* are *chiral* and their chiral components have different group *quantum numbers*. Hence, they are, as the *gauge vector bosons*, *massless* in absence of the *BEH mechanism*, i.e. in absence of the *scalar condensate*. The *condensate* $\langle \phi \rangle \neq 0$ gives *mass* to the W+, W–, Z bosons and to all *fermions* except the three *chiral neutrinos* which have no opposite chirality counterpart in the conventional *Standard Model*. The *photons* and the *gluons* remain *massless* but the latter become *short range* due in the conventional description to a highly non-perturbative vacuum (*resulting from a mechanism somehow dual to the BEH mechanism*).

The discovery of the Z and W *bosons* in 1983 and the precision experiments testing the quantum consistency of the *Standard Model* established the validity of the mechanism, but it was still unclear whether this was the result of a *dynamical symmetry breaking* or of a particle identifiable as an elementary *boson* at the energy scale considered.

4 THE DISCOVERY

In the *Standard Model*, there is one *real massive scalar boson* φ (also labelled H). It couples to the *massive* W and Z bosons. This follows from Figure 8 and the *couplings* are depicted in Figure 11 (a).

...

Figure 11. Coupling of the scalar boson φ to *massive gauge bosons* and to elementary *fermions*.

Its *coupling* to *elementary fermions* similarly follows from the *couplings* in Figure 9 as shown in Figure 11 (b). The *coupling* to the *massless photons* is a genuine *quantum effect* involving *loops*, even in the lowest order, as indicated in Figure 12.

...

Figure 12. Coupling of the *scalar boson* φ to *photons*.

The Large Hadron Collider (LHC) site circling under the French-Swiss border is schematically indicated in the picture of Figure 13. The 27 km circular tunnel containing two opposite beams of *protons* surrounded by guiding superconducting electromagnets cooled by superfluid helium is pictured in Figure 14. Figures 15 and 16 are pictures of the ATLAS and CMS detectors at diametrical opposite sides of the tunnel. There collisions occur and were used primarily to detect and identify the *scalar boson* of the *Standard Model* (and possibly other ones). At the end of 2012 *proton-proton collisions* occurred at the rate of nearly 10^9 s^{-1} and the *proton energy* reached 8 TeV. At these energies, all *quarks* of Figure 10 and the *gluons* connecting them may contribute to the production of the *scalar boson*. The leading production processes are represented in Figure 17.

Figure 13. Schematic location of the LHC.

Figure 14. The LHC dipole magnets.

Figure 15. The Atlas detector.

Figure 16. The CMS detector.

Figure 17. Production of the Standard Model scalar.

Figure 18. Decay of the scalar boson into 4 leptons from two Z,s.

As an example of the data gathered by CMS and ATLAS, Figure 18 presents the data obtained by the CMS group of observed decays into 4 *leptons* at the end of the 2012 run. The blue area is the expected background, namely those decays which would follow from the *Standard Model* if, at given *total mass*, there would be no contribution from the *scalar boson*. The red curve measures the contribution that could be due to the *scalar* decaying into two *Z vector bosons* which further decay into *leptons*, as was confirmed by further analysis. Note that one of the Z is real but the other is "virtual," meaning that this decay is forbidden by *energy conservation* but may contribute in the *quantum theory*. Consideration of other decay channels and *spin* analysis show that the particle detected is consistent with the *Standard Model scalar* with a *mass* $m_H \simeq 125$ GeV. The absence of new particles at comparable energies, as well as the success of Feynman graph analysis including *loops*, points towards an elementary particle, at least up to the energy range considered. *This is the first elementary spin zero particle ever detected.* It raises the interesting possibility of *supersymmetry broken at attainable energies*, although there is no indication of it so far.

The elementary character of the *scalar* already eliminates many dynamical models of *symmetry breaking* and raises interesting possibilities for extrapolation beyond presently known energies, up to those close to the Planck scale where *quantum gravity* effects might play a dominant role. The analysis of these speculations is beyond the scope of this talk.

Peter W. Higgs – 2013 Nobel Lecture, December 8, 2013. *Evading the Goldstone Theorem.*

Peter Higgs – Nobel Lecture. NobelPrize.org. https://www.nobelprize.org/prizes/physics/2013/higgs/lecture/.

In his Nobel Prize lecture Higgs provided an account of how he came to publish his paper.

———————————

My story begins in 1960, when I was appointed Lecturer in Mathematical Physics at the University of Edinburgh. Before I took up my appointment, I was invited to serve on the committee of the first Scottish Universities Summer School in Physics. I was asked to act as Steward at the School in July, my principal duty being to purchase and look after supplies of the wine which was to be served at dinner each evening. The students at the school included four who stayed up late into the night in the common room of Newbattle Abbey College (the crypt of a former abbey) discussing theoretical physics, and rarely got up in time for the first lecture of the following day. They were Dr. N. Cabibbo (Rome), Dr. S. L. Glashow (CERN), Mr. D. W. Robinson (Oxford) and Mr. M. J. G. Veltman (Utrecht). Many years later, Cabibbo told me that their discussions had been lubricated by bottles of wine collected after dinner and hidden inside the grandfather clock in the crypt. I did not take part in these discussions, since I had other things to do (such as conserving wine). Consequently, I did not learn about Glashow's paper on *electroweak unification*, which had already been written.

Broken Symmetries

During my first year as a lecturer, I was in search of a worthwhile research program. In the previous four years in London, I had rather lost my way in particle physics and had become interested in quantum gravity. Symmetry had fascinated me since my student days, and I was puzzled by the approximate symmetries (what are now called flavor symmetries) of particle physics.

Then in 1961, I read Nambu's and Goldstone's papers on models of symmetry breaking in particle physics based on an analogy with the theory of super conductivity. (Nambu's models were inspired by the Bardeen, Cooper & Schrieffer theory, based on Bose condensation of Cooper pairs of electrons: Goldstone used scalar fields, with a 'wine bottle' potential to induce Bose condensation, as in the earlier Ginzburg-Landau theory.) What I found very attractive was the concept of a *spontaneously* broken symmetry, one that is exact in the underlying dynamics but appears broken in the observed phenomena as a consequence of an asymmetric *ground state* ("*vacuum*" in *quantum field theory*).

Most particle theorists at the time did not pay much attention to the ideas of Nambu and Goldstone. Quantum field theory was out of fashion, despite its successes in *quantum electrodynamics*; it was failing to describe either the *strong* or the *weak interactions*. Besides, condensed matter physics was commonly viewed as another country. At a Cornell seminar in 1960, Victor Weisskopf remarked (as recalled by Robert Brout):

> "Particle physicists are so desperate these days that they have to borrow from the new things coming up in many body physics—like BCS [Bardeen, Cooper & Schrieffer theory]. Perhaps something will come of it."

The Goldstone Theorem

There was an obstacle to the success of the Nambu-Goldstone program. Nambu had shown how *spontaneous breaking* of a *chiral symmetry* could generate the *masses* of *spin-½ particles*, such as the *proton* and *neutron*, but his model predicted *massless spin-0 particles* (*pions?*), contrary to experimental evidence. (As noted by Weinberg, any such particles would dominate the radiation of energy from stars). Goldstone had argued that such massless particles would always be the result of excitations around the trough of the wine bottle potential.

In 1962 a paper entitled *"Broken Symmetries"* by Goldstone, Salam and Weinberg proved the *"Goldstone Theorem"*, that *"In a manifestly Lorentz-invariant quantum field theory*, if there is a continuous symmetry *under which the Lagrangian is invariant*, then either the *vacuum state* is also *invariant* or there must exist *spinless particles of zero mass"*. This theorem appeared to put an end to Nambu's program.

Can one evade the Goldstone Theorem?

In 1963 the condensed matter theorist Phil Anderson pointed out that in a superconductor the Goldstone mode becomes a massive *"plasmon"* mode due to long-range (Coulomb) forces, and that this mode is just the longitudinal partner of transverse *electromagnetic* modes, which are also massive. Anderson remarked "The Goldstone zero-mass difficulty is not a serious one, because we can probably cancel it off against an equal Yang-Mills zero-mass problem". However, he did not show that there was a flaw in the *Goldstone theorem* and he did not discuss any *relativistic* model, so particle theorists such as myself received his remark with skepticism.

In March 1964 Abe Klein and Ben Lee suggested that, even in *relativistic theories*, a certain equation which was crucial for the proof of the *Goldstone theorem* could be modified by the addition of an extra term, just as in condensed matter theories. But in June, Wally Gilbert (who was in transition from theoretical physics to molecular biology, for which he later won a Nobel Prize for Chemistry) ruled out this term as a violation of *Lorentz invariance*. It was at this point that my intervention took place.

How to evade the Goldstone Theorem

I read Gilbert's paper on July 16, 1964—it had been published a month earlier, but in those days the University of Edinburgh's copies of *Physical Review Letters* came by sea—and I was upset because it implied that there was no way to evade *Goldstone's theorem*.

> [*Goldstone's Theorem* states that "*in a manifestly Lorentz-invariant quantum field theory, if there is a continuous symmetry transformation under which the Lagrangian is invariant, then either the vacuum state is also invariant under the transformation, or there must exist spinless particles of zero mass*".]

But over the following weekend I began to recall that I had seen similar apparent violations of *Lorentz invariance* elsewhere, in no less a theory than *quantum electrodynamics*, as formulated by Julian Schwinger.

Quantum electrodynamics is invariant under *gauge transformations* and the *gauge* must be fixed before well-defined quantum formalism can be set up. The fashionable way to do this was to choose a *Lorentz gauge*, which was manifestly compatible with *relativity*. However, such a *gauge* had unsatisfactory features that led Schwinger to prefer a *Coulomb gauge*, which introduces an apparent *conflict with relativity*. Nevertheless, it was well known that this choice did not lead to any conflict between the predicted physics and relativity.

Schwinger had, as recently as 1962, written papers in which he demolished the folklore that it is *gauge invariance* alone that requires *photons* to be massless. He had provided examples of some properties of a *gauge theory* containing massive "*photons*", but without describing explicitly the underlying dynamics.

During the weekend of 18–19 July it occurred to me that Schwinger's way of formulating *gauge theories* undermined the axioms which had been used to prove the *Goldstone theorem*. So, *gauge theories* might save Nambu's program. During the following week I wrote a short paper about this. It was sent to *Physics Letters* on 24 July and was accepted for publication. [Higgs, P. W. (September, 1964). Broken symmetries, massless particles and gauge fields. *Physics Letters.*, 12, 2, 132–3; doi:10.1016/0031-9163(64)91136-9.]

By then I had written down the (classical) field equations of the simplest illustrative model that I could imagine, the result of introducing an *electromagnetic interaction* into Goldstone's simplest scalar model. It became obvious that in this model the Goldstone *massless mode* became the longitudinal polarization of a massive spin-1 "*photon*", just as

Anderson had suggested. My second short paper, consisting of a brief account of this model, was sent to *Physics Letters* on 31 July. *It was rejected.* The editor (at CERN) suggested that I develop my ideas further and write a full account for *Il Nuovo Cimento*.

I was indignant; it seemed that the referee had not seen the point of my paper. (Later, a colleague who returned from a month's visit to CERN told me that the theorists there did not think it had any relevance to particle physics.) Besides, it seemed odd that the earlier paper had been accepted but the more physical sequel had not.

I decided to augment the paper by some remarks on possible physical consequences, and to send the revised version across the Atlantic to *Physical Review Letters*. Among the additional material was the remark, "It is worth noting that an essential feature of this type of theory is the prediction of incomplete multiplets of scalar and vector bosons."

The revised paper was received by *Physical Review Letters* on 31 August and was accepted. [Higgs, P. W. (October, 1964). Broken Symmetries and the Masses of Gauge Bosons. See above.] The referee invited me to comment on the relation of my paper to that of Englert and Brout, whose paper (received on 22 June) had been published that day. Until then I had been unaware of their work, but I added a footnote to my paper as soon as I had received a copy of theirs. Twenty years later, at a conference in 1984, I met Nambu, who revealed that he had refereed both papers.

Postcript

It took some time for the work of Englert and Brout and myself (and of Guralnik, Hagen and Kibble, who published a little later) to gain acceptance.

My longer (1966) paper was written in autumn 1965 at Chapel Hill, North Carolina, where I was spending a sabbatical year at the invitation of Bryce DeWitt as a consequence of my interest in *quantum gravity*. [Higgs, P. W.* (May, 1966). Spontaneous Symmetry Breakdown without Massless Bosons. See below.] A preprint sent to Freeman Dyson received a positive response; he invited me to give a talk at I.A.S. Princeton. There, in March 1966, I faced an audience including axiomatic quantum field theorists who still believed that there could be no exceptions to the *Goldstone theorem*.

The next day I gave a talk at Harvard (arranged by Stanley Deser) to another skeptical audience, including Wally Gilbert. I survived this too. After the seminar Shelly Glashow complimented me on having invented 'a nice model', but he did not recognize its relevance to his *electroweak theory*—a missed opportunity!

Like Nambu, the six of us who published in 1964 expected to apply our ideas to the *broken flavor symmetries* of the *strong interactions*, but this did not work. So, it was left to Weinberg and Salam in 1967 to find the right application. [Salam, A. Weak and Electromagnetic Interactions; in (1968). *Elementary Particle Theory* (Ed. N Svartholm), Almqvist and Wiskell, pp. 367–377; Weinberg, S. (1967). A Model of Leptons. *Phys. Rev. Let.*, 19, 1264. See below.]

Four more years passed before Gerard 't Hooft, in an extension of Veltman's program, proved the *renormalizability* of such theories and another two before the discovery of *weak neutral currents* indicated that Glashow's *electroweak unification* was the correct one. And in 1976 Ellis, Gaillard and Nanopoulos at CERN encouraged experimentalists to look for the massive *spinless boson* that the theory predicted.

Abdus Salam (January 29, 1926–November 21, 1996).

Salam was a Pakistani theoretical physicist. He shared the 1979 Nobel Prize in Physics with Sheldon Glashow and Steven Weinberg "for their contributions to the theory of the *unified weak and electromagnetic interaction* between elementary particles, including, inter alia, the prediction of the *weak neutral current*". He was the first Pakistani to receive a Nobel Prize.

Salam was born on January 29, 1926 in the Punjab Province of British India (now in Pakistan). His father Choudhary Muhammad Hussain was a minor educational official and a teacher. Abdus Salam's father was stationed in a poor farming district in Jhang, where Abdus Salam spent his early years.

Salam very early established a reputation throughout Punjab for outstanding brilliance and academic achievement. At age 14, Salam scored the highest marks ever recorded for the entrance examination at the Punjab University. He won a full scholarship to the Government College University of Lahore. Salam was a versatile scholar, interested in Urdu and English literature in which he excelled. After a month in Lahore, he went to Bombay to study. In 1947, he came back to Lahore. But he soon picked up Mathematics as his concentration. Salam's mentor and tutors wanted him to become an English teacher, but Salam decided to stick with Mathematics. As a fourth-year student there, he published his work on Srinivasa Ramanujan's problems in mathematics, and took his B.A. in Mathematics in 1944.

While in Lahore, Salam went on to attend the graduate school of Government College University. He received his MA in Mathematics from the Government College University in 1946. That same year, he was awarded a scholarship to St John's College, Cambridge, where he completed a BA degree with Double First-Class Honors in Mathematics and Physics in 1949. In 1950, he received the Smith's Prize from Cambridge University for the most outstanding pre-doctoral contribution to Physics. After finishing his degrees, Fred Hoyle advised Salam to spend another year in the Cavendish Laboratory to do research in experimental physics, but Salam had no patience for carrying out long experiments in the laboratory. Salam returned to Jhang and renewed his scholarship and returned to the United Kingdom to do his doctorate.

He obtained a PhD degree in theoretical physics from the Cavendish Laboratory at Cambridge. His doctoral thesis titled "*Developments in quantum theory of fields*" contained comprehensive and fundamental work in *quantum electrodynamics*. By the time it was published in 1951, it had already gained him an international reputation and the Adams Prize. During his doctoral studies, his mentors challenged him to solve within one year an intractable problem which had defied such great minds as Paul Dirac and Richard Feynman. Within six months, Salam had found a solution for the *renormalization of meson theory*. As he proposed the solution at the Cavendish Laboratory, Salam had attracted the attention of Hans Bethe, J. Robert Oppenheimer and Dirac.

After receiving his doctorate in 1951, Salam returned to Lahore at the Government College University as a Professor of Mathematics where he remained till 1954. In 1952, he was appointed professor and Chair of the Department of Mathematics at the neighboring University of the Punjab. In 1953, Salam was unable to establish a research institute in Lahore, as he faced strong opposition from his peers. As a result of the 1953 Lahore riots, Salam went back to Cambridge and joined St John's College, and took a position as a professor of mathematics in 1954. In 1957, he was invited to take a chair at Imperial College, London, and he and Paul Matthews went on to set up the Theoretical Physics Group at Imperial College.

In 1959, Salam took a fellowship at the Princeton University, where he met with J. Robert Oppenheimer and to whom he presented his research work on *neutrinos*. Salam had worked on theory of the *neutrino* – an elusive particle that was first postulated by Wolfgang Pauli in the 1930s. Salam introduced *chiral symmetry* in the *theory of neutrinos*. The introduction of *chiral symmetry* played crucial role in subsequent development of the theory of *electroweak interactions*. Salam introduced the massive *Higgs bosons* to the theory of the *Standard Model*, where he later predicted the existence of *proton decay*. In 1963, Salam published his theoretical work on the *vector meson*. The paper introduced the *interaction* of *vector meson*, *photon* (*vector electrodynamics*), and the *renormalisation* of *vector mesons'* known *mass* after the *interaction*.

In 1961, Salam began to work with John Clive Ward on *symmetries* and *electroweak unification*. [Salam, A. & Ward, J. C. (1961). On a gauge theory of elementary interactions. *Il Nuovo Cimento*, 19, 1, 165–70; https://doi.org/10.1007/BF02812723.]

While at Imperial College, in their 1962 paper "*Broken Symmetries*", Goldstone, Salam and Weinberg proved the "*Goldstone Theorem*" that "*In a manifestly Lorentz-invariant quantum field theory, if there is a continuous symmetry under which the Lagrangian is invariant, then either the vacuum state is also invariant or there must exist spinless particles of zero mass*".

In 1964, Salam and Ward worked on the synthesis of the *weak* and *electromagnetic interaction*, obtaining a *gauge theory* based on the SU(2) × U(1) model. [Salam, A. & Ward, J. C. (November, 1964). Electromagnetic and Weak Interactions. See below.]

Physicists had believed that there were four fundamental forces of nature: the *gravitational force*, the *strong* and *weak nuclear forces*, and the *electromagnetic force*. Salam had worked on the unification of these forces from 1959 with Glashow and Weinberg. While at Imperial College London, Salam successfully showed that *weak nuclear* forces are not really different from *electromagnetic forces*, and two could inter-convert. Salam provided a theory that shows the unification of two fundamental forces of nature, *weak nuclear forces* and the *electromagnetic forces*, one into another. Glashow had also formulated the same work, and the theory was combined in 1966. In 1967, Salam proved the *electroweak unification theory* mathematically, and finally published the papers.

Following the publication of *Physics Review Letters (PRL) Symmetry Breaking* papers in 1964, Steven Weinberg and Salam were the first to apply the *Brout–Englert–Higgs* (BEH) *mechanism* to *electroweak symmetry breaking*. Salam provided a mathematical postulation for the *interaction* between the *Higgs boson* and the *electroweak symmetry theory*.

In 1967-8, Salam and Weinberg incorporated the *Brout–Englert–Higgs* (BEH) *mechanism* into Glashow's discovery, giving it a modern form in *electroweak theory*, and thus theorized half of the *Standard Model*. In 1968, together with Weinberg and Sheldon Glashow, Salam finally formulated the mathematical concept of their work.

In the 1970s Salam continued trying to unify forces by including the *strong interaction* in a *grand unified theory*. In 1972, Salam began to work with Indian-American theoretical physicist Jogesh Pati. Pati wrote to Salam several times expressing interest to work under Salam's direction, in response to which Salam eventually invited Pati to the ICTP seminar in Pakistan. Salam suggested to Pati that there should be some deep reason why the *protons* and *electrons* are so different and yet carry equal but opposite *electric charge*. *Protons* are composed of *quarks*, but the *electroweak theory* was concerned only with the *electrons* and *neutrinos*, with nothing postulated about *quarks*. If all of nature's ingredients could be brought together in one new symmetry, it might reveal a reason for the various features of these particles and the forces they feel. This led to the development of *Pati–Salam model* in particle physics. In 1973, Salam and Pati were the first to notice that since *quarks* and *leptons* have very similar SU(2) × U(1) representation content, they all may have similar entities. They provided a simple realization of the *quark-lepton symmetry* by postulating that the *lepton number* was a fourth *quark color*, dubbed "violet". [Pati, J. C. & Salam, A. (August, 1973). Unified Lepton-Hadron Symmetry and a Gauge Theory of the Basic Interactions. See below)]

Salam, Glashow, and Weinberg were awarded the 1979 Nobel Prize in Physics "for their contributions to the theory of the *unified weak and electromagnetic interaction* between elementary particles, including, inter alia, the prediction of the *weak neutral current*".

Salam was scientific advisor to the Ministry of Science and Technology in Pakistan from 1960 to 1974, a position from which he played a major and influential role in the development of the country's science infrastructure. Salam contributed to numerous developments in theoretical and particle physics in Pakistan. In 1974, Salam departed from his country in protest after the Parliament of Pakistan passed unanimously a parliamentary bill declaring members of the Ahmadiyya Muslim community, to which Salam belonged, non-Muslim.

Salam made a major contribution in *quantum field theory* and in the advancement of Mathematics at Imperial College London. With his student, Riazuddin, Salam made important contributions to the modern theory on *neutrinos*, *neutron stars* and *black holes*, as well as the work on modernizing *quantum mechanics* and *quantum field theory*.

Salam married twice; first time to a cousin, the second time as well in accordance with Islamic law. At his death, he was survived by three daughters and a son by his first wife, and a son and daughter by his second.

Salam, A. & Ward, J. C.* (November, 1964). Electromagnetic and weak interactions.

Physics Letters, 13, 2, 168–71; https://doi.org/10.1016/0031-9163(64)90711-5.

Imperial College, London. * Permanent address, John Hopkins University, Baltimore.

Received September 24, 1964.

In 1964, Salam and Ward worked on a *gauge theory* for the *weak* and *electromagnetic interaction*, subsequently obtaining the SU(2) × U(1) model. Salam was convinced that all the *elementary particle interactions* are actually the *gauge interactions*.

One of the recurrent dreams in *elementary particle* physics is that of a possible fundamental synthesis between *electro-magnetism* and *weak interactions*. The idea has its origin in the following shared characteristics:

1) *Both forces affect equally all forms of matter- leptons as well as hadrons.*

2) *Both are vector in character.*

3) *Both (individually) possess universal coupling strengths.*

Since *universality* and *vector* character are features of a *gauge-theory* these shared characteristics suggest that *weak* forces just like the *electromagnetic* forces arise from a *gauge principle*.

There of course also are profound differences:

1) *Electromagnetic coupling strength* is vastly different from the *weak*. Quantitatively one may state it thus: if *weak* forces are assumed to have been mediated by intermediate *bosons* (W), the *boson mass* would have to equal $137\,M_p$, in order that the (dimensionless) weak *coupling constant* $g_W^2/4\pi$ equals $e^2/4\pi$. In the sequel we assume just this. For the outrageous *mass* value itself ($M_W \sim 137\,M_p$) *we can offer no explanation*. We seek however for a synthesis in terms of a group structure such that the remaining differences, viz:

2) Contrasting *space-time behavior* (V for *electromagnetic* versus V and A for *weak*).

3) And *contrasting ΔS and ΔI behaviors* both appear as aspects of the same fundamental *symmetry*. Naturally for *hadrons* at least the group structure must be compatible with SU_3.

Lepton interactions define both the unit of the *electric charge* and (from μ-decay) the (bare) value of the *weak coupling constant*. *Leptons* therefore must be treated first. ...

Higgs, P. W.* (May, 1966). Spontaneous Symmetry Breakdown without Massless Bosons.

Phys. Rev., 145, 4, 1156; https://doi.org/10.1103/PhysRev.145.1156.

Department of Physics, University of North Carolina, Chapel Hill, North Carolina.

* On leave from the Tait Institute of Mathematical Physics, University of Edinburgh, Scotland.

Received December 27, 1965.

Abstract

We examine a simple *relativistic* theory of two *scalar fields*, first discussed by Goldstone, in which as a result of *spontaneous breakdown* of U(1) *symmetry* one of the *scalar bosons* is *massless*, in conformity with the *Goldstone theorem*. When the *symmetry group* of the Lagrangian is extended from *global* to *local* U(1) *transformations* by the introduction of *coupling* with a *vector gauge field*, the *Goldstone boson* becomes the *longitudinal state* of a *massive vector boson* whose *transverse states* are the *quanta* of the *transverse gauge field*. A *perturbative* treatment of the model is developed in which the major features of these phenomena are present in zero order. *Transition amplitudes* for *decay* and *scattering* processes are evaluated in lowest order, and it is shown that they may be obtained more directly from an equivalent Lagrangian in which the original *symmetry* is no longer manifest. When the system is coupled to other systems in a *U(1) invariant* Lagrangian, the other systems display an induced *symmetry breakdown*, associated with a partially conserved *current* which interacts with itself via the *massive vector boson*.

Steven Weinberg (May 3, 1933–July 23, 2021).

Weinberg was an American theoretical physicist who was awarded the 1979 Nobel Prize in Physics, together with, Abdus Salam and Sheldon Glashow "for their contributions to the theory of the *unified weak and electromagnetic interaction* between elementary particles, including, inter alia, the prediction of the *weak neutral current*".

Steven Weinberg was born in 1933 in New York City. His parents were Jewish immigrants; his father, Frederick, worked as a court stenographer, while his mother, Eva (Israel), was a housewife. Becoming interested in science at age 16 through a chemistry set handed down by a cousin, he graduated from Bronx High School of Science in 1950. He was in the same graduating class as Sheldon Glashow, whose research, independent of Weinberg's, resulted in their (and Abdus Salam's) sharing the 1979 Nobel in physics. Weinberg received his bachelor's degree from Cornell University in 1954.

In 1954 Weinberg married legal scholar Louise Goldwasser and they had a daughter, Elizabeth.

He then went to the Niels Bohr Institute in Copenhagen, where he started his graduate studies and research. After one year, Weinberg moved to Princeton University, where he earned his Ph.D. in physics in 1957, completing his dissertation, "*The role of strong interactions in decay processes*".

After completing his Ph.D., Weinberg worked as a postdoctoral researcher at Columbia University (1957–59) and University of California, Berkeley (1959) and then was promoted to the faculty at Berkeley (1960–66). He did research in a variety of topics of particle physics, such as the high energy behavior of quantum field theory, symmetry breaking, pion scattering, infrared photons and quantum gravity (soft graviton theorem). It was also during this time that he developed the approach to *quantum field theory* described in the first chapters of his book "*The Quantum Theory of Fields*" and started to write his textbook *Gravitation and Cosmology*, having taken up an interest in *general relativity* after the discovery of *cosmic microwave background radiation*. The Quantum Theory of Fields spanned three volumes and over 1,500 pages, and is often regarded as the leading book in the field.

In 1966, Weinberg left Berkeley and accepted a lecturer position at Harvard. In 1967 he was a visiting professor at MIT. It was in that year at MIT that Weinberg proposed his *model of unification of electromagnetism and nuclear weak forces* (such as those involved in *beta-decay* and *kaon-decay*), with the masses of the force-carriers of the weak part of the interaction being explained by *spontaneous symmetry breaking*. The paper by

Weinberg in which he presented this theory is one of the most cited works ever in high-energy physics. [Weinberg, S. (November, 1967). Model of Leptons. *Phys. Rev. Let.*, 19, 21, 1264–6. See below.]

One of its fundamental aspects was the *prediction of the existence of the Higgs boson.* Weinberg's model, now known as the *electroweak unification theory*, had the same *symmetry structure* as that proposed by Glashow in 1961: both included the then-unknown *weak interaction* mechanism between *leptons*, known as *neutral current* and mediated by the *Z boson*. The 1973 experimental discovery of *weak neutral currents* (mediated by this Z boson) was one verification of the *electroweak unification*. [Haidt, D. (2004). The discovery of the weak neutral currents. *CERN Courier*.]

After his 1967 seminal work on the *unification of weak and electromagnetic interactions*, Weinberg continued his work in many aspects of particle physics, quantum field theory, gravity, supersymmetry, superstrings and cosmology. In the years after 1967, the full *Standard Model* of elementary particle theory was developed through the work of many contributors. In it, *the weak and electromagnetic interactions already unified by the work of Weinberg, Salam and Glashow, are made consistent with a theory of the strong interactions between quarks, in one overarching theory.* In 1973, Weinberg proposed a modification of the *Standard Model* that did not contain that model's fundamental *Higgs boson*. Also, during the 1970s, he proposed a theory later known as technicolor, in which new *strong interactions* resolve the hierarchy problem.

Weinberg became Eugene Higgins Professor of Physics at Harvard University in 1973, a post he held until 1983. In 1979 he pioneered the modern view on the *renormalization* aspect of *quantum field theory* that considers all quantum field theories *effective field theories* and *changed the viewpoint of previous work (including his own in his 1967 paper) that a sensible quantum field theory must be renormalizable*. This approach allowed the development of *effective theory of quantum gravity*, *low energy quantum chromodynamics*, *heavy quark effective field theory* and other developments, and is a topic of considerable interest in current research.

In 1979, *some six years after the experimental discovery of the neutral currents*—i.e. the discovery of the inferred existence of the *Z boson*—and after the 1978 experimental discovery of the theory's predicted amount of *parity violation due to Z bosons' mixing with electromagnetic interactions,* [Charles Y. Prescott (June 30, 1978). Parity violation in inelastic scattering of polarized electrons (PDF). Sixth Trieste Conference on Particle Physics. *AIP Conference Proceedings. Vol. 51*. Trieste, Italy: American Institute of Physics. p. 202; doi:10.1063/1.31766] Weinberg was awarded the Nobel Prize in physics with Glashow and Salam, who had independently proposed a *theory of electroweak unification* based on *spontaneous symmetry breaking*.

In 1982 Weinberg moved to the University of Texas at Austin as the Jack S. Josey-Welch Foundation Regents Chair in Science, and started a theoretical physics group at the university that now has eight full professors and is one of the leading research groups in the field in the U.S.

Weinberg died on July 23, 2021, at age 88 at a hospital in Austin, where he had been undergoing treatment for several weeks.

Weinberg, S.[†] (November, 1967). Model of Leptons.[*]

Phys. Rev. Let., 19, 21-22, 1264–6; https://doi.org/10.1103/PhysRevLett.19.1264.

[*] This work is supported in part through funds provided by the U. S. Atomic Energy Commission under Contract No. AT(30-1)2098.

[†] On leave from the University of California, Berkeley, California.

Laboratory for Nuclear Science and Physics Department, Massachusetts Institute of Technology, Cambridge, Massachusetts.

Received October 17, 1967

In 1967 Steven Weinberg incorporated the *Brout–Englert–Higgs* (BEH) *mechanism* into Glashow's *electroweak interaction*, giving it its modern form. Weinberg proposed his *model of unification of electromagnetism and nuclear weak forces* with the *masses* of the force-carriers of the *weak* part of the *interaction* being explained by *spontaneous symmetry breaking*, in which the *symmetry* between the *electromagnetic* and *weak interactions* is *spontaneously broken*, but in which the *Goldstone bosons* are avoided by introducing the *photon* and the *intermediate boson fields* as *gauge fields*.

Leptons interact only with *photons*, and with the *intermediate bosons* that presumably mediate *weak interactions*. What could be more natural than to unite[1] these *spin-one bosons* into a *multiplet* of *gauge fields*?

[A *lepton* is an elementary particle of half-integer *spin* (*spin* ½) that does not undergo *strong interactions*. Two main classes of *leptons* exist: *charged leptons* (also known as the *electron-like leptons* or *muons*), including the *electron, muon,* and *tauon,* and *neutral leptons*, better known as *neutrinos*. *Charged leptons* can combine with other particles to form various composite particles such as atoms and *positronium*, while *neutrinos* rarely interact with anything, and are consequently rarely observed. The best known of all *leptons* is the *electron*.

There are six types of *leptons*, known as flavors, grouped in three generations. The *first-generation leptons*, also called electronic leptons, comprise the electron (e^-) and the electron neutrino (v_e); the second are the *muonic leptons*, comprising the muon (μ^-) and the *muon neutrino* (v_μ); and the third are the *tauonic leptons*, comprising the *tau* (τ^-) and the *tau neutrino* (v_τ). *Electrons* have the least mass of all the *charged leptons*. The heavier *muons* and *taus* will rapidly change into *electrons* and *neutrinos* through a process of particle decay: the transformation from a higher mass state to a lower mass state. Thus, *electrons* are stable and the most common charged *lepton* in the universe, whereas *muons* and *taus* can only be produced in high-energy collisions (such as those involving cosmic rays and those carried out in particle accelerators).

Leptons have various intrinsic properties, including *electric charge*, *spin*, and *mass*. Unlike *quarks*, however, *leptons* are not subject to the *strong interaction*, but they are subject to the other three fundamental interactions: *gravitation*, the *weak interaction*, and to *electromagnetism*, of which the latter is proportional to charge, and is thus zero for the electrically neutral *neutrinos*.

For every *lepton flavor*, there is a corresponding type of *antiparticle*, known as an *antilepton*, that differs from the *lepton* only in that some of its properties have equal magnitude but opposite sign. According to certain theories, *neutrinos* may be their own antiparticle. It is not currently known whether this is the case.

The first *charged lepton*, the *electron*, was theorized in the mid-19th century by several scientists and was discovered in 1897 by J. J. Thomson. The next *lepton* to be observed was the *muon*, discovered by Carl D. Anderson in 1936, which was classified as a meson at the time. After investigation, it was realized that the *muon* did not have the expected properties of a meson, but rather behaved like an *electron*, only with higher mass. It took until 1947 for the concept of "*leptons*" as a family of particles to be proposed. The term *lepton* was first used by physicist Léon Rosenfeld in 1948.

The first *neutrino*, the *electron neutrino*, was proposed by Wolfgang Pauli in 1930 to explain certain characteristics of *beta decay*. It was first observed in the Cowan–Reines *neutrino* experiment conducted by Clyde Cowan and Frederick Reines in 1956. The *muon neutrino* was discovered in 1962 by Leon M. Lederman, Melvin Schwartz, and Jack Steinberger, and the *tau* discovered between 1974 and 1977 by Martin Lewis Perl and his colleagues from the Stanford Linear Accelerator Center and Lawrence Berkeley National Laboratory. The *tau neutrino* remained elusive until July 2000, when the DONUT collaboration from Fermilab announced its discovery.

Leptons are an important part of the *Standard Model*. Electrons are one of the components of atoms, alongside protons and neutrons. Exotic atoms with muons and *taus* instead of *electrons* can also be synthesized, as well as *lepton–antilepton* particles such as *positronium*.]

[1] The history of attempts to unify *weak* and *electromagnetic interactions* is very long, and will not be reviewed here. Possibly the earliest reference is Fermi, E. (March, 1934) Versuch einer Theorie der β-Strahlen. I. (Attempt at a theory of β rays. I.) *Zeit. Phys.*, 88, 161–77 (1934). See above. A model similar to ours was discussed by Glashow, S. L. (February, 1961). Partial-symmetries of weak interactions. *Nuclear Physics*, 22, 4, 579–88, see above; the chief difference is that Glashow introduces *symmetry-breaking* terms into the Lagrangian, and therefore gets less definite predictions.

Standing in the way of this synthesis are the obvious differences in the *masses* of the *photon* and *intermediate mesons*, and in their *couplings*. We might hope to understand these differences by imagining that the *symmetries* relating the *weak* and *electromagnetic interactions* are exact *symmetries* of the Lagrangian but are broken by the *vacuum*. However, this raises the specter of unwanted *massless Goldstone bosons*[2].

[2] Goldstone, J. (January, 1961). Field theories with "Superconductor" solutions. *Nuovo Cimento*, 19, 154–64, see above; Goldstone, J., Salam, A. & Weinberg, S. (August, 1962). Broken Symmetries. *Phys. Rev.* 127, 965, see above.

This note will describe a model in which the *symmetry* between the *electromagnetic* and *weak interactions* is *spontaneously broken*, but in which the *Goldstone bosons* are avoided by introducing the *photon* and the *intermediate boson fields* as *gauge fields*[3].

[3] Higgs, P. W. (September, 1964). Broken symmetries, massless particles and gauge fields. *Physics Letters.*, 12, 2, 132–3; doi:10.1016/0031-9163(64)91136-9; Higgs, P. W. (October, 1964). Broken Symmetries and the Masses of Gauge Bosons. *Phys. Rev. Let.*, 13, 16, 508–9, see above; Higgs, P. W. (May, 1966). Spontaneous Symmetry Breakdown without Massless Bosons. *Phys. Rev.*, 145, 4, 1156, see above; Englert, F. & Brout, R. (August, 1964). Broken Symmetry and the Mass of Gauge Vector Mesons. *Phys. Rev. Lett.*, 13, 9, 321–3, see above; Guralnik, G., Hagen, C. & Kibble, T. (November, 1964). Global Conservation Laws and Massless Particles. *Phys. Rev. Lett.*, 13, 20, 585; doi:10.1103/PhysRevLett.13.585, see above.

The model may be *renormalizable*. We will restrict our attention to *symmetry groups* that connect the observed *electron-type leptons* only with each other, i.e., not with *muon-type leptons* or other unobserved *leptons* or *hadrons*. The *symmetries* then act on a left-handed *doublet*

$$L \equiv [\tfrac{1}{2}(1 + \gamma_5)] \binom{\nu_e}{e} \tag{1}$$

and on a right-handed *singlet*

$$R \equiv [\tfrac{1}{2}(1 + \gamma_5)]e. \tag{2}$$

The largest group that leaves *invariant* the *kinematic terms* ... of the Lagrangian consists of the *electronic isospin* $\vec{T}$ acting on L, plus the numbers N_L, N_R of left- and right-handed *electron-type leptons*. As far as we know, two of these *symmetries* are entirely unbroken: the *charge* $Q = T_3 - N_R - \tfrac{1}{2} N_L$, and the *electron number* $N = N_R + N_L$. But the *gauge field* corresponding to an *unbroken symmetry* will have zero *mass*[4], and there is no massless particle coupled to N, [5] so we must form our *gauge group* out of the *electronic isospin* $\vec{T}$ and the *electronic hypercharge* $Y = N_R + \tfrac{1}{2} N_L$.

[4] See particularly Kibble, T. W. B. (March, 1967). Symmetry Breaking in Non-Abelian Gauge Theories. *Phys. Rev.*, 155, 5, 1554; https://doi.org/10.1103/PhysRev.155.1554. A similar phenomenon occurs in the *strong interactions*; the *p-meson mass* in zeroth-order perturbation theory is just the bare *mass*, while the *meson* picks up an extra contribution from the spontaneous breaking of *chiral symmetry*. See Weinberg, S. (March, 1967). Precise Relations between the Spectra of Vector and Axial-Vector Mesons. *Phys. Rev. Lett.*, 18, 13, 507; https://doi.org/10.1103/PhysRevLett.18.507, especially footnote 7; Schwinger, J. (January, 1967). Chiral dynamics. *Phys. Letters*, 24B, 473; https://doi.org/ 10.1016/0370-2693(67)90277-8; Glashow, S., Schnitzer, H. & S. Weinberg, S. (July, 1967). Sum Rules for the Spectral Functions of SU(3) x SU(3). *Phys. Rev. Lett.*, 19, 3, 139; https://doi.org/10.1103/PhysRevLett.19.139, Eq. (13) et seq..

[5] Lee, T. D. & Yang, C. N. (June, 1955). Conservation of Heavy Particles and Generalized Gauge Transformations. *Phys. Rev.*, 98, 5, 101; https://doi.org/10.1103/ PhysRev.98.1501.

Therefore, we shall construct our Lagrangian out of L and B, plus *gauge fields* $\vec{A_\mu}$ and B^μ coupled to $\vec{T}$ and Y, plus a *spin-zero doublet*

$$\varphi = \begin{pmatrix} \varphi^0 \\ \varphi^- \end{pmatrix} \tag{3}$$

whose *vacuum expectation* value will break $\vec{T}$ and Y and give the *electron* its *mass*. The only *renormalizable* Lagrangian which is invariant under $\vec{T}$ and Y *gauge transformations* is

$$\mathscr{L} = \dots . \tag{4}$$

We have chosen the *phase* of the R field to make G_e real, and can also adjust the *phase* of the L and Q fields to make the *vacuum expectation value* $\lambda \equiv \langle \varphi^0 \rangle$ real. The "physical" φ *fields* are then φ^- and

$$\varphi_1 \equiv (\varphi^0 + \varphi^{0\dagger} - 2\lambda) \sqrt{2}, \qquad \varphi_2 \equiv (\varphi^0 - \varphi^{0\dagger}) i\sqrt{2}, \tag{5}$$

385

The condition that φ_1 have zero *vacuum expectation value* to all orders of *perturbation theory* tells us that $\lambda^2 \simeq M_1^2/2h$, and therefore the *field* φ_1 has *mass* M_1 while φ_2 and φ^- have *mass* zero. But we can easily see that the *Goldstone bosons* represented by φ_2 and φ^- have no physical *coupling*. The Lagrangian is *gauge invariant*, so we can perform a combined *isospin* and *hypercharge gauge transformation* which eliminates φ^- and φ_2 everywhere[6] without changing anything else.

[6] This is the same sort of transformation as that which eliminates the nonderivative $\pi^{\rightarrow}$ *couplings* in the σ model; see Weinberg, S. (January, 1967). Dynamical Approach to Current Algebra. *Phys. Rev. Lett.*, 18, 5, 188; https://doi.org/10.1103/PhysRevLett. 18.188. The $\pi^{\rightarrow}$ reappears with derivative *coupling* because the *strong-interaction* Lagrangian is not invariant under a *chiral gauge transformation*.

We will see that G_e is very small, and in any case M_1 might be very large[7], so the φ_1 *couplings* will also be disregarded in the following.

[7] For a similar argument applied to the σ *meson*, see Weinberg, Ref. 6.

The effect of all this is just to replace φ everywhere by its *vacuum expectation value*

$$\langle\varphi\rangle = \begin{matrix}(1)\\(0)\end{matrix} \tag{6}$$

The first four terms in $\mathscr{L}$ remain intact, while the rest of the Lagrangian becomes

$$\dots . \tag{7}$$

We see immediately that the *electron mass* is λG_e. The charged *spin-1 field* is

$$W_\mu \equiv 2^{-1/2} (A_\mu^1 + iA_\mu^2) \tag{8}$$

and has *mass*

$$M_W = \tfrac{1}{2} \lambda g. \tag{9}$$

The neutral *spin-1 fields* of definite *mass* are

$$Z_\mu = \dots , \tag{10}$$
$$A_\mu = \dots . \tag{11}$$

Their *masses* are

$$M_Z = \tfrac{1}{2} \lambda(g^2 + g'^2)^{1/2}, \tag{12}$$
$$M_A = 0, \tag{13}$$

so A_μ is to be identified as the *photon field*. The interaction between *leptons* and *spin-1 mesons* is

$$\ldots .\qquad(14)$$

We see that the rationalized *electric charge* is

$$E = gg'/(g^2 + g'^2)^{1/2}\qquad(15)$$

and, assuming that W_μ couples as usual to *hadrons* and *muons*, the usual *coupling constant* of *weak interactions* is given by

$$G_W/\sqrt{2} = g^2/8M_W^2 = 1/2\lambda^2.\qquad(16)$$

Note that then the e-p *coupling constant* is

$$G_e = M_e/\lambda = 2^{1/4}\, M_e G_W = 2.07\ 10–6.$$

The coupling of φ_1 to *muons* is stronger by a factor M_μ/M_e, but still very weak. Note also that (14) gives g and g' larger than e, so (16) tells us that $M_W > 40$ BeV, while (12) gives $M_Z > M_W$ and $M_Z > 80$ BeV.

The only unequivocal new predictions made by this model have to do with the *couplings* of the *neutral intermediate meson* Z_μ. If Z_μ does not couple to *hadrons* then the best place to look for effects of Z_μ is in *electron-neutron scattering*. Applying a *Fierz transformation* to the *W-exchange* terms, the total effective e-v *interaction* is

$$\ldots .$$

If g » e then g » g', and this is just the usual e-v *scattering matrix element* times an extra factor 3/2. If g ≃ e then g « g', and the *vector interaction* is multiplied by a factor ½ rather than 3/2. Of course, our model has too many arbitrary features for these predictions to be taken very seriously, but it is worth keeping in mind that the standard calculation[8] of the *electron-neutrino cross section* may well be wrong.

[8] Feynman, R. P. & Gell-Mann, M. (January, 1958). Theory of the Fermi Interaction. *Phys. Rev.*, 109, 1, 193; https://doi.org/10.1103/PhysRev.109.193.

Is this model *renormalizable*?

[*Renormalization* is a collection of techniques in *quantum field theory, statistical field theory*, and the *theory of self-similar geometric structures*, that are used to treat infinities arising in calculated quantities *by altering values of these quantities to compensate for effects of their self-interactions.* But even if no infinities arose in *loop diagrams* in *quantum field theory*, it could be shown that it would be necessary to *renormalize* the *mass* and *fields* appearing in the original Lagrangian.

For example, an *electron theory* may begin by postulating an *electron* with an initial *mass* and *charge*. In *quantum field theory* a cloud of virtual particles, such as *photons*, *positrons*, and others surround and interact with the initial *electron*. Accounting for the interactions of the surrounding particles (e.g. collisions at different energies) shows that the electron-system behaves as if it had a different *mass* and *charge* than initially postulated. *Renormalization*, in this example, mathematically replaces the initially postulated *mass* and *charge* of an *electron* with *the experimentally observed mass and charge.* ???!]

We usually do not expect non-Abelian *gauge theories* to be *renormalizable* if the *vector-meson mass* is not zero, but our Z_μ and W_μ *mesons* get their *mass* from the *spontaneous breaking* of the *symmetry*, not from a *mass* term put in at the beginning. Indeed, the model Lagrangian we start from is probably *renormalizable*, so the question is whether this *renormalizability* is lost in the reordering of the *perturbation theory* implied by our redefinition of the *fields*. And if this model is *renormalizable*, then what happens when we extend it to include the *couplings* of $\vec{A}_\mu$ and B_μ to the *hadrons*?

Gerardus (Gerard) 't Hooft (born July 5, 1946).

't Hooft is a Dutch theoretical physicist and professor at Utrecht University, the Netherlands. He is most famous for his contributions to the development of *gauge theories* in particle physics. The best known of these is the proof in his PhD thesis that (Yang–Mills) *gauge theories* are *renormalizable*, for which he shared the 1999 Nobel Prize in Physics with his thesis advisor Martinus J. G. Veltman "for elucidating the quantum structure of *electroweak interactions*". For this proof he introduced (with his adviser Veltman) the technique of dimensional regularization. His work concentrates on gauge theory, black holes, quantum gravity and fundamental aspects of quantum mechanics.

Gerard 't Hooft was born in Den Helder on July 5, 1946, but grew up in The Hague. He was the middle child of a family of three. He comes from a family of scholars. His great uncle was Nobel prize laureate Frits Zernike, and his grandmother was married to Pieter Nicolaas van Kampen, a professor of zoology at Leiden University. His uncle Nico van Kampen was an (emeritus) professor of theoretical physics at Utrecht University, and his mother married a maritime engineer. Following his family's footsteps, he showed interest in science at an early age. When his primary school teacher asked him what he wanted to be when he grew up, he replied, "a man who knows everything."

After primary school Gerard attended the Dalton Lyceum, a school that applied the ideas of the Dalton Plan, an educational method that suited him well. He excelled at science and mathematics courses. At the age of sixteen he won a silver medal in the second Dutch Math Olympiad. After Gerard 't Hooft passed his high school exams in 1964, he enrolled in the physics program at Utrecht University. He opted for Utrecht instead of the much closer Leiden, because his uncle was a professor there and he wanted to attend his lectures. Because he was so focused on science, his father insisted that he join the Utrechtsch Studenten Corps, a student association, in the hope that he would do something else besides studying. This worked to some extent; during his studies he was a coxswain with their rowing club "Triton" and organized a national congress for science students with their science discussion club "Christiaan Huygens".

In the course of his studies, he decided he wanted to go into what he perceived as the heart of theoretical physics, elementary particles. His uncle had grown to dislike the subject and in particular its practitioners, so when it became time to write his doctoraalscriptie (former name of the Dutch equivalent of a master's thesis) in 1968, 't Hooft turned to the newly appointed professor Martinus Veltman, who specialized in *Yang–Mills theory*, a relatively fringe subject at the time because it was thought that these could not be *renormalized*. His assignment was to study the Adler–Bell–Jackiw anomaly, a mismatch in the theory of the decay of *neutral pions*; formal arguments forbid the decay into *photons*, whereas practical calculations and experiments showed that this was the primary form of decay. The resolution of the problem was completely unknown at the time, and 't Hooft was unable to provide one.

In 1969, 't Hooft started on his doctoral research with Martinus Veltman as his advisor. He would work on the same subject Veltman was working on, the *renormalization of Yang–Mills theories*. In 1971 his first paper was published. ['t Hooft, G. (October, 1971). Renormalization of Massless Yang-Mills Fields. *Nucl. Phys.* B, 33, 1, 173-99; https://doi.org/10.1016/055090395-6.] In it he showed how to *renormalize massless Yang–Mills fields*, and was able to derive relations between amplitudes, which would be generalized by Andrei Slavnov and John C. Taylor, and become known as the Slavnov–Taylor identities.

The world took little notice, but Veltman was excited because he saw that the problem he had been working on was solved. A period of intense collaboration followed in which they developed the technique of *dimensional regularization*. Soon 't Hooft's second paper was ready to be published, in which he showed that *Yang–Mills theories with massive fields due to spontaneous symmetry breaking could be renormalized*. ['t Hooft, G. (December, 1971). Renormalizable Lagrangians for Massive Yang-Mills Fields. See below.] This paper earned them worldwide recognition, and would ultimately earn the pair the 1999 Nobel Prize in Physics.

These two papers formed the basis of 't Hooft's dissertation, *The Renormalization procedure for Yang–Mills Fields*, and he obtained his PhD degree in 1972. In the same year he married Albertha Schik (Betteke), a student of medicine in Utrecht, with whom he has two daughters.

After obtaining his doctorate 't Hooft went to CERN in Geneva, where he had a fellowship. He further refined his methods for Yang–Mills theories with Veltman (who went back to Geneva). In this time, he became interested in the possibility that the *strong interaction* could be described as a *massless Yang–Mills theory*, i.e. one of a type that he had just proved to be *renormalizable* and hence be susceptible to detailed calculation and comparison with experiment.

According to 't Hooft's calculations, this type of theory possessed just the right kind of scaling properties (asymptotic freedom) that this theory should have according to deep inelastic scattering experiments. *This was contrary to popular perception of Yang–Mills theories at the time*, that like *gravitation* and *electrodynamics*, their intensity should decrease with increasing distance between the interacting particles; such conventional behavior with distance was unable to explain the results of *deep inelastic scattering*, whereas 't Hooft's calculations could.

When 't Hooft mentioned his results at a small conference at Marseilles in 1972, Kurt Symanzik urged him to publish this result; but 't Hooft did not, and the result was eventually rediscovered and published by Hugh David Politzer, David Gross, and Frank Wilczek in 1973, which led to their earning the 2004 Nobel Prize in Physics.

In 1974, 't Hooft returned to Utrecht where he became assistant professor. In 1976, he was invited for a guest position at Stanford and a position at Harvard as Morris Loeb lecturer. His eldest daughter, Saskia Anne, was born in Boston, while his second daughter, Ellen Marga, was born in 1978 after he returned to Utrecht, where he was made full professor. In the academic year 1987–1988 't Hooft spent a sabbatical in the Boston University Physics Department along with Howard Georgi, Robert Jaffe and others arranged by the then new Department chair Lawrence Sulak.

In 1999 't Hooft shared the Nobel prize in Physics with his thesis adviser Veltman. On July 1, 2011 he was appointed Distinguished professor by Utrecht University.

't Hooft, G. (December, 1971). Renormalizable Lagrangians for Massive Yang-Mills Fields.

Nucl. Phys. B, 35, 1, 167-88; https://doi.org/10.1016/0550-3213(71)90139-8.

Institute for Theoretical Physics, University of Utrecht.

Received July 13, 1971.

Abstract.

Renormalizable models are constructed in which *local gauge invariance* is *broken spontaneously. Feynman rules* and *Ward identities* can be found by means of a path integral method, and they can be checked by algebra.

> [In quantum field theory, a *Ward–Takahashi identity* is an identity between correlation functions that follows from the global or gauge symmetries of the theory, and which remains valid after *renormalization*.]

In one of these models, which is studied in more detail, *local* SU(2) is broken in such a way that *local* U(1) remains as a *symmetry*. A *renormalizable and unitary theory results*, with *photons, charged massive vector particles*, and additional *neutral scalar particles*. It has three independent parameters. Another model has local SU(2) x U(1) as a *symmetry* and may serve as a *renormalizable* theory for ρ-*mesons* and *photons*. In such models, *electromagnetic mass-differences* are finite and can be calculated in *perturbation theory*.

---

1. INTRODUCTION

In a preceding article[1], henceforth referred to as I, it has been shown that, owing to their large *symmetry, massless Yang-Mills fields* may be *renormalized*, provided that a certain set of *Ward identities* is not violated by *renormalization* effects.

[1] 't Hooft, G. (October, 1971). Renormalization of Massless Yang-Mills Fields. *Nucl. Phys. B*, 33, 1, 173-99; https://doi.org/10.1016/055090395-6. Referred to as I.

With this we mean that anomalies like those of the *axial current Ward identities* in *nucleon-nucleon interactions*, which are due to an unallowed shift of integration variables in the "formal" proof, must not occur. In I it is proved that such anomalies are absent in diagrams with one closed loop, *if there are no parity-changing transformations in the local gauge group*. We do know an extension of this proof for diagrams with an arbitrary number of closed loops, but it is rather involved and we shall not present it here.

Thus, our prescription for the *renormalization* procedure is consistent, so the *ultraviolet problem* for *massless Yang-Mills fields* has been solved. A much more complicated problem is formed by the *infrared divergencies* of the system. Weinberg[5] has pointed out that, contrary to the *quantum electrodynamical* case, this problem cannot merely be solved by some closer contemplation of the measuring process.

[5] Weinberg, S. (October, 1965). Infrared Photons and Gravitons. *Phys. Rev.*, 140, B516; https://doi.org/10.1103/PhysRev.140.B516.

The disaster is such that the *perturbation expansion* breaks down in the *infrared* region, so we have no rigorous *field theory* to describe what happens.

However, although the Lagrangian is *invariant* under *local gauge transformations*, the physical solutions we are interested in may provide us with a certain preference *gauge*, in which these solutions take a simple form. If this is the case, then the *local gauge invariance* is hidden, and it is very well possible that all *Yang-Mills bosons* become *massive vector particles*. We do not know whether such a thing can happen with *massless Yang-Mills fields* alone, but it surely can happen in other models, of which we present some.

In all these models, additional *scalar fields* are introduced, which are *representations* of the *local gauge group*. If, in some *gauge*, these *fields* have a non-zero *vacuum expectation value*, then they may fix the *gauge*, either completely, or partly. In the latter case, *invariance* under transformations of a *local subgroup* of the original *invariance group* remains evident, and some of the *Yang-Mills bosons* remain *massless*.

The transition from a "*symmetric*" to a "*non-symmetric*" *representation* is done in a way analogous to the treatment of the σ-model by Lee and Gervais. The difference is of course that we have a *local invariance*, and we have no *symmetry breaking* term in the Lagrangian.

Our result is a large set of different models with *massive*, *charged* or *neutral*, *spin one bosons*, *photons*, and *massive scalar particles*. Due to the *local symmetry* our models are *renormalizable*, *causal*, and *unitary*. They all contain a small number of independent physical parameters.

A nice feature is that in certain models the *electromagnetic mass-differences* are finite and can be expressed in terms of the other parameters.

In sect. 2 we give a short review of the results in the preceding paper (I) on *massless Yang-Mills fields*. A general procedure appears to exist for deriving Feynman rules for models with a *local gauge invariance*. One statement must be made on our use of *path integrals* here: we only apply *path integral* techniques in order to get some idea of what the *Feynman rules* and *Ward-identities* might be. *Consistency* and *unitarity* of the *renormalized* expressions must always be checked later on. This has been done for the models described in this paper.

In sect. 3 we consider SU(2) *gauge fields* and an additional *scalar isospin one boson*. We show how the *vacuum expectation value* of this *boson field* can become non-zero due to dynamical effects, and how two of the *Yang-Mills bosons* become *massive, oppositely charged, vector particles*, while the third becomes an ordinary *photon*. Of the original *scalar fields* one component survives in the form of a *neutral spinless particle. Interaction* and *gauge* are formulated in such a way that the theory remains *renormalizable*. In sect. 4 a *renormalization* scheme is presented, but for a more elaborate description of the *renormalization* procedures we refer to I. In sect. 5 we prove that the model of sect. 3 is *unitary* and it is easily seen that the proof applies also to the other models.

In sect. 6 we describe an example where *local invariance* seems *broken*, while *global invariance* remains evident. All *Yang-Mills particles* get equal *mass*, and the model resembles very much the *massive Yang-Mills field* studied by other authors except for the presence of one extra *neutral scalar boson* with arbitrary *mass*. The model can be used to describe ρ-*mesons* as *elementary particles*.

In sect. 7 it is shown that our "ρ-*meson model*" can be enriched with *electromagnetic interactions* without destroying *renormalizability* or *unitarity*. ρ° - γ mixture leads to phenomena like *vector-dominance*.

In the appendix we formulate the *Feynman rules* for the various models.

Sheldon Lee Glashow, Abdus Salam and Steven Weinberg – the 1979 Nobel Prize in Physics.

The Nobel Prize in Physics 1979 was awarded jointly to Sheldon Lee Glashow, Abdus Salam and Steven Weinberg "for their contributions to the theory of the *unified weak and electromagnetic interaction* between *elementary particles*, including, inter alia, the prediction of the *weak neutral current*". [The Nobel Prize in Physics 1979. NobelPrize.org. https://www.nobelprize.org/prizes/physics/1979/summary/.]

The Royal Swedish Academy of Sciences has decided to award the 1979 Nobel Prize in physics to be shared equally between Professor Sheldon L. Glashow, Harvard University, USA, Professor Abdus Salam, International Centre for Theoretical Physics, Italy and Imperial College, Great Britain, and Professor Steven Weinberg, Harvard University, USA, for their contributions to the *theory of the unified weak and electromagnetic interaction between elementary particles*, including inter alla the prediction of the weak neutral current.

Physics, like other sciences, aspires to find common causes for apparently unrelated natural or experimental observations. A classical example is the force of gravitation introduced by Newton to explain such disparate phenomena as the apple falling to the ground and the moon moving around the earth.

Another example occurred in the 19th century when it was realized, mainly through the work of Oersted in Denmark and Faraday in England, that electricity and magnetism are closely related, and are really different aspects of the *electromagnetic force* or *interaction* between *charges*. The final synthesis was presented in the 1860's by Maxwell in England. His work predicted the existence of *electromagnetic waves* and interpreted light as an electromagnetic wave phenomenon.

The discovery of the *radioactivity* of certain heavy elements towards the end of last century, and the ensuing development of the physics of the atomic nucleus, led to the introduction of two new forces or interactions: the *strong* and the *weak nuclear forces*. Unlike *gravitation* and *electromagnetism* these forces act only at very short distances, of the order of nuclear diameters or less. While the *strong interaction* keeps *protons* and *neutrons* together in the *nucleus*, the *weak interaction* causes the so-called *radioactive beta-decay*. The typical process is the *decay* of the *neutron*: the *neutron*, with charge zero, is transformed into a positively charged *proton*, with the emission of a negatively charged *electron* and a neutral, massless particle, the *neutrino*.

Although the *weak interaction* is much weaker than both the *strong* and the *electromagnetic interactions*, it is of great importance in many connections. The actual strength of the *weak interaction* is also of significance. The energy of the sun, all-important for life on earth, is produced when hydrogen fuses or burns into helium in a chain of nuclear reactions occurring in the interior of the sun. The first reaction in this chain, the transformation of hydrogen into heavy hydrogen (deuterium), is caused by the *weak force*. Without this force solar energy production would not be possible. Again, had the *weak force* been much stronger, the life span of the sun would have been too short for life to have had time to evolve on any planet. The *weak interaction* finds practical application in the radioactive elements used in medicine and technology, which are in general *beta-radioactive*, and in the *beta-decay* of a carbon isotope into nitrogen, which is the basis for the carbon-14 method for dating of organic archaeological remains.

Theories of weak interaction

A first theory or *weak interaction* was put forward already in 1934 by the Italian physicist Fermi. However, a satisfactory description of the *weak interaction* between particles at low energy could be given only after the discovery in 1956 that the *weak force* differs from the other forces in *not being reflection symmetric*; in other words, the *weak force* makes a distinction between left and right. Although this theory was valid only for low energies and thus had a restricted domain of validity, it suggested a certain kinship between the *weak* and the *electromagnetic interactions*.

In a series of separate works in the 1960's this year's Nobel Prize winners, Glashow, Salam and Weinberg developed *a theory which is applicable also at higher energies, and which at the same time unifies the weak and electromagnetic interactions in a common formalism.* Glashow, Salam and Weinberg started from earlier contributions by other scientists. Of special importance was a generalization of the so-called *gauge principle* for the description of the *electromagnetic interaction*. This generalization was worked out around the middle of the 1950's by Yang and Mills in USA. After the fundamental work in the 1960's the theory has been further developed. An important contribution was made in 1971 by the young Dutch physicist van't Hooft.

The theory predicts among other things the existence of a new type of *weak interaction*, in which the reacting particles do not change their *charges*. This behavior is similar to what happens in the *electromagnetic interaction*, and one says that the *interaction* proceeds via a *neutral current*. One should contrast this with the *beta-decay* of the *neutron*, where the *charge* is altered when the *neutron* is changed into a *proton*.

First observation of the weak neutral current

The first observation of an effect of the new type of *weak interaction* was made in 1973 at the European nuclear research laboratory, CERN, in Geneva in an experiment where *nuclei* were bombarded with a beam of *neutrinos*. Since then, a series of *neutrino* experiments at CERN and at the Fermi Laboratory near Chicago have given results in good agreement with theory. Other laboratories have also made successful tests of effects of the *weak neutral current interaction*. Of special interest is a result, published in the summer of 1978, of an experiment at the *electron* accelerator at SLAC in Stanford, USA. In this experiment the scattering of high energy *electrons* on *deuterium nuclei* was studied and an effect due to a direct interplay between the *electronmagnetic* and *weak* parts of the unified *interaction* could be observed.

Interaction carried by particles

An important consequence of the theory is that the *weak interaction* is carried by particles having some properties in common – with the *photon*, which carries the *electromagnetic interaction* between *charged particles*. These so-called weak *vector bosons* differ from the massless *photon* primarily by having a *large mass*; this corresponds to the short range of the *weak interaction*. *The theory predicts masses of the order of one hundred proton masses*, but today's particle accelerators are not powerful enough to be able to produce these particles.

The contributions awarded this year's Nobel Prize in physics have been of great importance for the intense development of particle physics in this decade.

[NobelPrize.org. https://www.nobelprize.org/prizes/physics/1979/press-release/].

Steven Weinberg – Nobel Lecture, December 8, 1979. *Guage Unification of Fundamental Forces.*

Steven Weinberg – Nobel Lecture. NobelPrize.org. https://www.nobelprize.org/prizes/physics/1979/weinberg/lecture/

In his Nobel lecture, Weinberg attempted (but failed) to justify renormalization in the Standard Model.

Our job in physics is to see things simply, to understand a great many complicated phenomena in a unified way, in terms of a few simple principles. At times, our efforts are illuminated by a brilliant experiment, such as the 1973 discovery of *neutral current neutrino reactions*. But even in the dark times between experimental breakthroughs, there always continues a steady evolution of theoretical ideas, leading almost imperceptibly to changes in previous beliefs. In this talk, I want to discuss the development of two lines of thought in theoretical physics. One of them is the slow growth in our understanding of *symmetry*, and in particular, *broken or hidden symmetry*. The other is the old struggle to come to terms with the *infinities in quantum field theories*. To a remarkable degree, our present detailed theories of elementary particle interactions can be understood deductively, as consequences of *symmetry principles* and of a *principle of renormalizability* which is invoked to deal with the infinities. I will also briefly describe how the convergence of these lines of thought led to my own work on the unification of *weak* and *electromagnetic interactions*. For the most part, my talk will center on my own gradual education in these matters, because that is one subject on which I can speak with some confidence. With rather less confidence, I will also try to look ahead, and suggest what role these lines of thought may play in the physics of the future.

Symmetry principles made their appearance in twentieth century physics in 1905 with Einstein's identification of the invariance group of space and time. With this as a precedent, symmetries took on a character in physicists' minds as a priori principles of universal validity, expressions of the simplicity of nature at its deepest level. So, it was painfully difficult in the 1930's to realize that there are *internal symmetries*, such as *isospin conservation*, having nothing to do with space and time, symmetries which are far from self-evident, and that only govern what are now called the *strong interactions*. The 1950's saw the discovery of another *internal symmetry* - the *conservation of strangeness* - which is not obeyed by the *weak interactions*, and even one of the supposedly sacred symmetries of space-time - *parity* - was also found to be violated by *weak interactions*. Instead of moving toward unity, physicists were learning that different interactions are apparently governed by quite different symmetries. Matters became yet more confusing with the recognition in the early 1960's of a *symmetry group* - the *"eightfold way"* - which is not even an exact *symmetry* of the *strong interactions*.

These are all *"global"* symmetries, for which the symmetry transformations *do not depend on position in space and time*. It had been recognized in the 1920's that *quantum electrodynamics* has another *symmetry* of a far more powerful kind, a *"local"* symmetry under transformations in which the *electron field* suffers a *phase change* that can vary freely from point to point in space-time, and the *electromagnetic vector potential* undergoes a corresponding *gauge transformation*. Today this would be called a U(1) *gauge symmetry*, because a simple *phase change* can be thought of as multiplication by a 1 x 1 unitary matrix. The extension to more complicated groups was made by Yang and Mills[6] in 1954 in a seminal paper

> [6] Yang, C. N. & Mills, R. (October, 1954). Conservation of Isotopic Spin and Isotopic Gauge Invariance. See above.

in which they showed how to construct an SU(2) *gauge theory* of *strong interactions*. (The name "SU(2)" means that the *group of symmetry transformations* consists of 2 x 2 unitary matrices that are "special," in that they have determinant unity). But here again it seemed that the symmetry if real at all would have to be approximate, because at least on a naive level *gauge invariance* requires that *vector bosons* like the *photon* would have to be *massless*, and it seemed obvious that the *strong interactions* are not mediated by *massless particles*. The old question remained: if *symmetry principles* are an expression of the simplicity of nature at its deepest level, then how can there be such a thing as an approximate symmetry? Is nature only approximately simple?

At some time in 1960 or early 1961, I learned of an idea which had originated earlier in solid state physics and had been brought into particle physics by those like Heisenberg, Nambu, and Goldstone, who had worked in both areas. It was the idea of *"broken symmetry"*, *that the Hamiltonian and commutation relations of a quantum theory could possess an exact symmetry, and that the physical states might nevertheless not provide neat representations of the symmetry*. In particular, a *symmetry* of the Hamiltonian might turn out to be not a *symmetry of the vacuum*.

As theorists sometimes do, I fell in love with this idea. But as often happens with love affairs, at first, I was rather confused about its implications. I thought (as turned out, wrongly) that the *approximate symmetries - parity, isospin, strangeness*, the *eight-fold way* - might really be exact a priori symmetry principles, and that the observed violations of these symmetries might somehow be brought about by *spontaneous symmetry breaking*. It was therefore rather disturbing for me to hear of a result of Goldstone[7], that in at least one simple case the *spontaneous breakdown* of a *continuous symmetry* like *isospin* would necessarily entail the existence of a *massless spin zero particle* - what would today be called a "*Goldstone boson*".

[7] Goldstone, J. (January, 1961). Field theories with "Superconductor" solutions. See above.

It seemed obvious that there could not exist any new type of *massless* particle of this sort which would not already have been discovered.

I had long discussions of this problems with Goldstone at Madison in the summer of 1961, and then with Salam while I was his guest at Imperial College in 1961-62. The three of us soon were able to show that *Goldstone bosons* must in fact occur whenever a *symmetry* like *isospin* or *strangeness* is *spontaneously broken*, and that their *masses* then remain zero to all orders of perturbation theory. I remember being so discouraged by these zero masses that when we wrote our joint paper on the subject[8],

[8] Goldstone, J., Salam, A. & Weinberg, S. (August, 1962). Broken Symmetries. See above.

I added an epigraph to the paper to underscore the futility of supposing that anything could be explained in terms of a *non-invariant vacuum state*: it was Lear's retort to Cordelia, "Nothing will come of nothing: speak again." Of course, *The Physical Review* protected the purity of the physics literature, and removed the quote. Considering the future of the *non-invariant vacuum* in theoretical physics, it was just as well.

There was actually an exception to this proof, pointed out soon after wards by Higgs, Kibble, and others. They showed that if the *broken symmetry* is a *local, gauge symmetry*, like *electromagnetic gauge invariance*, then although the *Goldstone bosons* exist formally, and are in some sense real, they can be eliminated by a *gauge transformation*, so that they do not appear as physical particles. The missing *Goldstone bosons* appear instead as helicity *zero states* of the *vector particles*, which thereby acquire a *mass*.

I think that at the time physicists who heard about this exception generally regarded it as a technicality. This may have been because of a new development in theoretical physics which suddenly seemed to change the role of *Goldstone bosons* from that of unwanted intruders to that of welcome friends.

In 1964 Adler and Weisberger independently derived sum rules which gave the *ratio of axial-vector to vector coupling constants* in *beta decay* in terms of *pion-nucleon cross sections*. One way of looking at their calculation, (perhaps the most common way at the time) was as an analogue to the old dipole sum rule in atomic physics: a complete set of hadronic states is inserted in the commutation relations of the axial vector currents. This is the approach memorialized in the name of "*current algebra*". But there was another way of looking at the Adler-Weisberger sum rule. One could suppose that the *strong interactions* have an *approximate symmetry*, based on the group SU(2) x SU(2), and that this *symmetry* is spontaneously broken, giving rise among other things to the *nucleon masses*. The *pion* is then identified as (approximately) a *Goldstone boson*, with small non-zero *mass*, an idea that goes back to Nambu. Although the SU(2) x SU(2) *symmetry* is *spontaneously broken*, it still has a great deal of predictive power, but its predictions take the form of approximate formulas, which give the matrix elements for low energy *pionic* reactions. In this approach, the Adler-Weisberger sum rule is obtained by using the predicted *pion nucleon scattering lengths* in conjunction with a well-known sum rule, which years earlier had been derived from the dispersion relations for *pion-nucleon scattering*.

In these calculations one is really using not only the fact that the *strong interactions* have a *spontaneously broken* approximate SU(2) x SU(2) *symmetry*, but also that the *currents* of this *symmetry group* are, up to an overall constant, to be identified with the *vector* and *axial vector currents* of *beta decay*. (With this assumption g_A/g_V gets into the picture through the Goldberger-Treiman relation, which gives g_A/g_V in terms of the *pion decay constant* and the *pion nucleon coupling*.) Here, in this relation between the *currents* of the *symmetries* of the *strong interactions* and the physical currents of *beta decay*, there was a tantalizing hint of a deep connection between the *weak interactions* and the *strong interactions*. But this connection was not really understood for almost a decade.

I spent the years 1965-67 happily developing the implications of spontaneous symmetry breaking for the strong interactions. It was this work that led to my 1967 paper on *weak* and *electromagnetic unification*. But before I come to that I have to go back in history and pick up one other line of thought, having to do with the *problem of infinities in quantum field theory*. I believe that it was Oppenheimer and Waller in 1930 who independently first noted that *quantum field theory* when pushed beyond the lowest approximation yields ultraviolet divergent results for radiative self-energies. Professor Waller told me last night that when he described this result to Pauli, Pauli did not believe it. It must have seemed that these *infinities* would be a disaster for the *quantum field theory* that had just been developed by Heisenberg and Pauli in 1929-30. And indeed, these *infinites* did lead to a sense of discouragement about *quantum field theory*, and many attempts were made in the 1930's and early 1940's to find alternatives. The problem was solved (at least for *quantum electrodynamics*) after the war, by Feynman, Schwinger, Tomonaga and Dyson.

[See Underwood, T. G. (2023). *Quantum Electrodynamics – annotated sources*, Volume II.]

It was found that all infinities disappear if one identifies the observed finite values of the *electron mass* and *charge*, not with the parameters m and e appearing in the Lagrangian, but with the *electron mass* and *charge* that are calculated from m and e, when one takes into account the fact that the *electron* and photon are always surrounded with clouds of *virtual photons* and *electron-positron pairs*. Suddenly all sorts of calculations became possible, and gave results in spectacular agreement with experiment [???].

But even after this success, opinions differed as to the significance of the ultraviolet divergences in *quantum field theory*. Many thought - and some still do think - that what had been done *was just to sweep the real problems under the rug*.

> [Underwood, T. G. (2023). *Quantum Electrodynamics - annotated sources*, Volume II, pp. 34-35: "Schwinger, in the Preface of his 1958 book [*Selected Papers on Quantum electrodynamics*.], "questioned whether *renormalization* simply corrected a mathematical error that causes the divergencies, or whether *there is a serious flaw in the structure of field theory*". He concluded that "the observational basis of quantum electrodynamics is self-contradictory" and that "a convergent theory cannot be formulated consistently within the framework of present space-time concepts" … "It can never explain the observed value of the dimensionless *coupling constant* measuring the *electron charge* … a full understanding of the *electron charge* can exist only when the *theory of elementary particles* has come to a stage of perfection that is presently unimaginable".
>
> Tomonaga, in his 1965 Nobel prize speech, note that "In order to overcome the difficulty of an infinitely large *electromagnetic mass*, *Lorentz considered the electron not to be point-like but to have a finite size. It is very difficult, however, to incorporate a finite sized electron into the framework of relativistic quantum theory*. Many people tried various means to overcome this problem of infinite quantities, but nobody succeeded".
>
> Feynman, in his 1965 Nobel prize speech, described *renormalization* as "simply a way to sweep the difficulties of the divergences of electrodynamics under the rug".
>
> Dirac's final judgment on *quantum field theory*, in his last paper published in 1987 (The inadequacies of quantum field theory.), was that "These rules of *renormalization* give surprisingly, excessively good agreement with experiments. Most physicists say that these working rules are, therefore, correct. I feel that is not an adequate reason. Just because the results happen to be in agreement with observation does not prove that one's theory is correct.""]

And it soon became clear that there was only a limited class of so-called *"renormalizable"* theories in which the infinities could be eliminated by absorbing them into a *redefinition*, or a *"renormalization"*, of a finite number of physical parameters. (Roughly speaking, in *renormalizable* theories no coupling constants can have the dimensions of negative powers of mass. But every time we add a *field* or a *space-time derivative* to an *interaction*, we reduce the dimensionality of the associated *coupling constant*. So only a few simple types of interaction can be *renormalizable*.) In particular, *the existing Fermi theory of weak interactions clearly was not renormalizable*. (The Fermi coupling constant has the dimensions of $[\text{mass}]^{-2}$.) The sense of discouragement about *quantum field theory* persisted into the 1950's and 1960's.

I learned about *renormalization theory* as a graduate student, mostly by reading Dyson's papers[19].

> [19] Dyson, F. J. (February, 1949). The Radiation Theories of Tomonaga, Schwinger, and Feynman. *Phys. Rev.*, 75, 3, 486; https://doi.org/10.1103/PhysRev.75.486; (June, 1949). *Ibid.*, 75, 11, 1736; https://doi.org/10.1103/PhysRev.75.1736. Also in Underwood, T. G. (2023). *Quantum Electrodynamics - annotated sources*, Volume II.

From the beginning it seemed to me to be a wonderful thing that very few *quantum field theories* are *renormalizable*. Limitations of this sort are, after all, what we most want, not mathematical methods which can make sense of an infinite variety of physically irrelevant theories, but methods which carry constraints, because these constraints may point the way toward the one true theory. In particular, I was impressed by the fact that *quantum electrodynamics* could in a sense be derived from *symmetry principles* and the constraints of *renormalizability* [???]; the only Lorentz invariant and gauge invariant renormalizable Lagrangian for photons and electrons is precisely the original Dirac Lagrangian of *quantum electrodynamics*. Of course, that is not the way Dirac came to his theory. He had the benefit of the information gleaned in centuries of experimentation on electromagnetism, and in order to fix the final form of his theory he relied on ideas of simplicity (specifically, on what is sometimes called minimal *electromagnetic coupling*). But we have to look ahead, to try to make theories of phenomena which have not been so well studied experimentally, and we may not be able to trust purely formal ideas of simplicity. I thought that *renormalizability* might be the key criterion, which also in a more general context would impose a precise kind of simplicity on our theories and help us to pick out the one true physical theory out of the infinite variety of conceivable *quantum field theories*. As I will explain later, I would say this a bit differently today, but I am more convinced than ever that the use of *renormalizability* as a *constraint* [???] on our theories of the observed *interactions* is a good strategy. Filled with enthusiasm for renormalization theory, I wrote my Ph.D. thesis under Sam Treiman in 1957 on the use of a limited version of *renormalizability* to set constraints on the *weak interactions*, and a little later I worked out a rather tough little theorem which completed the proof by Dyson and Salam *that ultraviolet divergences really do cancel out to all orders in nominally renormalizable theories*. But none of this seemed to help with the important problem, of how to make a *renormalizable* theory of weak interactions.

Now, back to 1967. I had been considering the implications of the broken SU(2) x SU(2) *symmetry* of the *strong interactions*, and I thought of trying out the idea that perhaps the SU(2) x SU(2) symmetry was a "*local*", not merely a "*global*", symmetry. That is, the *strong interactions* might be described by something like a Yang-Mills theory, but in addition to the *vector ϱ mesons* of the Yang-Mills theory, there would also be axial vector Al *mesons*. To give the *ϱ mesons* a *mass*, it was necessary to insert a common *ϱ* and Al *mass* term in the Lagrangian, and the *spontaneous breakdown* of the SU(2) x SU(2) *symmetry* would then split the *ϱ* and Al by something like the Higgs mechanism, but since the theory would not be *gauge invariant* the *pions* would remain as physical *Goldstone bosons*. This theory gave an intriguing result, that the *mass ratio* should be $\sqrt{2}$ and in trying to understand this result without relying on *perturbation theory*, I discovered certain sum rules, the "*spectral function sum rules*", which turned out to have variety of other uses. But the SU(2) x SU(2) theory was not *gauge invariant*, and hence it could not be *renormalizable*, [24] so I was not too enthusiastic about it. Of course, if I did not insert the *mass* term in the Lagrangian, then the theory would be *gauge invariant* and *renormalizable*, and the Al would be massive. But then there would be no *pions* and the *mesons* would be massless, in obvious contradiction (to say the least) with observation.

At some point in the fall of 1967, I think while driving to my office at M.I.T., it occurred to me that I had been applying the right ideas to the wrong problem. It is not the *mesons* that is massless: it is the *photon*. And its partner is not the Al, but the massive intermediate *boson*, which since the time of Yukawa had been suspected to be the mediator of the *weak interactions*. The *weak* and *electromagnetic interactions* could then be described in a unified way in terms of an exact but *spontaneously broken gauge symmetry*. [Of course, not necessarily SU(2) x SU(2)]. And this theory would be *renormalizable* like *quantum electrodynamics* because it is *gauge invariant* like *quantum electrodynamics*.

It was not difficult to develop a concrete model which embodied these ideas. I had little confidence then in my understanding of *strong interactions*, so I decided to concentrate on *leptons*. There are two left-handed electron-type *leptons*, the v_{eL} and e_L, and one right-handed electron-type, the e_R so I started with the group U(2) x U(1): all unitary 2 x 2 matrices acting on the left-handed e-type *leptons*, together with all unitary 1 x 1 matrices acting on the right-handed e-type *lepton*. Breaking up U(2) into *unimodular transformations* and *phase transformations*, one could say that the group was SU(2) x U(1) x U(1). But then one of the U(l)'s could be identified with ordinary *lepton number*, and since the *lepton number* appears to be conserved and there is no massless *vector* particle coupled to it, I decided to exclude it from the group. This left the four-parameter group SU(2) x U(1). The *spontaneous breakdown* of SU(2) x U(1) to the U(1) of ordinary *electromagnetic gauge invariance* would give *masses* to three of the four *vector gauge bosons*: the *charged bosons* W±, and a *neutral boson* that I called the Z_0. The fourth *boson* would automatically remain massless, and could be identified as the *photon*. Knowing the strength of the ordinary *charged current weak interactions* like *beta decay* which are mediated by W±, the *mass* of the W± was then determined as about **40 GeV/sin Θ** where is **Θ** is the **γ-Z_0 mixing** angle.

To go further, one had to make some hypothesis about the mechanism for the breakdown of SU(2) x U(1). The only kind of *field* in a *renormalizable* SU(2) x U(1) theory whose *vacuum expectation values* could give the *electron* a *mass* is a *spin zero* SU(2) *doublet*, so for simplicity I assumed that these were the only *scalar fields* in the theory. The mass of the Z0 was then determined as about 80 GeV/sin 2Θ. This fixed the strength of the *neutral current weak interactions*. Indeed, just as in *quantum electrodynamics*, once one decides on the menu of *fields* in the theory all details of the theory are completely determined by *symmetry principles* and *renormalizability*, with just a few free parameters: the *lepton charge* and *masses*, the *Fermi coupling constant* of *beta decay*, the *mixing angle* and the *mass* of the *scalar particle*. (It was of crucial importance to impose the constraint [???] of *renormalizability*; otherwise *weak interactions* would receive contributions from SU(2) x U(I) - *invariant four-fermion couplings* as well as from *vector boson exchange*, and the theory would lose most of its predictive power.) The naturalness of the whole theory is well demonstrated by the fact that much the same theory was independently developed by Salam in 1968.

The next question now was *renormalizability*. The Feynman rules for Yang-Mills theories with unbroken *gauge symmetries* had been worked out by deWitt, Faddeev and Popov and others, and it was known that such theories are *renormalizable*. But in 1967 I did not know how to prove that this *renormalizability* was not spoiled by the *spontaneous symmetry breaking*. I worked on the problem on and off for several years, partly in collaboration with students, but I made little progress. With hindsight, my main difficulty was that in quantizing the *vector fields* I adopted a *gauge* now known as the *unitarity gauge*: this *gauge* has several wonderful advantages, it exhibits the true particle spectrum of the theory, but it has the disadvantage of making *renormalizability* totally obscure.

Finally, in 1971 't Hooft[31] showed in a beautiful paper how the problem could be solved.

[31] 't Hooft, G. (December, 1971). Renormalizable Lagrangians for Massive Yang-Mills Fields. See above.

He invented a *gauge*, like the *"Feynman gauge"* in *quantum electrodynamics*, in which the Feynman rules manifestly lead to only a finite number of types of *ultraviolet divergence*. It was also necessary to show that these *infinities* satisfied essentially the same constraints as the Lagrangian itself, so that they could be absorbed into a redefinition of the parameters of the theory. (This was plausible, but not easy to prove, because a *gauge invariant theory* can be quantized only after one has picked a specific *gauge*, so it is not obvious that the *ultraviolet divergences* satisfy the same *gauge invariance constraints* as the Lagrangian itself.) The proof was subsequently completed by Lee and Zinn-Justin and by 't Hooft and Veltman. More recently, Becchi, Rouet and Stora have invented an ingenious method for carrying out this sort of proof, by using a *global supersymmetry* of *gauge theories* which is preserved even when we choose a specific *gauge*.

I have to admit that when I first saw 't Hooft's paper in 1971, I was not convinced that he had found the way to' prove *renormalizability*. The trouble was not with 't Hooft, but with me: I was simply not familiar enough with the *path integral* formalism on which 't Hooft's work was based, and I wanted to see a derivation of the Feynman rules in 't Hooft's *gauge* from canonical quantization. That was soon supplied (for a limited class of *gauge theories*) by a paper of Ben Lee, and after Lee's paper I was ready to regard the *renormalizability of the unified theory* as essentially proved.

By this time, many theoretical physicists were becoming convinced of the general approach that Salam and I had adopted: that is, the *weak* and *electromagnetic interactions* are governed by some group of *exact local gauge symmetries*; this group is *spontaneously broken* to U(l), giving *mass* to all the *vector bosons* except the *photon*; and the theory is *renormalizable*. What was not so clear was that our specific simple model was the one chosen by nature. That, of course, was a matter for experiment to decide.

It was obvious even back in 1967 that the best way to test the theory would be by searching for *neutral current weak interactions*, mediated by the *neutral intermediate vector boson*, the Z_0. Of course, the possibility of *neutral currents* was nothing new. There had been speculations about possible *neutral currents* as far back as 1937 by Gamow and Teller, Kemmer, and Wentzel, and again in 1958 by Bludman and Leite-Lopes. Attempts at a unified *weak* and *electromagnetic theory* had been made[36] by Glashow and Salam and Ward in the early 1960's, and these had *neutral currents* with many of the features that Salam and I encountered in developing the 1967-68 theory.

[36] Glashow, S. L. (February, 1961). Partial-symmetries of weak interactions, see above; Salam, A. & Ward, J. C. (November, 1964). Electromagnetic and weak interactions, see above.

But since one of the predictions of our theory was a value for the *mass* of the Z_0, it made a definite prediction of the strength of the *neutral currents*. More important, now we had a comprehensive *quantum field theory* of the *weak* and *electromagnetic interactions* that was physically and mathematically satisfactory in the same sense as was *quantum electrodynamics* - a theory that treated *photons* and *intermediate vector bosons* on the same footing, that was based on an *exact symmetry principle*, and that allowed one to carry calculations to any desired degree of accuracy. To test this theory, it had now become urgent to settle the question of the existence of the *neutral currents*.

Late in 1971, I carried out a study of the experimental possibilities. The results were striking. Previous experiments had set upper bounds on the rates of *neutral current* processes which were rather low, and many people had received the impression that *neutral currents* were pretty well ruled out, but I found that in fact the 1967-68 theory *predicted* quite low rates, low enough in fact to have escaped clear detection up to that time. For instance, experiments a few years earlier had found an upper bound of 0.12 ± 0.06 on the *ratio of a neutral current process, the elastic scattering of muon neutrinos by protons, to the corresponding charged current process*, in which a *muon* is produced. I found a predicted ratio of 0.15 to 0.25, depending on the value of the Z_0 *mixing angle* was every reason to look a little harder. So, there was every reason to look a little harder.

As everyone knows, *neutral currents* were finally discovered[39] in 1973.

> [39] The first published discovery of *neutral currents* was at the Gargamelle Bubble Chamber at CERN: Hasert, F. J. et al. (1973). *Phys. Lett.*, 468, 121, 138. Also see Musset, P. (1973). *Jour. de Physique*, 11 /12 T34. Muonless events were seen at about the same time by the HPWF group at Fermilab, but when publication of their paper was delayed, they took the opportunity to rebuild their detector, and then did not at first find the same *neutral current* signal. The HPWF group published evidence for *neutral currents* in Benvenuti, A. et al. (1974). *Phys. Rev. Lett.*, 52, 800.

There followed years of careful experimental study on the detailed properties of the *neutral currents*. It would take me too far from my subject to survey these experiments, so I will just say that they have confirmed the 1967-68 theory with steadily improving precision for *neutrino-nucleon* and *neutrino-electron neutral current* reactions, and since the remarkable SLAC-Yale experiment last year, for the *electron nucleon neutral current* as well.

This is all very nice. But I must say that I would not have been too disturbed if it had turned out that the correct theory was based on some other *spontaneously broken gauge group*, with very different *neutral currents*. One possibility was a clever SU(2) theory proposed in 1972 by Georgi and Glashow, which has no *neutral currents* at all. The important thing to me was the idea of an *exact spontaneously broken gauge symmetry*, which connects the *weak* and *electromagnetic interactions*, and allows these interactions to be *renormalizable*. Of this I was convinced, if only because it fitted my conception of the way that nature ought to be.

There were two other relevant theoretical developments in the early 1970's, before the discovery of *neutral currents*, that I must mention here. One is the important work of *Glashow*, *Iliopoulos*, and *Maiani* on the *charmed quark*.

[43] Glashow, S. L., Iliopoulos, J. & Maiani, L. (October, 1970). Weak Interactions with Lepton-Hadron Symmetry. *Phys. Rev. D*, 2, 7, 1285; https://doi.org/10.1103/PhysRevD.2.1285. This paper was cited in ref. 37 as providing a possible solution to the problem of *strangeness changing neutral currents*. However, at that time I was skeptical about the *quark model*, so in the calculations of ref. 37 *baryons* were incorporated in the theory by taking the *protons* and *neutrons* to form an SU(2) *doublet*, with *strange particles* simply ignored.

Their work provided a solution to what otherwise would have been a serious problem, that of *neutral strangeness changing currents*. I leave this topic for Professor Glashow's talk. The other theoretical development has to do specifically with the *strong interactions*, but it will take us back to one of the themes of my talk, the theme of *symmetry*. …

…

Abdus Salam – Nobel Lecture, December 8, 1979. *Guage Unification of Fundamental Forces.*

https://www.nobelprize.org/prizes/physics/1979/salam/lecture/.

In his Nobel Prize lecture Abdus Salam provided a brief history of the development of *the gauge unification of the fundamental forces*.

> [There were 98 references in the printed copy of this lecture. Only the most important have been included.]

Introduction: In June 1938, Sir George Thomson, then Professor of Physics at Imperial College, London, delivered his 1937 Nobel Lecture. …

…

I. FUNDAMENTAL PARTICLES, FUNDAMENTAL FORCES AND GAUGE UNIFICATION

The Nobel lectures this year are concerned with a set of ideas relevant to the *gauge unification of the electromagnetic force with the weak nuclear force*. These lectures coincide nearly with the 100th death-anniversary of Maxwell, with whom the first unification of forces (*electric* with the *magnetic*) matured and with whom *gauge theories* originated. They also nearly coincide with the anniversary of the birth of Einstein - the man who gave us the vision of an ultimate unification of all forces.

The ideas of today started more than twenty years ago, as gleams in several theoretical eyes. They were brought to predictive maturity over a decade back. And they started to receive experimental confirmation some six years ago.

In some senses then, our story has a fairly long background in the past. In this lecture I wish to examine some of the theoretical gleams of today and ask the question if these may be the ideas to watch for maturity twenty years from now.

From time immemorial, man has desired to comprehend the complexity of nature in terms of as few elementary concepts as possible. Among his quests - in Feynman's words - has been the one for "wheels within wheels" - the task of natural philosophy being to discover the innermost wheels if any such exist. A second quest has concerned itself with the *fundamental forces* which make the wheels go round and enmesh with one another. *The greatness of gauge ideas - of gauge field theories - is that they reduce these two quests to just one*; *elementary particles* (described by relativistic quantum fields) are representations of certain *charge operators*, corresponding to gravitational mass, spin, flavor, color, electric charge and the like, while the *fundamental forces* are the forces of attraction or repulsion between these same *charges*. A third quest seeks for a *unification* between the *charges* (and thus of the *forces*) by searching for a single entity, of which the various *charges* are components in the sense that they can be transformed one into the other.

But are all fundamental forces gauge forces? Can they be understood as such, in terms of *charges* - and their corresponding *currents* - only? And if they are, how many *charges*? What unified entity are the *charges* components of? What is the nature of *charge*? Just as Einstein comprehended the nature of gravitational charge in terms of space-time curvature, can we comprehend the nature of the other charges - the nature of the entire unified set, as a set, in terms of something equally profound? This briefly is the dream, much reinforced by the verification of *gauge theory* predictions. But before I examine the new theoretical ideas on offer for the future in this particular context, I would like your indulgence to range over a one-man, purely subjective, perspective in respect of the developments of the last twenty years themselves. The point I wish to emphasize during this part of my talk was well made by G. P. Thomson in his 1937 Nobel Lecture. G. P. said ". . . The goddess of learning is fabled to have sprung full grown from the brain of Zeus, but it is seldom that a scientific conception is born in its final form, or owns a single parent. More often it is the product of a series of minds, each in turn modifying the ideas of those that came before, and providing material for those that come after."

II. THE EMERGENCE OF SPONTANEOUSLY BROKEN SU(2) x U(1) GAUGE THEORY

I started physics research thirty years ago as an experimental physicist in the Cavendish, experimenting with tritium-deuterium scattering. Soon I knew the craft of experimental physics was beyond me - it was the sublime quality of patience - patience in accumulating data, patience with recalcitrant equipment - which I sadly lacked. Reluctantly I turned my papers in, and started instead on *quantum field theory* with Nicholas Kemmer in the exciting department of P. A. M. Dirac.

The year 1949 was the culminating year of the Tomonaga-Schwinger Feynman-Dyson reformulation of *renormalized* Maxwell-Dirac *gauge theory*, and its triumphant experimental vindication. A *field theory* must be *renormalizable* and be capable of being made free of infinities - first discussed by Waller - if perturbative calculations with it are to make any sense. [???] More - a *renormalizable theory*, with no dimensional parameter in its *interaction term*, connotes somehow that the fields represent "structureless" elementary entities. With Paul Matthews, we started on an exploration of *renormalizability*

410

of *meson theories*. Finding that *renormalizability* held only for *spin-zero mesons* and that these were the only *mesons* that empirically existed then, (pseudoscalar *pions*, invented by Kemmer, following Yukawa) one felt thrillingly euphoric that with the triplet of *pions* (considered as the carriers of the *strong nuclear force* between the *proton-neutron doublet*) one might resolve the dilemma of the origin of this particular force which is responsible for fusion and fission. By the same token, the so-called *weak nuclear force* - the force responsible for β-radioactivity (and described then by Fermi's *non-renormalizable* theory) had to be mediated by some unknown *spin-zero mesons* if it was to be *renormalizable*. If *massive charged spin-one mesons* were to mediate this *interaction*, the theory would be *non-renormalizable*, according to the ideas then. Now this agreeably *renormalizable spin-zero* theory for the *pion* was a *field theory*, but not a *gauge field theory. There was no conserved charge* which determined the *pionic* interaction. As is well known, shortly after the theory was elaborated, it was found wanting. The (3/2, 3/2) resonance Δ effectively killed it off as a fundamental theory; we were dealing with a complex dynamical system, not "structureless" in the held-theoretic sense.

For me, personally, the trek to *gauge theories* as candidates for fundamental physical theories started in earnest in September 1956 - the year I heard at the Seattle Conference Professor Yang expound his and Professor Lee's ideas[1] on the possibility of the hitherto sacred principle of *left-right symmetry*, being violated in the realm of the *weak nuclear force*.

[1] Lee, T. D. & Yang, C. N., (October, 1956). Question of Parity Conservation in Weak Interactions. *Phys. Rev.*, 104, 254; https://journals.aps.org/pr/abstract/10.1103/ PhysRev. 104.254

Lee and Yang had been led to consider abandoning *left-right symmetry* for *weak nuclear interactions* as a possible resolution of the puzzle. I remember travelling back to London on an American Air Force (MATS) transport flight. Although I had been granted, for that night, the status of a Brigadier or a Field Marshal - I don't quite remember which - the plane was very uncomfortable; full of crying service-men's children - that is, the children were crying, not the servicemen. I could not sleep. I kept reflecting on why Nature should violate left-right symmetry in weak interactions. Now the hallmark of most *weak interactions* was the involvement in radioactivity phenomena of Pauli's *neutrino*. While crossing over the Atlantic, came back to me a deeply perceptive question about the *neutrino* which Professor Rudolf Peierls had asked when he was examining me for a Ph. D. a few years before. Peierls' question was: "The *photon mass* is zero because of Maxwell's principle of a *gauge symmetry* for *electromagnetism*; tell me, why is the *neutrino mass* zero?" I had then felt somewhat uncomfortable at Peierls. asking for a Ph. D. viva, a question of which he himself said he did not know the answer. But during that comfortless night the answer came. The analogue for the *neutrino*, of the *gauge symmetry* for the *photon* existed; it had to do with the *masslessness* of the *neutrino*, with *symmetry* under the γ_5 transformation (later christened "*chiral symmetry*")[2].

[2] Abdus Salam, A. (1957). On parity conservation and neutrino mass. *Nuovo Cimento*, 5, 299; https://doi.org/10.1007/BF02812841.

The existence of this *symmetry* for the massless *neutrino* must imply a combination $(1 + \gamma_5)$ or $(1 - \gamma_5)$ for the *neutrino interactions*. Nature had the choice of an aesthetically satisfying but a *left-right symmetry violating theory*, with a *neutrino* which travels exactly with the velocity of light; or alternatively a theory where *left-right symmetry* is *preserved*, but the *neutrino* has a tiny *mass* - some ten thousand times smaller than the *mass* of the *electron*.

It appeared at that time clear to me what choice Nature must have made. Surely, *left-right symmetry* must be sacrificed in all *neutrino interactions*. I got off the plane the next morning, naturally very elated. I rushed to the Cavendish, worked out the Michel parameter and a few other consequences of γ_5 *symmetry*, rushed out again, got into a train to Birmingham where Peierls lived. To Peierls I presented my idea; he had asked the original question; could he approve of the answer? Peierls' reply was kind but firm. He said "I do not believe *left-right symmetry* is violated in *weak nuclear forces* at all. I would not touch such ideas with a pair of tongs." Thus, rebuffed in Birmingham, like Zuleika Dobson, I wondered where I could go next and the obvious place was CERN in Geneva, with Pauli - the father of the *neutrino* - nearby in Zurich. At that time CERN lived in a wooden hut just outside Geneva airport. Besides my friends, Prentki and d'Espagnat, the hut contained a gas ring on which was cooked the staple diet of CERN - Entrecôte à la creme. The hut also contained Professor Villars of MIT, who was visiting Pauli the same day in Zurich. I gave him my paper. He returned the next day with a message from the Oracle; "Give my regards to my friend Salam and tell him to think of something better". This was discouraging, but I was compensated by Pauli's excessive kindness a few months later, when Mrs. Wu's, Lederman's and Telegdi's experiments were announced showing that *left-right symmetry was indeed violated* and ideas similar to mine about *chiral symmetry* were expressed independently by Landau and Lee and Yang. I received Pauli's first somewhat apologetic letter on 24 January 1957. Thinking that Pauli's spirit should by now be suitably crushed, I sent him two short notes I had written in the meantime. These contained suggestions to extend *chiral symmetry* to *electrons* and *muons*, assuming that their masses were a consequence of what has come to be known as *dynamical spontaneous symmetry breaking*. With *chiral symmetry* for *electrons*, *muons* and *neutrinos*, the only *mesons* that could mediate weak decays of the muons would have to carry spin one. Reviving thus the notion of *charged intermediate spin-one bosons*, one could then postulate for these a type of *gauge*

412

invariance which I called the "neutrino gauge". Pauli's reaction was swift and terrible. He wrote on 30th January 1957, then on 18 February and later on 11, 12 and 13 March: "I am reading (along the shores of Lake Zurich) in bright sunshine quietly your paper..." "I am very much startled on the title of your paper '*Universal Fermi interaction*'

... For quite a while I have for myself the rule if a theoretician says universal it just means pure nonsense. This holds particularly in connection with the *Fermi interaction*, but otherwise too, and now you too, Brutus, my son, come with this word. ..." Earlier, on 30 January, he had written "There is a similarity between this type of *gauge invariance* and that which was published by Yang and Mills . . . In the latter, of course, no γ_5 was used in the exponent." and he gave me the full reference of Yang and Mills' paper [Yang, C. N. & Mills, R. (October, 1954). Conservation of Isotopic Spin and Isotopic Gauge Invariance. (1954). See above.]. I quote from his letter: "However, there are dark points in your paper regarding the vector field B_l. If the rest mass is infinite (or very large), how can this be compatible with the *gauge transformation* and he concludes his letter with the remark: "Every reader will realize that you deliberately conceal here something and will ask you the same questions". Although he signed himself "With friendly regards", Pauli had forgotten his earlier penitence. He was clearly and rightly on the warpath.

Now the fact that I was using *gauge* ideas similar to the Yang-Mills (non-Abelian SU(2)-invariant) *gauge theory* was no news to me. This was because the *Yang-Mills theory*[9] (which married gauge ideas of Maxwell with the internal *symmetry* SU(2) of which the *proton-neutron* system constituted a doublet) had been independently invented by a Ph. D. pupil of mine, Ronald Shaw,[10] at Cambridge at the same time as Yang and Mills had written.

[9] Yang, C. N. & Mills, R. (October, 1954). Conservation of Isotopic Spin and Isotopic Gauge Invariance. *Phys. Rev.*, 96, 1, 191–5. See above.
[10] Shaw, R., "*The problem of particle types and other contributions to the theory of elementary particles*", Cambridge Ph. D. Thesis (1955), unpublished.

Shaw's work is relatively unknown; it remains buried in his Cambridge thesis. I must admit I was taken aback by Pauli's fierce prejudice against universalism - against what we would today call unification of basic forces but I did not take this too seriously. I felt this was a legacy of the exasperation which Pauli had always felt at Einstein's somewhat formalistic attempts at unifying gravity with electromagnetism - forces which in Pauli's phrase "cannot be joined - for God hath rent them asunder". But Pauli was absolutely right in accusing me of darkness about the problem of the *masses* of the *Yang-Mills fields*; one could not obtain a *mass* without wantonly destroying the *gauge symmetry* one had started with. And this was particularly serious in this context, because Yang and Mills had conjectured the desirable *renormalizability* of their theory with a proof which relied heavily and exceptionally on the *mass lessness* of their *spin-one intermediate mesons*. The problem was to be solved only seven years later with the understanding of what is now known as the *Higgs mechanism*, but I will come back to this later.

Be that as it may, the point I wish to make from this exchange with Pauli is that already in early 1957, just after the first set of *parity* experiments, many ideas coming to fruition now, had started to become clear. These are:

1. First was the idea of *chiral symmetry leading to a V-A theory*.

> [A *V–A theory* is a theory with a *vector minus axial vector* or left-handed Lagrangian for *weak interactions*. In this theory, the *weak interaction* acts only on *left-handed particles* (and right-handed antiparticles). Since the mirror reflection of a left-handed particle is right-handed, this explains the maximal violation of *parity*. The *V–A theory* was developed before the discovery of the *Z boson*, so it did not include the right-handed fields that enter in the *neutral current interaction*.]

In those early days my humble suggestion, of this was limited to *neutrinos, electrons* and *muons* only, while shortly after, that year, Sudarshan and Marshak, Feynman and Gell-Mann, and Sakurai had the courage to postulate *symmetry* for *baryons* [composite subatomic particle, including the *proton* and the *neutron*, that contains an odd number of *valence quarks*, conventionally three] as well as *leptons*, making this into a universal principle of physics[1].

> [1] Today we believe *protons* and *neutrons* are composites of *quarks*, so that *symmetry* is now postulated for the elementary entities of today - the *quarks*.

Concomitant with the V-A theory was the result that if weak interactions are mediated by intermediate mesons, these must carry spin one.

2. Second, was the idea of *spontaneous breaking of chiral symmetry to generate electron and muon masses*: though the price which those latter-day Shylocks, Nambu and Jona-Lasinio and Goldstone[15] exacted for this (i.e. the appearance of *massless scalars*), was not yet appreciated.

> [15] Nambu, Y. (April, 1960). Axial Vector Current Conservation in Weak Interactions. *Phys. Rev. Lett.*, 4, 7, 380-2, see above; Goldstone, J. (January, 1961). Field theories with "Superconductor" solutions. *Nuovo Cimento*, 19, 154–64, see above.

3. And finally, though the use of a *Yang-Mills-Shaw (non-Abelian) gauge theory for describing spin-one intermediate charged mesons* was suggested already in 1957, the

giving of masses to the intermediate bosons through spontaneous symmetry breaking, in a manner to preserve the *renormalizability* of the theory, was to be accomplished only during a long period of theoretical development between 1963 and 1971.

Once the Yang-Mills-Shaw ideas were accepted as relevant to the *charged weak currents* - to which the *charged intermediate mesons* were coupled in this theory - during 1957 and 1958 was raised the question of what was the third component of the SU(2) *triplet*, of which the *charged weak currents* were the two members. There were the two alternatives: the *electroweak unification* suggestion, where the *electromagnetic current* was assumed to be this third component; and the rival suggestion that the third component was a *neutral current* unconnected with *electroweak unification*. With hindsight, I shall call these the Klein (1938) and the Kemmer (1937) alternatives. …

…

To give you the flavor of, for example, the year 1960, there is a paper written that year of Ward and myself with the statement: "Our basic postulate is that it should be possible to generate *strong*, *weak* and *electromagnetic interaction terms* with all their correct *symmetry* properties (as well as with clues regarding their relative strengths) *by making local gauge trans formations on the kinetic energy terms in the free Lagrangian for all particles*. This is the statement of an ideal which, in this paper at least, is only very partially realized". I am not laying a claim that we were the only ones who were saying this, but I just wish to convey to you the temper of the physics of twenty years ago - qualitatively no different today from then. But what a quantitative difference the next twenty years made, first with new and far-reaching developments in theory-and then, thanks to CERN, Fermilab, Brook haven, Argonne, Serpukhov and SLAG in testing it!

So far as theory itself is concerned, it was the next seven years between 1961-67 which were the crucial years of quantitative comprehension of the phenomenon of *spontaneous symmetry breaking* and the emergence of the *SU(2) x U(1) theory* in a form capable of being tested. The story is well known and Steve Weinberg has already spoken about it. So, I will give the barest outline. First there was the realization that the two alternatives mentioned above, a pure *electromagnetic current* versus a pure *neutral current* - Klein Schwinger versus Kemmer-Bludman - were not alternatives; they were complementary. As was noted by Glashow and independently by Ward and myself (W', Z^0, both types of *currents* and the corresponding *gauge particles* and were needed in order to build a theory that could simultaneously accommodate *parity violation* for *weak* and *parity conservation* for the *electromagnetic* phenomena. Second, there was the influential paper of Goldstone[25] in 1961 which, utilizing a *non-gauge self-interaction* between scalar particles, showed that the price of *spontaneous breaking* of a *continuous internal symmetry* was the appearance of *zero mass scalars* - a result foreshadowed earlier by Nambu.

[25] Goldstone, J. (January, 1961). Field theories with "Superconductor" solutions. *Nuovo Cimento*, 19, 154–64. See above.

In giving a proof of this theorem[26] with Goldstone I collaborated with Steve Weinberg, who spent a year at Imperial College in London.

[26] Goldstone, J., Salam, A. & Weinberg, S. (August, 1962). Broken Symmetries. *Phys. Rev.* 127, 965. See above.

I would like to pay here a most sincerely felt tribute to him and to Sheldon Glashow for their warm and personal friendship.

I shall not dwell on the now well-known contributions of Anderson, Higgs, Brout & Englert, Guralnik, Hagen and Kibble starting from 1963, which showed the way how *spontaneous symmetry breaking* using *spin-zero fields* could generate *vector-meson masses*, defeating Goldstone at the same time. This is the so-called Higgs mechanism.

The final steps towards the *electroweak theory* were taken by Weinberg[31] and myself[32] (with Kibble at Imperial College tutoring me about the Higgs phenomena).

[31] Weinberg, S. (November, 1967). Model of Leptons. *Phys. Rev. Let.*, 19, 21, 1264–6. See above.
[32] Salam, A. (1968). *Proceedings of the 8th Nobel Symposium*, Ed. Svartholm, N., Almqvist and Wiksell, Stockholm.

We were able to complete the present formulation of the *spontaneously broken* SU(2) x U(1) theory so far as *leptonic weak interactions* were concerned - with one parameter $\sin^2\theta$ describing all *weak* and *electromagnetic* phenomena and with one *isodoublet Higgs multiplet*. An account of this development was given during the contribution[32] to the Nobel Symposium (organized by Nils Svartholm and chaired by Lamek Hulthén held at Gothenburg after some postponements, in early 1968). As is well known, we did not then, and still do not, have a prediction for the *scalar Higgs mass*.

Both Weinberg and I suspected that this theory was likely to be *renormalizable*[2].

[2] When I was discussing the final version of the SU(2) x U(1) theory and its possible *renormalizability* in Autumn 1967 during a post-doctoral course of lectures at Imperial College, Nino Zichichi from CERN happened to be present. I was delighted because Zichichi had been badgering me since 1958 with persistent questioning of what theoretical avail his precise measurements on (g-2) for the *muon* as well as those of the *muon lifetime* were, when not only the magnitude of the *electromagnetic* corrections to weak decays was

uncertain, but also conversely the effect of non-renormalizable *weak interactions* on "*renormalized*" electromagnetism was so unclear.

Regarding spontaneously broken Yang-Mills-Shaw theories in general this had earlier been suggested by Englert, Brout and Thiry. But this subject was not pursued seriously except at Veltman's school at Utrecht, where the proof of *renormalizability* was given by 't Hooft in 1971[33].

[33] 't Hooft, G., (1971). Renormalization of massless Yang-Mills fields. *Nucl. Phys. B*, 33, 173; https://webspace.science.uu.nl/~hooft101/gthpub/massless.pdf; (December, 1971). Renormalizable Lagrangians for massive Yang-Mills fields. *Ibid.*, 35, 167-88; see above.

This was elaborated further by that remarkable physicist the late Benjamin Lee, working with Zinn Justin, and by 't Hooft and Veltman. This followed on the earlier basic advances in Yang-Mills calculational technology by Feynman, DeWitt, Faddeev and Popov, Mandelstam, Fradkin and Tyutin, Boulware, Taylor, Slavnov, Strathdee and Salam. In Coleman's eloquent phrase "'t Hooft's work turned the Weinberg-Salam frog into an enchanted prince". Just before had come the GIM (Glashow, Iliopoulos and Maiani) mechanism, emphasizing that the existence of the fourth *charmed quark* (postulated earlier by several authors) was essential to the natural resolution of the dilemma posed by the absence of *strangeness--violating currents*. This tied in naturally with the understanding of the Steinberger-Schwinger-Rosenberg-Bell-Jackiw-Adler anomaly and its removal for SU(2) x U(1) by the parallelism of four *quarks* and four *leptons*, pointed out by Bouchiat, Iliopoulos and Meyer and independently by Gross and Jackiw.

If one has kept a count, I have so far mentioned around fifty theoreticians. As a failed experimenter, I have always felt envious of the ambience of large experimental teams and it gives me the greatest pleasure to acknowledge the direct or the indirect contributions of the "series of minds" to the *spontaneously broken* SU(2) x U(1) *gauge theory*. My profoundest personal appreciation goes to my collaborators at Imperial College, Cambridge, and the Trieste Centre, John Ward, Paul Matthews, Jogesh Pati, John Strathdee, Tom Kibble and to Nicholas Kemmer.

In retrospect, what strikes me most about the early part of this story is how uninformed all of us were, not only of each other's work, but also of work done earlier. For example, only in 1972 did I learn of Kemmer's paper written at Imperial College in 1937.

Kemmer's argument essentially was that Fermi's weak theory was not globally SU(2) *invariant* and should be made so - though not for its own sake but as a prototype for *strong interactions*. Then this year I learnt that earlier, in 1936, Kemmer's Ph.D. supervisor, Gregor Wentzel, had introduced (the yet undiscovered) analogues of *lepto-quarks*, whose mediation could give rise to *neutral currents* after a Fierz reshuffle. And only this summer, Cecilia Jarlskog at Bergen rescued Oscar Klein's paper from the anonymity of the *Proceedings of the International Institute of Intellectual Cooperation of Paris*, and we learnt of his anticipation of a theory similar to *Yang-Mills-Shaw's* long before these authors. As I indicated before, the interesting point is that Klein was using his triplet, of two *charged mesons* plus the *photon*, not to describe *weak interaction* but for *strong nuclear force* unification with the *electromagnetic* - something our generation started on only in 1972 - and not yet experimentally verified. Even in this recitation I am sure I have inadvertently left off some names of those who have in some way contributed to SU(2) x U(1). Perhaps the moral is that not unless there is the prospect of quantitative verification, does a qualitative idea make its impress in physics.

And this brings me to experiment, and the year of the Gargamelle. I still remember Paul Matthews and I getting off the train at Aix-en-Provence for the 1973 European Conference and foolishly deciding to walk with our rather heavy luggage to the student hostel where we were billeted. A car drove from behind us, stopped, and the driver leaned out. This was Musset whom I did not know well personally then. He peered out of the window and said: "Are you Salam?" I said "Yes". He said: "Get into the car. I have news for you. We have found *neutral currents*." I will not say whether I was more relieved for being given a lift because of our heavy luggage or for the discovery of *neutral currents*. At the Aix-en-Provence meeting that great and modest man, Lagarrigue, was also present and the atmosphere was that of a carnival - at least this is how it appeared to me. Steve Weinberg gave the rapporteur's talk with T. D. Lee as the chairman. T. D. was kind enough to ask me to comment after Weinberg finished. That summer Jogesh Pati and I had predicted *proton* decay within the context of what is now called grand unification and in the flush of this excitement I am afraid I ignored *weak neutral currents* as a subject which had already come to a successful conclusion, and concentrated on speaking of the possible decays of the *proton*. I understand now that *proton* decay experiments are being planned in the United States by the Brookhaven, Irvine and Michigan and the Wisconsin-Harvard groups and also by a European collaboration to be mounted in the Mont Blanc Tunnel Garage No. 17. The later quantitative work on *neutral currents* at CERN, Fermilab., Brookhaven, Argonne and Serpukhov is, of course, history, but a special tribute is warranted to the beautiful SLAC-Yale-CERN experiment of 1978 which exhibited the effective Z^0-*photon* interference in accordance with the predictions of the theory. This was foreshadowed by Barkov *et al*'s experiments at Novosibirsk in the USSR in their exploration of *parity violation* in the atomic potential for bismuth. There is the apocryphal story about Einstein, who was asked what he would have thought if experiment had not confirmed the light deflection predicted by him. Einstein is supposed to have said, "Madam, I would have thought the Lord has missed a most marvelous opportunity." I believe, however, that the following quote from Einstein's Herbert Spencer lecture of 1933 expresses his, my colleagues' and my own views more accurately. "Pure logical thinking cannot yield us any knowledge of the empirical world; all knowledge of reality starts from experience and ends in it." This is exactly how I feel about the Gargamelle-SLAC experience[3].

[3] "To my mind the most striking feature of theoretical physics in the last thirty-six years is the fact that not a single new theoretical idea of a fundamental nature has been successful. The notions of relativistic quantum theory have in every instance proved stronger than the revolutionary ideas of a great number of talented physicists. We live in a dilapidated house and we seem to be unable to move out. The difference between this house and a prison is hardly noticeable" - Res Jost (1963) in Praise of Quantum Field Theory (Siena European Conference).

III. THE PRESENT AND ITS PROBLEMS

Thus far we have reviewed the last twenty years and the emergence of SU(2) x U(1) with the twin developments of a *gauge theory* of basic interactions, linked with *internal symmetries*, and of the *spontaneous breaking* of these symmetries. I shall first summarize the situation as we believe it to exist now and the immediate problems. Then we turn to the future.

1. To the level of energies explored, we believe that the following sets of particles are "structureless" (in a field-theoretic sense) and, at least to the level of energies explored hitherto, constitute the *elementary entities* of which all other objects are made.

<u>SU(3) triplets</u>

Family I	*quarks*	(u_R, u_Y, u_B) (d_R, d_Y, d_B)	*leptons* (ν_e) SU(2) doublets (e)
Family II	*quarks*	(c_R, c_Y, c_B) (s_R, s_Y, s_B)	*leptons* (ν_μ) SU(2) doublets (μ)
Family III	*quarks*	(t_R, t_Y, t_B) (b_R, b_Y, b_B)	*leptons* (ν_τ) SU(2) doublets (τ)

Together with their *antiparticles* each family consists of 15 or 16 two-component *fermions* (15 or 16 depending on whether the *neutrino* is *massless* or not). The third family is still conjectural, since the *top quark* has not yet been discovered. Does this family really follow the pattern of the other two? Are there more families? Does the fact that the families are replicas of each other imply that Nature has discovered a dynamical stability about a system of 15 (or 16) objects, and that by this token there is a more basic layer of structure underneath?

2. Note that *quarks* come in three *colors*; Red (R), Yellow (Y) and Blue (B). Parallel with the *electroweak* SU(2) x U(1), a *gauge field theory* (SU(3)) of *strong (quark) interactions* (*quantum chromodynamics*, QCD) has emerged which gauges the three *colors*. The indirect discovery of the (eight) *gauge bosons* associated with QCD (*gluons*), has already been surmised by the groups at DESY.

3. All known *baryons* and *mesons* are singlets of *color* SU(3). This has led to a hypothesis that *color* is always confined. *One of the major unsolved problems of field theory is to determine if QCD - treated non-perturbatively - is capable of confining quarks and gluons.*

4. In respect of the *electroweak* SU(2) x U(1), all known experiments on *weak* and *electromagnetic* phenomena below 100 GeV carried out to date agree with the theory which contains one theoretically undetermined parameter $\sin'\vartheta = 0.230 + 0.009$. The predicted values of the associated *gauge boson* (W' and Z") masses are: $m_W \approx 77\text{-}84$ GeV, $m_Z \approx 89\text{-}95$ GeV, for $0.21 \geq \sin\vartheta \geq 0.21$.

5. Perhaps the most remarkable measurement in *electroweak* physics is that of the parameter $\rho = (m_W/m_Z \cos\vartheta)^2$. Currently this has been determined from the ratio of neutral to charged *current* cross-sections. The predicted value $\rho = 1$ for weak *iso-doublet Higgs* is to be compared with the experimental[4] $\rho = 1.00 + 0.02$.

> [4] The one-loop radiative corrections to ρ suggest that the maximum *mass* of *leptons* contributing to ρ is less than 100 GeV.

6. Why does Nature favor the simplest suggestion in SU(2) x U(1) theory of the *Higgs scalars* being *iso-doublet*?[5]

> [5] To reduce the arbitrariness of the Higgs couplings and to motivate their *iso-doublet* character, one suggestion is to use *supersymmetry*. Supersymmetry is a *Fermi-Bose symmetry*, so that *isodoublet leptons* like (ν_e, e) or (ν_μ, μ) in a *super-symmetric theory* must be accompanied in the same multiplet by *iso-doublet Higgs*.
>
> Alternatively, one may identify the *Higgs* as composite fields associated with bound states of a yet new level of elementary particles and new (so-called *technicolor*) forces of which, at present low energies, we have no cognizance and which may manifest themselves in the 1-100 TeV range. Unfortunately, both these ideas at first sight appear to introduce complexities, though in the context of a wider theory, which spans *energy scales* up to much higher *masses*, a satisfactory theory of the *Higgs* phenomena, incorporating these, may well emerge.

Is there just one physical *Higgs*? Of what *mass*? At present the *Higgs interactions* with *leptons*, *quarks* as well as their self-interactions are *non-gauge interactions*. For a three-family (6-*quark*) model, 21 out of the 26 parameters needed, are attributable to the *Higgs interactions*. Is there a basic principle, as compelling and as economical as the *gauge principle*, which embraces the *Higgs* sector? Alternatively, could the *Higgs* phenomenon itself be a manifestation of a dynamical breakdown of the *gauge symmetry*[5].

7. Finally there is the problem of the families; is there a distinct SU(2) for the first, another for the second as well as a third SU(2), with *spontaneous symmetry breaking* such that the SU(2) apprehended by present experiment is a diagonal sum of these "family" SU(2)'s? To state this in another way, how far in *energy* does the e-μ universality (for example) extend? Are there more Z^0 than just one, effectively differentially coupled to the e and the μ systems? (If there are, this will constitute mini-modifications of the theory, but not a drastic revolution of its basic ideas.)

In the next section I turn to a direct extrapolation of the ideas which went into the *electroweak* unification, so as to include *strong interactions* as well. Later I shall consider the more drastic alternatives which may be needed for the unification of all forces (including gravity) - ideas which have the promise of providing a deeper understanding of the *charge* concept. Regretfully, by the same token, 1 must also become more technical and obscure for the non-specialist. I apologize for this. The non-specialist may sample the flavor of the arguments in the next section (Sec. IV), ignoring the Appendices and then go on to Sec. V which is perhaps less technical.

IV. DIRECT EXTRAPOLATION FROM THE ELECTROWEAK TO THE ELECTRONUCLEAR

4.1 The three ideas

The three main ideas which have gone into the *electronuclear* - also called *grand - unification* of the *electroweak* with the *strong nuclear force* (and which date back to the period 1972-1974), are the following:

1. First: the psychological break (for us) of grouping *quarks* and *leptons* in the same *multiplet* of a *unifying group* G, suggested by Pati and myself in 1972. The group G must contain SU(2) x U(l) x SU(3); must be simple, if all quantum numbers (*flavor*, *color*, *lepton*, *quark* and *family numbers*) are to be automatically quantized and the resulting *gauge theory* asymptotically free.

2. Second: an extension, proposed by Georgi and Glashow which places not only (*left-handed*) *quarks* and *leptons* but also their *antiparticles* in the same *multiplet* of the *unifying group*.

Appendix I displays some examples of the *unifying groups* presently considered.

Now a *gauge theory* based on a "simple" (or with discrete symmetries, a "semi-simple") group G contains one basic *gauge constant*. This constant would manifest itself physically above the "*grand unification mass*" M, exceeding all particle *masses* in the theory - these themselves being generated (if possible) hierarchically through a suitable *spontaneous symmetry-breaking* mechanism.

3. The third crucial development was by Georgi, Quinn and Weinberg who showed how, using *renormalization group* ideas, one could relate the observed low-energy couplings ... to the magnitude of the *grand unifying mass* M and the observed value of $\sin^2 \vartheta(\mu)$. ...
... .

V. ELEMENTARITY: UNIFICATION WITH GRAVITY AND NATURE OF CHARGE

In some of the remaining parts of this lecture I shall be questioning two of the notions which have gone into the direct extrapolation of Sec. IV - first, *do quarks and leptons represent the correct elementary" fields, which should appear in the matter Lagrangian*, and which are structureless for renormalizability; second, could some of the presently considered *gauge fields* themselves be composite? This part of the lecture relies heavily on an address I was privileged to give at the European Physical Society meeting in Geneva in July this year.

5.1 The quest for elementarity, prequarks (preons and pre-peons)

If *quarks* and *leptons* are *elementary*, we are dealing with 3 x 15 = 45 *elementary* entities. The "natural" group of which these constitute the fundamental *representation* is SU(45) with 2,024 *elementary gauge bosons*. It is possible to reduce the size of this group to SU(11) for example (see Appendix I), with only 120 *gauge bosons*, but then the number of *elementary fermions* increases to 561, (of which presumably 3 x 15 = 45 objects are of low and the rest of Planckian *mass*). Is there any basic reason for one's instinctive revulsion when faced with these vast numbers of *elementary fields*.
...

422

Sheldon Glashow – 1979 Nobel Lecture, December 8, 1979. *Towards a Unified Theory – Threads in a Tapestry.*

Sheldon Glashow – Nobel Lecture. NobelPrize.org. https://www.nobelprize.org/prizes/physics/1979/glashow/lecture/.

> [The Nobel Prize in Physics 1979 was awarded jointly to Sheldon Lee Glashow, Abdus Salam and Steven Weinberg "for their contributions to the theory of the *unified weak and electromagnetic interaction* between elementary particles, including, inter alia, the prediction of the *weak neutral current*".
>
> According to modern physics, four fundamental forces exist in nature. *Electromagnetic interaction* is one of these. The *weak interaction*—responsible, for example, for the beta decay of nuclei—is another. Thanks to contributions made by Sheldon Glashow, Abdus Salam, and Steven Weinberg in 1968, these two interactions were unified to one single, called *electroweak*. The theory predicted, for example, that *weak interaction* manifests itself in "*neutral weak currents*" when certain elementary particles interact. This was later confirmed. [Sheldon Glashow – Facts. NobelPrize.org. https://www.nobelprize.org/prizes/physics/1979/glashow/facts/.]

In his Nobel Prize lecture Glashow provided a brief history of his work.

———————————

INTRODUCTION

In 1956, when I began doing theoretical physics, the study of elementary particles was like a patchwork quilt. *Electrodynamics*, *weak interactions*, and *strong interactions* were clearly separate disciplines, separately taught and separately studied. There was no coherent theory that described them all. Developments such as the observation of *parity violation*, the successes of *quantum electrodynamics*, the discovery of *hadron* resonances and the appearance of *strangeness* were well-defined parts of the picture, but they could not be easily fitted together.

Things have changed. Today we have what has been called a "*standard theory*" [subsequently referred to as the *Standard Model*] *of elementary particle physics* in which *strong*, *weak*, and *electromagnetic interactions* all arise from a *local symmetry principle*. It is, in a sense, a complete and apparently correct theory, offering a qualitative description of all particle phenomena and precise quantitative predictions in many instances. There is no experimental data that contradicts the theory. In principle, if not yet in practice, all experimental data can be expressed in terms of a small number of "fundamental" *masses* and *coupling constants*. The theory we now have is an integral work of art: the patchwork quilt has become a tapestry.

Tapestries are made by many artisans working together. The contributions of separate workers cannot be discerned in the completed work, and the loose and false threads have been covered over. So it is in our picture of particle physics. Part of the picture is the *unification of weak and electromagnetic interactions* and the prediction of *neutral currents*, now being celebrated by the award of the Nobel Prize. Another part concerns the reasoned evolution of the *quark hypothesis* from mere whimsy to established dogma. Yet another is the development of *quantum chromodynamics* into a plausible, powerful, and predictive theory of *strong interactions*. All is woven together in the tapestry; one part makes little sense without the other. Even the development of the *electroweak theory* was not as simple and straightforward as it might have been. It did not arise full blown in the mind of one physicist, nor even of three. It, too, is the result of the collective endeavor of many scientists, both experimenters and theorists.

Let me stress that I do not believe that the standard theory will long survive as a correct and complete picture of physics. All *interactions* may be *gauge interactions*, but surely, they must lie within a *unifying group*. This would imply the existence of a new and very *weak interaction* which mediates the *decay* of *protons*. All matter is thus inherently unstable, and can be observed to decay. Such a synthesis of *weak*, *strong*, and *electromagnetic interactions* has been called a "*grand unified theory*", but a theory is neither grand nor unified unless it includes a description of *gravitational* phenomena. We are still far from Einstein's truly grand design.

Physics of the past century has been characterized by frequent great but unanticipated experimental discoveries. If the *standard theory* is correct, this age has come to an end. Only a few important particles remain to be discovered, and many of their properties are alleged to be known in advance. Surely this is not the way things will be, for Nature must still have some surprises in store for us.

Nevertheless, the *standard theory* will prove useful for years to come. The confusion of the past is now replaced by a simple and elegant synthesis. The *standard theory* may survive as a part of the ultimate theory, or it may turn out to be fundamentally wrong. In either case, it will have been an important way-station, and the next theory will have to be better.

In this talk, I shall not attempt to describe the tapestry as a whole, nor even that portion which is the *electroweak* synthesis and its empirical triumph. Rather, I shall describe several old threads, mostly overwoven, which are closely related to my own researches. My purpose is not so much to explain who did what when, but to approach the more difficult question of why things went as they did. I shall also follow several new threads which may suggest the future development of the tapestry.

EARLY MODELS

In the 1920's, it was still believed that there were only two fundamental forces: *gravity* and *electromagnetism*. In attempting to unify them, Einstein might have hoped to formulate a universal theory of physics. However, the study of the *atomic nucleus* soon revealed the need for two additional forces: the *strong force* to hold the *nucleus* together and the *weak force* to enable it to decay. Yukawa asked whether there might be a deep analogy between these new forces and *electromagnetism*. All forces, he said, were to result from the *exchange* of *mesons*. His conjectured *mesons* were originally intended to mediate both the *strong* and the *weak interactions*: they were strongly coupled to *nucleons* and weakly coupled to *leptons*. This first attempt to unify *strong* and *weak interactions* was fully forty years premature. Not only this, but Yukawa could have predicted the existence of *neutral currents*. His *neutral meson*, essential to provide the charge independence of nuclear forces, was also weakly coupled to pairs of *leptons*.

Not only is *electromagnetism* mediated by *photons*, but it arises from the requirement of *local gauge invariance*. This concept was generalized in 1954 to apply to non-Abelian *local symmetry groups*[1].

[1] Yang, C. N. & Mills, R. (October, 1954). Conservation of Isotopic Spin and Isotopic Gauge Invariance. *Phys. Rev.*, 96, 1, 191-5, see above. Also, Shaw, R., unpublished.

It soon became clear that a more far-reaching analogy might exist between *electromagnetism* and the other forces. They, too, might emerge from a *gauge principle*.

A bit of a problem arises at this point. All *gauge mesons* must be massless, yet the *photon* is the only massless *meson*. *How do the other gauge bosons get their masses?* There was no good answer to this question until the work of Weinberg and Salam[2] as proven by 't Hooft[3] (for *spontaneously broken gauge theories*) and of Gross, Wilczek, and Politzer[4] (for *unbroken gauge theories*).

[2] Weinberg, S. (November, 1967). Model of Leptons, see above; Salam, A. (1968), in *Elementary Particle Physics*, ed. Svartholm, N., Almqvist and Wiksell, Stockholm.
[3] 't Hooft, G. (1971). *Nuclear Physics B*, 33, 173; (December, 1971). Renormalizable Lagrangians for Massive Yang-Mills Fields. *Ibid.*, 35, 167-88, see above; Lee, B. W. &

Zinn-Justin, J. (1972). *Phys. Rev. D*, 5, 3121-60; 't Hooft, G. & Veltman, M. (1972). *Nuclear Physics B*, 44, 189.

[4] Gross, D. J. & Wilczek, F. (1973). *Phys. Rev. Lett.*, 30, 1343; Politzer, H. D. (1973). *Phys. Rev. Lett.*, 30, 1346.

Until this work was done, *gauge meson masses* had simply to be put in ad hoc.

Sakurai suggested in 1960 that *strong interactions* should arise from a *gauge principle*[5].

[5] Sakurai, J. J. (September, 1960). Theory of Strong Interactions. See above.

Applying the Yang-Mills construct to the *isospin hypercharge symmetry group*, he predicted the existence of the *vector mesons* and ω. This was the first phenomenological SU(2) x U(1) *gauge theory*. It was extended to local SU(3) by Gell-Mann and Ne'eman in 1961[6].

[6] *Gell-Mann, M., and Ne'eman, Y., The Eightfold Way.* (1964). Benjamin, W. A., New York.

Yet, these early attempts to formulate a gauge theory of strong interactions were doomed to fail. In today's jargon, they used "*flavor*" as the relevant dynamical variable, rather than the hidden and then unknown variable "*color*". Nevertheless, this work prepared the way for the emergence of *quantum chromodynamics* a decade later.

Early work in *nuclear beta decay* seemed to show that the relevant *interaction* was a mixture of S, T, and P [???].

> [The *Standard Model* of particle physics has three related *natural near-symmetries*. These state that the universe in which we live should be indistinguishable from one where a certain type of change is introduced.
>
> *Charge conjugation symmetry* (*C-symmetry*), a universe where every *particle* is replaced with its *antiparticle*
>
> *Parity symmetry* (*P-symmetry*), a universe where everything is *mirrored* along the three physical axes, known as *chirality*.
>
> *Time reversal symmetry* (*T-symmetry*), a universe where the direction of time is reversed. *T-symmetry* is counterintuitive (the future and the past are not symmetrical) but explained by the fact that the *Standard Model* describes *local properties*, not *global* ones like entropy.]

Only after the discovery of *parity violation*, and the undoing of several wrong experiments, did it become clear that the *weak interactions* were in reality V-A.

> [In 1957, Robert Marshak and George Sudarshan and, somewhat later, Richard Feynman and Murray Gell-Mann proposed a V−A (*vector minus axial vector* or left-handed) Lagrangian for *weak interactions*. In this theory, the *weak interaction* acts only on *left-handed particles* (and *right-handed antiparticles*). Since the mirror reflection of a left-handed particle is right-handed, this explains the maximal violation of parity. The V−A theory was developed before the discovery of the Z *boson*, so it did not include the right-handed fields that enter in the *neutral current interaction*.]

The synthesis of Feynman and Gell-Mann and of Marshak and Sudarshan was a necessary precursor to the notion of a *gauge theory* of *weak interactions*[7].

[7] Feynman, R., & Gell-Mann, M. (1958). *Phys. Rev.*, 109, 193; Marshak, R. & Sudarshan, E. C. G. (1958). *Phys. Rev.*, 109, 1860.

Bludman formulated the first SU(2) *gauge theory* of *weak interactions* in 1958[8].

[8] Bludman, S. (1958). *Nuovo Cimento*, 10, 9, 433.

No attempt was made to include *electromagnetism*. The model included the conventional *charged-current interactions*, and in addition, a set of *neutral current couplings*. These are of the same strength and form as those of today's theory in the limit in which the weak mixing angle vanishes. Of course, a *gauge theory* of *weak interactions* alone cannot be made *renormalizable*. For this, the *weak* and *electromagnetic interactions* must be unified.

Schwinger, as early as 1956, believed that the *weak* and *electromagnetic interactions* should be combined together into a gauge theory[9].

[9] Schwinger, J. (November, 1958). A theory of the fundamental interactions. *Ann. Phys.*, 2, 5, 407-34: https://doi.org/10.1016/0003-4916(57)90015-5.

The *charged massive vector* intermediary and the *massless photon* were to be the *gauge mesons*. As his student, I accepted this faith. In my 1958 Harvard thesis, I wrote: "It is of little value to have a potentially *renormalizable* theory of *beta processes* without the possibility of a *renormalizable electrodynamics*. We should care to suggest that a fully acceptable theory of these *interactions* may only be achieved if they are treated together ....."[10].

[10] Glashow, S. L. (1958). *Harvard University Thesis*, p. 75.

We used the original SU(2) *gauge interaction* of Yang and Mills. Things had to be arranged so that the *charged current*, but not the *neutral (electromagnetic) current*, would violate *parity* and *strangeness*. Such a theory is technically possible to construct, but it is both ugly and experimentally false[11].

[11] Georgi, H. & Glashow, S. L. (1972). *Phys. Rev. Lett.*, 28, 1494.

We know now that *neutral currents* do exist and that the *electroweak gauge group* must be larger than SU(2).

Another *electroweak* synthesis without *neutral currents* was put forward by Salam and Ward in 1959[12].

[12] Salam, A. & Ward, J. (February, 1959). Weak and electromagnetic interactions. *Nuovo Cimento*, 11, 568–577 (1959); https://doi.org/10.1007/BF02726525

Again, they failed to see how to incorporate the experimental fact of *parity violation*. Incidentally, in a continuation of their work in 1961, they suggested a *gauge theory* of *strong*, *weak*, and *electromagnetic interactions* based on the *local symmetry group* SU(2) x SU(2)[13].

[13] Salam, A. & Ward, J. (January, 1961). On a gauge theory of elementary interactions. *Nuovo Cimento*, 19, 165-70; https://doi.org/10.1007/BF02812723.

This was a remarkable portent of the SU(3) x SU(2) x U(1) model which is accepted today.

We come to my own work[14] done in Copenhagen in 1960, and done independently by Salam and Ward[15].

[14] Glashow, S. L. (February, 1961). Partial-symmetries of weak interactions. See above.

[15] Salam, A., & Ward, J. (November, 1964). *Physics Letters*, 13, 168 (1964). See above.

We finally saw that a *gauge group* larger than SU(2) was necessary to describe the *electroweak interactions*. Salam and Ward were motivated by the compelling beauty of *gauge theory*. I thought I saw a way to a *renormalizable* scheme. I was led to the group SU(2) x U(1) by analogy with the approximate *isospin-hypercharge group* which characterizes *strong interactions*. In this model there were two electrically neutral intermediaries: the *massless photon* and a *massive neutral vector meson* which I called B but which is now known as Z. The weak mixing angle determined to what linear combination of SU(2) x U(1) *generators* B would correspond. The precise form of the predicted *neutral-current interaction* has been verified by recent experimental data. However, the strength of the *neutral current* was not prescribed, and the model was not in fact *renormalizable*. These glaring omissions were to be rectified by the work of Salam and Weinberg and the subsequent proof of *renormalizability*. Furthermore, the model was a model of *leptons* - it could not evidently be extended to deal with *hadrons*.

RENORMALIZABILITY

In the late 50's, *quantum electrodynamics* and *pseudoscalar meson theory* were known to be *renormalizable*, thanks in part to work of Salam. Neither of the customary models of *weak interactions - charged intermediate vector bosons* or *direct four-fermion couplings -* satisfied this essential criterion. My thesis at Harvard, under the direction of Julian Schwinger, was to pursue my teacher's belief in a unified *electroweak gauge theory*. I had found some reason to believe that such a theory was less singular than its alternatives. Feinberg, working with *charged intermediate vector mesons* discovered that a certain type of divergence would cancel for a special value of the *meson anomalous magnetic moment*[16].

[16] Feinberg, G. (1958). *Phys. Rev.*, 110, 1482.

It did not correspond to a *"minimal electromagnetic coupling"*, but to the *magnetic* properties demanded by a *gauge theory*. Tzou Kuo-Hsien examined the zero- mass limit of *charged vector meson electrodynamics*[17].

[17] Kuo-Hsien, T. (1957). *Comptes Rendus*, 245, 289.

Again, a sensible result is obtained only for a very special choice of the *magnetic dipole moment* and *electric quadrupole moment*, just the values assumed in a *gauge theory*. Was it just coincidence that the *electromagnetism* of a *charged vector meson* was least pathological in a *gauge theory*?

Inspired by these special properties, I wrote a notorious paper[18].

[18] Glashow, S. L. (1959). *Nuclear Physics*, 10, 107.

I alleged that a softly-broken *gauge theory*, with *symmetry breaking* provided by explicit *mass* terms, was *renormalizable*. It was quickly shown that this is false.

Again, in 1970, Iliopoulos and I showed that a wide class of divergences that might be expected would cancel in such a *gauge theory*[19].

[19] Glashow, S. L. & Iliopoulos J. (1971). *Phys. Rev. D*, 3, 1043.

We showed that the naive divergences of order $(\alpha\Lambda^4)^n$ were reduced to "merely" $(\alpha\Lambda^2)^n$ where Λ is a cut-off *momentum*. This is probably the most difficult theorem that Iliopoulos or I had even proven. Yet, our labors were in vain. In the spring of 1971, Veltman informed us that his student Gerhart 't Hooft had established the *renormalizability* of *spontaneously broken gauge theory*.

In pursuit of *renormalizability*, I had worked diligently but I completely missed the boat. The *gauge symmetry* is an exact *symmetry*, but it is hidden. One must not put in *mass* terms by hand. The key to the problem is the idea of *spontaneous symmetry breakdown*: the work of Goldstone as extended to *gauge theories* by Higgs and Kibble in 1964[20].

> [20] Many authors are involved with this work: Brout, R., Englert, F., Goldstone, J., Guralnik, G., Hagen, C., Higgs, P., Jona-Lasinio, G., Kibble, T., and Nambu, Y.

These workers never thought to apply their work on formal *field theory* to a phenomenologically relevant model. I had had many conversations with Goldstone and Higgs in 1960. Did I neglect to tell them about my SU(2) x U(1) model, or did they simply forget?

Both Salam and Weinberg had had considerable experience in formal *field theory*, and they had both collaborated with Goldstone on *spontaneous symmetry breaking*. In retrospect, it is not so surprising that it was they who first used the key. Their SU(2) x U(1) *gauge symmetry* was *spontaneously broken*. The *masses* of the W and Z and the nature of *neutral current* effects depend on a single measurable parameter, not two as in my un-renormalizable model. The strength of the *neutral currents* was correctly predicted. The daring Weinberg-Salam conjecture of *renormalizability* was proven in 1971. *Neutral currents* were discovered in 1973[21], but not until 1978 was it clear that they had just the predicted properties[22].

> [21] Hasert, F. J., *et al.* (1973). *Physics Letters*, 46B, 138; and (1974). *Nuclear Physics B*, 73, 1. Benvenuti, A., *et al.* (1974). *Phys. Rev. Lett.*, 32, 800.
> [22] Prescott, C. Y., *et al.* (1978). *Phys. Lett.*, B 77, 347.

THE STRANGENESS-CHANGING NEUTRAL CURRENT

I had more or less abandoned the idea of an *electroweak gauge theory* during the period 1961-1970. Of the several reasons for this, one was the failure of my naive foray into *renormalizability*. Another was the emergence of an empirically successful description of *strong interactions* - the *SU(3) unitary symmetry* scheme of Gell-Mann and Ne'eman. This theory was originally phrased as a *gauge theory*, with ϱ, ω and K* as *gauge mesons*. It was completely impossible to imagine how both *strong* and *weak interactions* could be *gauge* *theories*: there simply wasn't room enough for commuting structures of *weak* and *strong currents*. Who could foresee the success of the *quark model*, and the displacement of SU(3) from the arena of *flavor* to that of *color*? The predictions of *unitary symmetry* were being borne out - the predicted Ω^- was discovered in 1964. Current algebra was being successfully exploited. *Strong interactions* dominated the scene.

When I came upon the SU(2) x U(1) model in 1960, I had speculated on a possible extension to include *hadrons*. To construct a model of *leptons* alone seemed senseless: *nuclear beta decay*, after all, was the first and foremost problem. One thing seemed clear. The fact that the *charged current* violated *strangeness* would force the *neutral current* to violate *strangeness* as well. It was already well known that *strangeness-changing neutral currents* were either strongly suppressed or absent. I concluded that the Z^0 had to be made very much heavier than the W. This was an arbitrary but permissible act in those days: the *symmetry breaking* mechanism was unknown. I had "solved" the problem of *strangeness-changing neutral currents* by suppressing all *neutral currents*: the baby was lost with the bath water.

I returned briefly to the question of *gauge theories* of *weak interactions* in a collaboration with Gell-Mann in 1961[23].

[23] Glashow, S. L. & Gell-Mann, M. (September, 1961). Gauge Theories of Vector Particles. See above.

From the recently developing ideas of current algebra, we showed that a *gauge theory* of *weak interactions* would inevitably run into the problem of *strangeness-changing neutral currents*. We concluded that something essential was missing. Indeed, it was. Only after *quarks* were invented could the idea of the fourth *quark* and the *GIM mechanism* arise.

[The *GIM* or *Glashow–Iliopoulos–Maiani* mechanism is the mechanism through which *flavor-changing neutral currents* (FCNCs) are suppressed in loop diagrams. It also explains why *weak interactions* that change *strangeness* by 2 ($\Delta S = 2$ transitions) are suppressed, while those that change *strangeness* by 1 ($\Delta S = 1$ transitions) are allowed, but only in *charged current interactions*. The rare *leptonic* decay of the neutral Kaon was predicated on the *GIM mechanism*.

The mechanism was put forth in a famous paper by Glashow, Iliopoulos & Maiani (1970); at that time, only three *quarks* (*up*, *down*, and *strange*) were thought to exist. Bjorken & Glashow (1964) had previously predicted a fourth *quark*, but there was little evidence for its existence. The *GIM mechanism* however, required the existence of a fourth *quark*, and the prediction of the *charm quark* is usually credited to Glashow, Iliopoulos, & Maiani.]

431

From 1961 to 1964, Sidney Coleman and I devoted ourselves to the exploitation of the *unitary symmetry* scheme. In the spring of 1964, I spent a short leave of absence in Copenhagen. There, Bjorken and I suggested that the Gell-Mann-Zweig-system of three *quarks* should be extended to four[24].

[24] Bjorken, J. & Glashow, S. L. (August, 1964). Elementary particles and SU(4). *Physics Letters*, 11, 3, 255-257; https://doi.org/10.1016/0031-9163(64)90433-0.

(Other workers had the same idea at the same time). We called the fourth *quark* the *charmed quark*. Part of our motivation for introducing a fourth *quark* was based on our mistaken notions of *hadron* spectroscopy. But we also wished to enforce an analogy between the *weak leptonic current* and the *weak hadronic current*. Because there were two weak *doublets* of *leptons*, we believed there had to be two weak *doublets* of *quarks* as well. The basic idea was correct, but today there seem to be three *doublets* of *quarks* and three *doublets* of *leptons*.

The *weak current* Bjorken and I introduced in 1964 was precisely the *GIM current*. The associated *neutral current*, as we noted, conserved *strangeness*. Had we inserted these currents into the earlier *electroweak theory*, we would have solved the problem of *strangeness-changing neutral currents*. We did not. I had apparently quite forgotten my earlier ideas of *electroweak synthesis*. The problem which was explicitly posed in 1961 was solved, in principle, in 1964. No one, least of all me, knew it. Perhaps we were all befuddled by the chimera of *relativistic* SU(6), which arose at about this time to cloud the minds of theorists.

Five years later, John Iliopoulos, Luciano Maiani and I returned to the question of *strangeness-changing neutral currents*[26].

[26] Glashow, S. L., Iliopoulos, J. & Maiani, L. (October, 1970). Weak Interactions with Lepton-Hadron Symmetry. *Phys. Rev. D*, 2, 7, 1285; https://doi.org/10.1103/PhysRevD.2.1285.

It seems incredible that the problem was totally ignored for so long. We argued that unobserved effects (a large K_1K_2 *mass* difference; decays like $K \rightarrow \pi\nu\nu^-$; etc.) would be expected to arise in any of the known *weak interaction models*: four *fermion couplings*; *charged vector meson* models; or the *electroweak gauge theory*. We worked in terms of cut-offs, since no *renormalizable* theory was known at the time. We showed how the unwanted effects would be eliminated with the conjectured existence of a fourth *quark*. After languishing for a decade, the problem of the *selection rules* of the *neutral current* was finally solved. Of course, not everyone believed in the predicted existence of *charmed hadrons*.

This work was done fully three years after the epochal work of Weinberg and Salam, and was presented in seminars at Harvard and at M.I.T. Neither I, nor my coworkers, nor Weinberg, sensed the connection between the two endeavors. We did not refer, nor were we asked to refer, to the Weinberg-Salam work in our paper.

The relevance became evident only a year later. Due to the work of 't Hooft, Veltman, Benjamin Lee, and Zinn-Justin, it became clear that the Weinberg-Salam ansatz was in fact a *renormalizable* theory. With *GIM*, it was trivially extended from a *model of leptons* to a *theory of weak interactions*. The ball was now squarely in the hands of the experimenters. Within a few years, *charmed hadrons* and *neutral currents* were discovered, and both had just the properties they were predicted to have.

FROM ACCELERATORS TO MINES

Pions and *strange* particles were discovered by passive experiments which made use of the natural flux of *cosmic rays*. However, in the last three decades, most discoveries in particle physics were made in the active mode, with the artificial aid of particle accelerators. Passive experimentation stagnates from a lack of funding and lack of interest. Recent developments in theoretical particle physics and in astrophysics may mark an imminent rebirth of passive experimentation. The concentration of virtually all high energy physics endeavors at a small number of major accelerator laboratories may be a thing of the past.

This is not to say that the large accelerator is becoming extinct; it will remain an essential if not exclusive tool of high-energy physics. Do not forget that the existence of Z^0 at ~ 100 GeV is an essential but quite untested prediction of the *electroweak theory*. There will be additional dramatic discoveries at accelerators, and these will not always have been predicted in advance by theorists. The construction of new machines like LEP and ISABELLE is mandatory.

Consider the successes of the *electroweak synthesis*, and the fact that the only plausible theory of *strong interactions* is also a *gauge theory*. We must believe in the ultimate synthesis of *strong*, *weak*, and *electromagnetic interactions*. It has been shown how the *strong* and *electroweak gauge groups* may be put into a larger but simple gauge group[27].

[27] Georgi, H. & Glashow, S. L. (February,1974). Unity of All Elementary-Particle Forces. *Phys. Rev. Lett.*, 33, 8, 438; https://doi.org/10.1103/PhysRevLett.32.438.

Grand unification - perhaps along the lines of the original SU(5) theory of Georgi and me - must be essentially correct. This implies that the *proton*, and indeed all *nuclear matter*, must be inherently unstable. Sensitive searches for *proton* decay are now being launched.

If the *proton* lifetime is shorter than 10^{32} years, as theoretical estimates indicate, it will not be long before it is seen to decay.

Once the effect is discovered (and I am sure it will be), further experiments will have to be done to establish the precise modes of decay of nucleons. The *selection rules*, *mixing angles*, and *space-time structure* of a new class of effective four-fermion couplings must be established. The heroic days of the discovery of the nature of *beta decay* will be repeated. The first generation of *proton decay* experiments is cheap, but subsequent generations will not be. Active and passive experiments will compete for the same dwindling resources.

Other new physics may show up in elaborate passive experiments. Today's theories suggest modes of *proton decay* which violate both *baryon number* and *lepton number* by unity. Perhaps this $\Delta B = \Delta L = 1$ law will be satisfied. Perhaps $\Delta B = -\Delta L$ transitions will be seen. Perhaps, as Pati and Salam suggest, the *proton* will decay into three *leptons*. Perhaps two nucleons will annihilate in $\Delta B = 2$ transitions. The effects of *neutrino oscillations* resulting from *neutrino masses* of a fraction of an election volt may be detectable. "Superheavy isotopes" which may be present in the Earth's crust in small concentrations could reveal themselves through their multi-GeV decays. *Neutrino* bursts arising from distant astronomical catastrophes may be seen. The list may be endless or empty. Large passive experiments of the sort now envisioned have never been done before. Who can say what results they may yield?

PREMATURE ORTHODOXY

The discovery of J/Ψ in 1974 made it possible to believe in a system involving just four *quarks* and four *leptons*. Very quickly after this a third *charged lepton* (the *tau*) was discovered, and evidence appeared for a third $Q = -1/3$ *quark* (the b *quark*). Both discoveries were classic surprises. It became immediately fashionable to put the known *fermions* into *families* or *generations*:

$$(u \quad v_e) \quad (c \quad v_\mu) \quad (t \quad v_\tau)$$
$$(d \quad e) \quad (s \quad \mu) \quad (b \quad \tau)$$

The existence of a third $Q = 2/3$ *quark* (the t *quark*) is predicted. The Cabibbo-GIM scheme is extended to a system of six *quarks*. The three-family system is the basis to a vast and daring theoretical endeavor. For example, a variety of papers have been written putting experimental constraints on the four parameters which replace the Cabibbo angle in a *six-quark* system. The detailed manner of decay of particles containing a single b *quark* has been worked out. All that is wanting is experimental confirmation. A new orthodoxy has emerged, one for which there is little evidence, and one in which I have little faith.

The predicted t *quark* has not been found. While the *upsilon mass* is less than 10 GeV, the analogous tt particle, if it exists at all, must be heavier than 30 GeV. Perhaps it doesn't exist.

Howard Georgi and I, and other before us, have been working on models with no t *quark*[28].

[28] Georgi, H. & Glashow, S. L., *Harvard Preprint HUTP-79/A 053*.

We believe this unorthodox view is as attractive as its alternative. And, it suggests a number of exciting experimental possibilities.

We assume that b and τ share a *quantum number*, like *baryon number*, that is essentially exactly conserved. (Of course, it may be violated to the same extent that *baryon number* is expected to be violated.) Thus, the system is assumed to be distinct from the lighter four *quarks* and four *leptons*. There is, in particular, no mixing between b and d or s. The original GIM structure is left intact. An additional mechanism must be invoked to mediate *b decay*, which is not present in the SU(3) x SU(2) x U(1) *gauge theory*.

One possibility is that there is an additional SU(2) *gauge interaction* whose effects we have not yet encountered. It could mediate such decays of b as these

$$b \rightarrow \tau^+ + (e^- \text{ or } \mu^-) + (d \text{ or } s).$$

All decays of b would result in the production of a pair of *leptons*, including a τ^+ or its neutral partner. There are other possibilities as well, which predict equally bizarre decay schemes for *b-matter*. How the b *quark* decays is not yet known, but it soon will be.

The new SU(2) *gauge theory* is called upon to explain *CP violation* as well as *b decay*. In order to fit experiment, three additional *massive neutral vector bosons* must exist, and they cannot be too heavy. One of them can be produced in $e^+ e^-$ annihilation, in addition to the expected Z^0. Our model is rife with experimental predictions, for example: a second Z^0, a heavier version of b and of τ, the production of τ b in e p collisions, and the existence of heavy neutral unstable *leptons* which may be produced and detected in $e^+ e^-$ or in up collisions.

This is not the place to describe our views in detail. They are very speculative and probably false. The point I wish to make is simply that it is too early to convince ourselves that we know the future of particle physics. There are too many points at which the conventional picture may be wrong or incomplete. The SU(3) x SU(2) x U(1) *gauge theory* with three families is certainly a good beginning, not to accept but to attack, extend, and exploit. We are far from the end.

Makoto Kobayashi (born April 7, 1944, in Nagoya, Japan).

Kobayashi is a Japanese physicist known for his work on *CP-violation* who was awarded one-fourth of the 2008 Nobel Prize in Physics "for the discovery of the origin of the *broken symmetry* which predicts the existence of at least three families of *quarks* in nature."

> [*CP violation* is a violation of *CP-symmetry* (or *charge conjugation parity symmetry*): the combination of *C-symmetry* (*charge conjugation symmetry*) and *P-symmetry* (*parity symmetry*). *CP-symmetry* states that the laws of physics should be the same if a particle is interchanged with its *antiparticle* (*C-symmetry*) while its *spatial coordinates* are inverted ("mirror" or *P-symmetry*).]

Makoto Kobayashi was born in Nagoya, Japan in 1944. When he was two years old, Kobayashi's father Hisashi died. The Kobayashi family home was destroyed by the Bombing of Nagoya, so they stayed at his mother's (surnamed Kaifu) family house. One of Makoto's cousins, Toshiki Kaifu, the 51st Prime Minister of Japan, was living in the same place. His other cousin was an astronomer, Norio Kaifu.

After graduating from the School of Science of Nagoya University in 1967, he obtained a DSc degree from the Graduate School of Science of Nagoya University in 1972. During college years, he received guidance from Shoichi Sakata and others. After completing his doctoral research at Nagoya University in 1972, Kobayashi worked as a research associate on particle physics at Kyoto University. Together, with his colleague Toshihide Maskawa, he worked on explaining *CP-violation* within the *Standard Model* of particle physics. Kobayashi and Maskawa's theory required that there were at least three generations of *quarks*, a prediction that was confirmed experimentally four years later by the discovery of the *bottom quark*.

Kobayashi and Maskawa's 1973 article, explained *broken symmetry* within the framework of the *Standard Model*, but *required that the Model be extended to three families of quarks* to explain *CP violation*. [Kobayashi, M. & Maskawa, T. (February, 1973). CP-Violation in the Renormalizable Theory of Weak Interaction. See below.] The Cabibbo–Kobayashi–Maskawa matrix, which defines the mixing parameters between *quarks* was the result of this work. Kobayashi and Maskawa were jointly awarded half of the 2008 Nobel Prize in Physics for this work, with the other half going to Yoichiro Nambu.

He married Sachiko Enomoto in 1975; they had one son, Junichiro. After his first wife died, Kobayashi married Emiko Nakayama in 1990, they had a daughter, Yuka.

Toshihide Maskawa (February 7, 1940–July 23, 2021).

Maskawa was a Japanese theoretical physicist known for his work on *CP-violation* who was awarded one quarter of the 2008 Nobel Prize in Physics "for the discovery of the origin of the *broken symmetry* which predicts the existence of at least three families of *quarks* in nature."

Maskawa was born in Nagoya, Japan. After World War II ended, the Maskawa family operated as a sugar wholesaler. A native of Aichi Prefecture, Toshihide Maskawa graduated from Nagoya University in 1962 and received a Ph.D. degree in particle physics from the same university in 1967. His doctoral advisor was the physicist Shoichi Sakata.

From early life Maskawa liked trivia, also studied mathematics, chemistry, linguistics and various books. In high school, he loved novels, especially detective and mystery stories and novels by Ryūnosuke Akutagawa.

At Kyoto University in the early 1970s, he collaborated with Makoto Kobayashi on explaining *broken symmetry* (the *CP violation*) within the *Standard Model* of particle physics. Maskawa and Kobayashi's theory required that there be at least three generations of *quarks*, a prediction that was confirmed experimentally four years later by the discovery of the *bottom quark*.

Maskawa was director of the Yukawa Institute for Theoretical Physics from 1997 to 2003. He was special professor and director general of Kobayashi-Maskawa Institute for the Origin of Particles and the Universe at Nagoya University, director of Maskawa Institute for Science and Culture at Kyoto Sangyo University, and professor emeritus at Kyoto University.

On December 8, 2008, after Maskawa told the audience "Sorry, I cannot speak English", he delivered his Nobel lecture on "What Did CP Violation Tell Us?" in Japanese language, at Stockholm University. The audience followed the subtitles on the screen behind him.

Maskawa married Akiko Takahashi in 1967. The couple have two children, Kazuki and Tokifuji.

On 23 July 2021 at the same day as the opening ceremony of Tokyo Summer Olympic Games, Maskawa died of oral cancer at his home in Kyoto at the age of 81. He was cremated in October 2021 after the private funeral.

Kobayashi, M. & Maskawa, T. (February, 1973). CP-Violation in the Renormalizable Theory of Weak Interaction.

Progress of Theoretical Physics, 49, 2, 652–7; https://doi.org/10.1143/PTP.49.652.

Department of Physics, Kyoto University, Kyoto.

Received September 1, 1972.

Explained *broken symmetry* within the framework of the *Standard Model*, but *required that the Model be extended to three families of quarks to explain CP violation,* which ultimately led to the *six-quark model.*

[*CP violation* is a violation of *CP-symmetry* (or *charge conjugation parity symmetry*): the combination of *C-symmetry* (*charge conjugation symmetry*) and *P-symmetry* (*parity symmetry*). *CP-symmetry* states that the laws of physics should be the same if a particle is interchanged with its *antiparticle* (*C-symmetry*) while its *spatial coordinates* are inverted ("mirror" or *P-symmetry*).

A parity transformation (also called *parity inversion)* is the flip in the sign of one spatial coordinate.

Charge conjugation is a transformation that switches all particles with their corresponding *antiparticles*, thus changing the sign of all *charges*: not only *electric charge* but also the *charges* relevant to other forces.]

[James Watson Cronin, 1980 Nobel Lecture, December 8, 1980: "Recently, much attention has been given to the role that *CP violation* may play in the early stages of the evolution of the universe. A mechanism has been proposed with *CP violation* as one ingredient *which leads from matter--antimatter symmetry in the early universe to the small excess of matter observed in the universe at the present time.* The first published account of this mechanism, of which I am aware, was made by Sakharov in 1967[1].

[1] Sakharov, A. D., (January, 1967). Violation of CP invariance, C asymmetry, and Baryon asymmetry of the universe. *ZhETF Pis'ma*, 5, 32; English translation (1967). *JETP Letters (USSR)*, 5, 24-7; https://www.osti.gov/biblio/4449128.

He explicitly stated the three ingredients which form the foundation of the mechanism as it is presently discussed. These ingredients are: (1) *baryon instability*, (2) *CP violation*, and (3) appropriate *lack of thermal equilibrium*. The recent intense interest in this problem has risen because *baryon instability* is a natural consequence of the present ideas of unification of the *strong interactions* with the successfully unified *electromagnetic* and *weak interactions*. This latter unification was discussed in the 1979 Nobel lectures of Glashow, Salam, and Weinberg.]

[Makoto Kobayashi – Nobel Lecture, December 8, 2008. CP Violation and Flavor Mixing: "At that time only three *quarks* were widely accepted, but the *three-quark model* had some flaws in the *gauge theory*. Therefore, from a theoretical point of view, the *four-quark* model of the [*Glashow-Illiopoulos-Miani*] GIM type was considered preferable. However, it is impossible to accommodate *CP violation* in a model of the GIM type. We found that even if we relax the conditions for the GIM type, we cannot make any realistic model of *CP violation* with four *quarks*. This implies that there must be some unknown particles besides the fourth *quark*. I thought that this was quite strong and an important conclusion of our argument. Then we considered a few possible mechanisms of *CP violation* by introducing new particles. We proposed the *six-quark model* as one such possible mechanism."
… "We thought that this mechanism of *CP violation* is very interesting and elegant, *but we had no further reason to single out the six-quark model from the other possibilities*. The model was not so special, because if the system has sufficiently many particles, it is not difficult to violate *CP symmetry*. However, *the subsequent experimental development pushed up the six-quark model to a special position*.]

Abstract

In a framework of the *renormalizable* theory of *weak interaction*, problems of *CP-violation* are studied. It is concluded that no realistic models of *CP-violation* exist in the *quartet scheme* without introducing any other new fields. Some possible models of *CP-violation* are also discussed.

When we apply the *renormalizable* theory of *weak interaction*[1] to the *hadron* system, we have some limitations on the *hadron model*.

[1] Weinberg, S. (November, 1967). Model of Leptons. *Phys. Rev. Let.*, 19, 21, 1264–6, see above; (December, 1971). Physical Processes in a Convergent Theory of the Weak and Electromagnetic Interactions. *Idem.*, 27, 24, 1688; https://doi.org/10.1103/PhysRevLett.27.1688.

[Hadron [classification].

A *hadron* (from Ancient Greek ἁδρός (hadrós) 'stout, thick') is a *composite subatomic particle made of two or more quarks held together by the strong interaction.* They are analogous to molecules, which are held together by the electric force. Most of the *mass* of ordinary matter comes from two *hadrons*: the *proton* and the *neutron*, while most of the *mass* of the *protons* and *neutrons* is in turn due to the *binding energy* of their constituent *quarks*, due to the *strong force.*

Hadrons are categorized into two broad families: *baryons*, made of an odd number of *quarks* (usually three *quarks*) and *mesons*, made of an even number of *quarks* (usually two *quarks*: one *quark* and one *anti-quark*). *Protons* and *neutrons* (which make the majority of the *mass* of an atom) are examples of *baryons*; *pions* are an example of a *meson.*

"Exotic" *hadrons*, containing more than three *valence quarks*, have been discovered in recent years. A *tetraquark state* (an exotic *meson*), named the Z(4430)−, was discovered in 2007 by the Belle Collaboration and confirmed as a resonance in 2014 by the LHCb collaboration. Two *pentaquark states* (exotic *baryons*), named P+c(4380) and P+c(4450), were discovered in 2015 by the LHCb collaboration. There are several more exotic *hadron* candidates and other color-singlet *quark* combinations that may also exist.

Almost all "free" *hadrons* and *anti-hadrons* (meaning, in isolation and not bound within an atomic nucleus) are believed to be unstable and eventually decay into other particles. The only known possible exception is free *protons*, which appear to be stable, or at least, take immense amounts of time to decay (order of 10^{34}+ years). By way of comparison, free *neutrons* are the longest-lived unstable particle, and decay with a half-life of about 611 seconds, and have a mean lifetime of 879 seconds.

Hadron physics is studied by colliding *hadrons*, e.g. *protons*, with each other or the nuclei of dense, heavy elements, such as lead (Pb) or gold (Au), and detecting the debris in the produced particle showers. A similar process occurs in the natural environment, in the extreme upper-atmosphere, where *muons* and *mesons* such as *pions* are produced by the collisions of *cosmic rays* with rarefied gas particles in the outer atmosphere.

The term "*hadron*" is a new Greek word introduced by L.B. Okun in a plenary talk at the 1962 International Conference on High Energy Physics at CERN. He opened his talk with the definition of a new category term:

"Notwithstanding the fact that this report deals with *weak interactions*, we shall frequently have to speak of *strongly interacting* particles. These particles pose not only numerous scientific problems, but also a terminological problem. The point is that "*strongly interacting particles*" is a very clumsy term which does not yield itself to the formation of an adjective. For this reason, to take but one instance, decays into *strongly interacting particles* are called "*non-leptonic*". This definition is not exact because "*non-leptonic*" may also signify *photonic*. In this report I shall call strongly interacting particles "*hadrons*", and the corresponding decays "*hadronic*" (the Greek ἁδρός signifies "large", "massive", in contrast to λεπτός which means "small", "light"). I hope that this terminology will prove to be convenient." — L.B. Okun (1962).]

It is well known that there exists, in the case of the *triplet model*, a difficulty of the *strangeness* changing *neutral current* and that the *quartet model* is free from this difficulty. Furthermore, Maki and one of the present authors (T.M.) have shown[2] that, in the latter case, the *strong interaction* must be *chiral* SU(4) x SU(4) *invariant* as precisely as the conservation of the third component of the *isospin* I_8.

[2] Maki, Z. & Maskawa, T. (April, 1972). RIFP-146 (preprint).

In addition to these arguments, for the theory to be realistic, *CP-violating interactions* should be incorporated in a *gauge invariant* way. This requirement will impose further limitations on the *hadron* model and the *CP-violating interaction* itself. The purpose of the present paper is to investigate this problem. In the following, it will be shown that in the case of the above-mentioned *quartet model*, we cannot make a *CP-violating interaction* without introducing any other new fields when we require the following conditions: a) The *mass* of the fourth member of the *quartet*, which we will call ζ, is sufficiently large, b) the model should be consistent with our well-established knowledge of the *semi-leptonic* processes. After that some possible ways of bringing *CP-violation* into the theory will be discussed.

We consider the *quartet model* with a *charge* assignment of Q, Q–1, Q–1 and Q for p, n, λ and ζ, respectively, and we take the same underlying *gauge group* $SU_{weak}(2)$ x SU(1) and the *scalar doublet field* φ as those of Weinberg's original model[1]. Then, *hadronic* parts of the Lagrangian can be divided in the following way:

$$\mathscr{L}_{had} = \mathscr{L}_{kin} + \mathscr{L}_{mass} + \mathscr{L}_{strong} + \mathscr{L}',$$

where $\mathscr{L}_{kin}$ is the *gauge-invariant* kinetic part of the *quartet field*, q, so that it contains *interactions* with the *gauge fields*. $\mathscr{L}_{mass}$ is a generalized *mass* term of q, which includes *Yukawa couplings* to φ since they contribute to the *mass* of q through the *spontaneous*

441

breaking of *gauge symmetry*. $\mathscr{L}_{\text{strong}}$ is a *strong-interaction* part which conserves I_8 and therefore *chiral* SU (4) x SU(4) *invariant*[2]. We assume C- and P-*invariance* of $\mathscr{L}_{\text{strong}}$. The last term $\mathscr{L}'$ denotes residual *interaction* parts if they exist. Since $\mathscr{L}_{\text{mass}}$ includes *couplings* with φ, it has possibilities of violating *CP-conservation*. As is known as Higgs phenomena[3] three massless components of φ can be absorbed into the *massive gauge fields* and eliminated from the Lagrangian.

[3] Higgs, P. W. (1964). Broken symmetries, massless particles and gauge fields. *Physics. Letters*, 12, 2, 132-3; https://doi.org/10.1016/0031-9163(64)91136-9; (1964). Broken symmetries and the masses of gauge bosons. *Phys. Rev. Lett.*, 13, 16, 508–9, see above. Guralnik, G., Hagen, C. & Kibble, T. (November, 1964). Global Conservation Laws and Massless Particles. *Phys. Rev. Lett.*, 13, 20, 585, see above.

Even after this has been done, both *scalar* and *pseudoscalar* parts remain in $\mathscr{L}_{\text{mass}}$. For the *mass* term, however, we can eliminate such *pseudoscalar* parts by applying an appropriate constant *gauge transformation* on q, which does not affect on $\mathscr{L}_{\text{strong}}$ due to *gauge invariance*.

Now we consider possible ways of assigning the *quartet field* to *representations* of the SU$_{\text{weak}}$(2). *Since this group is commutative with the Lorentz transformation*, the left and right components of the *quartet field*, which are respectively defined as $q_L = ½ (1 + \gamma_5)q$ and $q_R = ½ (1 - \gamma_5)q$, do not mix each other under the *gauge transformation*. Then, *each* component has three possibilities:

A) **4 = 2 + 2**,
B) **4 = 2 + 1 + 1**,
C) **4 = 1 + 1 + 1 + 1**,

where on the r.h.s., **n** denotes an n-dimensional *representation* of SU(2). The present scheme of *charge assignment* of the *quartet* does not permit *representations* of n > 3. As a result, we have nine possibilities which we will denote by (A, A), (A, B), $\cdots$, where the former (latter) in the parentheses indicates the transformation properties of the left (right) component. Since all members of the *quartet* should take part in the *weak interaction*, and size of the *strangeness changing neutral current* is bounded experimentally to a very small value, the cases of (B, C), (C, B) and (C, C) should be abandoned. The models of (B, A) and (C, A) are equivalent to those of (A, B) and (A, C), respectively, except relative signs between *vector* and *axial vector* parts of the *weak current*. Since g_A/g_Y ratios are measured only for composite states, this difference of the relative signs would be reduced to a dynamical problem of the composite system. So, we investigate in detail the cases of (A, A), (A, B), (A, C) and (B, B).

i)   Case (A, C)

This is the most natural choice in the *quartet model*. Let us denote two (SU$_{weak}$(2)) *doublets* and four *singlets* by L_{d1}, L_{d2}, $R_{s1}^{(p)}$, $R_{s2}^{(p)}$, $R_{s1}^{(n)}$ and $R_{s2}^{(n)}$, where superscript p(n) indicates p-like (n-like) *charge states*. In this case, $\mathscr{L}_{mass}$ takes, in general, the following form:

$$\mathscr{L}_{mass} = \dots$$

...

Therefore, if $\mathscr{L}' = 0$, no *CP-violations* occur in this case. It should be noted, however, that this argument does not hold when we introduce one more *fermion* doublet with the same *charge* assignment. This is because all *phases* of elements of a 3 x 3 unitary matrix cannot be absorbed into the *phase* convention of six fields. This possibility of *CP-violation* will be discussed later on.

ii) Case (A, B)

This is a rather delicate case. We denote two left *doublets*, one right *doublet* and two *singlets* by L_{d1}, L_{d2}, R_s, $R_s^{(p)}$, and $R_s^{(n)}$, respectively. The general form of $\mathscr{L}_{mass}$ is given by

$$\mathscr{L}_{mass} = \dots ,$$

...

... Thus, it seems difficult to reconcile the hierarchy of *chiral symmetry breaking* with the experimental knowledge of the *semi-leptonic* processes.

iii) Case (B, B)

As a previous one, in this case also, occurrence of *CP-violation* is possible, but in order to suppress $|\Delta S| = 1$ *neutral currents*, coefficients of the *axial-vector* part of $\Delta S = 0$ and $|\Delta S| = 1$ *weak currents* must take signs opposite to each other. This contradicts again the experiments on the *baryon β-decay*.

iv) Case (A, A)

In a similar way, we can show that no *CP-violation* occurs in this case as far as $\mathscr{L}' = 0$. Furthermore this model would reduce to an exactly U(4) symmetric one.

Summarizing the above results, *we have no realistic models in the quartet scheme as far as $\mathscr{L}' = 0$*. Now we consider some examples of *CP-violation* through $\mathscr{L}'$. Hereafter we will consider only the case of (A, C). The first one is to introduce another *scalar doublet field* ψ. Then, we may consider an *interaction* with this new *field* ...

...

... So, this *interaction* can cause a *CP-violation*.

Another one is a possibility associated with the *strong interaction*. Let us consider a *scalar (pseudoscalar) field* S which mediates the *strong interaction*. For the *interaction* to be *renormalizable* and $SU_{weak}(2)$ *invariant*, it must belong to a (4, 4*) + (4*, 4) representation of *chiral* SU(4) x SU(4) and interact with φ through *scalar* and *pseudoscalar couplings*. It also interacts with φ and possible *renormalizable* forms are given as follows:

...

... It is easy to see that these *interaction* terms can violate *CP-conservation*.

Next, we consider a 6-plet model, another interesting model of CP-violation. Suppose that a 6-plet with charges (Q, Q, Q, Q–1, Q–1, Q–1) is decomposed into $SU_{weak}(2)$ *multiplets* as 2 + 2 + 2 and 1 + 1 + 1 + 1 + 1 + 1 for left and right components, respectively. Just as the case of (A, C), we have a similar expression for the *charged weak current* with a 3 x 3 instead of 2 x 2 unitary matrix in Eq. (5). As was pointed out, in this case we cannot absorb all *phases* of matrix elements into the *phase convention* and can take, for example, the following expression:

$$\ldots .\tag{13}$$

Then, we have *CP-violating* effects through the interference among these different *current* components. An interesting feature of this model is that the *CP-violating* effects of lowest order appear only in $\Delta S \neq 0$ *non-leptonic* processes and in the *semi-leptonic* decay of neutral *strange mesons* (we are not concerned with higher states with the new quantum number) and not in the other *semi-leptonic*, $\Delta S = 0$ *non-leptonic* and *pure-leptonic* processes.

So far, we have considered only the straightforward extensions of the original Weinberg's model. However, other schemes of underlying *gauge groups* and/or *scalar fields* are possible. Georgi and Glashow's model[4] is one of them.

[4] Georgi, H. & Glashow, S. L. (May, 1972). Unified Weak and Electromagnetic Interactions without Neutral Currents. *Phys. Rev. Lett.*, 28, 22, 1494; https://doi.org/10.1103/PhysRevLett.28.1494.

We can easily see that *CP-violation* is incorporated into their model without introducing any other fields than the (many) new fields which they have introduced already.

Yoichiro Nambu, Makoto Kobayashi and Toshihide Maskawa – the 2008 Nobel Prize in Physics.

The Nobel Prize in Physics 2008 was divided, one half awarded to Yoichiro Nambu "for the discovery of the mechanism of *spontaneous broken symmetry* in subatomic physics", the other half jointly to Makoto Kobayashi and Toshihide Maskawa "*for the discovery of the origin of the broken symmetry which predicts the existence of at least three families of quarks in nature*".

[The Nobel Prize in Physics 2008. NobelPrize.org.   https://www.nobelprize.org/prizes/physics/2008/summary/.]

Passion for symmetry

The fact that our world does not behave perfectly symmetrically is due to deviations from symmetry at the microscopic level.

As early as 1960, Yoichiro Nambu formulated his mathematical description of *spontaneous broken symmetry* in *elementary particle physics*. *Spontaneous broken symmetry* conceals nature's order under an apparently jumbled surface. It has proved to be extremely useful, and Nambu's theories permeate the *Standard Model* of *elementary particle physics*. The Model unifies the smallest building blocks of all matter and three of nature's four forces in one single theory.

The *spontaneous broken symmetries* that Nambu studied, differ from the *broken symmetries* described by Makoto Kobayashi and Toshihide Maskawa. These spontaneous occurrences seem to have existed in nature since the very beginning of the universe and came as a complete surprise when they first appeared in particle experiments in 1964. It is only in recent years that scientists have come to fully confirm the explanations that Kobayashi and Maskawa made in 1972. It is for this work that they are now awarded the Nobel Prize in Physics. They explained *broken symmetry* within the framework of the *Standard Model*, but *required that the Model be extended to three families of quarks*. These predicted, hypothetical new *quarks* have recently appeared in physics experiments. As late as 2001, the two particle detectors BaBar at Stanford, USA and Belle at Tsukuba, Japan, both detected *broken symmetries* independently of each other. The results were exactly as Kobayashi and Maskawa had predicted almost three decades earlier.

A hitherto unexplained *broken symmetry* of the same kind lies behind the very origin of the cosmos in the Big Bang some 14 billion years ago. If equal amounts of *matter* and *antimatter* were created, they ought to have annihilated each other. But this did not happen, there was a tiny deviation of one extra particle of *matter* for every 10 billion *antimatter* particles. It is this *broken symmetry* that seems to have caused our cosmos to survive. *The question of how this exactly happened still remains unanswered.* Perhaps the new particle accelerator LHC at CERN in Geneva will unravel some of the mysteries that continue to puzzle us. [Press release. NobelPrize.org. https://www.nobelprize.org/prizes/physics/2008/press-release/.]

Makoto Kobayashi – Nobel Lecture, December 8, 2008. CP Violation and Flavor Mixing.

Makoto Kobayashi – Nobel Lecture. NobelPrize.org. https://www.nobelprize.org/prizes/physics/2008/kobayashi/lecture/

In his Nobel Prize lecture Makoto Kobayashi provided a brief history of the development of the *six-quark model*.

---

1. INTRODUCTION

We know that ordinary *matter* is made of *atoms*. An *atom* consists of the *atomic nucleus* and *electrons*. The *atomic nucleus* is made of a number of *protons* and *neutrons*. A *proton* and a *neutron* are further made of two kinds of *quarks*, u and d. Therefore, the fundamental building blocks of ordinary matter are the *electron* and two kinds of *quarks*, u and d.

The *standard model*, which gives a comprehensive description of current understanding of the *elementary particle* phenomena, however, tells us that the number of species of *quarks* is six. The additional *quarks* are called s, c, b and t. The reason why we do not find them in ordinary *matter* is that they are unstable in the usual environment. Similarly, the *electron* belongs to a family of six members called *leptons*. Three types of *neutrinos* are included among these six.

Another important ingredient of the *standard model* is *fundamental interactions*. Three kinds of *interactions* act on the *quarks* and *leptons*. The *strong interaction* is described by *quantum chromodynamics* (QCD) and the *electro-magnetic* and *weak interactions* by the *Weinberg-Salam-Glashow theory* in a unified manner. All of them belong to a special type of *field theory* called *gauge theory*.

> …
> Figure 1. The Standard Model.

The *standard model* was established in the 1970s. It was triggered by the development of studies of *gauge theories*. In particular, it was proved that generalized *gauge theory* is *renormalizable*[1].

[1] 't Hooft, G., (December, 1971). Renormalizable Lagrangians for massive Yang-Mills fields. *Nucl. Phys. B*, 35, 167-88; see above; 't Hooft, G. & Veltman, M. J. G. (1972). Regularization and Renormalization of Gauge Fields. *Nucl. Phys.*, B 44, 189.

This opened the possibility that all the *interactions* of an *elementary particle* can be described by the *quantum field theory* without the difficulty of *divergence*. Before this time, such description was possible only for *electro-magnetic interaction*.

The discovery of the new *flavors* made in 1970s played an important role in the establishment of the *standard model*. In particular, the τ-*lepton* and c- and b-*quarks* were found in the 1970s. When we proposed the *six-quark model* to explain *CP violation* with Dr. Toshihide Maskawa in 1973[2], only three *quarks* were widely accepted, and a slight hint of the fourth *quark* was there, but no one thought there would be six *quarks*.

[2] Kobayashi, M. & Maskawa, T. (February, 1973). CP-Violation in the Renormalizable Theory of Weak Interaction. *Progress of Theoretical Physics*, 49, 2, 652–7. See above.

In the following, I will describe the development of the studies on *CP violation* and the *quark* and *lepton flavors*, putting some emphasis on contributions from Japan. The next section will be devoted to the pioneering works of the Sakata School, from which I learned many things. The work on *CP violation* will be discussed in Section 3. I will explain what we thought and what we found at that time. Section 4 will describe subsequent development related to our work. Experimental verification of the proposed model has been done by using accelerators called *B-factories*. A brief outline of those experiments will be given. Finally in Section 5, I will look briefly at *flavor mixing* in the *lepton* sector, because this is a phenomenon parallel to the *flavor mixing* in the *quark* sector, and Japan has made unique and important contributions in this field.

2. SAKATA SCHOOL

Both Dr. Maskawa and I graduated from and obtained our PhD's from Nagoya University. When I entered the graduate program, the theoretical particle physics group of Nagoya University was known for its unique research activity and was led by Professor Shoichi Sakata.

In the early 1950s, a number of *strange* particles were discovered, with the first evidence having been found in the *cosmic ray* events of 1947. In the current terminology, *strange* particles contain an *s-quark* or *anti-s-quark* as a constituent, while non-strange particles do not. But what we are about to consider is the era before the *quark model* appeared.

In 1956, Sakata[3] proposed a model which is known as the *Sakata model*.

[3] Sakata, S. (September, 1956). On a Composite Model for the New Particles. *Progr. Theor. Phys.*, 16, 6, 686-8. See above.

In this model, all *hadrons*, strange and non-strange, are supposed to be composite *states* of the *triplet* of *baryons*, the *proton* (p), the *neutron* (n), and the *lambda* (Λ). In other words, three *baryons*, p, n, and Λ are the fundamental building blocks of the *hadrons* in the model. Eventually, the *Sakata model* was replaced by the *quark model*, where the *triplet* of *quarks*, u, d, and s replace p, n, and Λ. But the root of the idea of fundamental *triplet* is in the *Sakata model*.

In the following, we focus on the *weak interactions* in the *Sakata model*. Usual *beta-decays* of the *atomic nucleus* are caused by the transition of a *neutron* into a *proton*. Similarly, we can consider the transition of a *lambda* into a *proton*. In the *Sakata model*, all the *weak interaction* of the *hadrons* can be explained by these two kinds of transitions among the fundamental *triplet*.

This pattern of the *weak interaction* is quite similar to the *weak interaction* of the *leptons*;
…

It should be noted that at that time, the *neutrino* was thought of as a single species. This similarity of the *weak interaction* between the *baryons* and the *leptons* was pointed out by Gamba, Marshak and Okubo.

In 1960, Maki, Nakagawa, Ohnuki and Sakata[5] developed the idea of *baryon-lepton* or *B-L symmetry* further and proposed the so-called *Nagoya model*.

[5] Maki, Z., Nakagawa, M., Ohnuki, Y. & Sakata, S. (June, 1960). A Unified Model for Elementary Particles. *Progr. Theor. Phys.*, 23, 6, 1174-80; https://doi.org/10.1143/PTP.23.1174.

They considered that the *triplet baryons*, p, n, and Λ are composite states of a hypothetical object called *B-matter* and the *neutrino*, the *electron*, and the *muon*, respectively;

$$p = \langle B^+ \nu \rangle, \quad n = \langle B^+ e \rangle, \quad \Lambda = \langle B^+ \mu \rangle,$$

where B-matter is denoted as B^+.

Although the composite picture of the *Nagoya model* did not lead to remarkable progress, some ideas in the *Nagoya model* developed in an interesting way. In 1962, it was discovered that there exist two kinds of *neutrinos*, corresponding to the *electron* and the *muon*, respectively. When the results of this discovery at Brookhaven National Laboratory were to come out, two interesting papers were published, one written by Maki, Nakagawa and Sakata[7] and the other by Katayama, Matsumoto, Tanaka and Yamada[8].

[7] Maki, Z., Nakagawa, M., & Sakata, S. (1962). Remarks on the Unified Model of Elementary Particles. *Progr. Theor. Phys.*, 28, 870.

[8] Katayama, Y., Matumoto, K., Tanaka, S. & Yamada, E. (1962.) Possible Unified Models of Elementary Particles with Two Neutrinos. *Progr. Theor. Phys.*, 28 (1962) 675.

Both papers discussed the modification of the *Nagoya model* to accommodate two *neutrinos* in the model.

In the course of the argument to associate *leptons* and *baryons*, Maki et al. discussed the *masses* of *neutrinos* and derived the relation describing the mixing of the *neutrino states*;

$$\nu_1 = \cos\vartheta\,\nu_e + \sin\vartheta\,\nu_\mu,$$
$$\nu_2 = \sin\vartheta\,\nu_e + \cos\vartheta\,\nu_\mu,$$

where ν_1 and ν_2 are the *mass eigenstates* of *neutrinos*, and they assumed that the *proton* is the composite state of the *B-matter* and ν_1. Although the last assumption is not compatible with the current experimental evidence, it is remarkable that they did present the correct formulation of *lepton flavor mixing*. To recognize their contribution, the *lepton flavor mixing matrix* is called the *MNS matrix* today.

Lepton flavor mixing gives rise to the phenomenon called *neutrino oscillation*. Many years later, *neutrino oscillation* was discovered in an unexpected manner. We will come back to this point later.

Another important outcome of this argument is the possible existence of the fourth fundamental particle associated with ν_2. This was discussed by Katayama et al. in some detail. At the time, the fundamental particles were still considered *baryons* but the structure of the *weak interaction* discussed here is the same as that of the *Glashow-Illiopoulos-Miani*[9] *scheme*.

[9] Glashow, S. L., Iliopoulos, J. & Maiani, L. (October, 1970). Weak Interactions with Lepton-Hadron Symmetry. *Phys. Rev. D*, 2, 7, 1285; https://doi.org/10.1103/PhysRevD.2.1285.

These works were revived in 1971, when Niu and his collaborators found new kind of events in emulsion chambers exposed to *cosmic rays*. One of the events they found is shown in Figure 3. In this event, we see kinks on two tracks, which indicate the decay of new particles produced in pairs. The estimated *mass* of the new particle was 2~3 GeV and the *life* was a few times 10^{-14} sec. under some reasonable assumptions.

When this result came to his attention, Shuzo Ogawa, a member of the Sakata group, immediately pointed out that this new particle might be related to the fourth element expected in the extended version of the *Nagoya model*. By that time, the Sakata model had already been replaced by the *quark model*, so that what he meant was that those new particles might be *charmed* particles, in the current terminology. Following this suggestion, several Japanese groups, including mine, began to investigate the *four-quark model*. At that time, I was a graduate student at Nagoya University.

...

Figure 3. A cosmic ray event.

So far, I have explained about the unique activities of the Sakata School. I mentioned the *four-quark model* in some detail. But I do not mean to imply that the *six-quark model* we proposed is a simple extension of the *four-quark model*. What was most important for me was the atmosphere of the particle physics group of Nagoya University. Although most of the work we discussed in this section had been done by Sakata and his group before I entered the graduate course, the spirit created by this work was still there. I learned the importance of capturing the entire picture, which is necessary for this kind of work.

3. SIX-QUARK MODEL

In 1971, the *renormalizability* of the non-Abelian gauge theory was proved[1]. This enabled a description of the *weak interactions* with the *quantum field theory* in a consistent manner, and the *Weinberg-Salam-Glashow*[13] *theory* began to attract attention.

[13] Weinberg, S. (November, 1967). Model of Leptons. *Phys. Rev. Lett.*, 19, 21, 1264–6, see above; Salam, A. (1968). Weak and Electromagnetic Interactions, originally printed in "*N. Svartholm: Elementary Particle Theory, Proceedings of the Nobel Symposium Held 1968 at Lerum, Sweden*", pp. 367–377; Glashow, S. L. (February, 1961). Partial-symmetries of weak interactions. *Nuclear Physics*, 22, 4, 579–88, see above.

In 1972, I obtained my PhD from Nagoya University and moved to Kyoto University. Then my work on *CP violation* started.

CP violation was first found in 1964 by Cronin, Fitch et al.[14] in the decay of the *neutral K-meson* [*Kaon*].

[14] Christenson, J. H., Cronin, J. W., Fitch, V. L. & Turlay, R. (July 1964). Evidence for the 2π Decay of the K_2^0 Meson. *Phys. Rev. Lett.*, 13, 4, 138–40; https://doi.org/10.1103/PhysRevLett.13.13.

[*Kaons* have played a distinguished role in our understanding of fundamental conservation laws: *CP violation*, a phenomenon generating the observed *matter–*

451

antimatter asymmetry of the universe, was discovered in the *kaon* system in 1964. This was acknowledged by the award of the 1980 Nobel Prize in Physics jointly to James Watson Cronin and Val Logsdon Fitch "for the discovery of violations of fundamental symmetry principles in the decay of neutral K-mesons".

CP violation is a violation of *CP-symmetry* (or *charge conjugation parity symmetry*): the combination of *C-symmetry* (*charge conjugation symmetry*) and *P-symmetry* (*parity symmetry*). *CP-symmetry* states that the laws of physics should be the same if a particle is interchanged with its *antiparticle* (*C-symmetry*) while its *spatial coordinates* are inverted ("mirror" or *P-symmetry*).

A parity transformation (also called *parity inversion)* is the flip in the sign of one spatial coordinate.

Charge conjugation is a transformation that switches all particles with their corresponding *antiparticles*, thus changing the sign of all *charges*: not only *electric charge* but also the *charges* relevant to other forces.]

CP violation means violation of *symmetry* between *particles* and *anti-particles*. The discovery of *CP violation* implies that there is an essential difference between *particles* and *anti-particles*.

We thought that if the *gauge theory* can describe the *interactions* of particles consistently, *CP violating interaction* should also be included in it. It was rather straightforward to solve the problem. We simply investigated conditions for *CP violation* in the *renormalizable gauge theory*. What we found then is summarized as follows[2].

At that time only three *quarks* were widely accepted, but the *three-quark model* had some flaws in the *gauge theory*. Therefore, from a theoretical point of view, the *four-quark* model of the *Glashow-Illiopoulos-Miani* (GIM) type was considered preferable. However, it is impossible to accommodate *CP violation* in a model of the GIM type. We found that even if we relax the conditions for the GIM type, we cannot make any realistic model of *CP violation* with four *quarks*. This implies that there must be some unknown particles besides the fourth *quark*. I thought that this was quite strong and an important conclusion of our argument.

Then we considered a few possible mechanisms of *CP violation* by introducing new particles. We proposed the *six-quark model* as one such possible mechanism.

Below we will discuss the *quark flavor mixing* in some detail, in order to understand why four *quarks* are not enough and six *quarks* are needed to accommodate *CP violation*.

In the frame work of the *gauge theory, flavor mixing* arises from a mismatch between *gauge symmetry* and *particle states. Gauge symmetry* lumps a certain number of particles into a group called a *multiplet*. However, each *multiplet* member is not necessarily identical to a single species of particles, but sometimes it is a *superposition of particles*. The *flavor mixing* is nothing but this *superposition*. In the present case, the relevant *gauge group* is SU(2) of the *Weinberg-Salam-Glashow theory* and the *multiplet* is a *doublet*.

Assuming that four *quarks* consist of two *doublets* of the SU(2) *group*, we can denote the most general form of them as

$$(u), \quad (c),$$
$$(d') \quad (s'),$$

where d' and s' are the *superposition* of real *quark states* d and s, described in a matrix form as

$$(d') \quad = \quad (V_{ud} \quad V_{us}) \quad (d),$$
$$(s') \qquad\qquad (V_{cd} \quad V_{cs}) \quad (s)$$

where the matrix describing the mixing should be what is called a *unitary matrix* in mathematics.

The next problem is what the condition for *CP violation* is. In *quantum field theory, CP violation* is related to *complex coupling constants*. To be more concrete in the present formulation, *CP violation* will occur if *irreducible* complex numbers appear in the elements of the *mixing matrix*. The matrix elements of a *unitary matrix* are complex numbers in general, but some of them can be made real by adjusting the *phase factor* of the *particle state* without changing the physics results. In such case, those complex numbers are called *reducible*, and otherwise, *irreducible*. Therefore, *one condition of CP violation is that there remain complex numbers which cannot be removed by the phase adjustment of the particle states.*

In the *four-quark model*, adjustment factors are described by two diagonal matrices whose elements are mere *phase factors*. It is easy to see that, if we choose them properly, we can make any 2 x 2 unitary matrix into a real matrix:

$$\dots .$$

Therefore, in this case, we cannot accommodate *CP violation*.

How does this argument change in the *six-quark model*? In this case, we can express the *flavor mixing* as follows:

(u),	(c),	(t),
(d')	(s')	(b'),

(d')	=	(V_{ud}	V_{us}	V_{ub})	(d),
(s')		(V_{cd}	V_{cs}	V_{cb})	(s)
(b')		(V_{td}	V_{ts}	V_{tb})	(b)

This time the mixing matrix is a 3 x 3 *unitary matrix*. In this case, however, we cannot remove all the *phase factors* of the matrix elements by adjusting the *phases* of the *quark states*. The best we can do by adjusting the *phases* is to express them by a certain standard form with four parameters. A popular parameterization is

$$V = \ldots .$$

where $c_{ij} = \cos \vartheta_{ij}$ and $s_{ij} = \sin \vartheta_{ij}$ with i, j = 1, 2, 3. *We note that, unless $\delta = 0$, the imaginary part remains in the matrix elements and therefore CP symmetry is violated.*

Taking into account the hierarchy of the actual values of the parameters, the following approximate parameterization is frequently used in phenomenological analyses.

$$V \approx \ldots .$$

In this parameterization, *if η is not zero, the system is violating CP symmetry.*

We thought that this mechanism of *CP violation* is very interesting and elegant, *but we had no further reason to single out the six-quark model from the other possibilities.* The model was not so special, because if the system has sufficiently many particles, it is not difficult to violate *CP symmetry.* However, *the subsequent experimental development pushed up the six-quark model to a special position.*

In 1974, the *J/ψ particle* was discovered, and soon it turned out that it is the *bound state* of the fourth *quark* c and its *anti-particle.* The discovery had a great impact on particle physics, but it had little effect on the *six-quark model.*

[*J/ψ (J/psi) meson [composite particle]*; charm quark and charm anti-quark [elementary particles]
In 1974, the *J/ψ (J/psi) meson* (a composite particle which is a *vector meson* (*quark* and *antiquark*) was discovered by groups headed by Burton Richter and Samuel Ting, both American physicists, demonstrating the existence of the *charm quark and charm anti-quark*, for which they shared the 1976 Nobel Prize for Physics. They discovered that they had found the same particle, and both announced their discoveries on November 11, 1974. [Aubert, J. J. et al. (1974). Experimental

Observation of a Heavy Particle J. *Phys. Rev. Let.*, 33, 23, 1404–6; https://doi.org/10.1103/PhysRevLett.33.1404; Augustin, J. -E. et al. (1974). Discovery of a Narrow Resonance in e+e− Annihilation. *Phys. Rev. Let.*, 33, 23, 1406–8; https://doi.org/10.1103/PhysRevLett.33.1406.]

Burton Richter (March 22, 1931–July 18, 2018) led the Stanford Linear Accelerator Center (SLAC) team, and Samuel Ting (born January 27, 1936) led the Brookhaven National Laboratory (BNL) team. The importance of this discovery is highlighted by the fact that the subsequent, rapid changes in high-energy physics at the time have become collectively known as the "*November Revolution*".

The *J/ψ meson* is a subatomic particle, a *flavor-neutral meson* consisting of a *charm quark* and a *charm antiquark*. *Mesons* formed by a bound state of a *charm quark* and a *charm anti-quark* are generally known as "*charmonium*" or *psions*. The J/ψ is the most common form of *charmonium*, due to its *spin* of 1 and its low *rest mass*. The J/ψ has a *rest mass* of 3.0969 GeV/c^2, just above that of the η_c (2.9836 GeV/c^2), and a *mean lifetime* of 7.2×10^{-21} s. *This lifetime was about a thousand times longer than expected.*

The *J/ψ meson* had been proposed by James Bjorken and Sheldon Glashow in 1964) [Bjørken, B. J. & Glashow, S. L. (1964). Elementary Particles and SU(4). *Phys. Let.*, 11, 3, 255–7; https://www.sciencedirect.com/science/article/abs/pii/0031916364904330.]

> [*Vector meson [classification]*
> A *vector meson* is a *meson* with *total spin* 1 and odd parity. *Vector mesons* have been seen in experiments since the 1960s, and are well known for their spectroscopic pattern of masses.
>
> The *vector mesons* contrast with the *pseudovector mesons*, which also have a *total spin* 1 but instead have *even parity*. The *vector* and *pseudovector mesons* are also dissimilar in that the spectroscopy of *vector mesons* tends to show nearly pure states of constituent *quark flavors*, whereas *pseudovector mesons* and *scalar mesons* tend to be expressed as composites of mixed states.]]

In 1975, the *τ-lepton* was discovered. This discovery had a significant effect on our model. The *τ-lepton* is the fifth member of *leptons*. Although it is a *lepton*, it suggested the existence of a third family in the *quark* sector, too. That was when people began to pay attention to our model. Early works which discussed the *six-quark model* include … .

[Tau [elementary particle]

In 1975, the *tau* discovered by a group headed by Martin Perl. [Perl, M. L. (1975). Evidence for Anomalous Lepton Production in e+–e− Annihilation. *Phys. Rev. Let.*, 35, 22, 1489–92; https://doi.org/10.1103/PhysRevLett.35.1489.] Their equipment consisted of SLAC's then-new electron–positron colliding ring, called SPEAR, and the LBL magnetic detector. They could detect and distinguish between *leptons*, *hadrons*, and *photons*. They did not detect the *tau* directly, but rather discovered anomalous events.

The search for *tau* started in 1960 at CERN by the Bologna-CERN-Frascati (BCF) group led by Antonino Zichichi. Zichichi came up with the idea of a new sequential *heavy lepton*, now called *tau*, and invented a method of search. He performed the experiment at the ADONE facility in 1969 once its accelerator became operational; however, the accelerator he used did not have enough energy to search for the *tau* particle.

The *tau* was independently anticipated in a 1971 article by Yung-su Tsai. [Tsai, Y-S. (November, 1971). Decay correlations of heavy leptons in e+ + e− → ℓ+ + ℓ−. *Phys. Rev. D.*, 4, 9, 2821; https://doi.org/10.1103/PhysRevD.4.2821.] Providing the theory for this discovery, the *tau* was detected in a series of experiments between 1974 and 1977 by Perl with his and Tsai's colleagues at the Stanford Linear Accelerator Center (SLAC) and Lawrence Berkeley National Laboratory (LBL) group.

The *tau*, also called the *tau lepton*, *tau particle*, *tauon* or *tau electron*, is an elementary particle similar to the *electron*, with negative *electric charge* and a *spin* of ½. Like the *electron*, the *muon*, and the three *neutrinos*, the *tau* is a *lepton*, and like all elementary particles with *half-integer spin*, the *tau* has a corresponding *antiparticle* of *opposite charge but equal mass and spin*. In the *tau*'s case, this is the "*anti-tau*" (also called the *positive tau*). *Tau leptons* have a *lifetime* of 2.9×10^{-13} s and a *mass* of 1776.9 MeV/c^2 (compared to 105.66 MeV/c^2 for *muons* and 0.511 MeV/c^2 for *electrons*). Since their *interactions* are very similar to those of the *electron, a tau can be thought of as a much heavier version of the electron. Because of their greater mass, tau particles do not emit as much bremsstrahlung (braking radiation) as electrons; consequently, they are potentially much more highly penetrating than electrons.*

Because of its *short lifetime*, the *range* of the *tau* is mainly set by its *decay length*, which is too small for *bremsstrahlung* to be noticeable. *Its penetrating power appears only at ultra-high velocity and energy (above petaelectronvolt energies),* when time dilation extends its otherwise very short path-length. As with the case of the other *charged leptons*, the *tau* has an associated *tau neutrino*.]

In 1977, the *upsilon particle* was discovered, and it turned out that it is a *bound state* of the fifth *quark* b and *anti*-b.

[*Upsilon meson* [*composite particle*]; bottom quark and bottom antiquark [elementary particle and elementary antiparticle]

In 1977, *the upsilon meson* was observed by a team at Fermilab led by Leon Lederman, demonstrating the existence of the *bottom quark*. This was a strong indicator of the *top quark*'s existence: without the *top quark*, the *bottom quark* would have been without a partner. [Herb, S. W. et al. (1977). Observation of a Dimuon Resonance at 9.5 GeV in 400-GeV Proton-Nucleus Collisions. *Phys. Rev. Let.*, 39, 5, 252–5; https://doi.org/10.1103/PhysRevLett.39.252.] The evidence for the *bottom quark* was first obtained by the Fermilab E288 experiment team led by Leon M. Lederman, when *proton-nucleon* collisions produced bottomonium decaying to pairs of *muons*. The discovery was confirmed about a year later by the PLUTO and DASP2 Collaborations at the electron-positron collider DORIS at DESY.

The *upsilon meson* is a quarkonium state (i.e. flavorless *meson*) formed from a *bottom quark* and its *antiparticle*. It was the first particle containing a *bottom quark* to be discovered because it is the lightest that can be produced without additional massive particles. It has a *lifetime* of 1.21×10^{-20} s and a *mass* about 9.46 GeV/c^2 in the ground state.

The *bottom quark* is an elementary particle of the *third generation*. It is a *heavy quark* with a *charge* of − 1/3 e. All *quarks* are described in a similar way by *electroweak interaction* and *quantum chromodynamics*, but the *bottom quark* has exceptionally low rates of transition to lower-mass quarks. The *bottom quark* is also notable because *it is a product in almost all top quark decays*, and is a frequent decay product of the *Higgs boson*.

The *bottom quark* was proposed by Kobayashi and Maskawa in 1973, for which they shared half of the 2008 Nobel Prize in Physics "for [their 1973] discovery of *the origin of the broken symmetry which predicts the existence of at least three families of quarks in nature*", with the other half going to Yoichiro Nambu "for the discovery of *the mechanism of spontaneous broken symmetry* in subatomic physics". [Kobayashi, M. & Maskawa, T. (February, 1973). CP-Violation in the

Renormalizable Theory of Weak Interaction. *Progress of Theoretical Physics*, 49, 2, 652–7; https://doi.org/10.1143/PTP.49.652. See above.]]

The discovery of the last *quark*, t, occurred as recently as in 1995, but before that time the *six-quark model* became a standard one.

[Top quark [elementary particle]
It was not until 1995 that the *top quark* was finally observed at Fermilab [Abe, F. et al. (CDF collaboration) (1995). Observation of Top quark production in p–p Collisions with the Collider Detector at Fermilab. *Phys. Rev. Let.*, 74, 14, 2626–31; arXiv:hep-ex/9503002; doi:10.1103/PhysRevLett.74.2626; S. Arabuchi et al. (D0 collaboration) (1995). Observation of the Top Quark. *Phys. Rev. Let.*, 74, 14, 2632–7; arXiv:hep-ex/9503003; doi:10.1103/ PhysRevLett.74.2632.]

[The DØ experiment (sometimes written D0 experiment) was a worldwide collaboration of scientists conducting research on the fundamental nature of matter. DØ was one of two major experiments (the other was the CDF experiment) both located at the Tevatron Collider at Fermilab in Batavia, Illinois. The Tevatron was the world's highest-energy accelerator from 1983 until 2009, when its energy was surpassed by the Large Hadron Collider. The DØ experiment stopped taking data in 2011, when the Tevatron shut down, but data analysis is still ongoing.]

It had a mass much larger than expected, almost as large as that of a gold atom.

The *top quark* is *the most massive of all observed elementary particles*. It derives its mass from its *coupling to the Higgs boson*. This coupling is very close to unity; in the *Standard Model* of particle physics, *it is the largest (strongest) coupling at the scale of the weak interactions and above*. Like all other quarks, the *top quark* is a fermion with *spin ½* and *participates in all four fundamental interactions: gravitation, electromagnetism, weak interactions, and strong interactions*. It has an *electric charge* of $+2/3$ e. It has a mass of 172.76 ± 0.3 GeV/c2, which is close to the rhenium atom mass. The *antiparticle* of the *top quark* is the *top antiquark* (sometimes called *antitop quark* or simply *antitop*), which differs from it only in that some of its properties have equal magnitude but opposite sign.

The *top quark* interacts with *gluons* of the *strong interaction* and is typically produced in *hadron* colliders via this interaction. However, once produced, the *top* (or antitop) can decay only through the *weak force*. It decays to a *W boson* and either a *bottom quark* (most frequently), a *strange quark*, or, on the rarest of occasions, a *down quark*.

The *Standard Model* determines the *top quark*'s mean *lifetime* to be roughly 5×10^{-25} s. This is about a twentieth of the timescale for *strong interactions*, and therefore *it does not form hadrons*, giving physicists a unique opportunity to study a "bare" *quark* (all other *quarks* hadronize, meaning that they combine with other *quarks* to form *hadrons* and can only be observed as such).

Because the *top quark* is so massive, *its properties allowed indirect determination of the mass of the Higgs boson*. As such, the *top quark*'s properties are extensively studied as a means to discriminate between competing theories of new physics beyond the *Standard Model. The top quark is the only quark that has been directly observed* due to its decay time being shorter than the hadronization time.]

Meanwhile, it was pointed out that we could expect large *CP asymmetry* in the *B-meson* system[21].

[21] Carter, A. B. & Sanda, A. I. (1981). CP Violation in B Meson Decays. *Phys. Rev. D*, 23, 1567; Bigi, I. I. Y. & Sanda, A. I. (1981). Notes on the Observability of CP Violations in B Decays. *Nucl. Phys. B*, 193, 85.

This opened the possibility to test the model with *B-factories*. *B-meson* implies a meson containing b- or *anti*-b as a constituent, and *B-factory* means an accelerator, which produces a lot of *B-mesons* like a factory.

4. EXPERIMENTAL VERIFICATION AT B-FACTORIES

In order to verify the *six-quark model* experimentally, two B-factories, KEKB at KEK in Japan and PEPII at SLAC in the US, were built. Those accelerators are unusual ones. Colliding *electrons* and *positrons* have different energies, so that the *B-mesons* produced are boosted. Both experimental groups, Belle (KEKB) and BaBar (PEPII), are large international teams organized with participation from many countries.

They were approved and started experiments almost at the same time. PEPII/BaBar ceased operation this year, while KEKB/Belle is still running. They achieved luminosities more than 10^{34} cm^{-2}s^{-1}, which are record high. Luminosity is a key parameter representing the performance of the colliding accelerator.

…

In the light of the B-factory results, the present status of *CP violation* may be summarized as follows.

- B-factory results show that *quark mixing* of the *six-quark model* is the dominant source of the observed *CP violation*.
- B-factory results, however, allow small room for additional source from new physics beyond the *standard model*.
- And *matter dominance* of the universe seems to require new sources of *CP violation*, because it appears that *CP violation* of the *six-quark model* is too small to explain matter dominance.

It has been proposed that the last point may be related to *lepton flavor mixing*, which is the counterpart of *quark mixing*. In regard to *lepton flavor mixing*, very important contributions have been made in Japan, which will be discussed in the next section.

5. LEPTON FLAVOUR MIXING

The most important achievement is the discovery of *neutrino oscillation* at Super Kamiokande, which is a huge water tank detector built in the Kamioka mine in central Japan. They were observing *neutrinos* produced by *cosmic rays* in the atmosphere surrounding the earth. Since *neutrinos* penetrate the earth, those *neutrinos* come to the detector also from below. The *neutrino oscillation* implies the species of *neutrino* changes during its flight. So, if the *neutrino oscillation* takes place while *neutrinos* are travelling the distance from the other side of the earth, the observed number of the particular kind of *neutrino* will be reduced …

… The results show a clear deficit of observed *neutrinos* and are completely consistent with *neutrino oscillation*.

This great discovery was led by Yoji Totsuka (1942-2008). To our deep regret, he passed away in this last July. The *neutrino oscillation* was further confirmed by two experiments using man-made *neutrinos*. One is the K2K experiment. In this experiment, *neutrinos* were produced by the *proton synchrotron* in the KEK laboratory and those *neutrinos* were observed by the Super-Kamiokande. … Data show a clear oscillation pattern.

The other experiment is the KamLAND experiment. The KamLAND detector uses liquid scintillator instead of water, and it is also located in the Kamioka mine. They observed *neutrinos* produced in the nuclear reactors in the surrounding area. Data show a clean agreement with the oscillation. …

PART II   Unified Gravity (update; May 2, 2025)

A promising recent attempt below to integrate *gravity* into what is currently the *Standard Model* is founded on introducing a more appropriate way of representing the *spin of the electron* in formulating *Maxwell's equations* describing how *non-relativistic* electric and magnetic fields are generated by charges, currents, and changes of the fields and how fluctuations in electromagnetic fields (waves) propagate at a constant speed in a vacuum.

The first paper notes that "advances in optics and photonics technologies enabled detailed experimental investigations of the spin and orbital angular momenta of light and the related physics of structured light fields and chiral quantum optics. Classical [*non-relativistic*] *Maxwell's equations* and [*non-relativistic*] QED provided solid basis for theoretical understanding of the related physical phenomena". *They describe how electric and magnetic fields are generated by charges, currents, and changes of the fields* and may be combined to demonstrate how fluctuations in *electromagnetic fields* (waves) propagate at a constant speed in vacuum, c.

The most interesting innovation was that, in order to develop their relativistic theory, they first introduced an *eight-component [non-relativistic] spinorial Maxwell equation*, the solutions of which are called *electromagnetic spinors*. This approach leads to the appearance of *8 × 8 matrices in the spinorial Maxwell equation* and to the electromagnetic spinors having eight components.

> [*Spinors* were first applied to mathematical physics by Wolfgang Pauli in 1927, when he introduced his *spin matrices*. The following year, Paul Dirac introduced his *relativistic theory of electron spin* by showing the connection between *spinors* and the *Lorentz group*.
>
> In geometry and physics, *spinors* (pronounced "spinner") are elements of a complex vector space that can be associated with Euclidean space. A *spinor* transforms linearly when the Euclidean space is subjected to a slight (infinitesimal) rotation, but unlike geometric vectors and tensors, a *spinor* transforms to its negative when the space rotates through 360°. It takes a rotation of 720° for a *spinor* to go back to its original state. This property characterizes *spinors*: spinors can be viewed as the "square roots" of vectors (although this is inaccurate and may be misleading; they are better viewed as "square roots" of sections of vector bundles). In the 1920s physicists discovered that *spinors* are essential to describe the *intrinsic angular momentum*, or "*spin*", of the electron and other subatomic particles.

461

Spinors are characterized by the specific way in which they behave under *rotations*. They change in different ways depending not just on the overall final rotation, but the details of how that rotation was achieved (by a continuous path in the rotation group). In mathematical terms, spinors are described by a double-valued projective representation of the rotation group SO(3).

Although *spinors* can be defined purely as elements of a representation space of the spin group (or its Lie algebra of infinitesimal rotations), they are typically defined as elements of a vector space that carries a linear representation of the ***Clifford algebra***. The ***Clifford algebra*** is an associative algebra that can be constructed from Euclidean space and its inner product in a basis-independent way. Both the spin group and its Lie algebra are embedded inside the Clifford algebra in a natural way, and in applications the Clifford algebra is often the easiest to work with. A ***Clifford space*** operates on a ***spinor space***, and the elements of a ***spinor space*** are ***spinors***.

A ***spinor representation*** is a linear representation of the *Clifford algebra*. It is also defined as a spin representation of the orthogonal Lie algebra. Spinors are typically defined as elements of a vector space that carries a linear representation of the Clifford algebra.

A ***Clifford algebra*** is an algebra generated by a vector space with a quadratic form, and is a unital associative algebra with the additional structure of a distinguished subspace. The theory of Clifford algebras is intimately connected with the theory of quadratic forms and orthogonal transformations. Clifford algebras have important applications in a variety of fields including geometry, theoretical physics and digital image processing.

In the papers below, a coupling is developed between the ***electromagnetic field***, the ***Dirac electron-positron field***, and the ***gravitational field*** based on an ***eight-component spinorial representation*** of the ***electromagnetic field***. In the first paper this is used to reformulate ***quantum electrodynamics*** (QED) and in the second paper to extend the ***Standard Model*** to include gravity in a theory of ***unified gravity***.]

Section II introduced the ***electromagnetic spinor*** and formulated the [***non-relativistic***] ***spinorial Maxwell equation*** by introducing an ***electromagnetic spinor*** Ψ and a ***charge-current spinor*** Φ, which both have eight components. The advantage of using an eight-dimensional electromagnetic spinor was that it was ***gauge-independent*** [***non-relativistic***]. "The eight-component spinors describe the physics of *all four* [*non-relativistic*] Maxwell's

462

equations including the charge and current densities: one component is related to **Gauss's law of magne**tism, three components to the **Ampère Maxwell law**, one component to **Gauss's law of electricity**, and three components to **Faraday's law**. In any inertial frame, the components of Ψ(t,r) are composed of the *electric and magnetic fields*, E(t,r) and B(t,r), and the components of Φ(t, r) are composed of the *total electric charge* and *current densities* ρ_c(t, r) and J_c(t,r)". Omitting the time and position arguments, …

$$\Psi = \sqrt{\varepsilon_0}/2 \begin{bmatrix} 0 \\ \mathbf{E} \\ 0 \\ i c \mathbf{B} \end{bmatrix}, \qquad \Phi = \sqrt{\varepsilon_0}/2 \begin{bmatrix} 0 \\ \mu_0 c \mathbf{J}_e \\ \rho_e/\varepsilon_0 \\ \mathbf{0} \end{bmatrix}, \qquad (1)$$

Here ε_0, μ_0, and c = $1/\sqrt{\varepsilon_0 \mu_0}$ are the **permittivity**, **permeability**, and the **speed of light in a vacuum**, respectively. The zero elements of the *electromagnetic spinor* Ψ are related to the property of all electromagnetic spinors to be eigenstates of the spin operator squared, having a well-defined spin S = 1. The zero elements of the charge-current spinor Φ are related to the nonexistence of magnetic monopoles as elementary particles. …

In order to generate a **relativistic** formulation of the electrodynamics equations they wrote the spinorial **Maxwell equation** in a form similar to the [*relativistic*] Dirac equation describing spin-1/2 particles in the case when the mass of the particle is set to zero and when an external source term is added. The [*relativistic*] spinorial **Maxwell equation**, which describes all of classical electromagnetism, is written as

$$\gamma^a{}_B \partial_a \Psi = - \Phi. \qquad (2)$$

In analogy to the **Dirac equation** for electrons, the [*relativistic*] spinorial **Maxwell equation** is a first-order differential equation that was covariant in **Lorentz transformations** between inertial frames. One particular advantage of their 8×8 gamma matrices was that all Maxwell's equations could be presented as a single equation allowing then to rewrite QED in the eight-spinor notation.

When the [*non-relativistic*] expressions of the electromagnetic spinor and the charge-current spinor from Eq. (1)

$$[\Psi = \sqrt{\varepsilon_0}/2 \begin{bmatrix} 0 \\ \mathbf{E} \\ 0 \\ i c \mathbf{B} \end{bmatrix}, \qquad \Phi = \sqrt{\varepsilon_0}/2 \begin{bmatrix} 0 \\ \mu_0 c \mathbf{J}_e \\ \rho_e/\varepsilon_0 \\ \mathbf{0} \end{bmatrix}, \qquad (1)$$

were substituted into the [*relativistic*] spinorial **Maxwell equation** in Eq. (2),

$$[\gamma^a{}_B \partial_a \Psi = -\, \Phi. \qquad\qquad (2)\,]$$

They obtained the full set of four [*non-relativistic*] **Maxwell's equations** in the conventional form as

$\nabla \cdot \mathbf{B} = 0$	(7)	Gauss's Law for Magnetism
$\nabla \times \mathbf{B} = -\,\mu_0\, \mathbf{J}_e + 1/c^2\, \partial\mathbf{E}/\partial t)$	(8)	The Maxwell-Ampère Law with Maxwell's addition.
$\nabla \cdot \mathbf{E} = \rho/\varepsilon_0 = 0$	(9)	Gauss's Law
$\nabla \times \mathbf{E} = -\, \partial\mathbf{B}/\partial t$	(10)	The Maxwell–Faraday version of Faraday's Law of Induction

[Underwood, T.G. (2024). *Electricity & Magnetism*, p. 300: "The Heaviside formulation of Maxwell's Equations (1884) in a region with no charges, such as in a vacuum, reduces to:

$\nabla \cdot \mathbf{B} = 0$	Gauss's Law for Magnetism
$\nabla \times \mathbf{B} = -\,\mu_0\,(\mathbf{J} + \varepsilon_0\, \partial\mathbf{E}/\partial t) = -\,\mu_0\varepsilon_0\, \partial\mathbf{E}/\partial t$	The Maxwell-Ampère Law with Maxwell's addition.
$\nabla \cdot \mathbf{E} = \rho/\varepsilon_0 = 0$	Gauss's Law
$\nabla \times \mathbf{E} = -\, \partial\mathbf{B}/\partial t$	The Maxwell–Faraday version of Faraday's Law of Induction."

In 1884 Oliver Heaviside recast Maxwell's mathematical analysis from its original cumbersome form (they had already been recast as *quaternions*) to its modern vector terminology, thereby *reducing twelve of the original twenty equations in twenty unknowns down to the four differential equations in two unknowns we now know as Maxwell's equations*. Heaviside presented these equations *in modern vector format* using the *nabla operator* (∇) devised by William Rowan Hamilton in 1837.]

In the electric and magnetic field formulation there are four equations that determine the fields for given charge and current distribution. A separate law of nature, the *Lorentz force law*, describes how, conversely, the electric and magnetic fields act on charged particles and currents. A version of this law was included in the original equations by Maxwell but, by convention, is included no longer. The four re-formulated Maxwell's equations describe the nature of *electric charges* (both static and moving), *magnetic fields*, and the relationship between the two, namely *electromagnetic fields*.

464

This vector calculus formalism has become standard. It is manifestly rotation invariant, and therefore mathematically much more transparent than Maxwell's original 20 equations in x, y, z components.

Symbols in bold represent vector quantities, and symbols in italics represent scalar quantities, unless otherwise indicated. The equations introduce the *electric field*, **E**, a vector field, and the *magnetic field*, **B**, a pseudovector field, each generally having a time and location dependence. The sources are

- the total *electric charge density* (total charge per unit volume), ρ, and
- the total *electric current density* (total current per unit area), **J**.

The universal constants appearing in the equations are:

- the *permittivity of a vacuum*, ε_0, and
- the *permeability of a vacuum*, μ_0, and
- the speed of light, $c = 1/\sqrt{(\varepsilon_0 \mu_0)}$.

...

In the differential equations,

- the *nabla* symbol, ∇, denotes the three-dimensional *gradient operator*, del,
- the $\nabla \cdot$ symbol (pronounced "del dot") denotes the *divergence operator*,
- the $\nabla \times$ symbol (pronounced "del cross") denotes the *curl operator*.

[In *vector calculus*, the **gradient** of a scalar-valued differentiable function f of several variables is the vector field (or vector-valued function) ∇f (*del* or *nabla* f) whose value at a point p is the "direction and rate of fastest increase".

Del, or ***nabla***, is an operator used in mathematics (particularly in vector calculus) as a vector differential operator, usually represented by the ***nabla symbol*** ∇. When applied to a function defined on a one-dimensional domain, it denotes the standard derivative of the function as defined in calculus. When applied to a field (a function defined on a multi-dimensional domain), it may denote any one of three operations depending on the way it is applied: the *gradient* or (locally) steepest slope of a scalar field (or sometimes of a vector field); the *divergence* of a vector field; or the *curl* (rotation) of a vector field.

The ***divergence*** of a vector field is a *scalar* field. The divergence is generally denoted by "*div*". The ***divergence*** of a vector field can be

465

calculated by taking the **scalar product** of the vector operator applied to the vector field. I.e., $\nabla \cdot F(x, y)$.

The **curl** of a vector field is a *vector* field. The **curl** of a vector field is obtained by taking the **vector product** of the vector operator applied to the vector field $F(x, y, z)$. I.e., $Curl\ F(x, y, z) = \nabla \times F(x, y, z)$.]

Thus, the very compact representation of the **spinorial Maxwell equation** in Eq. (2) together with the **electromagnetic** and **charge-current spinors** in Eq. (1) is equivalent to the full (*non-relativistic*) theory of classical electromagnetism. Again, the fields and charge and current densities in Eqs. (7) – (10) can be either real or complex valued.

The rest of the first paper is devoted to developing a **relativistic** theory. **Section III** is devoted to the description of Lorentz and Poincaré transformations and their generators, which provide a natural way to introduce a *relativistically consistent* spin structure to the theory of light. The quantum operators in the first quantization of the field are presented in **Sec. IV**. The spinorial photon eigenstates and their density expectation values and charge-parity-time (CPT) symmetry are described in **Sec. V**. The conventional [*relativistic*] QED Lagrangian density is reformulated using eight-spinors, and Euler-Lagrange equations are investigated in **Sec. VI**. **Section VII** briefly presents the foundations of the second quantization using eight-spinors. This section is provided as a technical tool for interested readers to outline how the key properties of QED emerge from the conventional quantization procedure when eight-spinors are used. **Section VIII** shows how the eight-spinor formalism of QED enables the description of the generating Lagrangian density of gravity that acts as the basis for the [relativistic] Yang-Mills gauge theory of **unified gravity**. Related elegant special-unitary symmetry-based derivations of the electromagnetic and Dirac field SEM tensors are also presented. Finally, conclusions are drawn in **Sec. IX**.

The second paper was focused on using the **eight-spinor formalism** to derive a **gauge theory of gravity** using compact, finite-dimensional symmetries in a way that resembled the formulation of the fundamental interactions of the Standard Model. For their eight-spinor representation of the Lagrangian, a quantity, called the **space-time dimension field**, was defined which enabled the extraction of four-dimensional space-time quantities from the eight-dimensional spinors. Four U(1) symmetries of the components of the space-time dimension field were used to derive *a gauge theory, called* **unified gravity**. The **stress-energy-momentum tensor** source term of **gravity** follows directly from these symmetries. The **metric tensor** entered in **unified gravity** through geoedetric conditions.

The paper concluded that whilst **unified gravity** is a powerful and mathematically transparent framework with the potential of being an ultimate quantum theory of all fundamental forces of nature, in the end, a physical theory must be grounded on experimental verification. The classical limit of ***unified gravity*** was equivalent to ***general relativity***, and thus, consistent with the observations on gravitational interaction. ***However, the predictability of unified gravity in the explanation of quantum gravity phenomena is yet to be proven by future experiments.***

Partanen, M.[1] & Tulkki, J.[2] (March, 2024). QED based on an eight-dimensional spinorial wave equation of the electromagnetic field and the emergence of quantum gravity.

Phys. Rev., 109, 032224; https://doi.org/10.1103/PhysRevA,109.03224.

1 Photonics Group, Department of Electronics and Nanoengineering, Aalto University, PO Box 13500, 00076 Aalto, Finland
2 Engineered Nanosystems Group, School of Science, Aalto University, PO Box 12200, 00076 Aalto, Finland

Received October 2023; revised February 2024; accepted February 2024.

Abstract

Quantum electrodynamics (QED) is the most accurate of all experimentally verified physical theories. How QED and other theories of fundamental interactions couple to gravity through special unitary symmetries, on which the standard model of particle physics is based, is, however, still unknown. ***Here we develop a coupling between the electromagnetic field, Dirac electron-positron field, and the gravitational field based on an eight-component spinorial representation of the electromagnetic field.*** Our spinorial representation is analogous to the well-known representation of particles in the Dirac theory but it is given in terms of *8 × 8 bosonic gamma matrices*. In distinction from earlier works on the spinorial representations of the electromagnetic field, ***we reformulate QED using eight-component spinors. This enables us to introduce the generating Lagrangian density of gravity based on the special unitary symmetry of the eight-dimensional spinor space***. The generating Lagrangian density of gravity plays, in the definition of the gauge theory of gravity and its symmetric stress energy-momentum tensor source term, a similar role as the conventional Lagrangian density of the free Dirac field plays in the definition of the gauge theory of QED and its electric four-current density source term. The fundamental consequence, the ***Yang-Mills gauge theory of unified gravity***, is studied in a separate work [below; (May, 2025). *Rep. Prog. Phys.*, 88, 057802; https://arxiv.org/abs/2310.01460], where the theory is also extended to cover the other fundamental interactions of the standard model. We devote ample space for details of the eight-spinor QED to provide solid mathematical basis for the present work and the related work on the *Yang-Mills gauge theory of unified gravity.*

I. INTRODUCTION

***Quantum electrodynamics* (QED)** agrees with experiments to an exceedingly high accuracy.

> [This applies to ***non-relativistic quantum electrodynamics*** but not to ***relativistic*** quantum dynamics, on which this article is focused. See Underwood, T. G. (2023). *Quantum Electrodynamics – annotated sources*, Vol. II.]

In this regard, it can be considered as the most successful theory of physics ever developed. The present formulation of [relativistic] quantum electrodynamics (QED) has, however, not enabled coupling of the electromagnetic and Dirac electron positron fields to the gravitational field through the special unitary symmetries on which the standard model of particle physics is based. Therefore, alternative approaches, such as ***string theory*** and ***loop quantum gravity***, are being developed.

We approach the problem from a different point of view: ***It is proposed that identifying a special-unitary symmetry-based coupling of the present standard model to the gravitational field leads to the [relativistic] Yang-Mills gauge theory of unified gravity, the theory of all known fundamental interactions of nature***. Thus, it is our interest to study whether QED, the simplest quantum field theory of the standard model interactions, can be reformulated in alternative ways using novel mathematical structures that may help us to identify the special-unitary-symmetry-based coupling to gravity. This is the ultimate goal of the present work. Due to the very specific goal of the present work, the review of the relation of the present *spinorial representation of the electromagnetic field* to earlier spinorial representations will be left as a topic of further work.

Advances in optics and photonics technologies enable detailed experimental investigations of the spin and orbital angular momenta of light and the related physics of structured light fields and chiral quantum optics. Classical [*non-relativistic*] ***Maxwell's equations*** and [*non-relativistic*] **QED** provide solid basis for theoretical understanding of the related physical phenomena.

> [***Maxwell's equations*** are a set of coupled partial differential equations that, together with the ***Lorentz force law***, form the foundation of classical [*non-relativistic*] electromagnetism, classical optics, electric and magnetic circuits. *They describe how electric and magnetic fields are generated by charges, currents, and changes of the fields.* The equations are named after the physicist and mathematician James Clerk Maxwell, who, in 1861 and 1862, published an early form of the equations that included the Lorentz force law. Maxwell's equations may

be combined to demonstrate how fluctuations in **electromagnetic fields** (waves) propagate at a constant speed, c, in a vacuum.]

However, in spite of overall success, theories of light are also associated with well-known enigma that exist independently of the question of the gravitational coupling.

In [*relativistic*] *quantum electrodynamics* (QED), the usual approach to quantize the electromagnetic field is based on the [*relativistic*] transverse-vector-potential eigenstates describing the transverse *photons*. In optical spectroscopy, only radiative processes, which can be associated with transverse photons, are discovered. One enigmatic feature of the transverse-vector-potential eigenstates is that their [*relativistic*] *Lorentz transformation* between inertial frames leads to a nonzero scalar potential and a nonzero longitudinal component in the transformed inertial frame. Thus, the transformed state does not belong to the set of [*relativistic*] transverse-vector-potential eigenstates, which excludes scalar and longitudinal photon states. This is associated with the vector potential being gauge dependent.

> [A *gauge theory* is a type of *relativistic field theory* in which the *Lagrangian*, and hence the dynamics of the system itself, do not change under *local* transformations according to certain smooth families of operations (Lie groups).]

The related enigma of the definition of the *photon wave function* has been a subject of a continuing dialogue. The same applies to the definition of the *relativistic quantum spin structure of light*. The enigma is conventionally avoided only in the formalism of *second quantization*.

For comparison, the wave functions of spin-1/2 particles and their antiparticles are described by the [*relativistic*] *Dirac equation* whose solutions are four-component spinors. The *Dirac spinors* transform between inertial frames by their own spinorial [*relativistic*] *Lorentz transformation*, which is not equivalent to the Lorentz transformation of four-vectors. The Lorentz transformed spinors belong to the set of eigenstates in contrast with the case of the transverse-four-potential eigenstates of photons as discussed above.

Furthermore, *spin* emerges from the Dirac theory through the generators of [*relativistic*] Lorentz transformations on the Dirac spinors. The origin of spin in the Dirac theory suggests that a similar Lorentz-transformation-based origin may be found for the spin of photons.

> [*Lorentz transformations* are linear transformations from a coordinate frame in spacetime to another frame that moves at a constant velocity relative to the former in which observers moving at different velocities may measure different distances,

elapsed times, and even different orderings of events, but always *such that the speed of light is the same in all inertial reference frames*. The invariance of the speed of light is one of the postulates of *special relativity*.]

A strict *relativistic* description of the *spin of photons* does not emerge within the principles of classical physics from Maxwell's equations, from the electromagnetic field tensor, or from the electromagnetic four-potential. The conventional [*relativistic*] formulations of the quantum field theory of photons neither lead to a natural emergence of the covariant and gauge-invariant spin structure of light. The previous two-component spinorial representations of the electromagnetic field are covariant but they have been formulated so that the electromagnetic spinors in a vacuum satisfy the *Weyl equation*.

[The *Weyl equation* is a *relativistic* wave equation formulated by Hermann Weyl in 1929, which describes massless spin-1/2 particles, known as *Weyl fermions*.]

Regarding the spin properties of the theory, the *Weyl equation* using Pauli spin matrices is more natural in the description of massless spin-1/2 particles, known as Weyl fermions, as discussed by Perkins. The rank-two bispinor formulation of Kiessling *et al.* satisfies the massless Dirac equation, which is likewise typically used to describe spin-1/2 particles.

The initial goal of the present work was to shed light on the wave function and spin properties of light *by constructing a [relativistic] covariant and gauge-invariant spinorial wave function* and its Lorentz transformation for photons with a natural emergence of the spin structure of light. However, we discovered later that the theory of light based on the spinorial electromagnetic field leads to much more far-reaching consequences on the gravitational coupling as discussed below.

First, we introduce *the eight-component [non-relativistic] spinorial Maxwell equation*, the solutions of which are called *electromagnetic spinors*. This approach leads to the appearance of *8 × 8 matrices in the spinorial Maxwell equation* and to the electromagnetic spinors having eight components.

The *relativistic* description of the quantum spin of light emerges from this theory through the generators of Lorentz transformations on electromagnetic spinors in analogy to the emergence of the spin in the Dirac theory. Since the electromagnetic spinors are gauge independent [*non-relativistic*], they also avoid the gauge dependence problem of the four-potential eigenstates.

The *spinorial representation* of the [*non-relativistic*] electromagnetic field enables rewriting the conventional [*relativistic*] quantum electrodynamics (QED) and its Lagrangian density without changes in the physical predictions of the theory.

471

[***Lagrangian mechanics**** is a formulation of classical [*non-relativistic*] mechanics founded on the d'Alembert principle of virtual work. It was introduced by the Italian-French mathematician and astronomer Joseph-Louis Lagrange in his presentation to the Turin Academy of Science in 1760 culminating in his 1788 grand opus, *Mécanique analytique*. Lagrangian mechanics describes a mechanical system as a pair (M, L) consisting of a configuration space M and a smooth function L within that space called a ***Lagrangian***. For many systems, L = T − V, where T and V are the ***kinetic*** and ***potential energy*** of the system, respectively. The stationary ***action principle*** requires that the action functional of the system derived from L must remain at a stationary point (specifically, a maximum, minimum, or saddle point) throughout the time evolution of the system. This constraint allows the calculation of the equations of motion of the system using Lagrange's equations.

Newton's laws and the concept of ***forces*** are the usual starting point for teaching about mechanical systems. This method works well for many problems, but for others the approach is nightmarishly complicated. Lagrangian mechanics adopts ***energy*** rather than ***force*** as its basic ingredient, leading to more abstract equations capable of tackling more complex problems. Lagrange's approach was to set up independent generalized coordinates for the position and speed of every object, which allows the writing down of a general form of the ***Lagrangian*** (*total kinetic energy minus potential energy of the system*) and summing this over all possible paths of motion of the particles yielded a formula for the '***action***', which he minimized to give a generalized set of equations. This summed quantity is minimized along the path that the particle actually takes. This choice eliminates the need for the constraint force to enter into the resultant generalized system of equations. There are fewer equations since one is not directly calculating the influence of the constraint on the particle at a given moment.

The ***Lagrangian density*** is a function that describes the dynamics of a field in physics. It measures the distribution of the *Lagrangia*n over space and is crucial in formulating the equations of motion for fields in both classical and quantum field theories. In the context of field theories, the *Lagrangian density* is used to derive the ***Euler-Lagrange equations***, which govern the behavior of the fields. The ***action***, which is the integral of the Lagrangian density over spacetime, plays a fundamental role in the ***principle of least action***.

The ***Euler-Lagrange equations***, expressed as
$$d/dt\,(\partial f/\partial \dot{x}) - \partial f/\partial x = 0.$$
are fundamental in the calculus of variations. These equations are used to find the path that minimizes or maximizes a functional f.]

472

However, as a fundamental consequence, the ***eight-spinor formulation of the theory [of quantum electrodynamics* (QED)]** enables the description of the *generating Lagrangian density of gravity* based on the special unitary symmetry of the eight-dimensional spinor space. The generating **Lagrangian density of gravity** leads to an elegant derivation of the symmetric *stress-energy-momentum* (SEM) tensors of the electromagnetic and Dirac electron-positron fields in accordance with **Noether's theorem**.

[**Noether's theorem** states that every continuous symmetry of the *action* of a physical system with conservative forces has a corresponding conservation law. The *action* of a physical system is the integral over time of a *Lagrangian function*, from which the system's behavior can be determined by the *principle of least action*. This theorem applies to continuous and smooth symmetries of physical space. Noether's theorem is used in theoretical physics and the calculus of variations. It reveals the fundamental relation between the symmetries of a physical system and the conservation laws. It also made modern theoretical physicists much more focused on symmetries of physical systems.

The ***stress-energy-momentum (SEM) tensor***, a symmetric, rank-2 tensor indexed by pairs of space-time indices, *describes the flux of energy-momentum across regions of space-time*. In a different context, general relativity, the manifestly symmetric SEM tensor is also the source of the gravitational field. The ***electromagnetic stress–energy tensor*** contains the negative of the classical **Maxwell stress tensor** that governs the electromagnetic interactions.

The **Maxwell stress tensor** is a symmetric second-order tensor in three dimensions that is used in classical *non-relativistic* electromagnetism to represent the interaction between *electromagnetic forces* and *mechanical momentum*. In simple situations, such as a point charge moving freely in a homogeneous magnetic field, it is easy to calculate the forces on the charge from the **Lorentz force law.**

The **Lorentz force law** describes how electrically charged particles behave in electromagnetic fields. It states that a particle of **charge q** moving with velocity v in an electric field **E** and a magnetic field **B** experiences a force given by $\mathbf{F} = q\mathbf{E} + q\mathbf{v} \times \mathbf{B}$. This fundamental principle in electromagnetism is crucial for understanding how charged particles move under the influence of electric and magnetic fields.

When the situation becomes more complicated, this ordinary procedure can become impractically difficult, with equations spanning multiple lines. It is therefore convenient to collect many of these terms in the *Maxwell stress tensor*, and to use tensor arithmetic to find the answer to the problem at hand.

In the *relativistic* formulation of *electromagnetism*, the nine components of the *Maxwell stress tensor* appear, negated, as components of the *electromagnetic stress–energy tensor*, which is the electromagnetic component of the total stress–energy tensor. The latter describes the *density* and *flux of energy and momentum* in spacetime.]

Previously, the *SEM tensors* have been derived only by utilizing the *external* space-time symmetry. Furthermore, the canonical *SEM tensors* of the conventional QED are asymmetric, being as such *incompatible with general relativity*. This suggests that the *generating Lagrangian density of gravity* provides the basis for the derivation of the quantum field theory of gravity. Accordingly, the fundamental consequence of the present eight-spinor theory, the [*relativistic*] *Yang-Mills gauge theory of unified gravity*, is investigated in a separate work [below]. In this work, we also extend the theory to cover, not just QED, but all fundamental interactions of the *standard model*. …

We present a self-contained description of the theory and reserve plenty of space for the detailed study of the foundations so that the reader can become convinced that the mathematical and physical formulations are technically sound in every detail. This work is organized as follows: **Section II** introduces the *electromagnetic spinor* and formulates the [*non-relativistic*] *spinorial Maxwell equation*, which is the basis of the present work. **Section III** is devoted to the description of *Lorentz and Poincaré transformations* and their generators, which provide a natural way to introduce a *relativistically consistent* spin structure to the theory of light. The quantum operators in the first quantization of the field are presented in **Sec. IV**. The *spinorial photon eigenstates* and their *density expectation* values and *charge-parity-time* (CPT) *symmetry* are described in **Sec. V**. The conventional QED *Lagrangian density* is reformulated using eight-spinors, and *Euler-Lagrange equations* are investigated in **Sec. VI**. **Section VII** briefly presents the foundations of the *second quantization* using eight-spinors. This section is provided as a technical tool for interested readers to outline how the key properties of QED emerge from the conventional quantization procedure when eight-spinors are used. **Section VIII** shows how the eight-spinor formalism of QED enables the description of the generating *Lagrangian density of gravity* that acts as the basis for the [*relativistic*] *Yang-Mills gauge theory of unified gravity*. Related elegant special-unitary symmetry-based derivations of the electromagnetic and Dirac field SEM tensors are also presented. Finally, conclusions are drawn in **Sec. IX**.

474

II. SPINORIAL MAXWELL EQUATION

A. Electromagnetic and charge-current spinors

We start the formulation of the theory by introducing an *electromagnetic spinor* **Ψ** and a *charge-current spinor* **Φ**, which both have eight components. As we show below, the advantage of using an eight-dimensional electromagnetic spinor is that it is ***gauge-independent [non-relativistic]***. The eight-component spinors describe the physics of *all four [non-relativistic] Maxwell's equations* including the charge and current densities: one component is related to *Gauss's law of magnetism*, three components to the *Ampère Maxwell law*, one component to *Gauss's law of electricity*, and three components to *Faraday's law*. In any inertial frame, the components of Ψ(t,r) are composed of the *electric and magnetic fields*, E(t,r) and B(t,r), and the components of Φ(t, r) are composed of the *total electric charge* and *current densities* ρ_c(t, r) and J_c(t,r). Omitting the time and position arguments, we write

$$\Psi = \sqrt{\varepsilon_0}/2 \begin{bmatrix} 0 \\ \mathbf{E} \\ 0 \\ i c\mathbf{B} \end{bmatrix}, \qquad \Phi = \sqrt{\varepsilon_0}/2 \begin{bmatrix} 0 \\ \mu_0 c \mathbf{J_e} \\ \rho_e/\varepsilon_0 \\ \mathbf{0} \end{bmatrix}, \qquad (1)$$

Here ε_0, μ_0, and $c = 1/\sqrt{\varepsilon_0 \mu_0}$ are the ***permittivity***, ***permeability***, and the ***speed of light in a vacuum***, respectively. The zero elements of the ***electromagnetic spinor*** Ψ are related to the property of all electromagnetic spinors to be eigenstates of the spin operator squared, having a well-defined *spin* S = 1. The zero elements of the ***charge-current spinor*** Φ are related to the nonexistence of magnetic monopoles as elementary particles. …

…

B. Spinorial Maxwell equation and the associated gamma and sigma matrices

Our formulation of electrodynamics equations below is closely analogous to the [*relativistic*] approach of Dirac starting from the second-order [*relativistic*] ***Klein-Gordon equation***,

$$[1/c^2 \, \partial^2/\partial t^2 \, \psi - \nabla^2 \, \psi + m^2 c^2/\hbar^2 \, \psi = 0,$$

where ψ is the *wave function* as a function of x_1, x_2, x_3, and t; m is the mass of the electron; c is the speed of light, and $\hbar$ is the reduced Planck's constant,]

to derive a first-order equation, the [*relativistic*] ***Dirac equation***.

$$\{p_0 + \rho_1 \, (\boldsymbol{\sigma}, \mathbf{p}) + \rho_3 mc\} \, \psi = 0;$$

where **p** is the *momentum* vector, p_1, p_2, p_3 are the components of the momentum, $p_0 = ih/c\ \partial/\partial t$, $p_r = -\ ih\ \partial/\partial x_r$, $r = 1, 2, 3$; **σ** denotes the vector $(\sigma_1, \sigma_2, \sigma_3)$ where σ_1, σ_2, σ_3 are the matrices

$$
\sigma_1 = \begin{pmatrix} 0 & 1 & 0 & 0 \\ 1 & 0 & 0 & 0 \\ 0 & 0 & 0 & 1 \\ 0 & 0 & 1 & 0 \end{pmatrix} \quad
\sigma_2 = \begin{pmatrix} 0 & -i & 0 & 0 \\ i & 0 & 0 & 0 \\ 0 & 0 & 0 & -i \\ 0 & 0 & i & 0 \end{pmatrix} \quad
\sigma_3 = \begin{pmatrix} 1 & 0 & 0 & 0 \\ 0 & -1 & 0 & 0 \\ 0 & 0 & 1 & 0 \\ 0 & 0 & 0 & -1 \end{pmatrix}];
$$

and ρ_1 and ρ_3 are the matrices

$$
\rho_1 = \begin{pmatrix} 0 & 0 & 1 & 0 \\ 0 & 0 & 0 & 1 \\ 1 & 0 & 0 & 0 \\ 0 & 1 & 0 & 0 \end{pmatrix} \quad
\rho_3 = \begin{pmatrix} 1 & 0 & 0 & 0 \\ 0 & 1 & 0 & 0 \\ 0 & 0 & -1 & 0 \\ 0 & 0 & 0 & -1 \end{pmatrix}.]
$$

Dirac, P.A.M (1928). The quantum theory of the electron, *Proc. R. Soc. A*, 117, 610.
Dirac, P.A.M. (1958). *The Principles of Quantum Mechanics*, Oxford University Press, Oxford.

Dirac's clever idea was to take a square root of the *wave operator*, which is equivalent to finding matrices γ^a, which satisfy $I_n \partial^a \partial_a = (\gamma^a \partial_a)^2$, where the Einstein summation convention is used and I_n is the $n \times n$ identity matrix.

In this work, the Latin indices $a, b, c, d \in \{0, x, y, z\}$ range over the four dimensions of the Minkowski space-time, the Latin indices $i\ j, k \in \{x, y, z\}$ range over the three spatial dimensions, and the Greek indices $\mu, \nu, \rho, \sigma \in \{x^0, x^1, x^2, x^3\}$ range over the four general space-time dimensions.

Dirac found that $I_4 \partial^a \partial_a = (\gamma^a_F \partial_a)^2$ is satisfied with 4×4 matrices γ^a_F, which became known as the ***Dirac gamma matrices***. Here we use the subscript F to indicate that these gamma matrices are associated with fermionic fields in distinction from the gamma matrices of bosonic fields introduced below.

[However, Dirac followed this up with a paper which noted that ***relativity was of no importance in the consideration of atomic and molecular structure and ordinary chemical reactions where quantum entanglement between the spins of the electrons was more important,*** [Dirac, P. A. M. (April, 1929). Quantum Mechanics of Many-Electron Systems. *Roy. Soc. Proc., A*, 123, 792, 714-33], in which he stated "§ 1. *Introduction. The general theory of quantum mechanics is* *now almost complete,* **the imperfections that still remain being in connection with the exact fitting in of the theory with relativity ideas.** These give rise to difficulties only when high-speed particles are involved, and **are therefore of no importance in the consideration of atomic and molecular structure and ordinary chemical reactions,** in which it is, indeed, usually *sufficiently accurate if one neglects relativity variation of mass with velocity and assumes only Coulomb forces between the various electrons and atomic nuclei.*

…

Already before the arrival of quantum mechanics there existed a theory of atomic structure, based on Bohr's ideas of quantized orbits, which was fairly successful in a wide field. *To get agreement with experiment it was found necessary to introduce the spin of the electron,* giving a doubling in the number of orbits of an electron in an atom. With the help of this spin and *Pauli's exclusion principle, a satisfactory theory of multiplet terms was obtained when one made the additional assumption that the electrons in an atom all set themselves with their spins parallel or antiparallel.*

…

If s denoted the magnitude of the resultant *spin angular momentum,* this s was combined vectorially with the resultant *orbital angular momentum l* to give a multiplet of multiplicity 2s + 1. *The fact that one had to make this additional assumption was, however, a serious disadvantage, as no theoretical reasons to support it could be given.* **It seemed to show that there were large forces coupling the spin vectors of the electrons in an atom,** much larger forces than could be accounted for as due to the interaction of the *magnetic moments* of the electrons. *The position was thus that there was empirical evidence in favor of these large forces, but that their theoretical nature was quite unknown.*

…

The solution of this difficulty in the explanation of multiplet structure is provided by the **exchange (austausch) interaction of the electrons,** *which arises owing to the electrons being indistinguishable one from another.* Two electrons may change places without our knowing it, and the proper allowance for the possibility of quantum jumps of this nature, which can be made in a treatment of the problem by quantum mechanics, gives rise to the new kind of interaction. The energies involved, the so-called *exchange energies,* are quite large. In fact, it is these *exchange energies* between electrons in different atoms that give rise to homopolar valency bonds, as shown by Heitler and London[#].

[#] Heitler, W. & London, F. (June, 1927). Wechselwirkung neutraler Atome und homöopolare Bindung nach der Quantenmechanik. (Interaction of

Neutral Atoms and Homopolar Binding in Quantum Mechanics.) *Zeit. Phys.*, 44, 2, 455-72; http://dx.doi.org/ 10.1007/ BF01397394.]

Our goal is to define gamma matrices, which enable writing the well-known second-order *wave equation* of the four-potential, given by $\partial^a \partial_a A^b = \mu_0 J^b_e$, as a first-order equation for the electromagnetic spinor in Eq. (1), which, unlike the Dirac equation, also includes the source term expressed by the charge-current spinor Φ. We next show that it is possible to construct 8×8 gamma matrices, which operate on the electromagnetic spinor Ψ in Eq. (1) leading to Maxwell's equations and to having desired properties. In analogy to the Dirac theory, our gamma matrices γ^a_B satisfy $I_8 \partial^a \partial_a = (\gamma^a_B \partial_a)^2$. Here we use the subscript B to indicate that these gamma matrices are associated with bosonic fields in distinction from the gamma matrices of fermionic fields in the Dirac theory.

We write the spinorial *Maxwell equation* in a form that closely reminds us of the [*relativistic*] *Dirac equation* describing spin-1/2 particles in the case when the mass of the particle is set to zero and when an external source term is added. The [*relativistic*] spinorial Maxwell equation, which, as shown in Sec. IIC, describes all of classical electromagnetism, is written as

$$\gamma^a_B \partial_a \Psi = - \Phi. \qquad (2)$$

In analogy to the **Dirac equation** for electrons, the [*relativistic*] **spinorial Maxwell equation** is a first-order differential equation that is covariant in Lorentz transformations between inertial frames. ...

...

The form of the *bosonic gamma matrices* γ^a_B in Eq. (3) recalls the conventional definition of the 4×4 *fermionic gamma matrices* γ^a_F in the Dirac theory [59], but the dimensions and the corresponding representations of the sigma matrices are different. In the Dirac theory, the 2×2 sigma matrices σ^i_F are known as **Pauli matrices**. The 8×8 gamma matrices of the present theory can also be seen as the generalization of the 6×6 gamma matrices that have been used together with the six-component wave-function-like concept in previous literature [67,68]. In comparison with the 6×6 gamma matrices, one particular advantage of our 8×8 gamma matrices is that all **Maxwell's equations** can be presented as a single equation allowing us to rewrite the QED in the eight-spinor notation as presented in Sec.VI. The bosonic gamma matrices in Eq. (3) can be used to define the spinorial Lorentz transformations as will be shown in Sec. III. This makes the electromagnetic spinor in Eq. (1) a true spinor with spin 1 in the full physical meaning of the term.

C. Dynamical equations in terms of the fields, charges, and currents

When the [*non-relativistic*] expressions of the electromagnetic spinor and the charge-current spinor from Eq. (1)

$$[\Psi = \sqrt{\varepsilon_0}/2 \begin{bmatrix} 0 \\ \mathbf{E} \\ 0 \\ i c \mathbf{B} \end{bmatrix}, \qquad \Phi = \sqrt{\varepsilon_0}/2 \begin{bmatrix} 0 \\ \mu_0 c \mathbf{J_e} \\ \rho_e/\varepsilon_0 \\ \mathbf{0} \end{bmatrix}, \qquad (1)$$

are substituted into the [*relativistic form of the*] **spinorial Maxwell equation** in Eq. (2),

$$[\gamma^a{}_B \partial_a \Psi = -\Phi. \qquad (2)]$$

we obtain the full set of four [*non-relativistic*] **Maxwell's equations** in the conventional form as

$\nabla \cdot \mathbf{B} = 0$	Gauss's Law for Magnetism (7)
$\nabla \times \mathbf{B} = -\mu_0 \mathbf{J_e} + 1/c^2 \, \partial\mathbf{E}/\partial t)$	The Maxwell-Ampère Law (8) with Maxwell's addition.
$\nabla \cdot \mathbf{E} = \rho/\varepsilon_0 = 0$	Gauss's Law (9)
$\nabla \times \mathbf{E} = -\partial\mathbf{B}/\partial t$	The Maxwell–Faraday (10) version of Faraday's Law of Induction

[Underwood, T.G. (2024). *Electricity & Magnetism*, p. 300: "The Heaviside formulation of Maxwell's Equations (1884) in a region with no charges, such as in a vacuum, reduces to:

$\nabla \cdot \mathbf{B} = 0$	Gauss's Law for Magnetism
$\nabla \times \mathbf{B} = -\mu_0 (\mathbf{J} + \varepsilon_0 \, \partial\mathbf{E}/\partial t) = -\mu_0\varepsilon_0 \, \partial\mathbf{E}/\partial t$	The Maxwell-Ampère Law with Maxwell's addition.
$\nabla \cdot \mathbf{E} = \rho/\varepsilon_0 = 0$	Gauss's Law
$\nabla \times \mathbf{E} = -\partial\mathbf{B}/\partial t$	The Maxwell–Faraday version of Faraday's Law of Induction."

Maxwell's equations, or the *Maxwell–Heaviside equations*, are a set of coupled partial differential equations that, together with the Lorentz force law, form the foundation of **classical electromagnetism**, classical optics, and electric circuits. The equations provide a mathematical model for electric, optical, and radio technologies, such as power generation, electric motors, wireless communication, lenses, radar, etc. They describe how electric and magnetic fields are generated by

479

charges, currents, and changes of the fields. The equations are named after the physicist and mathematician James Clerk Maxwell, who, in 1861 and 1862, published an early form of the equations that included the Lorentz force law. Maxwell first used the equations to propose that light is an electromagnetic phenomenon.

Formulation in terms of electric and magnetic fields (microscopic or in vacuum version)

Maxwell's equations may be combined to demonstrate how fluctuations in electromagnetic fields (waves) propagate at a constant speed in a vacuum, c (299,792,458 m/s). Known as *electromagnetic radiation*, these waves occur at various wavelengths to produce a spectrum of radiation from radio waves to gamma rays.

In 1884 Oliver Heaviside recast Maxwell's mathematical analysis from its original cumbersome form (they had already been recast as *quaternions*) to its modern vector terminology, thereby *reducing twelve of the original twenty equations in twenty unknowns down to the four differential equations in two unknowns we now know as Maxwell's equations*. Heaviside presented these equations *in modern vector format* using the *nabla operator* (**∇**) devised by William Rowan Hamilton in 1837.

> [Hunt, B. J. (2012). Oliver Heaviside: A first-rate oddity. *Physics Today*, 65, 11, 48–54: "… His most important step was to derive what he called the *second circuital law*, which relates the curl of **E** to the rate of change of **H**. Starting with Maxwell's relation $\mathbf{E} = -\partial \mathbf{A}/\partial t - \nabla \Psi$, Heaviside took the curl of both sides:
>
> $$\nabla \times \mathbf{E} = \nabla(-\partial \mathbf{A}/\partial t) - \nabla \times \nabla \Psi.$$
>
> Since the curl of any gradient is zero, the last term vanishes. By switching the order of the space and time differentiations, Heaviside obtained $\nabla \times \mathbf{E} = -\partial(\nabla \times \mathbf{A})/\partial t$. Since $\nabla \times \mathbf{A} = \mu \mathbf{H}$, this yielded the *second circuital law*, $\nabla \times \mathbf{E} = -\mu\, \partial \mathbf{H}/\partial t$.
>
> Heaviside then combined this with equations drawn from Maxwell's *Treatise* to obtain his new set of four "Maxwell's equations":
>
> $$\nabla \cdot \varepsilon \mathbf{E} = \rho$$
> $$\nabla \times \mathbf{E} = -\mu\, \partial \mathbf{H}/\partial t$$
> $$\nabla \cdot \mu \mathbf{H} = 0$$
> $$\nabla \times \mathbf{H} = k\mathbf{E} + \varepsilon\, \partial \mathbf{E}/\partial t,$$

where ε is the *permittivity*; μ the *permeability*; ρ the *charge density*; and k the *conductivity*.

Heaviside used what he called rational units, which eliminated the factors of 4π that otherwise appeared in so many electromagnetic equations.]

In the ***electric and magnetic field formulation*** there are four equations that determine the fields for given charge and current distribution. A separate law of nature, the ***Lorentz force law***, describes how, conversely, the electric and magnetic fields act on charged particles and currents. A version of this law was included in the original equations by Maxwell but, by convention, is included no longer. The four re-formulated ***Maxwell's equations*** describe the nature of *electric charges* (both static and moving), *magnetic fields*, and the relationship between the two, namely *electromagnetic fields*.

This ***vector calculus*** formalism has become standard. It is manifestly rotation invariant, and therefore mathematically much more transparent than Maxwell's original 20 equations in x, y, z components.

Symbols in bold represent vector quantities, and symbols in italics represent scalar quantities, unless otherwise indicated. The equations introduce the *electric field*, **E**, a vector field, and the *magnetic field*, **B**, a pseudovector field, each generally having a time and location dependence. The sources are

- the total *electric charge density* (total charge per unit volume), ρ, and
- the total *electric current density* (total current per unit area), **J**.

The universal constants appearing in the equations are:

- the *permittivity of a vacuum*, ε_0, and
- the *permeability of a vacuum*, μ_0, and
- the speed of light, $c = 1/\sqrt{(\varepsilon_0\mu_0)}$.

...

In the differential equations,

- the ***nabla symbol***, ∇, denotes the three-dimensional ***gradient operator***, del,
- the $\nabla\cdot$ ***symbol*** (pronounced "del dot") denotes the ***divergence operator***,
- the $\nabla\times$ ***symbol*** (pronounced "del cross") denotes the ***curl operator***.

[In ***vector calculus***, the *gradient* of a scalar-valued differentiable function f of several variables is the vector field (or vector-valued function)

∇f (*del* or *nabla* f) whose value at a point p is the "direction and rate of fastest increase".

Del, or *nabla,* is an operator used in mathematics (particularly in vector calculus) as a *vector differential operator*, usually represented by the *nabla symbol* ∇. When applied to a function defined on a one-dimensional domain, it denotes the standard derivative of the function as defined in calculus. When applied to a field (a function defined on a multi-dimensional domain), it may denote any one of three operations depending on the way it is applied: the *gradient* or (locally) steepest slope of a scalar field (or sometimes of a vector field); the *divergence* of a vector field; or the *curl* (rotation) of a vector field.

The *divergence* of a vector field is a *scalar* field. The divergence is generally denoted by "*div*". The *divergence* of a vector field can be calculated by taking the *scalar product* of the vector operator applied to the vector field. I.e., $\nabla \cdot F(x, y)$.

The *curl* of a vector field is a *vector* field. The *curl* of a vector field is obtained by taking the *vector product* of the vector operator applied to the vector field $F(x, y, z)$. I.e., *Curl* $F(x, y, z) = \nabla \times F(x, y, z)$.]

Thus, the very compact representation of the *spinorial Maxwell equation* in Eq. (2) together with the *electromagnetic* and *charge-current spinors* in Eq. (1) is equivalent to the full (*non-relativistic*) theory of *classical electromagnetism*. Again, the *fields* and *charge and current densities* in Eqs. (7) – (10) can be either real or complex valued.

Operating on Eq. (2) side by side with an operator $\gamma^a{}_B \partial a$ and using $(\gamma^a{}_B \partial a)^2 = I_8 \partial^a \partial_a$ on the left-hand side, we obtain $\partial^a \partial_a = -\gamma^a{}_B \partial_a \Phi$. This equation is equivalent to the continuity equation of the electric four-current density $J^a{}_e = (c\rho_e, \mathbf{J}_e)$ and the inhomogeneous wave equations of the electric and magnetic fields as

$$\partial \rho_e / \partial t + \nabla \cdot \mathbf{J}_e = 0, \tag{11}$$

$$\nabla^2 \mathbf{E} - 1/c^2 \, \partial^2/\partial t^2 \, \mathbf{E} = \mu_0 \, \partial/\partial t \, \mathbf{J}_e + 1/\varepsilon_0 \, \nabla \rho_e, \tag{12}$$

$$\nabla^2 \mathbf{B} - 1/c^2 \, \partial^2/\partial t^2 \, \mathbf{B} = -\mu_0 \nabla \times \mathbf{J}_e. \tag{13}$$

Therefore, in addition to the conventional Maxwell's equations, the *spinorial Maxwell equation* also compactly describes the physics related to the *conservation of the electric four-current density*. If the hypothetical magnetic charge and current densities were added

482

in the theory through substituting them in the zero elements of the charge-current spinor in Eq. (1), the conservation law of the magnetic four-current density would also follow.

…

III. LORENTZ AND POINCARÉ TRANSFORMATIONS AND INVARIANTS

In the present theory, there are two types of Lorentz transformations for eight-spinors: the Lorentz transformation of spinors generated by four-vectors, called *four-vector spinors*, such as the *charge-current* spinor and the *potential* spinor; and the Lorentz transformation of spin-1 *field spinors*, such as the *electromagnetic* spinor.

These transformations preserve the forms of the electromagnetic and charge-current spinors in Eq. (1) and the potential spinor in Eq. (14) in all inertial frames. The transformations are analogous to the Lorentz transformation of Dirac spinors in their construction through the Lorentz generators in a way that they differ from the Lorentz transformation of four-vectors. For completeness, the Lorentz transformation of the conventional spin-1/2 Dirac field spinors is also presented. In addition, we construct the Lorentz transformation of spin-2 fields, which can be presented using 8×8 matrices in the present theory. The Lorentz transformations are described in detail in the sections below. In the last three sections, we present the complete Lorentz transformations of fields, which involve the transformations of the coordinate arguments; the Poincaré transformations, which also involve space-time translations; and the construction of Lorentz invariants.

…

IV. QUANTUM OPERATORS IN THE FIRST QUANTIZATION

In this section, we determine the key quantum-mechanical operators for *electromagnetic spinors* given in Eq. (1). These operators are in the literature called the operators in the *first quantization*. For the quantum field of many photons, one will need the corresponding *second quantization* operators described in Sec. VII.

…

V. SPINORIAL PHOTON EIGENSTATES

In this section, we present selected well-known special cases of *photon* eigenstates in the spinorial formulation of the electromagnetic field. To be able to define the second quantized electromagnetic field, we need a complete set of solutions of the free-field spinorial Maxwell equation in Eq. (67), which we call *photon spinors*. Eigenvalue equations of quantum operators, presented below, ***cannot be satisfied for the electromagnetic spinor in Eq. (1) if it is made of classical real-valued fields***. Therefore, the eigenvalue equations of quantum operators require the determination of complex-valued photon spinors. The real-valued physical quantities are recovered in this formalism as expectation values.

...

VI. QED LAGRANGIAN DENSITY AND EULER-LAGRANGE EQUATIONS

In this section, we investigate the derivation of the present spinorial electromagnetic theory *from the Lagrangian density*. Thus, we follow the conventional approach of deriving field theories. There are formal differences in our spinorial representation compared with the conventional four-potential based QED, but the resulting theory is equivalent to the conventional QED as discussed in more detail below.

A. Lagrangian density of QED.

Here we present the conventional **Lagrangian density** of the coupled system of the **electromagnetic field** and the **Dirac field**. ...

...

VII. SECOND QUANTIZATION OF SPINORIAL ELECTROMAGNETIC FIELD

The goal of this section is to give the reader a concise idea of how the definitions of the key quantum operators and their matrix elements in the spinorial form transform into the pertinent quantities in the conventional QED. The quantization of the electromagnetic spinor field form $\Psi_{\mathscr{R}}$, made of real-valued fields, follows trivially from the conventional QED. It is obtained by substituting the conventional Hermitian electric and magnetic-field operators in places of the electric and magnetic fields in the electromagnetic spinor in Eq. (1). Similarly, the charge-current spinor is quantized by substituting the conventional Dirac field operator into Eq. (115) or the corresponding components of the electric four-current density into Eq. (1).

This section is devoted to the quantization of the complex valued electromagnetic spinor field form Ψ. ...

VIII. GENERATING LAGRANGIAN DENSITY OF GRAVITY, SPECIAL UNITARY SYMMETRY, AND THE STRESS-ENERGY-MOMENTUM TENSOR

In the sections above, we have presented how the conventional QED is expressed in the **eight-spinor formalism**. Next, we introduce the generating Lagrangian density of gravity that plays, in the definition of the gauge theory of gravity, a similar role as the conventional Lagrangian density of the free Dirac field plays in the definition of the gauge theory of QED. The generating Lagrangian density of gravity is associated with a special unitary symmetry of the quantum fields in the standard model, and it enables an elegant derivation of the symmetric **SEM tensors** as described in detail in the sections below. The **Yang-Mills gauge theory of unified gravity** that follows is presented in a separate work.

A. Generating Lagrangian density of gravity

As the starting point for defining the generating **Lagrangian density of gravity**, we use the analogy with standard gauge theory as it is applied in deriving the full QED from the **Lagrangian density** of the **free Dirac field**, $L_{QED,0} = L_{QED,D}$, in Eq. (111). We have presented a concise summary of the conventional gauge theory of QED in Appendix B. As shown in Appendix B, the unitary transformation $\psi \rightarrow e^{i\theta}\psi$ with parameter θ and the electromagnetic-gauge-covariant derivative $D^{\rightarrow}_a = \partial^{\rightarrow}_a + i\, q_e/\hbar\, A_{\mathscr{R}a}$ in Eq. (113) *generate* from the **Lagrangian density** $\mathscr{L}_{QED,0}$ the full QED. In Appendix B, we also show that the variation of the generating **Lagrangian density** of QED, $\delta\mathscr{L}_{QED,0} = -\hbar/q_e\, J^a_{\mathscr{R}a}\, \partial_a\theta$, is proportional to the **electric four-current density**, which is the source term of the **electromagnetic field**. Based on the known properties of gravitational interaction, we must set certain conditions that we require the generating **Lagrangian density of gravity** and the resulting **full gauge theory of unified gravity** to satisfy:

(1) The theory must satisfy the **global Lorentz invariance and the general covariance**, which means the form invariance of physical laws under general differentiable coordinate transformations. More strongly, we require diffeomorphism invariance.

(2) The **SEM tensor** must act as the source term of the gravitational field. It follows that the gravitational field is a *tensor gauge field* in contrast with the vector gauge fields of the standard model.

(3) Instead of the Abelian U(1) gauge theory of QED, we must use the non-Abelian **Yang-Mills gauge theory** analogous to the theories of weak and strong interactions. This is because four symmetry generators are needed for the description of the tensor gauge field.

(4) To enable **unification of gravity** with the fundamental interactions of the standard model, the **gauge theory of gravity** must be based on an internal special unitary symmetry or sub-symmetry of the quantum fields in the **standard model**.

(5) The theory must contain a **new coupling constant** g_g, called the **coupling constant of unified gravity**. The variation of the generating Lagrangian density of gravity with respect to the symmetry transformation parameters must be directly proportional to the SEM tensor divided by g_g.

485

(6) The **gauge theory of gravity** must enable writing the dynamical equations for the gravitational field through **the Euler-Lagrange equations**. The gravitational field equations must reproduce the experimentally verified predictions of **general relativity**. [???]

(7) Through the **Euler-Lagrange equations**, we must also obtain the generalized equations of motion containing **gravitational coupling** for all the fundamental interactions of the standard model. In the Minkowski metric limit of the gravitational gauge field, these equations must reproduce the dynamical equations of the **standard model**.

The **generating Lagrangian density of gravity** can be seen as the fundamental hypothesis of the theory that unifies the standard model and gravity. All quantum fields of the standard model can be included on equal footing, but, for simplicity, here we present the theory using the Dirac electron-positron field and the electromagnetic field only. ...

...

B. SU(8)$_{4D}$ symmetry

While the theories of the electromagnetic, weak, and strong interactions utilize U(1), SU(2), and SU(3) symmetries, here we utilize a **four-dimensional sub-symmetry of SU(8)**, defined below and denoted by SU(8)$_{4D}$. In analogy to the description of gauge symmetries in the conventional quantum field theory, we consider the SU(8)$_{4D}$ symmetry transformation under which the **generating Lagrangian density** in Eq. (147) is globally invariant and whose space-time-dependent variation without introducing the gauge field is related to the source term of the gauge field. In our case, this source term is the **SEM tensor**, while in the case of the U(1) symmetry of QED it is the **electric four-current density** as briefly presented in Appendix B. The **generating Lagrangian density of gravity** is invariant with respect to any global, i.e., space-time-independent, symmetry transformation of $I_g = I_8$, since such transformations do not make $\partial^{\rightarrow}{}_\nu I_g$ and $\partial^{\rightarrow}\nu\ I^{\rightarrow\dagger}{}_g$ nonzero. Thus, the **generating Lagrangian density of gravity** is trivially globally invariant with respect to the SU(8)$_{4D}$ symmetry.

The SU(8)$_{4D}$ symmetry differs from the U(1) symmetry of the conventional QED: While the U(1) symmetry operates on the *Dirac field* and generates the **electromagnetic field** as a gauge field, the SU(8)$_{4D}$ symmetry operates through its influence on I_g on both the **Dirac** and **electromagnetic fields** and generates the **gravitational field** as a gauge field. Such an operation is necessary since gravity is known to affect all fields and matter.

Here we limit our study of the SU(8)$_{4D}$ symmetry to its relation to the **SEM tensors** of the **Dirac and electromagnetic fields** without introducing the **gravitational gauge field**. ...

486

We define the SU(8)$_{4D}$ symmetry transformation as

$$\mathbf{I}_g \rightarrow \mathbf{U}\mathbf{I}_g, \text{ where } \mathbf{U} = \exp(i\varphi_a\mathbf{t}^a). \quad\quad\quad (148)$$

The symmetry transformation matrix $\mathbf{U}$ has determinant 1. Thus, it is an element of the *special unitary group SU(8)*. ...

... The group of transformations defined by $\mathbf{U}$ corresponds to a four-dimensional subgroup of SU(8), which is isomorphic to SU(2) $\otimes$ U(1), and which we denote SU(8)$_{4D}$. For comparison with the present SU(8)$_{4D}$ generators $\mathbf{t}_a$, in the case of the SU(2) symmetry of the *electroweak interaction*, the generators are the three *Pauli matrices*, and in the case of the SU(3) symmetry of *quantum chromodynamics* (QCD), the generators are the eight *Gell-Mann matrices*.

C. Symmetric stress-energy-momentum tensor

In this section, we generalize the *global SU(8)$_{4D}$ symmetry transformation* of Eq. (148) into a *local symmetry transformation* by making the symmetry transformation parameters φ_a space-time dependent as $\varphi_a = \varphi(x^0,x^1,x^2,x3)$. Would we follow the *Yang-Mills gauge theory*, we should simultaneously introduce the gauge-covariant derivative that makes the derivative terms of the *generating Lagrangian density of gravity* in Eq. (147) invariant. ... Here we consider the variation of the *generating Lagrangian density of gravity* without adding the gauge field. We recover the profound relationship between the SU(8)$_{4D}$ symmetry and the symmetric *SEM tensor* source term of gravity. This relationship is completely analogous to the relation between the U(1) symmetry of QED and the electric four-current density source term of electromagnetism, as briefly presented in Appendix B. ...

D. Summary of the generating Lagrangian density of gravity and its special unitary symmetry

We expect that the eight-spinor formulation of the standard model is necessary for the unification of gravity into it. This conclusion is based on the surprising simplicity of how all the fermionic and bosonic fields of the standard model can be coupled to the gravity in the gauge theory arising from the SU(8)$_{4D}$ symmetry presented in this work. The decades of work on this problem without an unambiguous solution also suggest that the formulation of the *quantum field theory of gravity* using the standard model without eight-spinor structures is extremely difficult if not impossible.

In the derivation of the SEM tensor using the *generating Lagrangian density of gravity*, the symmetric SEM tensor follows directly from the internal symmetry of the generating

Lagrangian density of gravity. This is in accordance with **Noether's theorem**, which states that each generator of a continuous symmetry is associated with a conserved current. In the present case, the conserved currents of the four symmetry generators t_a combine to a single **SEM tensor**. Using the **internal special unitary symmetry** SU(8)$_{4D}$ instead of the **external space-time symmetry** differs from the conventional Lagrangian derivation of the **SEM tensor**. In the conventional derivation, the **SEM tensor** follows from the external space-time symmetry of the **action** and it is asymmetric if no additional symmetrization procedures are introduced, such as the Belinfante-Rosenfeld symmetrization. This observation is one of the foundations for the development of the **Yang-Mills gauge theory of unified gravity** ... .

We have shown how the **SEM tensor** acting as the source term in the **Yang-Mills gauge theory of unified gravity** arises from the special unitary symmetry, but have left the description of the tensor gauge field to a separate work. Here we briefly note that the tensor gauge field is a Lorentz-invariant tensor whose representation as a 4 ×4 matrix is invariant in the Lorentz transformation of second-rank tensors, and whose representation in terms of 8 ×8 matrices is invariant in the Lorentz transformation of spin-2 fields, given in Sec. IIIE.

The different Lorentz transformation properties of vector and tensor gauge fields are one indication why the **Yang-Mills gauge theory of unified gravity** cannot be derived from the conventional *Lagrangian density*, but the **generating Lagrangian density of gravity** is needed. The conventional Lagrangian density follows from the gravitational-gauge-invariant form of the Lagrangian density of the Yang-Mills gauge theory of unified gravity in the Minkowski metric limit.

IX. CONCLUSIONS

In conclusion, we have presented QED based on the **eight-dimensional spinorial Maxwell equation**. The spinorial Maxwell equation is equivalent to the full set of Maxwell's equations. Consequently, it is equivalent to several formulations of electrodynamics, such as the most conventional three-vector-calculus, electromagnetic field tensor, exterior algebra of differential forms, space-time algebra, quaternions, two-component spinors, and rank-two bispinors. It provides an elegant representation of classical electrodynamics and QED, but it does not produce new physics if no other elements are added to the theory. In comparison to other known formulations of electrodynamics, the present formulation of the spinorial Maxwell equation is the most analogous to the Dirac equation.

Here we interpret the analogy in such a way that, in our case of *photons*, the pertinent gamma matrices must reflect the properties of a spin-1 field instead of the spin-1/2 field of the Dirac theory. Instead of the conventional 4 × 4 Dirac gamma matrices, our spinorial

Maxwell equation is given in terms of 8×8 bosonic gamma matrices satisfying the Dirac algebra. In contrast with 6 × 6 gamma matrices that have been studied in some previous works, the gamma matrices of the present theory allow the description the physics of all Maxwell's equations by a single equation, and accordingly, complete reformulation of QED using eight-spinors.

Many properties of the Dirac field are directly transferable to the eight-spinor electromagnetic theory. For example, we have formulated the Lorentz transformations of eight-spinors in analogy to the Lorentz transformation of Dirac spinors. As a result, the *relativistic* quantum spin operators of light emerge naturally through the generators of Lorentz transformations on eight-component electromagnetic spinors. Therefore, our work provides a well-defined electromagnetic spinor and field operator representations with a natural emergence of the *relativistic* quantum spin structure of light.

The **spinorial Maxwell equation** leads to the formulation of the **Lagrangian density of QED** using eight-spinors. *It enables the generating Lagrangian density of gravity with internal special-unitary-symmetry-based coupling of the quantum fields of the standard model to gravity.* The internal special unitary symmetry of the **generating Lagrangian density of gravity** was shown to enable an elegant derivation of the symmetric **SEM tensors** of the **electromagnetic** and Dirac fields. The possible space-time-dependent form of the quantity I_g in the generating Lagrangian density of gravity was left as a topic of further work. Due to the analogous forms of the underlying theories, the present eight-spinor formalism can be extended to describe the other fundamental interactions and the related fermionic and bosonic fields of the **standard model**. The related fundamental consequence, the **Yang-Mills gauge theory of unified gravity**, is studied in a separate work [see below].

Partanen, M.[1] & Tulkki, J.[2] (May, 2025). Gravity generated by four one-dimensional unitary gauge symmetries and the Standard Model.

Rep. Prog. Phys., 88, 057802 (65pp); https://doi.org/10.1088/1361-6633/adc82e; also at https://arxiv.org/abs/2310.01460.

[1] Photonics Group, Department of Electronics and Nanoengineering, Aalto University, PO Box 13500, 00076 Aalto, Finland.
[2] Engineered Nanosystems Group, School of Science, Aalto University, PO Box 12200, 00076 Aalto, Finland.

Received October 2024; revised March 2025; accepted for publication April 2025.

Abstract

The **Standard Model** of particle physics describes *electromagnetic, weak, and strong interactions*, which are three of the four known fundamental *forces* of nature. The unification of the fourth interaction, *gravity*, with the **Standard Model** has been challenging due to incompatibilities of the underlying theories—*general relativity* and *quantum field theory*. While *quantum field theory* utilizes *compact, finite-dimensional symmetries* associated with the *internal* degrees of freedom of quantum fields, *general relativity* is based on *noncompact, infinite-dimensional external space-time symmetries*. The present work aims at deriving the *gauge theory of gravity* using *compact, finite-dimensional symmetries* in a way that resembles the formulation of the fundamental interactions of the *Standard Model*. For our *eight-spinor representation of the Lagrangian*, we define a quantity, called the *space-time dimension field*, which enables extracting four-dimensional space-time quantities from the eight-dimensional spinors. Four U(1) symmetries of the components of the space-time dimension field are used to derive a gauge theory, called *unified gravity*. The *stress-energy-momentum tensor source term of gravity* follows directly from these symmetries. The metric tensor enters in *unified gravity* through geometric conditions. We show how the teleparallel equivalent of *general relativity* in the Weitzenböck gauge is obtained from *unified gravity* by a gravity-gauge-field-dependent geometric condition.

Unified gravity also enables a gravity-gauge-field-independent geometric condition that leads to an exact description of gravity in the Minkowski metric. *This differs from the use of metric in general relativity, where the metric depends on the gravitational field by definition.* Based on the Minkowski metric, unified gravity allows us to describe gravity within a single coherent mathematical framework together with the quantum fields of all fundamental interactions of the Standard Model. We present the Feynman rules for *unified gravity* and study the renormalizability and radiative corrections of the theory at one-loop order.

The *equivalence principle* is formulated by requiring that the renormalized values of the inertial and gravitational masses are equal. In contrast to previous gauge theories of gravity, *all infinities that are encountered in the calculations of loop diagrams can be absorbed by the redefinition of the small number of parameters of the theory in the same way as in the gauge theories of the Standard Model.* This result and our observation that *unified gravity* fulfills the Becchi–Rouet–Stora–Tyutin (BRST) symmetry and its coupling constant is dimensionless suggest that *unified gravity* can provide the basis for a complete, renormalizable theory of *quantum gravity*.

Contents

...

1. Introduction

Quantum field theory is a theoretical framework, which synthesizes classical field theory, quantum mechanics, and *special relativity*. The *Standard Model* of particle physics arises from this framework through *unitary symmetries of fermionic and Higgs fields* related to invariances of a physical system. [For problems with quantum field theory, see Underwood, T.G. (2023). Quantum Electrodynamics, Vol. II.]

> [A *fermion* is a subatomic particle that follows Fermi–Dirac statistics. Fermions have a *half-integer spin* (spin 1/2, spin 3/2, etc.) and obey the Pauli exclusion principle, so *only one fermion can occupy a particular quantum state at a given time*. These particles include all quarks and leptons and all composite particles made of an odd number of these, such as all baryons and many atoms and nuclei. Some fermions are elementary particles (such as electrons), and some are composite particles (such as protons). Fermions possess conserved baryon or lepton quantum numbers, so what is usually referred to as the spin-statistics relation is, in fact, a spin statistics-quantum number relation. Particles with *integer spin* are *bosons* which obey Bose–Einstein statistics.

In the *Standard Model*, the *Higgs particle* is a massive scalar boson with zero spin, even (positive) parity, no electric charge, and no color charge that couples to (interacts with) mass. It is also very unstable, decaying into other particles almost immediately upon generation. The Higgs field is a scalar field with two neutral and two electrically charged components that form a complex doublet of the weak isospin SU(2) symmetry. Its "Sombrero potential" leads it to take a nonzero value everywhere (including otherwise empty space), which breaks the weak isospin

491

symmetry of the electroweak interaction and, via the ***Higgs mechanism***, gives a rest mass to all massive elementary particles of the Standard Model, including the Higgs boson itself. The existence of the Higgs field became the last unverified part of the *Standard Model* of particle physics, and for several decades was considered "the central problem in particle physics".]

[The *symmetry group* of a geometric object is the group of all *transformations under which the object is invariant*, endowed with the group operation of composition.]

[***Symmetries*** in quantum mechanics describe features of spacetime and particles which are unchanged under some transformation, in the context of quantum mechanics, relativistic quantum mechanics and quantum field theory, and with applications in the mathematical formulation of the standard model and condensed matter physics. In general, symmetry in physics, invariance, and conservation laws, are fundamentally important constraints for formulating physical theories and models. In practice, they are powerful methods for solving problems and predicting what can happen. While conservation laws do not always give the answer to the problem directly, they form the correct constraints and the first steps to solving a multitude of problems.

Many powerful theories in physics are described by ***Lagrangians*** that are invariant under some symmetry transformation groups. They are said to have a ***global symmetry*** when they are invariant under a ***symmetry transformation identically*** performed at every point in the ***spacetime*** in which the physical processes occur. A ***local symmetry*** is one that keeps a property invariant when a ***possibly different symmetry transformation*** is applied at each point of spacetime. It is a stronger constraint and the cornerstone of ***gauge theories***. In fact, a global symmetry is just a local symmetry whose group's parameters are fixed in spacetime

[A ***gauge theory*** is a type of field theory in which the Lagrangian, and hence the dynamics of the system itself, do not change under *local transformations* according to certain smooth families of operations (*Lie groups*). Formally, the Lagrangian is invariant.

The term ***gauge*** refers to any specific mathematical formalism to regulate redundant degrees of freedom in the Lagrangian of a physical system. The transformations between possible *gauges*, called ***gauge transformations***, form a ***Lie group***—referred to as the *symmetry group* or the *gauge group* of the theory. Associated with any *Lie group* is the ***Lie algebra*** of *group generators*. For each ***group generator*** there necessarily arises a corresponding *field* (usually a *vector field*) called the **gauge field. Gauge fields** *are included in the Lagrangian to ensure its invariance under the local group transformations (called gauge invariance).* When such a theory is quantized, the *quanta* of the **gauge fields** are called **gauge bosons**.

A *local symmetry* is one that keeps a property invariant when a possibly different symmetry transformation is applied at each point of spacetime; specifically, a *local symmetry transformation* is parameterized by the spacetime co-ordinates, whereas a *global symmetry* is not. According to the **Standard Model**, there are three *local symmetries* in the universe in which we live, which should be indistinguishable where a certain type of change is introduced. The three *local symmetries* addressed by the *Standard Model* are:

> C-*symmetry* (**charge symmetry**), a universe where every particle is replaced with its *antiparticle*.
> P-*symmetry* (**parity symmetry**), a universe where everything is mirrored along the three physical axes. This excludes *weak interactions*.
> T-*symmetry* (**time reversal symmetry**), a universe where the direction of time is reversed.]

Gauge theories are important as the successful *field theories* explaining the dynamics of elementary particles. **Quantum electrodynamics is an abelian gauge theory** with the **symmetry group U(1)** and has one **gauge field, the** *electromagnetic four-potential, with the photon being the gauge boson.*

The *electromagnetic interaction* is mediated by photons, which have no electric charge. The *electromagnetic tensor* has an *electromagnetic four-potential field* possessing **gauge symmetry**. In **quantum electrodynamics**, the local symmetry group is U(1) and is abelian.

The *unitary group* of degree n, denoted U(n), is the group of n × n **unitary matrices,** with the **group operation of matrix multiplication.** In the simple case, n = 1, the group U(1) corresponds to the **circle group**, consisting of all *complex numbers* with absolute value 1, under multiplication.

The simplest unitary group is U(1), which is just the complex numbers of modulus 1. This one-dimensional matrix entry is of the form:

$$U = e^{-i\theta}$$

in which θ is the parameter of the group. The group is *Abelian* since one-dimensional matrices always commute under matrix multiplication.

> [An *abelian gauge theory* is a quantum field theory that explains the dynamics of elementary particles. It is a *gauge theory* with the symmetry group U(1) and has one *gauge field*, the *electromagnetic four-potential*, with the *photon* being the *gauge boson*. If the symmetry group is *non-commutative*, then the *gauge theory* is referred to as a *non-abelian gauge theory*, the usual example being the *Yang–Mills theory*.]

Lagrangians in *quantum field theory* for complex scalar fields are often invariant under U(1) transformations. If there is a quantum number a associated with the U(1) symmetry, for example baryon and the three lepton numbers in electromagnetic interactions, we have:

$$U = e^{-i\,a\,\theta}.$$

The general form of an element of a U(2) element is parametrized by two complex numbers a and b:

$$U = \begin{pmatrix} a & b \\ -b^* & a^* \end{pmatrix}$$

Important subgroups of each **U(*N*)** are those unitary matrices which have unit determinant: these are called the *special unitary groups* and are denoted SU(N).

The *special unitary group* of degree n, denoted SU(n), is the *Lie group of* **n × n** *unitary matrices* with *determinant* (real) **1**. The matrices of the more general unitary group may have complex determinants with absolute value 1, rather *than real 1 in the special case*. The *group operation* is *matrix multiplication*. The special unitary group is a normal subgroup of the unitary group U(n), consisting of all n × n unitary matrices.

The SU(n) groups find wide application in the *Standard Model* of particle physics, especially SU(2) in the *electroweak interaction* and SU(3) in the *strong interaction* (*quantum chromodynamics*).

The Standard Model is a non-abelian gauge theory with the symmetry group U(1) × SU(2) × SU(3), and has a total of twelve *gauge bosons:* the *photon,* three *weak bosons* and eight *gluons.*

The **Pauli matrices** are the generators of the **special unitary group** in two dimensions, denoted **SU(2)**. Their commutation relation is the same as for orbital angular momentum, aside from a factor of 2. The group **SU(2)** is a simple Lie group consisting on all 2 x 2 matrices of *determinant* 1. The group **SU(3)** is a simple Lie group consisting on all 3 x 3 matrices of *determinant* 1.

The eight **Gell-Mann matrices** are the generators for the **SU(3)** group in *quantum chromodynamics,* which is still of practical importance in nuclear physics. In *quantum chromodynamics*, the local symmetry group is **SU(3)** and is non-abelian. The **strong (color) interaction** is mediated by **gluons**, which can have eight *color charges*. There are **eight gluon field strength tensors** with a corresponding **gluon four potentials field**, each possessing gauge symmetry.]

The **gauge invariance** [*symmetry group* U(1)] of quantum electrodynamics (QED), related to the Abelian phase rotation transformations of fermions, is the most trivial example of such a symmetry [U(1)]. The **Yang–Mills theory** extends the gauge theory to non-Abelian special unitary symmetries, which enable mutually interacting force carriers. It describes the behavior of the other fundamental interactions of the **Standard Model** being at the core of the unification of electrodynamics to **weak** [SU(2)] and **strong** [SU(3)] interactions. The theories of these interactions are called the **electroweak theory** and **quantum chromodynamics** (QCD). The Yang–Mills gauge symmetries operate as matrix transformations of the **Higgs field** and doublets and triplets of fermionic fields. The symmetry groups of the **Standard Model** are all **compact** and **finite dimensional**. A similar compact and finite-dimensional unitary-symmetry-based approach to the description of gravity as a gauge field has remained unknown. Therefore, alternative approaches, such as string theory, loop quantum gravity, asymptotic safety, noncommutative geometry, and causal dynamical triangulation, are being developed. [See Part III.] There are also discussions on whether gravitational interaction should be quantized at all.

The current standard understanding of gravity [???] is based on *general relativity*, which describes gravitational interaction through the *curvature of space-time* that is similarly experienced by all objects. [But see Underwood, T.G. (2024). *Gravity – 2ⁿᵈ Edition*, Gravity in Three and Four Dimensions.] This universality is dictated by Einstein's equivalence principle. It follows that the space-time symmetries of gravity appear fundamentally different from the symmetries of the Standard Model. [No.] In the modern understanding of gravity, put forward by Einstein (1928) and Cartan [1925], it is recognized that, in addition to the curvature of space-time, the gravitational interaction can also be equivalently described by different metric-affine geometries using torsion or nonmetricity. Thus, curvature, torsion, and non-metricity provide three seemingly different representations of the same underlying theory of **general relativity** as special cases of a wider class of gauge theories of gravity.

The theory, where only **torsion** is nonzero, is called the **teleparallel equivalent of general relativity** (TEGR). TEGR is considered to be a natural way to understand the gauge field aspects of gravity since it can be formulated as the **gauge theory** of the translation group, and it enables the Lorentz-covariant definition of the **stress-energy-momentum (SEM) tensor of gravity**. Modifications of TEGR have also been widely studied.

The present work investigates the possibility of deriving the *gauge theory of gravity* by using *compact, finite dimensional gauge symmetry groups* instead of the *noncompact, infinite-dimensional translation gauge group of TEGR*. We use the **eight-spinor formalism** and the associated unitary symmetries in a way that closely resembles the formulation of the interactions of the **Standard Model**.

[**Spinors** (pronounced "spinner") are elements of a complex vector space that can be associated with **Euclidean space**. A spinor transforms linearly when the Euclidean space is subjected to a slight (infinitesimal) rotation, but unlike geometric vectors and tensors, a spinor transforms to its negative when the space rotates through 360°. It takes a rotation of 720° for a spinor to go back to its original state. This property characterizes spinors: spinors can be viewed as the "square roots" of vectors (although this is inaccurate and may be misleading; they are better viewed as "square roots" of sections of vector bundles).

It is also possible to associate a substantially similar notion of **spinor** to Minkowski space, in which case the Lorentz transformations of *special relativity* play the role of rotations.

[**Minkowski space** (or **Minkowski spacetime**) is the main mathematical description of spacetime in the absence of gravitation. It combines inertial space and time manifolds into a four-dimensional model. The model helps show how a spacetime interval between any two events is independent of the inertial frame of reference in which they are recorded. Mathematician Hermann Minkowski developed it from the work of Hendrik Lorentz, Henri Poincaré, and others.

Minkowski space is closely associated with Einstein's theories of special relativity and general relativity and is the most common mathematical structure by which special relativity is formalized. While the individual components in Euclidean space and time might differ due to length contraction and time dilation, in Minkowski spacetime, all frames of reference will agree on the total interval in spacetime between events. *Minkowski space differs from four-dimensional Euclidean space insofar as it treats time differently from the three spatial dimensions.*

In 3-dimensional Euclidean space, the isometry group (maps preserving the regular Euclidean distance) is the ***Euclidean group***. It is generated by rotations, reflections and translations. When time is appended as a fourth dimension, the group of all these transformations is called the ***Poincaré group***. *Minkowski's model follows special relativity, where motion causes time dilation changing the scale applied to the frame in motion and shifts the phase of light.* The group of transformations for ***Minkowski space*** that preserves the spacetime interval (as opposed to the spatial Euclidean distance) is the ***Lorentz group*** (as opposed to the ***Galilean group***).]

In the 1920s physicists discovered that *spinors* are essential to describe the intrinsic angular momentum, or "*spin*", of the electron and other subatomic particles.

Spinors are characterized by the specific way in which they behave under ***rotations***. They change in different ways depending not just on the overall final rotation, but the details of how that rotation was achieved (by a continuous path in the rotation group). In mathematical terms, spinors are described by a double-valued projective representation of the rotation group SO(3).

Although spinors can be defined purely as elements of a representation space of the spin group (or its Lie algebra of infinitesimal rotations), they are typically defined as elements of a vector space that carries a linear representation of the ***Clifford algebra***. The ***Clifford algebra*** is an associative algebra that can be constructed from Euclidean space and its inner product in a basis-independent way. Both the spin group and its Lie algebra are embedded inside the Clifford algebra in a natural way, and in applications the Clifford algebra is often the easiest to work with. A Clifford space operates on a spinor space, and the elements of a spinor space are spinors.

A spinor representation is a linear representation of the Clifford algebra. It is also defined as a spin representation of the orthogonal Lie algebra. Spinors are typically defined as elements of a vector space that carries a linear representation of the Clifford algebra.

In mathematics, a Clifford algebra is an algebra generated by a vector space with a quadratic form, and is a unital associative algebra with the additional structure of a distinguished subspace. The theory of Clifford algebras is intimately connected with the theory of quadratic forms and orthogonal transformations. Clifford algebras have important applications in a variety of fields including geometry, theoretical physics and digital image processing.

The ***eight-spinor formalism*** is a spinorial representation of the *electromagnetic field*. It has advantages over other formulations, such as representing all Maxwell's equations with a single spinorial equation. The eight-spinor formalism is used to describe the electric four-current density in the context of quantum electromagnetism. It involves the charge-current spinor, which is a four-vector-type eight-spinor. The formalism is used for complex-valued momenta and is related to spinor helicity. It simplifies complex amplitude computations by leveraging symmetry and dimensional analysis.]

Thus, the goal of the present work is to bring the ***gauge theory of gravity*** as close as possible to the *gauge theory formulation* of the ***Standard Model***, and thereby to contribute to improved understanding of the relations of all four fundamental interactions of nature.

[***Gauge theories*** are quantum theories of a vector field whose interactions with each other and with other fields follows from a local symmetry.]

Many authors have approached the problem of unifying the Standard Model and gravity by attempting to reformulate space-time symmetries in a way compatible with the gauge symmetries of the ***Standard Model***. The difference between external space-time symmetries and symmetries related to internal degrees of freedom, which govern the dynamics of quantum fields via creations and annihilations of field quanta, however, represents a challenge for this gauge theory approach to gravity especially at high energies

In previous literature, there are at least two different ways to interpret whether a symmetry is internal or external.

The first interpretation is based on observing how the given symmetry transformation operates on objects in the Lagrangian density.

[The ***Lagrangian density*** is a function that describes the dynamics of a field in physics. It measures the distribution of the Lagrangian over space and is crucial in formulating the equations of motion for fields in both classical and quantum field theories. In the context of field theories, the Lagrangian density is used to derive the ***Euler-Lagrange equations***, which govern the behavior of the fields.]

Internal symmetries operate via scalar and matrix multiplications, which do not depend on how the fields vary around the given space-time point. Examples are the multiplication of the Dirac field by a complex phase factor in QED and the color and weak isospin rotations of fermion field triplets and doublets and the Higgs field doublet in the strong and weak interactions of the Standard Model.

External symmetries, such as ***space-time translations***, are generated by differential operators.

The ***second interpretation*** of determining the internal or external nature of a symmetry is based on the well-known Coleman Mandula theorem. This theorem states that the symmetry group of a theory that can be described by an S-matrix is locally isomorphic to the direct product of the Poincaré group and internal symmetry groups. Therefore, any symmetries associated with the Poincaré symmetry structure of the space-time are clearly not internal symmetries from the point of view of the Coleman–Mandula theorem. Consequently, symmetries associated with gravity can be interpreted internal only according to the first interpretation discussed above.

The four ***U(1) symmetries of gravity***, to be revealed in the present work, are based on the eight-spinor representation of the Lagrangian density, and they are ***internal*** according to the ***first interpretation*** above. To avoid misunderstanding, we, however, call these symmetries the ***U(1) symmetries of gravity*** instead of internal symmetries.

The main challenge of the conventional ***gauge theory*** approach of ***gravity***, which emerges from the nature of the ***space-time symmetries***, is the ***non-renormalizability*** of the resulting theory without an infinite number of counter-terms. In contrast, all gauge theories of the Standard Model are ***renormalizable***, which means that their ultraviolet divergences can be reabsorbed into the redefinition of a finite number of parameters. The renormalization procedure then leads to the running of the coupling constants as a function of the energy scale. ***The non-renormalizability of conventional theories of gravity makes it impossible to use the quantized gauge theory of gravity to make predictions at high energies***. However, the quantum field theory treatment of general relativity can be argued to be successful as a low energy effective field theory. The main idea of the ***effective field theory*** is that the low-energy degrees of freedom organize themselves as quantum fields in such a way that one can make predictions without knowledge of the full high-energy theory. This also indicates that ***fundamental breakthroughs are needed to formulate a predictive quantum theory of gravity applicable to all energy scales***. Such a theory can finally answer ultimate questions on the structure of the Universe in circumstances of extremely high energy densities, such as those inside black holes and at the possible beginning of time.

On the experimental side, ***general relativity*** has so far passed all tests planned to probe gravitational interaction. The well-known classical tests involve the precession of the perihelion of Mercury, the bending of light by the Sun, and the gravitational redshift of light. New experiments are continuously developed. The waveforms recorded in recent measurements of gravitational waves are in good agreement with *general relativity*. The first image of a black hole is consistent with the shadow of a Kerr black hole predicted by the theory. Other recent measurements involve the study of the effect of gravity on the motion of antimatter and the measurement of gravity with milligram-scale masses and with bending beam resonators.

In a recent work [see above], we have reformulated QED based on an eight-component spinorial representation of the electromagnetic field. The eight-spinor formulation reveals a profound connection between the unitary symmetries of the Lagrangian density in the eight-dimensional spinor space and the symmetric SEM tensor of the Dirac and electromagnetic fields. Since the SEM tensor is the well-known source of the gravitational field in general relativity, it becomes obvious that the gauge theory obtained by studying the appropriate Lagrangian density under unitary symmetry transformations describes gravitational interaction. In such a theory, the SEM tensor appears analogously to the pertinent source terms of the quantum fields of the Standard Model. This is the basis for the development of the gauge theory of gravity, called **unified gravity**, in the present work. The present work also essentially generalizes some of the key mathematical concepts of our preliminary study.

In this work, we extend the eight-spinor formulation to cover the full Standard Model and derive the gauge theory of unified gravity. Our theory is based on introducing the concept of the ***space-time dimension field***, a geometric object which, by definition, enables extracting four-dimensional space-time quantities from the eight-dimensional spinor space. Introducing the space-time dimension field enables identifying compact, finite-dimensional unitary symmetries in the Lagrangian density, which can be utilized to form a gauge theory in analogy to how gauge symmetries are used in the ***Standard Model***. While the interaction symmetries of the Standard Model are based on symmetry transformations of fermionic and Higgs fields as discussed above, our extension of the Standard Model to cover gravity is based on four U(1) symmetry transformations, which act on the components of the space-time dimension field. Therefore, the symmetries of the space-time dimension field form a hierarchy separate from the symmetries of the ***Standard Model***.

The U(1) symmetry transformations of the space-time dimension field components allow us to couple the quantum fields of the ***Standard Model*** to a tensor gauge field in a way that is formally analogous to the gauge couplings of the fields in the ***electromagnetic***, ***weak***, and ***strong interactions***. Once the gauge field and its Lagrangian density are introduced, we obtain ***unified gravity*** and its dynamical equations, which are shown to describe the behavior of the gravitational field. The space-time metric tensor enters ***unified gravity*** through geometric conditions. We are allowed to use geometric conditions, in which the space-time metric tensor is independent of the gravity gauge field. This leads us to study *unified gravity* in the Minkowski metric (UGM) in an exact way. This differs from the use of metric in general relativity, where the metric depends on the gravitational field by definition, and whose effective quantization requires expansion of the metric about the flat or smooth background with an assumption that the deviation is small. In this respect, the conventional translation gauge formulation of *teleparallel equivalent of general relativity (TEGR)* is not significantly different from *general relativity* since a similar expansion of the tetrad is needed. Alternatively, in *unified gravity*, we are allowed to use geometric conditions, in which the space-time metric tensor depends on the gravity gauge field, as in *general relativity* and in TEGR.

We show that, within a particular Weitzenböck gauge fixing approach, the representation of *unified gravity* becomes equivalent to the known representation of **TEGR** in the Weitzenböck gauge **(TEGRW)**, where the teleparallel spin connection vanishes. This shows that *unified gravity* is in perfect agreement with the known nonlinear field equations of *general relativity*.

The harmonic gauge fixing of *unified gravity* in the *Minkowski metric* **(UGM)** is analogous to the Feynman gauge fixing of QED. It enables us to determine the Feynman rules for unified gravity. These rules are also written in a more general form for an arbitrary gauge fixing parameter. As a gauge theory similar to those of the Standard Model, *unified gravity* is subject to quantization. The quanta of the gauge field, the *gravitons*, are spin-2 tensor bosons. These quanta are to be added in the spectrum of the known elementary particles extending the Standard Model to describe gravity. However, *this can be done only after the non-renormalizability problem of quantum gravity has been fully resolved* making the quantum theory predictive at all energies. In this work, we study the renormalizability and the radiative corrections of unified gravity in the Minkowski metric (UGM) at *one-loop order*. We show that, in contrast to previous gauge theories of gravity, all infinities that are encountered in the calculations of loop diagrams can be absorbed by the redefinition of the small number of parameters of the theory in the same way as in the gauge theories of the Standard Model. Furthermore, *Einstein's equivalence principle* is formulated by requiring that the renormalized values of the gravitational and inertial masses are equal. The direct relation between the equivalence principle and renormalization is obviously absent in previous studies of the equivalence principle in the quantum regime. Based on the dimensionless coupling constant and the fulfillment of the Becchi–Rouet–Stora–Tyutin (BRST) symmetry, we expect that *unified gravity* is renormalizable to all loop orders. *The complete proof extending our one-loop results to all loop orders is, however, left as a topic of further works.*

501

Gravity couples to all fields and matter. Therefore, one cannot exclude any field or matter from the complete dynamical description of gravity. However, to make our theory of *unified gravity* more transparent and easier to understand for non-expert readers, we limit, in the first part of this work, our study to the coupling between gravity and electrodynamics. *The system of the electromagnetic field, Dirac electron–positron field, and the gravitational field provides all the insight needed for obtaining a unified description of gravity in a coherent frame work with the other known fundamental forces of nature.* The extension of *unified gravity* to cover the full **Standard Model** is presented in the later part of the present work.

This work is organized as follows:

Section 2 describes the theoretical concepts and conventions, including the eight-spinor formulation, ….

The kernel matrices, the space-time dimension field, and the generating Lagrangian density of gravity are presented in **Section 3**. We also formulate the equivalence principle in *unified gravity*. Furthermore, the generating Lagrangian density of gravity is shown to be equal to the Lagrangian density of QED in flat space-time.

Section 4 discusses the symmetries of the generating Lagrangian density of gravity and derives the conservation law of the SEM tensor of the Dirac and electromagnetic fields.

The gravity gauge field is introduced through the gauge-covariant derivative in **Section 5**. The gravity gauge field strength tensor is also discussed.

In **Section 6**, we present the locally gauge-invariant Lagrangian density of *unified gravity*. This section also provides the scaled representation of *unified gravity* to allow easier comparison with the gauge theories of the **Standard Model**.

Section 7 discusses the representation of *unified gravity* in the Minkowski metric in unified gravity in the Minkowski metric **(UGM)** and derives the corresponding dynamical equations.

Section 8 presents the Feynman rules and their selected applications in *unified gravity* in the Minkowski metric **(UGM)**.

Section 9 proves the renormalizability of *unified gravity* at one-loop order and determines the values of the related renormalization factors using the conventional on-shell renormalization scheme and dimensional regularization.

In **Section 10**, radiative corrections to the Coulomb and Newtonian potentials and to the anomalous magnetic moment of the electron are calculated to exemplify the calculation of radiative corrections using ***unified gravity***.

Section 11 derives teleparallel equivalent of ***general relativity*** in the Weitzenböck gauge **(TEGRW)** from ***unified gravity*** by applying the Weitzenböck gauge fixing approach. The dynamical equations for the Dirac, electromagnetic, and gravitational fields in TEGRW are also presented.

The extension of the theory to cover all quantum fields of the Standard Model is developed in **Section 12**.

The results are discussed and compared with previous theories in **Section 13**.

Finally, conclusions are drawn in **Section 14**.

2. Theoretical concepts and conventions

This section presents the theoretical concepts and conventions used in the present work. These include the index conventions, tetrads, metric tensors, connection coefficients, torsions, contortions, derivative operators, and the eight-spinor formalism ….

2.1. Index conventions, tetrads, and metric tensors

The Latin indices $a,b,c,d \in \{0,x,y,z\}$ in this work range over four Cartesian Minkowski space-time coordinates. The Latin indices starting from i range over the three spatial dimensions, i.e. $i,j,k \in \{x,y,z\}$, or over other values separately specified. The Greek indices denote the general space time indices, which range over the four general space-time dimensions $\mu,\nu,\rho,\sigma \in \{x^0,x^1,x^2,x^3\}$. The Latin Cartesian Minkowski space-time indices are lowered and raised by the Cartesian Minkowski metric tensor η_{ab} and its inverse η^{ab}. We use the sign convention $\eta^{00} = 1$ and $\eta^{xx} = \eta^{yy} = \eta^{zz} = -1$. The Greek general space-time indices are raised and lowered by the general space-time metric $g_{\mu\nu}$ and its inverse $g^{\mu\nu}$. The determinant of the general space-time metric tensor is denoted by $g = \det(g_{\mu\nu})$.

The Latin Cartesian Minkowski space-time indices and any Greek general space-time indices can be converted into each other by the tetrad $e_a{}^\mu$ and the inverse tetrad $e^a{}_\mu$. In *flat* space-time, i.e. in the absence of the gravitational field, and more generally, in a Minkowski manifold with the torsion and the spin connection equal to zero, the tetrad and the inverse tetrad are given by $e_{\circ a}{}^\mu = \partial^\mu x_a$ and $e_{\circ}{}^a{}_\mu = \partial_\mu x^a$, where x^a is a four-vector of Minkowski space-time coordinates, e.g. $x^a = (ct,x,y,z)$ in Cartesian coordinates and

$x^a = (ct, r\sin\theta\cos\phi, r\sin\theta\sin\phi, r\cos\theta)$ in spherical coordinates. Respectively, we use the symbol $e\bullet_a{}^\mu$ to highlight the tetrad of teleparallel equivalent of general relativity in the Weitzenböck gauge **(TEGRW)**, discussed in Section 11. The generic tetrad symbol $e_a{}^\mu$ is used to indicate that an equation is independent of the definition of the tetrad and one is not restricted to using the tetrad of the Minkowski manifold or that of TEGRW. In any definition of the tetrad, the general space-time met ric tensor is given in terms of the Cartesian Minkowski metric tensor and inverse tetrads as $g_{\mu\nu} = \eta_{ab}e^a{}_\mu e^b{}_\nu$. In the special case of the Minkowski space-time in Cartesian coordinates, used in **unified gravity** in the Minkowski metric **(UGM)**, the Latin and Greek indices are identical and the tetrad and the metric tensor are trivial as $e\circ_\mu{}^\nu = \delta^\nu{}_\mu$ and $g_{\mu\nu} = \eta_{\mu\nu}$, where $\delta^\nu{}_\mu$ is the Kronecker delta. Throughout this work, with a few exceptions, we use the Einstein convention for the summation over repeated indices. Exceptions to the summation convention are separately discussed and indicated by parentheses around the indices.

2.2. Connection coefficients, torsions, and contortions

The *Levi–Civita connection coefficients* of standard *general relativity*, i.e. the Christoffel symbols $\Gamma\circ^\mu{}_{\sigma\nu}$, are associated with all Greek space-time indices. They are used to define the Levi–Civita coordinate-covariant derivative as presented in section 2.3. The Christoffel symbols can always be written in terms of the metric tensor ...

$$\Gamma\circ^\mu{}_{\sigma\nu} = \tfrac{1}{2}\, g^{\mu\rho}(\partial_\sigma g_{\rho\nu} + \partial_\nu g_{\rho\sigma} - \partial_\rho g_{\sigma\nu}). \tag{1}$$

The Christoffel symbol contraction $\Gamma\circ^\sigma{}_{\rho\sigma}$, used in many equations of this work, can be given in terms of the determinant of the metric tensor as

$$\Gamma\circ^\sigma{}_{\rho\sigma} = 1/\sqrt{-g}\; \partial_\rho(\sqrt{-g})\,. \tag{2}$$

In a Minkowski manifold with the torsion and the spin connection equal to zero, the Christoffel symbols can be written in terms of tetrads as

$$\Gamma\circ^\mu{}_{\sigma\nu} = e\circ_a{}^\mu \partial_\nu e\circ^a{}_\sigma = -e\circ^a{}_\sigma \partial_\nu e\circ_a{}^\mu. \tag{3}$$

The condition of zero torsion tensor, $T\circ^\rho{}_{\mu\nu}$, is written as

$$T\circ^\rho{}_{\mu\nu} = \Gamma\circ^\rho{}_{\mu\nu} - \Gamma\circ^\rho{}_{\nu\mu} = 0. \tag{4}$$

The related contortion tensor, $K\circ^{\mu\nu\rho}$, is trivially zero as

$$K\circ^{\mu\nu\rho} = 1/2\,(T\circ^{\nu\mu\rho} + T\circ^{\mu\rho\nu} - T\circ^{\rho\nu\mu}) = 0. \tag{5}$$

504

In teleparallel equivalent of general relativity in the Weitzenböck gauge (**TEGRW**), where the spin connection is zero but the torsion is generally nonzero, the relation of the Christoffel symbols is given by

$$\Gamma^{\circ\mu}_{\sigma\nu} = \Gamma^{\bullet\mu}_{\sigma\nu} - K^{\bullet\mu}_{\sigma\nu}. \qquad (6)$$

...

2.4. Eight-spinor formalism

The *eight-spinor formulation* of QED, presented above, is not the first spinorial representation of the electromagnetic field. However, the *eight-spinor formalism* has certain advantages over other known formulations. For example, it enables the representation of all Maxwell's equations by a single spinorial equation. As discussed below, the term eight-spinor is used in a meaning that covers different types of eight-component physical quantities whose certain components can be defined to be zero and whose Lorentz transformation properties depend on the type of the eight-spinor. Here we briefly review the key quantities of the eight-spinor formulation.

The *eight-spinor theory* is formulated in terms of four 8×8 bosonic gamma matrices γ^a_B and $\gamma^5_B = i\gamma^0_B\gamma^x_B\gamma^y_B\gamma^z_B$ [55]. These matrices and their equivalence transformations are explicitly presented in section 1 of the supplementary material. The matrices γ^a_B satisfy the **Dirac algebra**, i.e. the Clifford algebra $C\ell_{1,3}(C)$. The defining property of the *Dirac algebra* of γ^a_B is *the anticommutation relation* $\{\gamma^a_B,\gamma^b_B\} = 2\eta^{ab}I_8$, where I_8 is the 8×8 identity matrix.

The conventional *4 × 4 Dirac gamma matrices* are denoted by γ^a_F. Here we use the subscript F in distinction to the subscript B used for bosonic gamma matrices discussed above. The Clifford algebra relation for the Dirac gamma matrices is given by $\{\gamma^a_F,\gamma^b_F\} = 2\eta^{ab}I_4$, where I_4 is the 4×4 identity matrix. In some places, we use the Feynman slash notation for the Dirac gamma matrices to write equations more compactly. For example, one can write $\gamma^\rho_F\partial_\rho = \partial\!\!\!/$ and $\gamma^\rho_F p_\rho = p\!\!\!/$.

...

3. Space-time dimension field and the generating Lagrangian density of gravity

In this section, we present the concept of the *space-time dimension field* and use it to formulate the *generating Lagrangian density of gravity*. Together with the *symmetry transformation* in section 4, the *space-time dimension field* forms the foundations of *unified gravity*. While the Standard Model symmetries are associated with *fermionic fields*, their doublets and triplets, and the Higgs field doublet, the symmetries of *unified gravity* are associated with the *components of the space-time dimension field* as discussed

505

in section 4. It follows from these foundations that ***unified gravity*** does not explicitly include internal degrees of freedom of quantum fields and their symmetry properties in the description of gravity. This limitation is common to all space-time-based formulations of general relativity.

In flat space-time, the ***generating Lagrangian density of gravity*** is equivalent to the known Lagrangian density of QED as shown in section 3.7. Due to this equivalence, the space-time dimension field is a mathematical, precisely defined tool, which is associated with space-time but does not assume any specific definition of the space-time metric tensor. The relation to the space-time metric tensor is obtained only after applying separate geometric conditions, to be discussed in this section and in sections 7.1 and 11.1. The gravitational interaction, which determines the structure of space-time in general relativity and teleparallel equivalent of general relativity (TEGR), is shown to arise from symmetries of the space-time dimension field as discussed in section 4 and in sections thereafter.

We have ended up at the concept of the ***space-time dimension field*** in a process of trial and error. The main goal has been to enable the gauge-theory description of gravity using a compact, finite-dimensional gauge group, similar to those of the fundamental interactions of the Standard Model. Without an explicit, separate quantity, like the space-time dimension field, the Lagrangian density of QED satisfies only the well-known U(1) phase rotation symmetry of QED and external space-time symmetries, which notoriously have a ***noncompact, infinite-dimensional gauge group. Therefore, the space-time dimension field is added in the Lagrangian density to enable additional symmetries***. The ***SEM tensor is the well-known source of the gravitational field in general relativity***. Thus, ***our goal is to construct the generating Lagrangian density***, for which the variation with respect to the symmetry transformation parameters gives the ***SEM tensor***. Through this procedure, the SEM tensor appears analogously to the pertinent source terms of the quantum fields of the ***Standard Model***. The eight-spinor representations of the SEM tensors of the Dirac and electromagnetic fields in terms of the kernel matrices, found above, provided us the hint to use the kernel matrices, discussed in section 3.1, in the construction of the ***space-time dimension field***. After investigating several alternatives, we have found a single meaningful definition of the space-time dimension field, to be given below.

We finally point out that ***the fundamental novelty of unified gravity is the addition of the space-time dimension field to the Lagrangian density of the Standard Model*** without introducing any free physical parameters. Thus, the fundamental hypothesis of ***unified gravity*** should sooner be considered to be the symmetry properties of the ***space-time dimension field*** and the way how it appears in the ***generating Lagrangian density of gravity***.

...

13.5. Remaining challenges in quantum gravity

13.5.1. Lack of experimental data on quantum gravity.

The development of the Standard Model has been deeply intertwined with experimental discoveries, which have played a crucial role in shaping and validating the theory. Initially, the Standard Model emerged as a theoretical framework to describe the fundamental particles and their interactions. However, it was through a series of groundbreaking experiments that its predictions were tested, and the Standard Model itself obtained its present form. In this respect, ***the lack of experimental data on quantum gravity, due to the weakness of the gravitational interaction, has so far been a notable challenge for the development of a well-functioning theory of quantum gravity***. Experimental advances can, however, take place in the following years. Any experiments must be planned carefully to clearly distinguish between the classical and quantum effects. For example, the gravitational Aharonov–Bohm effect has already been measured, but it can be explained semi-classically using the classical gravitational potential without requiring the full quantization of the gravitational field.

13.5.2. Possibility of divergences in high-order loop diagrams.

In the conventional ***effective field theory approach*** to ***quantum gravity***, the loop diagrams are problematic because they lead to divergences that cannot be renormalized in the usual sense. In the case of ***unified gravity***, one can speculate with the possibility that some high-order loop diagrams could not be renormalized. Even though the complete proof of the renormalizability of ***unified gravity*** to all loop orders remains a topic for future work, we have strong arguments against this scenario. One of these arguments is the fact that the known gauge theories of the Standard Model are based on similar unitary or special unitary groups and also have dimensionless coupling constants, and they have turned out to be renormalizable. Another, closely related, argument is the Becchi–Rouet–Stora–Tyutin (BRST) symmetry invariance.

13.5.3. Nonperturbative regime of the theory.

One challenge for ***unified gravity*** is provided by its nonperturbative regime at high energies. Previous quantum field theories, such as QCD at low energies, have shown that the nonperturbative regime is theoretically challenging to approach. This is primarily because, in the nonperturbative regime, the coupling constant is large, and perturbative methods fail. Therefore, alternative approaches, such as lattice gauge theory simulations or functional methods, are required. These methods are computationally very intensive. Even after complete unification of all fundamental interactions of nature, a comprehensive

507

understanding of the nonperturbative regimes of quantum field theories may still remain one of the most challenging and important goals in theoretical physics.

13.5.4. Eventual fundamental limitations of unified gravity.

Unified gravity is a powerful and mathematically transparent framework with the potential of being an ultimate quantum theory of all fundamental forces of nature. In the end, a physical theory must be grounded on experimental verification. *Unified gravity* does not contain a single free parameter that has not been measured in previous experiments. Since *unified gravity* contains the *Standard Model*, it is equally predictive in related phenomena. The classical limit of *unified gravity* is equivalent to *general relativity*, and thus, consistent with the observations on gravitational interaction. However, *the predictability of unified gravity in the explanation of quantum gravity phenomena is yet to be proven by future experiments*.

14. Conclusion

We have investigated the possibility of formulating a *gauge theory of gravity* using compact, finite-dimensional symmetry groups instead of the noncompact, infinite-dimensional translation gauge group of conventional theories of gravity. The resulting gauge theory, *unified gravity*, was made possible without a single free parameter by introducing the concept of the *space-time dimension field* and utilizing the recent *eight-spinor formulation of* QED [above] extended to cover the full Standard Model. Four U(1) symmetries of the components of the *space-time dimension field* lead to *unified gravity* in a way that resembles the gauge theories of the Standard Model. Thus, our theory differs from conventional theories of gravity, which are typically based on the *external* translation symmetry of the Lagrangian density. Compactness of the gauge group of unified gravity represents a fundamental change in the understanding of the structure of space-time and the emergence of gravity. The key steps in the emergence of unified gravity are summarized in table 5.

... (*Page 60 in article format; page 100 in this format.*)

508

PART III   Supersymmetry.

Supersymmetry is a theoretical framework in physics that suggests the existence of a *symmetry* between particles with *integer spin* (*bosons*) and particles with *half-integer spin* (*fermions*). It proposes that for every known particle, there exists a partner particle with different *spin* properties. *There have been multiple experiments on supersymmetry that have failed to provide evidence that it exists in nature.* If evidence is found, *supersymmetry* could help explain certain phenomena, such as the nature of *dark matter* and the *hierarchy problem* in particle physics.

[***There is no experimental evidence that either supersymmetry or misaligned supersymmetry holds in our universe***, and *many physicists have moved on from supersymmetry and string theory entirely due to the non-detection of supersymmetry at the Large Hadron Collider* (LHC).]

A *supersymmetric theory* is a theory in which the *equations for force* and the *equations for matter* are identical. In theoretical and mathematical physics, any theory with this property has the *principle of supersymmetry*. Dozens of supersymmetric theories exist. In theory, *supersymmetry* is a type of *spacetime symmetry* between two basic classes of particles: *bosons*, which have an integer-valued *spin* and follow Bose–Einstein statistics, and *fermions*, which have a half-integer-valued *spin* and follow Fermi–Dirac statistics. [Haber, H. *Supersymmetry. Part I (Theory).* Reviews, Tables and Plots. Particle Data Group.] *The names of bosonic partners of fermions are prefixed with s-, because they are scalar particles.*

In *supersymmetry*, each particle from the class of *fermions* would have an associated particle in the class of *bosons*, and vice versa, known as a *superpartner*. *The spin of a particle's superpartner is different by a half-integer.* For example, if the *electron* exists in a *supersymmetric theory*, then there would be a particle called a *selectron* (superpartner electron), a *bosonic* partner of the *electron*. In the simplest supersymmetry theories, with perfectly "unbroken" supersymmetry, each pair of *superpartners* would share the same *mass* and *internal quantum numbers* besides *spin*. *More complex supersymmetry theories have a spontaneously broken symmetry*, allowing *superpartners* to differ in *mass*.

Supersymmetry has various applications to different areas of physics, such as quantum mechanics, statistical mechanics, quantum field theory, condensed matter physics, nuclear physics, optics, stochastic dynamics, astrophysics, quantum gravity, and cosmology. *Supersymmetry* has also been applied to *high energy physics, where a supersymmetric extension of the Standard Model is a possible candidate for physics beyond the Standard*

Model. However, *no supersymmetric extensions of the Standard Model have been experimentally verified.*

History

A *supersymmetry* relating *mesons* and *baryons* was first proposed, in the context of hadronic physics, by Hironari Miyazawa in 1966. This *supersymmetry* did not involve *spacetime*, that is, it concerned *internal symmetry*, and was broken badly. Miyazawa's work was largely ignored at the time. [Miyazawa, H. (1966). Baryon Number Changing Currents. *Prog. Theor. Phys.*, 36, 6, 1266–76; https://doi.org/10.1143/PTP.36.1266; (1968). Spinor Currents and Symmetries of Baryons and Mesons. *Phys. Rev.*, 170, 5, 1586–90; https://doi.org/10.1103/PhysRev.170.1586.]

J. L. Gervais and B. Sakita (in 1971), Yu. A. Golfand and E. P. Likhtman (in 1973-4), and D. V. Volkov and V. P. Akulov (1972), independently rediscovered *supersymmetry* in the context of *quantum field theory, a radically new type of symmetry of spacetime and fundamental fields,* which establishes a relationship between *elementary particles* of different quantum nature, *bosons* and *fermions,* and *unifies spacetime and internal symmetries of microscopic phenomena.* [Gervais, J.-L. & Sakita, B. (1971). Field theory interpretation of supergauges in dual models. *Nuclear Physics B.* 34, 2, 632–9. https://doi.org/10.1016/0550-3213(71)90351-8; Volkov, D.V. & Akulov, V.P. (1973). Is the neutrino a goldstone particle? *Physics Letters B,* 46, 1, 109–10; https://doi.org/10.1016/0370-2693(73)90490-5; Akulov, V. P. & Volkov, D. V. (1974). Goldstone fields with spin 1/2. *Theoretical and Mathematical Physics,* 18, 1, 28–35. https://doi.org/10.1007/BF01036922.]

Supersymmetry with a consistent Lie-algebraic graded structure on which the Gervais–Sakita rediscovery was based directly first arose in 1971 in the context of an early version of *string theory* by Pierre Ramond, John H. Schwarz and André Neveu. [Ramond, P. (1971). Dual Theory for Free Fermions. *Phys. Rev. D,* 3, 10, 2415–8; https://doi.org/10.1103/PhysRevD.3.2415; Neveu, A.; Schwarz, J. H. (1971). Factorizable dual model of pions. *Nuclear Physics B,* 31, 1, 86–112; https://doi.org/10.1016/0550-3213(71)90448-2.]

In 1974, Julius Wess and Bruno Zumino identified the characteristic *renormalization* features of *four-dimensional supersymmetric field theories,* which identified them as remarkable *quantum field theories,* and they and Abdus Salam and their fellow researchers introduced early particle physics applications. The mathematical structure of *supersymmetry* (graded Lie superalgebras) has subsequently been applied successfully to other topics of physics, ranging from nuclear physics, critical phenomena, quantum mechanics to statistical physics, and *supersymmetry* remains a vital part of many proposed theories in many branches of physics.

In particle physics, *the first realistic supersymmetric version of the Standard Model was proposed in 1977 by Pierre Fayet and is known as the Minimal Supersymmetric Standard Model.* [Fayet, P. (1977). Spontaneously broken supersymmetric theories of weak, electromagnetic and strong interactions. *Physics Letters B*, 69, 4, 489-94; https://doi.org/10.1016/0370-2693(77)90852-8.]

It was proposed to solve, amongst other things, the *hierarchy problem*.

[The *hierarchy problem* is the problem concerning the large discrepancy between aspects of the *weak force* and *gravity*. There is no scientific consensus on why, for example, the *weak force* is 10^{24} times stronger than *gravity*. A *hierarchy problem* occurs when the fundamental value of some physical parameter, such as a *coupling constant* or a *mass*, in some Lagrangian is vastly different from its effective value, which is the value that gets measured in an experiment. This happens because the effective value is related to the fundamental value by a prescription known as *renormalization*, which applies corrections to it [???].]

Supersymmetry was coined by Abdus Salam and John Strathdee in 1974 as a simplification of the term *super-gauge symmetry* used by Wess and Zumino, although Zumino also used the same term at around the same time. The term *supergauge* was in turn coined by Neveu and Schwarz in 1971 when they devised *supersymmetry* in the context of *string theory*.

Extension of possible symmetry groups

One reason that physicists explored *supersymmetry* is because it offers an extension to the more familiar symmetries of *quantum field theory*. These *symmetries* are grouped into the *Poincaré group* and *internal symmetries* and the Coleman–Mandula theorem showed that under certain assumptions, *the symmetries of the S-matrix must be a direct product of the Poincaré group with a compact internal symmetry group or if there is not any mass gap, the conformal group with a compact internal symmetry group.* In 1971 Golfand and Likhtman were the first to show that the Poincaré algebra can be extended through introduction of four *anticommuting spinor generators* (in four dimensions), which later became known as *supercharges*. In 1975, the Haag–Łopuszański–Sohnius theorem analyzed all possible superalgebras in the general form, including those with an extended number of the *supergenerators* and *central charges*. This extended *super-Poincaré algebra* paved the way for obtaining a very large and important class of *supersymmetric field theories*.

The supersymmetry algebra

Traditional symmetries of physics are generated by objects that transform by the tensor *representations* of the Poincaré group and *internal symmetries. Supersymmetries, however, are generated by objects that transform by the spin representations.* According to the *spin-statistics theorem, bosonic fields commute while fermionic fields anti-commute.* Combining the two kinds of fields into a single algebra requires the introduction of a Z_2-grading under which the *bosons* are the even elements and the *fermions* are the odd elements. Such an algebra is called a *Lie superalgebra.*

There are *representations* of a *Lie superalgebra* that are analogous to *representations* of a *Lie algebra*. Each *Lie algebra* has an associated *Lie group* and a *Lie superalgebra* can sometimes be extended into *representations* of a *Lie supergroup.*

Supersymmetric quantum mechanics

Supersymmetric quantum mechanics adds the *supersymmetry* superalgebra to *quantum mechanics* as opposed to *quantum field theory. Supersymmetric quantum mechanics* often becomes relevant when studying the dynamics of supersymmetric *solitons,* and due to the simplified nature of having *fields* which are only functions of *time* (rather than *space-time*), a great deal of progress has been made in this subject and it is now studied in its own right.

> [A *soliton* is a nonlinear, self-reinforcing, localized *wave packet* that is strongly stable, in that it preserves its shape while propagating freely, at constant velocity, and recovers it even after collisions with other such localized *wave packets.* Its remarkable stability can be traced to a balanced cancellation of nonlinear and dispersive effects in the medium. (*Dispersive effects* are a property of certain systems where the speed of a wave depends on its frequency.) *Solitons* provide stable solutions of a wide class of weakly nonlinear dispersive partial differential equations describing physical systems.]

Supersymmetric quantum mechanics involves pairs of Hamiltonians which share a particular mathematical relationship, which are called *partner Hamiltonians.* (The *potential energy* terms which occur in the Hamiltonians are then known as *partner potentials.*) An introductory theorem shows that for every *eigenstate* of one Hamiltonian, its *partner Hamiltonian* has a corresponding *eigenstate* with the same *energy.* This fact can be exploited to deduce many properties of the *eigenstate spectrum.*

Supersymmetry in quantum field theory

In *quantum field theory, supersymmetry* is motivated by solutions to several theoretical problems, for generally providing many desirable mathematical properties, and for ensuring sensible behavior at high energies. *Supersymmetric quantum field theory* is often much easier to analyze, as many more problems become mathematically tractable. When *supersymmetry* is imposed as a *local symmetry*, Einstein's theory of *general relativity* is included automatically, and the result is said to be a *theory of supergravity*.

While *supersymmetry has not been discovered at high energy, supersymmetry* was found to be effectively realized at the intermediate energy of *hadronic* physics where *baryons* and *mesons* are *superpartners*. An exception is the *pion* that appears as a zero mode in the *mass* spectrum and thus protected by the *supersymmetry*: it has no *baryonic* partner.

Supersymmetry in string theory

Supersymmetry is an integral part of *string theory*, a possible theory of everything. There are two types of *string theory, supersymmetric string theory* or *superstring theory*, and *non-supersymmetric string theory*. By definition of *superstring theory, supersymmetry* is required at some level. However, even in *non-supersymmetric string theory*, a type of *supersymmetry* called *misaligned supersymmetry* is still required in the theory in order to ensure no physical *tachyons* appear.

[A *tachyon* is a hypothetical particle that always travels faster than light.]

Any *string theories* without some kind of *supersymmetry*, such as *bosonic string theory* and the $SU(16)$, and *heterotic string theories*, will have a *tachyon* and therefore the spacetime vacuum itself would be unstable and would decay into some tachyon-free string theory usually in a lower spacetime dimension.

There is no experimental evidence that either supersymmetry or misaligned supersymmetry holds in our universe, and *many physicists have moved on from supersymmetry and string theory entirely due to the non-detection of supersymmetry at the Large Hadron Collider* (LHC). Despite the null results for *supersymmetry* so far, some particle physicists have nevertheless moved to *string theory* in order to resolve the *naturalness crisis for certain supersymmetric extensions of the Standard Model.*

Supersymmetric extensions of the Standard Model

In *particle physics*, a *supersymmetric extension* of the *Standard Model* is a possible candidate for undiscovered particle physics, and seen by some physicists as an elegant solution to many current problems in particle physics if confirmed correct, which could resolve various areas where current theories are believed to be incomplete and where limitations of current theories are well established. If a *supersymmetric extension of the Standard Model* is correct, *superpartners* of the existing *elementary particles* would be new and undiscovered particles and *supersymmetry* is expected to be *spontaneously broken*.

Incorporating supersymmetry into the Standard Model requires doubling the number of particles since there is no way that any of the particles in the *Standard Model* can be *superpartners* of each other. With the addition of new particles, there are many possible new *interactions*. The simplest possible supersymmetric model consistent with the *Standard Model* is the *Minimal Supersymmetric Standard Model* which can include the necessary additional new particles that are able to be *superpartners* of those in the *Standard Model*.

One of the original motivations for the *Minimal Supersymmetric Standard Model* was that is the simplest supersymmetric extension of the *Standard Model* that could resolve major *hierarchy problems* within the *Standard Model*, by guaranteeing that *quadratic divergences* of all orders will cancel out in *perturbation theory*. Due to the quadratically *divergent* contributions to the *Higgs mass* squared in the *Standard Model, the quantum mechanical interactions of the Higgs boson cause a large renormalization of the Higgs mass* and unless there is an accidental cancellation, the natural size of the *Higgs mass* is the greatest scale possible. Furthermore, the *electroweak* scale receives *enormous Planck-scale quantum corrections. The observed hierarchy between the electroweak scale and the Planck scale must be achieved with extraordinary fine tuning.* This problem is known as the *hierarchy problem.*

Supersymmetry close to the *electroweak* scale, such as in the *Minimal Supersymmetric Standard Model*, would solve the hierarchy problem that afflicts the *Standard Model. It would reduce the size of the quantum corrections by having automatic cancellations between fermionic and bosonic Higgs interactions, and Planck-scale quantum corrections cancel between partners and superpartners* (owing to a minus sign associated with fermionic loops). The hierarchy between the *electroweak scale* and the *Planck scale* would be achieved in a natural manner, without extraordinary fine-tuning. If *supersymmetry* were restored at the *weak* scale, then the *Higgs mass* would be related to *supersymmetry breaking* which can be induced from small non-perturbative effects explaining the vastly different scales in the *weak interactions* and *gravitational interactions*. However, the *negative results* from the LHC since 2010 have already ruled out some *supersymmetric extensions to the Standard Model*, and *many physicists believe that the Minimal Supersymmetric Standard Model, while not ruled out, is no longer able to fully resolve the hierarchy problem. There is no experimental evidence that a supersymmetric extension to the Standard Model is correct, or whether or not other extensions to current models might be more accurate.* It is only since around 2010 that particle accelerators specifically designed to study physics beyond the *Standard Model* have become operational (i.e. the Large Hadron Collider (LHC)), and *it is not known where exactly to look, nor the energies required for a successful search.*

Another motivation for the *Minimal Supersymmetric Standard Model* came from *grand unification*, the idea that the *gauge symmetry groups* should unify at *high-energy*. *In the Standard Model, however, the weak, strong and electromagnetic gauge couplings fail to unify at high energy.* In particular, the *renormalization* group evolution of the three *gauge-coupling constants* of the *Standard Model* is somewhat sensitive to the present particle content of the theory. These *coupling constants* do not quite meet together at a common energy scale if we run the *renormalization group* using the *Standard Model*. After incorporating minimal *supersymmetry* at the *electroweak* scale, the running of the *gauge couplings* are modified, and joint convergence of the gauge coupling constants is projected to occur at approximately 10^{16} GeV. The modified running also provides a natural mechanism for radiative *electroweak symmetry breaking*.

In many *supersymmetric* extensions of the *Standard Model*, such as the *Minimal Supersymmetric Standard Model*, there was a heavy stable particle (such as the neutralino) which could serve as a *weakly interacting massive particle* dark matter candidate. The existence of a *supersymmetric dark matter candidate* is related closely to *R-parity*. *Supersymmetry* at the *electroweak* scale (augmented with a discrete symmetry) typically provides a candidate *dark matter particle* at a mass scale consistent with thermal relic abundance calculations.

The standard paradigm for incorporating *supersymmetry* into a realistic theory is to have the underlying dynamics of the theory be *supersymmetric*, but the ground state of the theory does not respect the symmetry and supersymmetry is broken spontaneously. *The supersymmetry break cannot be done permanently by the particles of the Minimal Supersymmetric Standard Model as they currently appear.* This means that there is a new sector of the theory that is responsible for the breaking. The only constraint on this new sector is that it must break *supersymmetry* permanently and must give *superparticles* TeV scale masses. There are many models that can do this and most of their details do not matter. In order to parameterize the relevant features of *supersymmetry breaking*, arbitrary soft *supersymmetric* breaking terms are added to the theory which temporarily break *supersymmetry* explicitly but could never arise from a complete theory of *supersymmetry breaking*.

Searches and constraints for supersymmetry

Supersymmetric extensions of the *Standard Model* are constrained by a variety of experiments, including measurements of low-energy observables – for example, the anomalous magnetic moment of the muon at Fermilab; the *weakly interacting massive particle* dark matter density measurement and direct detection experiments – for example, XENON-100 and LUX; and by particle collider experiments, including B-physics, Higgs phenomenology and direct searches for *superpartners* (sparticles), at the Large Electron–Positron Collider, Tevatron and the LHC. In fact, *CERN publicly states that if a supersymmetric model of the Standard Model "is correct, supersymmetric particles should appear in collisions at the LHC"*.

Historically, the tightest limits were from direct production at colliders. From 2003 to 2015, *weakly interacting massive particle's* and Planck's *dark matter density* measurements have strongly constrained *supersymmetric* extensions of the *Standard Model*, which, if they explain *dark matter*, have to be tuned to invoke a particular mechanism to sufficiently reduce the *neutralino* density.

The first runs of the LHC surpassed existing experimental limits from the Large Electron–Positron Collider and Tevatron and *partially excluded the aforementioned expected ranges*. In 2011–12, the LHC discovered a *Higgs boson* with a mass of about 125 GeV, and with couplings to *fermions* and *bosons* which are consistent with the *Standard Model*. The *Minimal Supersymmetric Standard Model* predicts that the mass of the lightest *Higgs boson* should not be much higher than the mass of the *Z boson*, and, in the absence of fine tuning (with the *supersymmetry breaking scale* on the order of 1 TeV), should not exceed 135 GeV. *The LHC found no previously unknown particles* other than the *Higgs boson* which was already suspected to exist as part of the *Standard Model*, and therefore *no evidence for any supersymmetric extension of the Standard Model*.

Indirect methods include the search for a permanent *electric dipole moment* in the known *Standard Model* particles, which can arise when the *Standard Model* particle interacts with the *supersymmetric* particles. The current best constraint on the *electron electric dipole moment* put it to be smaller than 10^{-28} e·cm, equivalent to a sensitivity to new physics at the TeV scale and matching that of the current best particle colliders. A permanent *electric dipole moment* in any fundamental particle points towards *time-reversal* violating physics, and therefore also *CP-symmetry* violation via the *CPT theorem*. Such *electric dipole moment* experiments are also much more scalable than conventional particle accelerators and offer a practical alternative to detecting physics beyond the *Standard Model* as accelerator experiments become increasingly costly and complicated to maintain. *The current best limit for the electron's electric dipole moment has already reached a sensitivity to rule out so called 'naive' versions of supersymmetric extensions of the Standard Model.* [Baron, J., Campbell, W. C., Demille, D., Doyle, J. M., Gabrielse, G., et al. (2014). Order of Magnitude Smaller Limit on the Electric Dipole Moment of the Electron. *Science*, 343, 6168, 269–72; arXiv:1310.7534; https://doi.org/10.1126/ science.1248213.]

Research in the late 2010s and early 2020s from experimental data on the cosmological constant, LIGO noise, and pulsar timing, suggests it is very unlikely that there are any new particles with *masses* much higher than those which can be found in the *Standard Model* or the LHC. However, this research has also indicated that *quantum gravity* or perturbative quantum field theory will become strongly coupled before 1 PeV, leading to other new physics in the TeVs.

Current status

The negative findings in the experiments disappointed many physicists, who believed that supersymmetric extensions of the Standard Model (and other theories relying upon it) were by far the most promising theories for "new" physics beyond the Standard Model, and had hoped for signs of unexpected results from the experiments. In particular, *the LHC result seems problematic for the Minimal Supersymmetric Standard Model*, as the value of 125 GeV is relatively large for the model and can only be achieved with large radiative loop corrections from *top squarks*, which many theorists consider to be "unnatural". [Draper, P., Meade, P., Reece, M., Shih, D. (December ,2011). Implications of a 125 GeV Higgs for the MSSM and Low-Scale SUSY Breaking. *Phys. Rev. D*, 85, 9, 095007; arXiv:1112.3068; https://doi.org/10.1103/PhysRevD.85.095007.]

In response to the so-called "naturalness crisis" in the Minimal Supersymmetric Standard Model, some researchers have abandoned naturalness and the original motivation to solve the hierarchy problem naturally with supersymmetry, while other researchers have moved on to other supersymmetric models such as split supersymmetry. Still *others have moved to string theory* as a result of the naturalness crisis. Former enthusiastic supporter Mikhail Shifman went as far as urging the theoretical community to search for new ideas and *accept that supersymmetry was a failed theory in particle physics*. However, some researchers suggested that this "naturalness" crisis was premature because various calculations were too optimistic about the limits of masses which would allow a supersymmetric extension of the *Standard Model* as a solution.

Extended supersymmetry

It is possible to have more than one kind of *supersymmetry transformation*. Theories with more than one *supersymmetry transformation* are known as *extended supersymmetric theories*. The more *supersymmetry* a theory has, the more constrained are the *field* content

517

and *interactions*. Typically, the number of copies of a *supersymmetry* is a power of 2 (1, 2, 4, 8...). In four dimensions, a *spinor* has four degrees of freedom and thus the minimal number of *supersymmetry generators* is four in four dimensions and having eight copies of *supersymmetry* means that there are 32 *supersymmetry generators*.

The maximal number of supersymmetry generators possible is 32. Theories with more than 32 *supersymmetry generators* automatically have massless fields with *spin* greater than 2. It is not known how to make massless fields with spin greater than two interact, so the maximal number of *supersymmetry generators* considered is 32. This corresponds to an N = 8 *supersymmetry theory*. Theories with 32 *supersymmetries* automatically have a *graviton*.

PART IV   String Theory.

String theory is a theoretical framework in which the point-like particles of particle physics are replaced by one-dimensional objects called strings. String theory describes how these *strings* propagate through space and *interact* with each other. On distance scales larger than the string scale, a *string* looks just like an ordinary particle, *with its mass, charge, and other properties determined by the vibrational state of the string.* In *string theory*, one of the many *vibrational states* of the string corresponds to the *graviton, a quantum mechanical particle* that carries the *gravitational force.* Thus, *string theory is a theory of quantum gravity.*

String theory is a broad and varied subject that attempts to address a number of deep questions of fundamental physics. *String theory* has contributed a number of advances to mathematical physics, which have been applied to a variety of problems in black hole physics, early universe cosmology, nuclear physics, and condensed matter physics, and it has stimulated a number of major developments in pure mathematics. Because *string theory* potentially provides a unified description of *gravity* and *particle physics*, it is a candidate for a theory of everything, a self-contained mathematical model that describes all fundamental forces and forms of matter. Despite much work on these problems, *it is not known to what extent string theory describes the real world* or how much freedom the theory allows in the choice of its details.

History

String theory was first studied in the late 1960s as a never completely successful theory of *hadrons*, the subatomic particles like the *proton* and *neutron* that feel the *strong interaction (strong nuclear force), before being abandoned in favor of quantum chromodynamics.* In the 1960s, Geoffrey Chew and Steven Frautschi discovered that the *mesons* make families called *Regge trajectories* with *masses* related to *spins* in a way that was later understood by Yoichiro Nambu, Holger Bech Nielsen and Leonard Susskind to be the relationship expected from *rotating strings*. Chew advocated making a theory for the *interactions* of these trajectories that did not presume that they were composed of any fundamental particles, but would construct their *interactions* from self-consistency conditions on the *S-matrix*. The *S-matrix* approach was started by Werner Heisenberg in the 1940s as a way of constructing a theory that did not rely on the local notions of space and time, which Heisenberg believed break down at the nuclear scale. [Heisenberg, W. (July, 1943). Die beobachtbaren Größen in der Theorie der Elementarteilchen. I. (The "observable quantities" in the theory of elementary particles. I.) *Zeit. Phys.*, 120, 7–10, 513-38; https://doi.org/10.1007/bf01329800; abstract in Underwood, T. G. (2023). *Quantum Electrodynamics – annotated sources.* Volume II, pp. 244-5.]

[*S-matrix theory* was proposed as a principle of *particle interactions* by Heisenberg in 1943, following John Archibald Wheeler's 1937 introduction of the *S-matrix*. It was a proposal for replacing *local quantum field theory* as the basic principle of elementary particle physics. It avoided the notion of space and time by replacing it with abstract mathematical properties of the *S-matrix*. In *S-matrix theory*, the *S-matrix* relates the infinite past to the infinite future in one step, without being decomposable into intermediate steps corresponding to time-slices.

This program was very influential in the 1960s, because it was a plausible substitute for *quantum field theory, which was plagued with the zero-interaction phenomenon at strong coupling*. Applied to the *strong interaction*, it led to the development of *string theory*.

S-matrix theory was largely abandoned by physicists in the 1970s, as *quantum chromodynamics* was recognized to solve the problems of *strong interactions* within the framework of *field theory*. But in the guise of *string theory*, *S-matrix theory* is still a popular approach to the problem of *quantum gravity*.]

While the scale was off by many orders of magnitude, the approach he advocated was ideally suited for a theory of *quantum gravity*.

Working with experimental data, R. Dolen, D. Horn and C. Schmid developed some sum rules for *hadron exchange*. When a *particle* and *antiparticle* scatter, virtual particles can be exchanged in two qualitatively different ways. In the *s-channel*, the two particles annihilate to make temporary intermediate states that fall apart into the final state particles. In the *t-channel*, the particles *exchange intermediate states* by emission and absorption. In *field theory*, the two contributions add together, one giving a continuous background contribution, the other giving peaks at certain energies. In the data, it was clear that the peaks were stealing from the background—the authors interpreted this as saying that the *t-channel* contribution was dual to the *s-channel* one, meaning both described the whole *amplitude* and included the other.

The result was widely advertised by Murray Gell-Mann, leading Gabriele Veneziano to construct a *scattering amplitude* that had the property of Dolen–Horn–Schmid duality, later renamed *world-sheet duality*. The *amplitude* needed *poles* where the particles appear, on straight-line trajectories, and there is a special mathematical function whose *poles* are evenly spaced on half the real line—the *gamma function*— which was widely used in *Regge theory*. By manipulating combinations of *gamma functions*, Veneziano was able to find a consistent *scattering amplitude* with *poles* on straight lines, with mostly positive residues, which obeyed duality and had the appropriate *Regge scaling* at high energy. The *amplitude* could fit near-beam scattering data as well as other *Regge* type fits and had a suggestive integral *representation* that could be used for generalization.

Over the next years, hundreds of physicists worked to complete the bootstrap program for this model, with many surprises. Veneziano himself discovered that for the *scattering amplitude* to describe the scattering of a particle that appears in the theory, an obvious self-consistency condition, the lightest particle must be a *tachyon*. Miguel Virasoro and Joel Shapiro found a different *amplitude* now understood to be that of *closed strings*, while Ziro Koba and Holger Nielsen generalized Veneziano's integral *representation* to multiparticle scattering. Veneziano and Sergio Fubini introduced an operator formalism for computing the *scattering amplitudes* that was a forerunner of *world-sheet conformal theory*, while Virasoro understood how to remove the *poles* with wrong-sign residues using a constraint on the *states*. Claud Lovelace calculated a *loop amplitude*, and noted that *there is an inconsistency unless the dimension of the theory is 26*. Charles Thorn, Peter Goddard and Richard Brower went on to prove that there are no wrong-sign propagating *states* in dimensions less than or equal to 26.

The earliest version of *string theory*, *bosonic string theory*, incorporated only the class of particles known as *bosons*. It later developed into *superstring theory*, which posits a connection called *supersymmetry* between *bosons* and the class of particles called *fermions*.

In the 1970s, many physicists became interested in *supergravity theories*, which combine *general relativity* with *supersymmetry*. Whereas *general relativity* makes sense in any number of dimensions, *supergravity* places an upper limit on the number of dimensions. In 1978, work by Werner Nahm showed that the maximum spacetime dimension in which one can formulate a consistent *supersymmetric theory* is eleven. In the same year, Eugene Cremmer, Bernard Julia, and Joël Scherk of the École Normale Supérieure showed that *supergravity* not only permits up to eleven dimensions but is in fact most elegant in this maximal number of dimensions.

Initially, many physicists hoped that *by compactifying eleven-dimensional supergravity, it might be possible to construct realistic models of our four-dimensional world*. The hope was that such models would provide a unified description of the four fundamental forces of nature: *electromagnetism*, the *strong* and *weak nuclear forces*, and *gravity*. Interest in *eleven-dimensional supergravity soon waned as various flaws in this scheme were discovered. One of the problems was that the laws of physics appear to distinguish between clockwise and counterclockwise, a phenomenon known as chirality.* Edward Witten and others observed this *chirality* property cannot be readily derived by compactifying from eleven dimensions.

In 1969–70, Yoichiro Nambu, Holger Bech Nielsen, and Leonard Susskind recognized that the theory could be given a description in space and time in terms of *strings*. The *scattering amplitudes* were derived systematically from the *action* principle by Peter Goddard, Jeffrey Goldstone, Claudio Rebbi, and Charles Thorn, giving a space-time picture to the *vertex operators* introduced by Veneziano and Fubini and a geometrical interpretation to the Virasoro conditions.

In 1971, Pierre Ramond added *fermions* to the model, which led him to formulate a two-dimensional *supersymmetry* to cancel the wrong-sign states. John Schwarz and André Neveu added another sector to the *fermion theory* a short time later. In the *fermion theories*, the critical dimension was 10. Stanley Mandelstam formulated a *world sheet conformal theory* for both the *bose* and *fermi* case, giving a two-dimensional field theoretic path-integral to generate the operator formalism. Michio Kaku and Keiji Kikkawa gave a different formulation of the *bosonic* string, as a *string field theory*, with infinitely many particle types and with fields taking values not on points, but on loops and curves.

In 1974, Tamiaki Yoneya discovered that all the known string theories included a *massless spin-two particle* that obeyed the correct *Ward identities* to be a *graviton*. John Schwarz and Joël Scherk came to the same conclusion and made the bold leap to suggest that *string theory was a theory of gravity, not a theory of hadrons*. They reintroduced Kaluza–Klein theory as a way of making sense of the extra dimensions. *At the same time, quantum chromodynamics was recognized as the correct theory of hadrons*, shifting the attention of physicists and apparently leaving the bootstrap program in the dustbin of history.

Subsequently, it was realized that the very properties that made *string theory* unsuitable as a theory of nuclear physics made it a promising candidate for a *quantum theory of gravity*. *String theory eventually made it out of the dustbin, but for the following decade, all work on the theory was completely ignored.* Still, the theory continued to develop at a steady pace thanks to the work of a handful of devotees. Ferdinando Gliozzi, Joël Scherk, and David Olive realized in 1977 that the original Ramond and Neveu Schwarz-strings were separately inconsistent and needed to be combined. The resulting theory did not have a *tachyon* and was proven to have *space-time supersymmetry* by John Schwarz and Michael Green in 1984. The same year, Alexander Polyakov gave the theory a modern *path integral formulation*, and went on to develop conformal *field theory* extensively. In 1979, Daniel Friedan showed that the *equations of motions of string theory*, which are *generalizations of the Einstein equations of general relativity*, emerge from the *renormalization* group equations for the *two-dimensional field theory*. Schwarz and Green discovered *T-duality*, and constructed two superstring theories—IIA and IIB related by *T-duality*, and type I theories with *open strings*. The consistency conditions had been so strong, that the entire theory was nearly uniquely determined, with only a few discrete choices.

In the early 1980s, Edward Witten discovered that most theories of *quantum gravity* could not accommodate *chiral fermions* like the *neutrino*. This led him, in collaboration with Luis Álvarez-Gaumé, to study violations of the conservation laws in gravity theories with anomalies, concluding that *type I string theories were inconsistent*. Green and Schwarz discovered a contribution to the anomaly that Witten and Alvarez-Gaumé had missed, which restricted the *gauge group* of the *type I string theory* to be SO(32). In coming to understand this calculation, Edward Witten became convinced that *string theory* was truly a consistent theory of gravity, and he became a high-profile advocate. Following Witten's lead, between 1984 and 1986, hundreds of physicists started to work in this field, and this is sometimes called the *first superstring revolution*.

During this period, David Gross, Jeffrey Harvey, Emil Martinec, and Ryan Rohm discovered *heterotic strings*. The gauge group of these closed strings was two copies of E8, and *either copy could easily and naturally include the Standard Model*. Philip Candelas, Gary Horowitz, Andrew Strominger and Edward Witten found that the Calabi–Yau manifolds are the compactifications that preserve a realistic amount of *supersymmetry*, while Lance Dixon and others worked out the physical properties of orbifolds, distinctive geometrical singularities allowed in *string theory*. Cumrun Vafa generalized *T-duality* from circles to arbitrary manifolds, creating the mathematical field of *mirror symmetry*. Daniel Friedan, Emil Martinec and Stephen Shenker further developed the covariant quantization of the *superstring* using conformal field theory techniques. David Gross and Vipul Periwal discovered that *string perturbation theory was divergent*. Stephen Shenker showed it diverged much faster than in *field theory* suggesting that new non-perturbative objects were missing.

Branes

Another approach to reducing the number of dimensions is the so-called *brane-world* scenario. In this approach, physicists *assume that the observable universe is a four-dimensional subspace of a higher dimensional space*. In such models, the force-carrying *bosons* of particle physics arise from *open strings* with endpoints attached to the four-dimensional subspace, while *gravity* arises from *closed strings* propagating through the larger ambient space. This idea plays an important role in attempts to develop models of real-world physics based on *string theory*, and it provides a natural explanation for the weakness of gravity compared to the other fundamental forces. [Randall, L. & Sundrum, R. (1999). An alternative to compactification. *Phys. Rev. Lett.*, 83, 23, 4690–3; arXiv:hep-th/9906064; https://doi.org/10.1103/PhysRevLett.83.4690.]

In 1987, Eric Bergshoeff, Ergin Sezgin, and Paul Townsend showed that *eleven-dimensional supergravity includes two-dimensional branes*. [Bergshoeff, E., Sezgin, E. &

Townsend, P. (1987). Supermembranes and eleven-dimensional supergravity. *Physics Letters B*, 189, 1, 75–8; https://doi.org/10.1016/0370-2693(87)91272-X.] Intuitively, these objects look like *sheets* or *membranes* propagating through the eleven-dimensional *spacetime*.

[*A brane is a physical object that generalizes the notion of a point particle to higher dimensions.* For instance, *a point particle can be viewed as a brane of dimension zero*, while *a string can be viewed as a brane of dimension one*. It is also possible to consider higher-dimensional *branes*. In dimension p, these are called *p-branes*. The word *brane* comes from the word "membrane" which refers to a two-dimensional brane.

Branes are dynamical objects which can propagate through spacetime according to the rules of quantum mechanics. They have *mass* and can have other attributes such as *charge*. A *p-brane* sweeps out a (p + 1)-dimensional volume in spacetime called its *worldvolume*. Physicists often study fields analogous to the *electromagnetic field* which live on the *worldvolume* of a *brane*.

In *string theory*, *D-branes* are an important class of *branes* that arise when one considers *open strings*. *As an open string propagates through spacetime, its endpoints are required to lie on a D-brane.* The letter "D" in *D-brane* refers to a certain mathematical condition on the system known as the *Dirichlet boundary condition*. The study of D-branes in *string theory* has led to important results such as the *AdS/CFT correspondence*, which has shed light on many problems in *quantum field theory*.]

In the 1990s, Joseph Polchinski discovered that the theory requires *higher-dimensional objects*, called *D-branes* and identified these with the *black-hole* solutions of *supergravity*. These were understood to be the new objects suggested by the perturbative divergences, and they opened up a new field with rich mathematical structure. It quickly became clear that *D-branes* and other *p-branes*, not just *strings*, formed the *matter content* of the *string theories*, and *the physical interpretation of the strings and branes was revealed—they are a type of black hole*. Leonard Susskind had incorporated the *holographic principle* of Gerardus 't Hooft into *string theory*, identifying the long highly excited *string states* with ordinary thermal *black hole states*. As suggested by 't Hooft, the fluctuations of the *black hole horizon*, the *world-sheet* or *world-volume theory*, describes not only the degrees of freedom of the *black hole*, but all nearby objects too.

Shortly after this discovery, Michael Duff, Paul Howe, Takeo Inami, and Kellogg Stelle considered a particular *compactification* of *eleven-dimensional supergravity* with one of

524

the dimensions curled up into a circle. In this setting, one can imagine the *membrane* wrapping around the circular dimension. If the radius of the circle is sufficiently small, then this *membrane* looks just like a *string* in ten-dimensional spacetime. Duff and his collaborators showed that this construction reproduces exactly the *strings* appearing in *type IIA superstring theory*. [Duff, M., Howe, P., Inami, T. & Stelle, K. (1987). Superstrings in D=10 from supermembranes in D=11. *Nuclear Physics B*, 191, 1, 70–4; https://doi.org/10.1016/0370-2693(87)91323-2.]

M-theory

Prior to 1995, theorists believed that there were five consistent versions of *superstring theory* (*type I, type IIA, type IIB*, and two versions of *heterotic string theory*). This understanding changed in 1995 when Edward Witten suggested that the five theories were just special limiting cases of an *eleven-dimensional theory* called *M-theory*. [Witten, Edward (1995). String theory dynamics in various dimensions. See below.] Witten's conjecture was based on the work of a number of other physicists, including Ashoke Sen, Chris Hull, Paul Townsend, and Michael Duff. His announcement led to a flurry of research activity now known as the *second superstring revolution*.

In 1995, at the annual conference of string theorists at the University of Southern California (USC), Edward Witten gave a speech on *string theory* that in essence *united the five string theories that existed at the time*, giving birth to a *new 11-dimensional theory called M-theory*. [Witten, Edward (1995). String theory dynamics in various dimensions.] *M-theory* was also foreshadowed in the work of Paul Townsend at approximately the same time. Witten's announcement drew together all of the previous results on *S- and T-duality* and the appearance of higher-dimensional *branes* in *string theory*. In the months following Witten's announcement, hundreds of new papers appeared on the Internet confirming different parts of his proposal. The flurry of activity that began at this time is sometimes called the *second superstring revolution*.

Initially, some physicists suggested that the new theory was a fundamental theory of membranes, but *Witten was skeptical of the role of membranes in the theory*. In a paper from 1996, Hořava and Witten wrote "As it has been proposed that the *eleven-dimensional theory* is a *supermembrane theory* but there are some reasons to doubt that interpretation, we will non-committally call it the *M-theory*, leaving to the future the relation of M to membranes." In the absence of an understanding of the true meaning and structure of M-theory, Witten has suggested that the M should stand for "magic", "mystery", or "membrane" according to taste, and the true meaning of the title should be decided when a more fundamental formulation of the theory is known.

BFSS matrix model

In 1997, Tom Banks, Willy Fischler, Stephen Shenker and Leonard Susskind formulated *matrix theory*, a full *holographic* description of *M-theory* using IIA D_0 *branes*. This was the first definition of *string theory* that was fully *non-perturbative* and a concrete mathematical realization of the *holographic principle*. It is an example of a *gauge-gravity duality* and is now understood to be a special case of the *AdS/CFT correspondence*. This theory describes the behavior of a set of nine large matrices. In their original paper, these authors showed, among other things, that the low energy limit of this matrix model is described by *eleven-dimensional supergravity*. These calculations led them to propose that *the BFSS matrix model is exactly equivalent to M-theory*. The *BFSS matrix model* could therefore be used as a prototype for a correct formulation of *M-theory* and a tool for investigating the properties of *M-theory* in a relatively simple setting. [Banks, T., Fischler, W., Schenker, S., & Susskind, L. (1997). M theory as a matrix model: A conjecture. *Phys. Rev. D*, 55, 8, 5112–28; arXiv:hep-th/9610043; https://doi.org/10.1103/ physrevd. 55.5112.]

The development of the matrix model formulation of *M-theory* led physicists to consider various connections between *string theory* and a branch of mathematics called *noncommutative geometry*. This subject is a generalization of ordinary geometry in which mathematicians define new geometric notions using tools from *noncommutative algebra*. In a paper from 1998, Alain Connes, Michael R. Douglas, and Albert Schwarz showed that some aspects of matrix models and *M-theory* are described by a *noncommutative quantum field theory*, a special kind of physical theory in which *spacetime* is described mathematically using *noncommutative geometry*. This established a link between *matrix models* and *M-theory* on the one hand, and *noncommutative geometry* on the other hand.

AdS/CFT correspondence

In late 1997, theorists discovered an important relationship called the *anti-de Sitter/conformal field theory* (AdS/CFT) *correspondence*), which relates *string theory* to *quantum field theory*. This is a theoretical result which implies that *string theory* is in some cases equivalent to a *quantum field theory*. In addition to providing insights into the mathematical structure of *string theory*, the *AdS/CFT correspondence* has shed light on many aspects of *quantum field theory* in regimes where traditional calculational techniques are ineffective.

The *AdS/CFT correspondence* was first proposed by Juan Maldacena in late 1997. [Maldacena, J. (1998). The Large N limit of superconformal field theories and supergravity. *Advances in Theoretical and Mathematical Physics*, 2, 2, 231–52; arXiv:hep-th/9711200; https://doi.org/10.4310/ATMP.1998. V2.N2.A1.] Maldacena noted that the low energy excitations of a theory near a *black hole* consist of objects close to the horizon, which for extreme charged *black holes* looks like an *anti-de Sitter space*. He noted that in this limit *the gauge theory describes the string excitations near the branes*. So, *he hypothesized that string theory on a near-horizon extreme-charged black-hole geometry, an anti-de Sitter space times a sphere with flux, is equally well described by the low-energy limiting gauge theory, the N = 4 supersymmetric Yang–Mills theory.*

Important aspects of the *correspondence* were elaborated in articles by Steven Gubser, Igor Klebanov, and Alexander Markovich Polyakov, and by Edward Witten. [Gubser, S., Klebanov, I. & Polyakov, A. (1998). Gauge theory correlators from non-critical string theory. *Physics Letters B*, 428, 1–2, 105–14; arXiv:hep-th/9802109; https://doi.org/10.1016/S0370-2693(98)00377-3; Witten, E. (1998). Anti-de Sitter space and holography. *Advances in Theoretical and Mathematical Physics*, 2, 2, 253–91; arXiv:hep-th/9802150; https://doi.org/10.4310/ATMP.1998.v2.n2.a2.] It is a concrete realization of the *holographic principle*, which has far-reaching implications for *black holes*, *locality* and *information* in physics, as well as the nature of the *gravitational interaction*. Through this relationship, *string theory has been shown to be related to gauge theories like quantum chromodynamics* and this has led to a more quantitative understanding of the behavior of *hadrons*, bringing *string theory* back to its roots.

Extra dimensions

String theories require extra dimensions of spacetime for their mathematical consistency. In *bosonic string theory*, spacetime is 26-dimensional, while in *superstring theory* it is 10-dimensional, and in *M-theory* it is 11-dimensional. In order to describe real physical phenomena using *string theory, one must therefore imagine scenarios in which these extra dimensions would not be observed in experiments.*

Compactification is one way of modifying the number of dimensions in a physical theory. In *compactification*, some of the extra dimensions are assumed to "close up" on themselves to form circles. In the limit where these curled up dimensions become very small; one obtains a theory in which *spacetime* has effectively a lower number of dimensions. *Compactification* can be used to construct models in which *spacetime* is effectively four-dimensional.

Dualities

One of the main developments of the past several decades in *string theory* was the discovery of certain '*dualities*', mathematical transformations that identify one physical theory with another. Physicists studying *string theory* have discovered a number of these *dualities* between different versions of *string theory*, and this has led to the conjecture that all consistent versions of *string theory* are subsumed in a single framework known as *M-theory*.

One of the relationships that can exist between different *string theories* is called *S-duality*. This is a relationship that says that *a collection of strongly interacting particles in one theory can, in some cases, be viewed as a collection of weakly interacting particles in a completely different theory*. Roughly speaking, a collection of particles is said to be *strongly interacting* if they combine and decay often and *weakly interacting* if they do so infrequently. *Type I string theory* turns out to be equivalent by *S-duality* to the SO(32) heterotic *string theory*. Similarly, *type IIB string theory* is related to itself in a nontrivial way by *S-duality*.

Another relationship between different *string theories* is *T-duality*. Here one considers strings propagating around a circular extra dimension. *T-duality states that a string propagating around a circle of radius R is equivalent to a string propagating around a circle of radius 1/R in the sense that all observable quantities in one description are identified with quantities in the dual description*. For example, a *string* has *momentum* as it propagates around a circle, and it can also wind around the circle one or more times. The number of times the string winds around a circle is called the *winding number*. If a *string* has *momentum* p and *winding number* n in one description, it will have *momentum* n and *winding number* p in the dual description. For example, *type IIA string theory* is equivalent to *type IIB string theory* via *T-duality*, and the two versions of *heterotic string theory* are also related by *T-duality*.

One of the challenges of string theory is that the full theory does not have a satisfactory definition in all circumstances. Another issue is that the theory is thought to describe an enormous landscape of possible universes, which has complicated efforts to develop theories of particle physics based on *string theory*. These issues have led some in the community to criticize these approaches to physics, and to question the value of continued research on *string theory unification*.